Probabilistic Methods in Discrete Mathematics

PROGRESS IN PURE AND APPLIED DISCRETE MATHEMATICS, Vol. 1

PROBABILISTIC METHODS IN DISCRETE MATHEMATICS

Proceedings of the Third International Petrozavodsk Conference

Petrozavodsk, Russia, May 12 - 15, 1992

Editors

V. F. Kolchin, V. Ya. Kozlov,
Yu. L. Pavlov, and Yu. V. Prokhorov

///VSP///
Utrecht, The Netherlands
Tokyo, Japan

TVP
Science Publishers
Moscow, Russia

TVP Science Publishers
Vavilov st. 42
117966 Moscow GSP-1
Russia

VSP BV
P.O. Box 346
3700 AH Zeist
The Netherlands

First published in 1993
ISBN 90-6764-158-8
UDC 519.1, 519.2, 519.7

Printed in Russia by Novosti, Moscow

CONTENTS

PREFACE

This volume contains the Proceedings of the Third Petrozavodsk Conference "Probabilistic Methods in Discrete Mathematics" which was held at 12–15 May 1992 in Petrozavodsk in Russia. The conference was organized by the Steklov Mathematical Institute of the Russia Academy of Sciences, the Karelian Scientific Centre of Russia Academy of Sciences and the Karelian State University.
The pioneering works carried out by V. L. Goncharov, published in 40's, are apparently the first where combinatorial objects were systematically investigated by modern probabilistic methods. These works contain deep results on random permutations and runs in random $(0,1)$-sequences. During the last three decades the interest in probabilistic problems of combinatorics has grown steadily both in our country and abroad. In 1983 the First All-Union conference "Probabilistic Methods in Discrete Mathematics" and in 1988 the Second conference were held in Petrozavodsk. The themes of the Petrozavodsk conferences cover almost all areas of probabilistic combinatorics including probabilistic problems of discrete mathematics, statistical problems in discrete mathematics and graph theory. The participants of the two first conferences came up with the proposal to extend the topics of the conference and give more attention to actual applied probabilistic problems of discrete mathematics. In line with these suggestions, for the first time the section "Mathematical methods of information security" was organized along with the traditional sections "Probabilistic problems of combinatorics", "Statistical methods in combinatorics", "Random graphs".
During the three working days of the III Petrozavodsk conference, 12 plenary and 33 sectional lectures were read. At the first plenary session the lectures were presented by B. Harris (Madison), V. F. Kolchin (Moscow), V. A. Vatutin (Moscow), Yu. L. Pavlov (Petrozavodsk); at the second plenary session the lectures were read by S. Golomb (Los-Angeles), V. G. Mikhailov (Moscow), A. A. Grusho (Moscow), V. A. Plaksin (Petrozavodsk); in the final day the conference heard the plenary lectures by I. N. Kovalenko and A. A. Levitskaya (Kiev), G. V. Balakin (Moscow), A. F. Ronzhin (Moscow). The plenary lectures prepared for publication are presented in the Proceedings in the above mentioned order. All sectional reports are arranged in alphabetic order of authors names.
The papers presented in the Proceedings reflect the current state of the art in probabilistic combinatorics and contain the information which will be interesting to those who work in theoretical and applied areas of discrete mathematics.
The Organizing Committee has the pleasant duty to thank the sponsors of the III Petrozavodsk conference without whose financial support the conference could not be held. These are TVP Science Publishers, Russian National Committee of Bernoulli Society, SCB Progress. The organizing duties were taken up by the small venture Elmus (Petrozavodsk).

We are deeply indebted to TVP Science Publishers who agreed to publish the Proceedings, we wish also to thank their highly skilled personal who carried out all their work at a top level.
The Editors are grateful to those who submitted their papers on time, to the referees who carried out their tasks thoroughly. We especially value the efforts of Elena Dyakonova who coordinated all the editorial activity. We wish also to express our gratitude to all participants of the Conference and invite all of them to take part in the next Petrozavodsk conference which will be held in 1996.

Vice-Chairman of the Organizing Committee
Professor V. F. Kolchin

Probabilistic Methods in Discrete Mathematics, pp. 1 – 22
V. F. Kolchin et al. (Eds.)
1993 TVP/VSP

A SURVEY OF THE EARLY HISTORY OF THE THEORY OF RANDOM MAPPINGS

B. HARRIS
Department of Statistics, University of Wisconsin, 1210 Dayton Street, Madison, Wisconsin 53706, U.S.A.

ABSTRACT

The early history (prior to 1965) of the theory of random mappings is surveyed. The paper describes the types of questions that were studied during that period and the results that were obtained. In particular, some unpublished works with significant results are described. The appendices to the paper contain a catalog of the various results that were obtained by type of result and by reference.

1. INTRODUCTION AND SUMMARY

In this report, the early history (prior to 1965) of the theory of random mappings is studied. In this report, the types of questions studied at that time and the methodology that was utilized are examined. In particular, a substantial fraction of the research described is unpublished and it is hoped that this study will serve the purpose of recognizing these contributions to the theory and establishing their place in the history of the subject.

The theory of random mappings, as described in this paper, concerns the following types of questions.

Let $\mathfrak{X}_n$ be a set with $|\mathfrak{X}_n| = n$. Let J_n be a subset of the mappings of $\mathfrak{X}_n$ into $\mathfrak{X}_n$. Let $\mathbf{P}_{J_n}$ be a probability measure on the subsets of J_n. With no loss of generality, it will be convenient to let $\mathfrak{X}_n = \{1, 2, \ldots, n\}$. Let $\alpha \in J_n$. Then since α is random, various characteristics of α are random variables and the presence of α in specified subsets of J_n are random events.

The theory of random mappings is concerned with determining the probability distributions and characteristics (such as moments) of these random variables and the probabilities of these random events. A substantial proportion of the principal results of the theory deals with the asymptotic behaviour of these quantities as $n \to \infty$. In the early history, $\mathbf{P}_{J_n}$ has been generally chosen as the uniform probability measure on J_n, that is $\mathbf{P}_{J_n} = |J_n|^{-1}$.

The following representation of the mapping $\alpha \in J_n$ is frequently employed. Let $x, y \in \mathfrak{X}_n$. If $\alpha(x) = y$, then a directed arc is drawn from x to y. This procedure is continued for each element $x \in \mathfrak{X}_n$ resulting in the representation of α by a directed graph (with labelled vertices) with the property that exactly one edge emanates from each vertex. This representation of a mapping as a directed graph appears to have been first utilized by A. Suschkewitsch (Suschkewitsch, 1928).

This representation provides a convenient method for visualizing the various characteristics of interest, including those defined below.

Fix $x \in \mathfrak{X}_n$ and consider the set $S_\alpha(x) = \{x, \alpha(x), \alpha^2(x), \ldots\}$. $S_\alpha(x)$ is called the set of successors of x under α. Similarly, let $P_\alpha(x) = \{x, \alpha^{-1}(x), \alpha^{-2}(x), \ldots\}$. $P_\alpha(x)$ is the set of predecessors of x under α. That is, $S_\alpha(x)$ is the set of elements in $\mathfrak{X}_n$ which "can be reached" from x in the directed graph representing α, and $P_\alpha(x)$ is the set of elements in $\mathfrak{X}_n$ from which x "can be reached" in the directed graph representing α.

Since $\mathfrak{X}_n$ is a finite set, there is a least positive integer m such that for some $r \geq 0$, $\alpha^{m+r}(x) = \alpha^r(x)$. The set $C_\alpha(x) = \{\alpha^r(x), \alpha^{r+1}(x), \ldots, \alpha^{r+m-1}(x)\}$ is the cycle reached from x under α. Clearly, $C_\alpha(x) \subset S_\alpha(x)$, m is the length of this cycle. The totality of cyclic elements determined by α is

$$C_\alpha = \bigcup_{x \in \mathfrak{X}_n} C_\alpha(x).$$

$\mathfrak{X}_n$ can be decomposed into equivalence classes by α as follows. Let $x, y \in \mathfrak{X}_n$. Then if there exist two non-negative integers r, s such that $\alpha^r(x) = \alpha^s(y)$, x and y are equivalent under α. It is easy to see that this is an equivalence relation. The equivalence classes are called components of α (more specifically, the components of the graph representing α). If α is a permutation, the components are the cycles of the permutation. We denote the component containing x by $K_\alpha(x)$. The following questions are of interest with respect to the components of a mapping: the number of components, denoted by K_α, the number of elements in the various components, and $|K_\alpha|$, the size of the component reached from a specified point $x \in \mathfrak{X}_n$; if there is only one component, the mapping is said to be connected.

For every $x \in \mathfrak{X}_n$, there is a least non-negative integer j, such that $\alpha^j(x) \in C_\alpha$. Then, x is said to be in $S_{j,\alpha}$, the jth stratum of $\mathfrak{X}_n$ under α. Specifically,

$$S_{j,\alpha} = \{x \in \mathfrak{X}_n\colon\ \alpha^j(x) \in C_\alpha,\ \alpha^{j-1}(x) \notin C_\alpha\}.$$

$S_{j,\alpha}$ is the set of elements whose distance from C_α is equal to j. C_α coincides with $S_{0,\alpha}$. The height of α is defined by $\max_j\{S_{j,\alpha} \neq \emptyset\}$.

In Section 2 of the paper, the work on random permutations during this period is discussed. More general mappings are treated in Section 3. A tabulation of the theorems and formulas obtained in the various references is given in the appendices to this paper.

2. RANDOM PERMUTATIONS

It had been well known that the distribution of the number of cycles in a random permutation on $\mathfrak{X}_n$ (with respect to the uniform probability measure) is given by

$$\mathbf{P}\{K_\alpha = j\} = \frac{|\alpha_{jn}|}{n!}, \qquad j = 1, 2, \ldots, n, \tag{1}$$

where K_α is the number of cycles (components) and α_{jn} denotes the Stirling numbers of the first kind. Such results have a very long history and can easily be traced as far as back as works of Cauchy.

However, the result that provided substantial impetus not only to the theory of random permutations, but also to the theory of random mappings is the following celebrated result of V. L. Goncharov (Goncharov, 1944):

$$\lim_{n\to\infty} \mathbf{P}\left\{\frac{K_\alpha - \log n}{\sqrt{\log n}} \leq x\right\} = \frac{1}{\sqrt{2\pi}}\int_{-\infty}^{x} e^{-t^2/2}dt. \tag{2}$$

That is, the expected value and variance of the number of cycles in a random permutation are both asymptotically equal to $\log n$ and the limit distribution of the number of cycles (when normalized as in (2)) is the standard normal distribution. Goncharov employed generating function methods to obtain the above results. R. E. Greenwood (Greenwood, 1953) calculated the mean and variance of K_α from the factorial moment generating function

$$M_n(t) = (t + n - 1)^{(n)}, \tag{3}$$

where $v^{(n)} = v(v-1)\cdots(v-n+1)$. However, it is evident from the definition of the Stirling numbers of the first kind that (3) is a generating function of the absolute values of these quantities (see (1)). In a technical report of the Jet Propulsion Laboratory, S. W. Golomb, L. R. Welch and R. M. Goldstein (Golomb et al., 1959) used random permutations to model the behaviour of nonlinear shift registers. They derived the mean and variance of the number of cycles and noted the connection to the Stirling numbers of the first kind. They also determined the distribution and expected length of the longest cycle. B. Harris (Harris, 1960) showed that the probability distribution of number of points in the cycle determined by a random point x is given by

$$\mathbf{P}\{|P_\alpha(x)| = j\} = \mathbf{P}\{|K_\alpha(x)| = j\} = \mathbf{P}\{|C_\alpha(x)| = j\} = \mathbf{P}\{|S_\alpha(x)| = j\} = \frac{1}{n}, \tag{4}$$

and employed this result to obtain the mean and variance of the number of cycles. He showed that the expected number of cycles with exactly j elements is j^{-1}. The analogous results were obtained for derangements (permutations with no fixed points). In such mappings, the mean and variance of the number of cycles and the distribution of the number of cycles was given. The reader should consult Appendix 1 for a tabulation of results on random permutations with references.

3. RANDOM MAPPINGS

The study of random mappings appears to have started with a brief article by N. Metropolis and S. Ulam (Metropolis and Ulam, 1953) in which random mappings were suggested as a procedure for studying the generation of pseudo-random numbers. They conjectured that the expected number of components is of the order of $\log n$. As an experiment, they chose $\mathfrak{X}_n = \{0, 1, \ldots, 4095\}$, representing each x in $\mathfrak{X}_n$ as a binary number. The mapping α was chosen as follows. Let $\gamma(x) = x^2$. Then delete the middle twelve binary digits. The result is $\alpha(x)$. They verified by computation that the number of components obtained is 7. Also, they verified experimentally that the distribution of the immediate pre-images of each point, $\alpha^{-1}(x)$, has an approximate Poisson distribution with expected value unity. (It is easy to show that the distribution is binomial with parameters n, n^{-1}, that is, $B(n, 1/n)$.) They also remarked that it is apparently quite difficult to obtain the expected value of the number of components. The Metropolis and Ulam paper is important, not by virtue of its mathematical content, but for its introduction of problems in this area and for motivating the work of other scientists. In particular, the problem of determining the expected number of components was solved by M. D. Kruskal (Kruskal, 1954). Further investigations in this area were made by L. Katz (Katz, 1955) who determined the probability that a mapping is connected for two models, the uniform probability measure on all mappings and the uniform probability measure on the set of mappings with no fixed points (which were called hollow mappings in the paper). A student of L. Katz, J. E. Folkert (Folkert, 1955) investigated the distribution of the number of components for these two models.

A very extensive study of random mappings was made by H. Rubin and R. Sitgreaves (Rubin and Sitgreaves, 1954) in a Stanford University Technical Report. Rubin and Sitgreaves studied the distribution of the number of components, the distribution of $|K_\alpha(x)|$, the size of the component determined by a specified point, the distribution of $|S_\alpha(x)|$ and the distribution of $|P_\alpha(x)|$, the number of successors and number of predecessors of a specified point. They also established the distribution of the size of components.

Noteworthy is their extension of these results to the uniform probability distribution over specified subsets of mappings. Specifically, they considered the set of mappings for which every element in $\mathfrak{X}_n$ has either 0 or k immediate predecessors ($k = 2$ is commonly encountered in many applications in computer science and communication theory), and the case in which every element of $\mathfrak{X}_n$ has no more than k immediate predecessors.

B. Harris (Harris, 1960) considered the set of all mappings and the set of mappings with no fixed points. For these two models the distributions of $|S_\alpha(x)|$, $|P_\alpha(x)|$ and $|C_\alpha(x)|$ were determined. The distribution of $|C_\alpha|$ was also obtained. Generating function methods were used to study the distribution of the number of components.

J. Riordan (Riordan, 1962) studied the use of generating functions in enumerating connected mappings with and without fixed points. He also produced generating functions for the number of components and the number of points on cycles. He exhibited a method for extending these results to mappings with restrictions on cycles, such as mappings with no cycles of length r ($r = 1$ is mappings with no fixed points). Riordan did not employ any probabilistic methods but only considered enumeration questions, however his results can readily be normalized and put in a probabilistic context.

Many properties of random mappings were studied in an extensive unpublished manuscript by A. Lenard (Lenard, 1964). This work contained many results which were previously known, but he also obtained many new results. Also noteworthy in his work is the extensive use of analytic function methods, rather than the use of probabilistic techniques, in the derivation of asymptotic results. A brief summary of his manuscript follows.

In his study of the components of a random mapping, Lenard obtained a formula for the probability that a mapping is connected (this is a different formula from that given by Katz). He employed this result to derive the distribution of the number of components. He also obtained the distribution of the size of the component starting from a specified point. In addition, he calculated the distribution of the largest component.

In his study of the cycles of a random mapping, Lenard derived an expression for the distribution of the number of points on cycles and the distribution of the number of points on the cycle which is reached from a specified point. He also considered the distribution of the number of points on "isolated cycles" (that is, cycles with no trees rooted at the cyclic points).

Lenard calculated the distribution of the successors of a random point and the distribution of the noncyclic predecessors of a random point. The point distribution of the in-degrees (number of immediate predecessors) of the various points was also obtained by Lenard as well as the probability that there are a specified number of points of a given degree. Specifically, the points of degree zero are the boundary points (terminal vertices) of the mapping. Lenard concludes his manuscript with the study of the distribution of the distance to the boundary from a random vertex.

REFERENCES

Folkert, J. E. (1955). *The Distribution of the Number of Components of a Random Mapping Function.* Ph. D. Dissertation, Michigan State University, East Lansing, Michigan.

Golomb, S. W., Welch, L. R. and Goldstein, R. M. (1959). Cycles from nonlinear shift registers. In: *Progress Report*, 20–389, *Jet Propulsion Laboratory,* California Institute of Technology, Pasadena, California.

Goncharov, V. L. (1944). Du domaine d'analyse combinatoire. *Bull. Acad. Sci. URSS, Ser. Math.* **8**, 3–48.

Greenwood, R. E. (1953). The number of cycles associated with the elements of a permutation group. *Amer. Math. Monthly* **60**, 407–409.

Harris, B. (1960). Probability distribution related to random mappings. *Ann. Math. Statist.* **31**, 1045–1062.

Katz, L. (1955). Probability of indecomposability of a random mapping function. *Ann. Math. Statist.* **26**, 512–517.

Kruskal, M. D. (1954). The expected number of components under a random mapping function. *Amer. Math. Monthly* **6**, 392–397.

Lenard, A. (1964). *Unpublished Manuscript.* Institute for Advanced Studies, Princeton University, Princeton, New Jersey, U.S.A.

Metropolis, N. and Ulam, S. (1953). A property of randomness of an arithmetical function. *Amer. Math. Monthly* **60**, 252–253.

Riordan, J. (1962). Enumeration of linear graphs for mappings of finite sets. *Ann. Math. Statist.* **33**, 178–185.

Rubin, H. and Sitgreaves, R. (1954). Probability distributions related to random transformations of a finite set. In: *Technical Report* 19A, Applied Mathematics and Statistics Laboratory, Stanford University, Stanford, California, U.S.A. Stanford, California, U.S.A.

Suschkewitsch, A. (1928). Untersuchungen über Verallgemeinerte Substitutionen. In: *Atti del Congresso Internazionale dei Mathematici*, Bologna, pp. 147–157.

APPENDIX 0
DESCRIPTION OF APPENDICES

In the appendices what follow, the various formulae obtained in the reference cited have been tabulated and indexed by the characteristic of the mapping that is being studied and the reference from which it has been obtained. The following notation will be employed:

$|A|$, for any set A denotes its cardinality

$K_\alpha(x)$, the component containing a specified (or randomly selected) point $x \in \mathfrak{X}_n$ under $\alpha \in J_n$

$a_{j,n}$, $j = 1,2,\ldots,n$; $n = 1,2,\ldots$, $j \leq n$, Stirling numbers of the first kind

γ, Euler's constant

c_m, number of cycles of length m, number of components of size m

m, length of cycle

S_n, symmetric group of n elements

$[a/b]$, greatest integer $\leq a/b$

D_n, nth derangement number (nearest integer to $n!/e$, $n > 0$, $D_0 = 1$)

$\sim$, asymptotically equal to (as $n \to \infty$)

$S_\alpha(x)$, successors of x under α, $x \in \mathfrak{X}_n$, $\alpha \in J_n$

$C_\alpha(x)$, cyclic successors of x under α, $x \in \mathfrak{X}_n$, $\alpha \in J_n$

$P_\alpha(x)$, predecessors of x under α, $x \in \mathfrak{X}_n$, $\alpha \in J_n$

C_α, cyclic elements of α, $\alpha \in J_n$

$\Phi(x)$, $\frac{1}{\sqrt{2\pi}} \int_{-\infty}^{x} e^{-t^2/2} dt$

$S_j^{(\alpha)}$, S_j, set of elements in $\mathfrak{X}_n$ with $\alpha^j(x) \in C_\alpha$, $\alpha^{j-1}(x) \notin C_\alpha$ ($S_0 = C_\alpha$)

I_j, $\mathbf{P}\{\alpha \in J_j \text{ is connected}\} \cdot j^j$

J_j, $\mathbf{P}\{\alpha \text{ is connected}\} \cdot (j-1)^j$

O_α, points on components with only cyclic elements

Q_k, vertices of degree k

Q_0, terminal vertices (boundary points)

$w(z)$, $\sum_{n=1}^{\infty} n^{-1} z^n/n!$

r, rk is maximum possible value of $|K_\alpha(x)|$ (Appendices 2.3, 2.4)

β^*, positive real root of $\sum_{j=2}^{r} (j-1)\beta^j/j! = 1$

APPENDIX 1.1
RANDOM PERMUTATIONS

Table 1. Components (cycles)

Characteristics	Formulae	References
$\mathbf{P}\{K_\alpha = j\}$	$\frac{\lvert\alpha_{ij}\rvert}{n!}, \quad j = 1, \dots, n$	(Harris, 1960)
$\mathbf{E}\{K_\alpha\}$	$\sum_{m=1}^{n} \frac{1}{m} \sim \log n + \gamma + o(1)$	(Golomb et al., 1959; Goncharov, 1944; Greenwood, 1953; Harris, 1960)
$\mathrm{Var}\{K_\alpha\}$	$\sum_{m=1}^{n} \frac{1}{m} - \sum_{m=1}^{n} \frac{1}{m^2}$ $\sim \log n + \gamma - \frac{\pi^2}{6} + o(1)$	(Goncharov, 1944; Greenwood, 1953; Harris, 1960)
asymptotic density of $(K_\alpha - \log n)/\sqrt{\log n}$	$\frac{1}{\sqrt{2\pi}} e^{-t^2/2}$	(Goncharov, 1944)
$\mathbf{P}\{\lvert K_\alpha(x)\rvert = j\}$	$\frac{1}{n}, \quad j = 1, \dots, n$	(Golomb et al., 1959; Harris, 1960)
$\mathbf{E}\{\lvert K_\alpha(x)\rvert\}$	$\frac{n+1}{2}$	
$\mathrm{Var}\{\lvert K_\alpha(x)\rvert\}$	$\frac{(n+1)(n-1)}{2}$	

Table 2. Length of cycles

Characteristics	Formulae	References
$\mathbf{E}(m)$	$\int_0^1 \frac{\Gamma(n+t)\,dt}{\Gamma(n)\Gamma(t+1)} \sim \frac{n}{\log n}$	(Goncharov, 1944)
$\mathbf{E}(c_m)$	$\frac{1}{m}$	(Goncharov, 1944)
$\mathrm{Var}(c_m)$	$\frac{1}{m}$	(Goncharov, 1944)
asymptotic distribution of c_m	$\frac{e^{-1/m}(1/m)^t}{t!} - P_0(m^{-1})$	(Goncharov, 1944)
asymptotic density of $\max\limits_{c_m \neq 0} \frac{m}{n}$	$\frac{1}{t}\sum_{h=0}^{\lambda-1} \frac{(-1)^h}{h!} I_h(t, 1-t)$, where $I_h(m,n) = \int \cdots \int_{\substack{t_i > m \\ \sum t_i < n}} (t_1, \ldots, t_h)^{-1}\, dt_1 \cdots dt_h$, $h \geq 1$; it is given by a sequence of analytic curves on $\frac{1}{j+1} \leq t \leq \frac{1}{j}$, $j = 1, 2, \ldots$, specifically $\frac{1}{t},\ \frac{1}{2} \leq t \leq 1,$ $\frac{1}{t}\left(1 - \log \frac{1-t}{t}\right),\ \frac{1}{3} \leq t \leq \frac{1}{2}$	(Goncharov, 1944)
$\mathbf{E}\left\{\max\limits_{c_m \neq 0} \frac{m}{n}\right\} = u_n$	$\frac{2}{3} + \frac{1}{3}\log\frac{27}{32} \leq u_n \leq 1 - e^{-1}$	(Golomb et al., 1959)
joint distribution of $(c_1, \ldots, c_n)$	$\prod_{r=1}^{n} r^{c_r} r! \Big/ \prod_{s=1}^{n} s^{c_s} c_s!,\quad \sum r c_r = n$	(Goncharov, 1944)
least probable configuration	$c_1 = n,\ c_j = 0,\ j \neq 1$	(Goncharov, 1944)
most probable configuration	$c_1 = 1,\ c_{n-1} = 1,\ c_j = 0,$ $j \neq 1,\ n-1$	(Goncharov, 1944)

Table 3. Generating functions for probabilities or expected values of characteristics of random parmutations

Characteristic	Coefficient of	Function	References
$(c_1, \ldots, c_n)$	$x_1^{c_1} \cdots x_n^{c_n}$	$\sum_{j=1}^{\infty} \frac{x_j}{j}$	(Goncharov, 1944)
cycle partitions of S_n indexed by c_j, $1 \leq j \leq n$	t^n	$\exp\left\{\sum_{j=1}^{\infty} \frac{t^j x_j}{j}\right\}$	(Goncharov, 1944)
K_α	$t^m x^n$	$\exp\left\{t \sum_{j=1}^{\infty} \frac{x^j}{j}\right\} = (1-x)^t$	(Goncharov, 1944)
K_α	t^m	$(n!)^{-1} t(t+1) \cdots (t+n-1)$	(Goncharov, 1944; Greenwood, 1953; Riordan, 1962)
K_α	x^n	$\frac{1}{m!} \log^m (1-x)^{-1}$	(Goncharov, 1944)
c_m	t^m	$\sum_{j=0}^{[n/m]} (j)^{-1} \left(\frac{t-1}{m}\right)^j$	(Goncharov, 1944)

APPENDIX 1.2
RANDOM PERMUTATIONS WITH NO FIXED POINTS

Table 1. Components

Characteristics	Formulae	References
$\mathrm{P}\{K_\alpha = k\}$	$\frac{n!}{D_n}\sum_{j=0}^{n}\frac{(-1)^j \alpha_{k-j,n-j}}{(n-j)!\,j!}$	(Harris, 1960)
$\mathrm{E}\{K_\alpha\}$	$\frac{n!}{D_n}\sum_{s=2}^{n}\frac{D_{n-s}}{s(n-s)!} \sim \log n + O(1)$	(Harris, 1960)
$\mathrm{Var}\{\|K_\alpha\}$	$\sim (2\gamma - 1)\log n + O(1)$	(Harris, 1960)
$\mathrm{P}\{\|K_\alpha(x)\| = j\}$	$\frac{(n-1)!\,D_{n-j}}{(n-j)!\,D_n}, \quad j = 1, \ldots, n$	(Harris, 1960)
$\mathrm{P}\{\|K_\alpha(x)\| = n-1\}$	$\sim \frac{1}{n}, \quad j \geq 2$	(Harris, 1960)
$\mathrm{P}\{\|K_\alpha(x)\| = n-2\}$	$\sim \frac{e}{2n}$	(Harris, 1960)
$\mathrm{P}\{\|K_\alpha(x)\| = n-3\}$	$\sim \frac{e}{3n}$	(Harris, 1960)

Table 2. Length of cycles

Characteristics	Formulae	References
joint distribution of $(c_2, \ldots, c_n)$	$\frac{n!}{D_n \prod_{j=2}^{n}(j^{c_j} c_j!)}, \quad \sum_{j=2}^{n} j c_j = n$	(Harris, 1960)

Table 3. Generating functions for probabilities or expected values of characteristics of random parmutations with no fixed points

Characteristic	Coefficient of	Function	References
$(c_1, \ldots, c_n)$	$x_1^{c_1} \cdots x_n^{c_n}$	$\frac{n!}{D_n}\exp\left\{\sum_{i=2}^{\infty}\frac{x_i}{i}\right\}$	(Harris, 1960)
K_α	$t^m x^n$	$\frac{n!}{D_n}\frac{e^{-tx}}{(1-x)^t}$	(Harris, 1960)

APPENDIX 2.1
RANDOM MAPPINGS

Table 1. Components

Characteristics	Formulae	References
$\mathbf{P}\{K_\alpha = j\}$	$n^{-n}\sum_{\mu=j}^{n}\frac{\alpha_{j,\mu}}{\mu!} \times \sum_{\substack{k_1,\ldots,k_n \\ k_i>0,\ \sum k_i=n}} \frac{n!}{k_1!\cdots k_n!}k_1^{k_1}\cdots k_\mu^{k_\mu}$	(Katz, 1955)
	$n^{-n}\frac{n!}{j!}\sum_{\mu=j}^{n}\alpha_{j,\mu} \times \sum_{\substack{k_1,\ldots,k_n \\ \sum k_i=\mu,\ \sum ik_i=n}} \frac{\mu!}{k_1!\cdots k_n!}\prod_{i=1}^{n}\left(\frac{i^i}{i!}\right)^{k_i}$	(Harris, 1960)
	$\sum_{i=1}^{n}\frac{(n-1)!\,i}{(n-i)!\,n^i}\lvert\alpha_{j,i}\rvert$	(Rubin and Sitgreaves, 1954)
$\mathbf{E}\{K_\alpha\}$	$\sum_{m=1}^{n}\frac{n!}{(n-m)!\,mn^m} \sim \frac{1}{2}(\log 2n+\gamma)+o(1)$	(Kruskal, 1954; Lenard, 1964; Riordan, 1962)
$\mathbf{P}\{\lvert K_\alpha(x)\rvert = j\}$	$\frac{(n-1)!}{(n-j)!(j-1)!}\frac{j^n(n-j)^{n-j}}{n^n} \times \sum_{l=1}^{j}\frac{(j-1)!}{(j-l)!\,j^l}$	(Rubin and Sitgreaves, 1954)
asymptotic density of $K_\alpha(x)/n$	$\frac{1}{2}(1-t)^{-1/2}, \quad 0 \le t \le 1$	(Lenard, 1964; Rubin and Sitgreaves, 1954)
$\mathbf{P}\{\alpha \text{ is connected}\}$ $= \mathbf{P}\{K_\alpha = 1\}$ $= \mathbf{P}\{K_\alpha(x) = n\}$	$\frac{(n-1)!}{n^n}\sum_{m=0}^{n-1}\frac{n^m}{m!} \sim \left(\frac{\pi}{2n}\right)^{1/2}$	(Harris, 1960; Lenard, 1964; Riordan, 1962)
	$\frac{n!}{n^n}\sum\frac{1}{S_0}\prod_{j=1}^{S_0}\left(\frac{(k_j+1)^{k_j-1}}{k_j!}\right)$, $S_0 \ge 1,\ k_i \ge 0,\ S_0+\sum_{i=1}^{n-1}k_i = n$	(Lenard, 1964)

Table 2. Sizes of components

Characteristics	Formulae	References
joint distribution of $(c_1, \ldots, c_n)$	$\dfrac{n! \prod_{j=1}^{n} I_j^{c_j}}{\prod_{j=1}^{n} (j!)^{c_j} c_j! \, n^n}$, $c_j > 0, \quad \sum_{j=1}^{n} j c_j = n$	(Harris, 1960; Lenard, 1964)
$\mathbf{E}\{c_k\}$	$\dfrac{n!}{n^n} \dfrac{(n-k)^{n-k}}{(n-k)!} \dfrac{I_k}{k!}$ $\sim I_k \dfrac{e^{-k}}{k!}$	(Lenard, 1964)
$\mathbf{E}\{c_k\}$	$\sim \dfrac{1}{2} \log \dfrac{b}{a}, \quad a\sqrt{n} < k < b\sqrt{n}$	(Lenard, 1964)
$\sum_{k>t} \mathbf{E}\{c_k\}$	$\sim \dfrac{1}{2} \log \dfrac{1 + \sqrt{1 - t/n}}{1 - \sqrt{1 - t/n}}$	(Lenard, 1964)
$\mathbf{E} \max_{c_m \neq 0} m/n = u$	$u \sim 0.758$ (conjectured)	(Lenard, 1964)
	$u \geq \dfrac{2}{3}$	(Lenard, 1964)

Table 3. Successors of x, predecessors of x and cyclic successors of x

Characteristics	Formulae	References
$\mathbf{P}\{\lvert S_\alpha(x)\rvert = j, \lvert C_\alpha(x)\rvert = k\}$	$\frac{(n-1)!}{(n-j)!\,n^j}, \quad 1 \le k \le j \le n$	(Harris, 1960; Lenard, 1964; Rubin and Sitgreaves, 1954)
$\mathbf{P}\{\lvert S_\alpha(x)\rvert = j\}$	$\frac{(n-1)!}{(n-j)!\,n^j}, \quad 1 \le j \le n$	(Harris, 1960; Lenard, 1964; Rubin and Sitgreaves, 1954)
$\mathbf{P}\{\lvert C_\alpha(x)\rvert = k\}$	$\sum_{j=k}^{n} \frac{(n-1)!}{(n-j)!\,n^j}, \quad 1 \le k \le n$	(Harris, 1960; Lenard, 1964; Rubin and Sitgreaves, 1954)
asymptotic density $\varphi(t,y)$ of $\left(\frac{\lvert S_\alpha(x)\rvert}{\sqrt{n}}, \frac{\lvert C_\alpha(x)\rvert}{\sqrt{n}}\right)$	$e^{-t^2/2}, \quad 0 < y \le t < \infty$	(Harris, 1960; Lenard, 1964)
asymptotic density of $\frac{\lvert S_\alpha(x)\rvert}{\sqrt{n}}$	$te^{-t^2/2}, \quad t > 0$	(Harris, 1960; Lenard, 1964; Rubin and Sitgreaves, 1954)
$\mathbf{E}\{\lvert C_\alpha(x)\rvert\}$	$= \mathbf{E}\left\{\frac{\lvert S_\alpha(x)\rvert + 1}{2}\right\} \sim \frac{(2\pi n)^{1/2}}{4}$	(Harris, 1960)
$\mathrm{Var}\{\lvert C_\alpha(x)\rvert\}$	$\sim n\left[\frac{2}{3} - \frac{2\pi}{16}\right]$	(Harris, 1960)
asymptotic density of $\frac{C_\alpha(x)}{\sqrt{n}}$	$\sqrt{2\pi}(1 - \Phi(t))$	(Harris, 1960; Lenard, 1964)
$\mathbf{P}\{\lvert P_\alpha(x)\rvert = j\}$	$\frac{(n-1)!\,j^{j-2}(n-j)^{n-j}}{(n-j)!(j-1)!\,n^{n-1}}$	(Harris, 1960; Rubin and Sitgreaves, 1954)
$\mathbf{P}\{\lvert S_\alpha(x)\rvert = q, \lvert(P_\alpha(x) \setminus \{x\}) \cap \overline{C_\alpha(x)}\rvert = k\}$	$(n-1)!(k+1)^{k-1} \times \frac{q(n-k-1)^{n-k-q}}{n^n k!(n-k-q)!}$	(Lenard, 1964)
$\mathbf{P}\{\lvert(P_\alpha(x) \setminus \{x\}) \cap \overline{C_\alpha(x)}\rvert = k\}$	$\frac{(k+1)^{k-1}q(n-1)!}{k!n^n} \times \sum_{q=1}^{n-k} \frac{(n-k+1)^{n-k-q}}{(n-k-q)!} \sim \frac{(k+1)^{k-1}e^{-(k+1)}}{k!}$	(Lenard, 1964)
$\mathbf{E}\{\lvert(P_\alpha(x) \setminus \{x\}) \cap \overline{C_\alpha(x)}\rvert\}$	$\sim \sqrt{\pi n/8}$	(Lenard, 1964)

Table 4. Cycles of α, strata of α

Characteristics	Formulae	References
probability of $\lvert S_0\rvert = j, \lvert S_1\rvert = n_1, \ldots, \lvert S_{n-1}\rvert = n_{n-1}, n_i \geq 0, \sum n_i = n$	$\dfrac{n!\, j!\, j^{n_1} n_1^{n_2} \cdots n_{n-2}^{n-1}}{j!\, n_1! \cdots n_{n-1}!}$	(Harris, 1960)
C_α	$\dfrac{(n-1)!\, j}{(n-j)!\, n^j}$	(Harris, 1960; Riordan, 1962; Rubin and Sitgreaves, 1954)
$\mathbf{P}\{\lvert C_\alpha\rvert \geq k\}$	$\dfrac{(n-1)!}{n^{k-1}(n-k)!} = \mathbf{P}\{\lvert S_\alpha(x)\rvert = j\}$	(Lenard, 1964)
asymptotic density of $\lvert C_\alpha\rvert/\sqrt{n}$	$te^{-t^2/2}$, $t > 0$ (coincides with that for $\lvert S_\alpha(x)\rvert/\sqrt{n}$)	(Harris, 1960; Lenard, 1964; Riordan, 1962)
	$\mathbf{E}\{\lvert S_\alpha(x)\rvert\} = \mathbf{E}\{\lvert P_\alpha(x)\rvert\} = \mathbf{E}\{\lvert C_\alpha\rvert\}$,	(Harris, 1960; Lenard, 1964)
	$\mathbf{E}\{\lvert S_\alpha(x)\rvert\}/n = \mathbf{E}\{\lvert S_\alpha(x)\rvert^{-1}\} = \mathbf{E}\{K_\alpha = 1\} = \mathbf{P}\{\lvert C_\alpha(x)\rvert = 1\} = \mathbf{E}\{\lvert P_\alpha(x)\rvert\}/n = \mathbf{E}\{\lvert C_\alpha\rvert\}/n$	(Harris, 1960; Lenard, 1964; Riordan, 1962)
$\mathbf{P}\{\lvert O_\alpha\rvert = j\}$	$(n-j)! \sum \prod_{i=2}^{n-j} \dfrac{[I_i - (n-1)!]^{c_i}}{c_i!\, i!}$, $\sum_{i=2}^{n-j} ic_i = n-j$	(Lenard, 1964)
	$\prod_{i=1}^{j} \left(1 - \dfrac{i}{n}\right)^{n-j} - \prod_{i=1}^{j+1} \left(1 - \dfrac{i}{n}\right)^{n-j-1}$	(Lenard, 1964)

Table 5. Degrees of vertices

Characteristics	Formulae	References
$\mathbf{P}\{\lvert\alpha^{-1}(x)\rvert = j\}$	$\binom{n}{j}\frac{(n-1)^{n-j}}{n^j n^{n-j}} \sim \frac{e^{-1}}{j!}$	(Harris, 1960; Lenard, 1964)
$\mathbf{P}\{\lvert Q_k\rvert = q\}$	$\frac{n!}{n^n}\binom{n}{q}\frac{1}{k!\,q} \times \sum_{\substack{k_i \neq k \\ \sum_{i=1}^{n-q} k_i + qk = n}} k_1! \cdots k_{n-q}!,$	(Lenard, 1964)
$\mathbf{E}\{\lvert Q_k\rvert\}$	$\binom{n}{k}\frac{(n-1)^{n-k}}{n^{n-1}}$	(Lenard, 1964)
asymptotic density of $\frac{(\lvert Q_k\rvert - n/(ek!))}{\sqrt{n}} \times \left(\frac{e^{-1}}{k!}\left(1 + \frac{2k-2-k^2}{ek!}\right)\right)$	$\frac{1}{\sqrt{2\pi}}e^{-t^2/2}$	(Lenard, 1964)
asymptotic density of $\frac{\lvert Q_0\rvert - n/e}{\sqrt{(n/e)(1-2/e)}}$	$\frac{1}{\sqrt{2\pi}}e^{-t^2/2}$	(Lenard, 1964)
asymptotic density of $\lvert Q_k\rvert$, $k, n \to \infty$, $n/(e(k+1)!) \to u > 0$	$e^{-u}, \quad u > 0$	(Lenard, 1964)
$\mathbf{P}\{$ distance from x to terminal vertex $\leq k\}$, $k \to \infty$	$\sim 1 - \frac{3}{k} + \frac{2}{3k^2}(\log k + \gamma)$	(Lenard, 1964)

Table 6. Generating functions for probabilities or expected values of characteristics of random mappings

Characteristic	Coefficient of	Function	References
$(c_1, \ldots, c_n)$	$x^{c_1} \cdots x_n^{c_n}$	$\frac{n!}{n^n} \exp\left\{\sum_{j=1}^{\infty} \frac{I_j x_j}{j!}\right\}$	(Harris, 1960)
component partition of J_n indexed by c_j, $1 \le j \le n$	t^n	$\frac{n!}{n^n} \exp\left\{\sum_{j=1}^{\infty} \frac{I_j t^j x_j}{j!}\right\}$	(Harris, 1960)
K_α	$t^k x^n$	$\frac{n!}{n^n} \exp\left\{\sum_{j=1}^{\infty} \frac{I_j x^j}{j!}\right\}$	(Harris, 1960)
		$\frac{n!}{n^n} \left(\sum \frac{j^j x^j}{j!}\right)^t$	(Harris, 1960)
$K_\alpha = 1$, α is connected	z^n	$w(z)$	(Lenard, 1964)
		$\frac{n!}{n^n}(-\log(1 - w(z)))$	(Lenard, 1964)
K_α	x^k	$\sum_{k=0}^{n-1} \frac{\binom{n-1}{k} x(x+1)\cdots(x+k)}{n^{k+1}}$	(Riordan, 1962)
$\mathrm{E}\{K_\alpha\}$	z^n	$\frac{n!}{n^n} \frac{1}{1 - w(z)} \log \frac{1}{1 - w(z)}$	(Lenard, 1964)
$\lvert O_\alpha \rvert$	z^{n-k}	$\frac{1 - z}{1 - w(z)}$	(Lenard, 1964)

APPENDIX 2.2
RANDOM MAPPINGS WITH NO FIXED POINTS

Table 1. Components

Characteristics	Formulae	References
α is connected, $K_\alpha = 1$	$\frac{(n-1)!}{(n-1)^n} \sum_{m=0}^{n-2} \frac{n^m}{m!} \sim e\left(\frac{\pi}{2(n-1)}\right)^{1/2}$	(Harris, 1960; Katz, 1955)
$\mathbf{P}\{K_\alpha = k\}$	$(n-1)^{-n} \sum_{\mu=k}^{[n/2]} \frac{\alpha_{k\mu}}{\mu!} \sum_{k_1,\ldots,k_\mu} \frac{n!}{k_1! \cdots k_\mu!} \times \prod_{i=1}^{\mu} (k_i - 1)^{k_i}$	(Folkert, 1955)
$\mathbf{P}\{K_\alpha = k\}$	$\frac{n!}{(n-1)^n} \sum_{\mu=k}^{[n/2]} \alpha_{k\mu} \times \sum_{\substack{k_2,\ldots,k_n \geq 0 \\ \sum k_i = \mu}} \prod \left(\frac{(i-1)^i}{i!}\right)^{k_i}$	(Harris, 1960)

Table 2. Sizes of components

Characteristics	Formulae	References
$(c_1, \ldots, c_n)$	$n! \prod_{j=2}^{n} J_j^{c_j} \Big/ \left((n-1)^n \prod_{j=2}^{n} (j!^{c_j} c_j!)\right)$, $c_j > 0$, $\sum_{j=2}^{n} jc_j = n$	(Harris, 1960)

Table 3. Successors of x, predecessors of x and cyclic successors of x

Characteristics	Formulae	References
$P\{\lvert S_\alpha(x)\rvert = j, \lvert C_\alpha(x)\rvert = k\}$	$\frac{(n-2)!}{(n-1)^{j-1}(n-j)!}$, $2 \le k \le j \le n$	(Harris, 1960)
$P\{\lvert S_\alpha(x)\rvert = j\}$	$\frac{(n-2)!(j-1)}{(n-1)^{j-1}(n-j)!}$, $2 \le j \le n$	(Harris, 1960)
$P\{\lvert C_\alpha(x)\rvert = k\}$	$\sum_{j=k}^{n} \frac{(n-2)!}{(n-1)^{j-1}(n-j)!}$, $2 \le k \le n$	(Harris, 1960)
asymptotic joint density of $\left(\frac{\lvert S_\alpha(x)\rvert}{\sqrt{n}}, \frac{\lvert C_\alpha(x)\rvert}{\sqrt{n}}\right)$	$e^{-t^2/2}$, $0 < t < y < \infty$	(Harris, 1960)
$P\{\lvert P_\alpha(x)\rvert = j\}$	$\frac{(n-1)!(n-j-1)^{n-j} j^{j-2}}{(n-1)^{n-1}(n-j)!(j-1)!}$, $j = 1, \ldots, n-2$ 0, $j = n-1$ $\frac{n!}{(n-1)^n} \sum_{i=2}^{n} \frac{n^{n-i-2}}{(n-i)!}$, $j = n$	(Harris, 1960)

Table 4. Cycles

Characteristics	Formulae	References
$P\{\lvert C_\alpha\rvert = j\}$	$\frac{n^{n-j} D_j \binom{n-1}{j-1}}{(n-1)^n}$	(Harris, 1960)
asymptotic density of $\frac{\lvert C_\alpha\rvert}{\sqrt{n}}$	$e^{-t^2/2}$	(Harris, 1960)

Table 5. Vertices

Characteristics	Formulae	References
$P\{\lvert \alpha^{-1}(x)\rvert = j\}$	$\binom{n-1}{j}\left(\frac{1}{n-1}\right)^j\left(\frac{n-2}{n-1}\right)^{n-j}$	(Harris, 1960)

Table 6. Generating functions for probabilities or expected values of characteristics of random mappings

Characteristic	Coefficient of	Function	References
$(c_1, \ldots, c_n)$	$x_1^{c_1} \cdots x_n^{c_n}$	$\frac{n!}{(n-1)^n} \exp\left\{\sum_{j=2}^{\infty} \frac{J_j x_j}{j!}\right\}$	(Harris, 1960)
component partition of J_n indexed by c_j, $2 \leq j \leq n$	t^n	$\frac{n!}{(n-1)^n} \exp\left\{\sum_{j=2}^{\infty} \frac{J_j t^j x_j}{j!}\right\}$	(Harris, 1960)
K_α	$t^k x^n$	$\frac{n!}{(n-1)^n} \exp\left\{t \sum_{j=2}^{\infty} \frac{J_j x^j}{j!}\right\}$ $= \frac{n!}{(n-1)^n} \sum \left[\frac{(i-1)^i x^i}{i!}\right]^t$	(Harris, 1960)

APPENDIX 2.3
RANDOM MAPPINGS WITH DEGREE OF VERTICES EITHER 0 OR k

Table 1. Components

Characteristics	Formulae	References
$\mathbf{P}\{K_\alpha = i\}$	$\sum_{j=1}^{r} \frac{k^{j-1} j(k-1)(r-1)!(kr-j-1)!\lvert\alpha_{ij}\rvert}{(kr-1)(r-j)!}$	(Rubin and Sitgreaves, 1954)
$\mathbf{P}\{K_\alpha = 1\}$, α is connected	$\sum_{j=1}^{r} \frac{k^{j-1}(k-1)(r-1)!(kr-j-1)!}{(kr-1)!(r-j)!}$	(Rubin and Sitgreaves, 1954)
$\mathbf{P}\{\lvert K_\alpha(x)\rvert = jk\}$	$\frac{(k(r-j))!(r-1)!r(k-1)!}{(kr)!(j(k-1)-1)!(r-j)!((r-j)(k-1))!} \sum_{t=1}^{j} \frac{(kj-t-1)^j k^{t-1}}{(j-t)!}$	(Rubin and Sitgreaves, 1954)
$\mathbf{E}\{K_\alpha\} \sim \mathrm{Var}\{K_\alpha\}$	$\sim \frac{1}{2} \log \frac{kr}{k-1}$	(Rubin and Sitgreaves, 1954)
asymptotic density of $\frac{K_\alpha - \mathbf{E}\{K_\alpha\}}{\sqrt{\mathrm{Var}\{K_\alpha\}}}$	$\frac{1}{\sqrt{2\pi}} e^{-t^2/2}$	(Rubin and Sitgreaves, 1954)
asymptotic density of $\frac{\lvert K_\alpha(x)\rvert}{kr}$	$\frac{1}{2(1-t)^{1/2}}$	(Rubin and Sitgreaves, 1954)

Table 2. Cycles

Characteristics	Formulae	References
$\mathbf{P}\{\lvert C_\alpha\rvert = j\}$	$\frac{k^{j-1} j(k-1)!(r-1)!(kr-j-1)!}{(kr-1)!(r-j)!}$, $1 \le j \le r$	(Rubin and Sitgreaves, 1954)
asymptotic density of $\frac{\lvert C_\alpha\rvert (k-1)^{1/2}}{kr}$	$te^{-t^2/2}$	(Rubin and Sitgreaves, 1954)

Table 3. Successors of x, predecessors of x and cyclic successors of x

Characteristics	Formulae	References
$\mathbf{P}\{\lvert S_\alpha(x)\rvert = j,$ $\lvert C_\alpha(x)\rvert = q, q \neq j\}$	$\frac{(k-1)k^{j-2}(r-1)!(kr-j+1)!}{(kr)!(r-j+1)!}$, $2 \leq j \leq r+1$	(Rubin and Sitgreaves, 1954)
$\mathbf{P}\{\lvert S_\alpha(x)\rvert = j,$ $\lvert C_\alpha(x)\rvert = j\}$	$\frac{k^{j-1}(r-1)!(kr-1)!}{(kr)!(r-j)!}$, $1 \leq j \leq n$	(Rubin and Sitgreaves, 1954)
$\mathbf{P}\{\lvert S_\alpha(x)\rvert = 1\}$	$\frac{1}{kr}$	(Rubin and Sitgreaves, 1954)
$\mathbf{P}\{\lvert S_\alpha(x)\rvert = j \neq 1\}$	$\frac{(k-1)(j-1)k^{j-2}(r-1)!(kr-j+1)!}{(kr)!(r-j+1)!}$, $2 \leq j \leq r+1$	(Rubin and Sitgreaves, 1954)
$\mathbf{P}\{\lvert C_\alpha(x)\rvert = q\}$	$\frac{k^q(r-1)!(kr-q)!}{(kr)!(r-q)!} \times \sum_{j=q+1}^{r} \frac{(k-1)k^{j-1}(r-1)!(kr-j)!}{(kr)!(r-j)!}$	(Rubin and Sitgreaves, 1954)
$\mathbf{P}\{\lvert P_\alpha(x)\rvert = 1\}$	$\frac{k-1}{k}$	(Rubin and Sitgreaves, 1954)
$\mathbf{P}\{\lvert P_\alpha(x)\rvert = jk\}$	$\frac{(r-1)!(r(k-1))!(kj-1)!(k(r-j))!}{(j-1)!(j(k-1))!(r-j)!((r-j)(k-1))!(rk)!}$	(Rubin and Sitgreaves, 1954)
$\mathbf{P}\{\lvert P_\alpha(x)\rvert = jk+1\}$	$\frac{(k(r-j))!(jk)!r!(r(k-1))!}{(r-j)!((k-1)(r-j)-1)!j!(j(k-1)+1)!(rk)!rk}$	(Rubin and Sitgreaves, 1954)
asymptotic density of $\frac{\lvert S_\alpha(x)\rvert - 1}{\sqrt{kr}}$	$(k-1)te^{-(k-1)t^2/2}$	(Rubin and Sitgreaves, 1954)
$\mathbf{P}\{\lvert C_\alpha(x)\rvert = q\}$	$\sim q\left(\frac{2\pi(k-1)}{kr}\right)^{1/2}\Phi\left(\sqrt{\frac{k-1}{kr}}\right)$	(Rubin and Sitgreaves, 1954)
$\mathbf{P}\{\lvert P_\alpha(x)\rvert = jk\}$	$\to 0, \quad r \to \infty$	(Rubin and Sitgreaves, 1954)
$\mathbf{P}\{\lvert P_\alpha(x)\rvert = jk+1\}$	$\sim (2\pi j^{3/2})^{-1/2}[(k-1)k]^{-1/2}$	(Rubin and Sitgreaves, 1954)
asymptotic density of $\frac{\lvert C_\alpha(x)\rvert}{\sqrt{n}}$	$(2\pi(k-(k-1)\beta^*))^{1/2}\Phi(k(k-(k-1)\beta^*)^{1/2})$	(Rubin and Sitgreaves, 1954)

APPENDIX 2.4
RANDOM MAPPINGS WITH DEGREE OF VERTICES NOT EXCEEDING k

Characteristics	Formulae	References
$\mathbf{E}\{K_\alpha\}$	$\sim \frac{\log n}{2}$	(Rubin and Sitgreaves, 1954)
$\mathrm{Var}\{K_\alpha\}$	$\sim \frac{\log n}{2}$	(Rubin and Sitgreaves, 1954)
asymptotic density of $\frac{K_\alpha - \mathbf{E}\{K_\alpha\}}{\sqrt{\mathrm{Var}\{K_\alpha\}}}$	$\frac{1}{\sqrt{2\pi}} e^{-t^2/2}$	(Rubin and Sitgreaves, 1954)
asymptotic density of $\frac{\lvert S_\alpha(x)\rvert - 1}{\sqrt{n}}$	$[k-(k-1)\beta^*]te^{-(t^2/2)(k-(k-1)\beta^*)}$	(Rubin and Sitgreaves, 1954)
asymptotic density of $\frac{\lvert C_\alpha(x)\rvert}{\sqrt{n}}$	$(2\pi(k-(k-1)\beta^*))^{1/2}\Phi(k(k-(k-1)\beta^*)^{1/2})$	(Rubin and Sitgreaves, 1954)
asymptotic density of $\frac{\lvert K_\alpha(x)\rvert}{\sqrt{n}}$	$\sim \frac{1}{2(1-t)^{1/2}}, \quad 0 < t < 1$	(Rubin and Sitgreaves, 1954)

Probabilistic Methods in Discrete Mathematics, pp. 23 – 27
V. F. Kolchin et al. (Eds.)
1993 TVP/VSP

ON GONCHAROV'S WORKS IN THE FIELD OF COMBINATORICS

V. F. KOLCHIN
Steklov Mathematical Institute, Vavilov Street 42, 117966 Moscow GSP-1, Russia

ABSTRACT

The papers of V. L. Goncharov in the field of combinatorics were the first papers where the probabilistic approach and methods based on generating and characteristic functions are systematically used in studies of combinatorial objects such as permutations and (0,1)-sequences. The importance of these works for the development of probabilistic combinatorics cannot be overestimated, besides, now they are not easily available, therefore it seems reasonable to give here a short review of the results proved in these papers.

Probabilistic combinatorics in our country has deep-rooted traditions which are based mainly on the very high level of the Russian school of probability theory. Combinatorics has played an important role in the development of probability theory at times of its formation. Now, when the interest in discrete models is regenerated, the probability theory which started with rigorous solutions of various combinatorial problems, begins to pay off a debt to combinatorics and suggests new approaches and methods for solving problems of discrete mathematics. Application of the analytic methods well developed in probability theory, and partucularly methods of asymptotic analysis, leads to success in complicated problems of combinatorial nature.
The development of probabilistic combinatorics in our country is related with the names of S. N. Bernstein, N. V. Smirnov, V. E. Stepanov and others. But an outstanding contribution to probabilistic combinatorics was made by Vasilii Leonidovich Goncharov (1896–1955) whose influence on the development of this area of mathematics cannot be overestimated.
In his works the probabilistic methods based on the use of generating and characteristic functions were applied in studies of the cyclic structure of permutations and sequences of zeros and ones. The results obtained by Goncharov became classic and had a great influence on the later investigations in this area of mathematics. Presenting examples of investigations of top quality, these papers brought the requirements to results acceptable for publication to a very high level and thereby eliminated the possibility to accept and publish papers of low quality.
Although the paper (Goncharov, 1944) originally published in 1944 was translated into English (Goncharov, 1962), the papers (Goncharov, 1942, 1943, 1944) are not easily available, so we found it reasonable to give here a short review of the results

obtained by Goncharov. Moreover, these results did not loss their significance and given here without editing they have a quite modern form.
The best known Goncharov's results in combinatorics are the results on the cyclic structure of permutations.
The probabilistic approach which now is a commonplace in studies of combinatorial objects was firstly formulated in an explicit form and applied in the investigations of the set S_n of all permutations of degree n. On the set S_n one defines the uniform distribution which assigns equal probabilities to each of $n!$ permutations of the set. Let α_r be the number of cycles of length r in a random permutation from S_n. It is shown that for any non-negative integers $m_1, \ldots, m_n$ such that $m_1 + 2m_2 + \cdots + nm_n = n$

$$\mathbf{P}\{\alpha_1 = m_1, \ldots, \alpha_n = m_n\} = \prod_{r=1}^{n} \frac{1}{m_r! r^{m_r}}$$

and the generating function

$$\sum_{n=0}^{\infty} \sum_{m_1+2m_2+\cdots+nm_n=n} \mathbf{P}\{\alpha_1 = m_1, \cdots, \alpha_n = m_n\} t^{m_1} \cdots t_n^{m_n}$$

as a formal power series is equal to

$$F(t_1, t_2, \ldots) = \exp\left\{\sum_{k=1}^{\infty} \frac{t_k}{k}\right\}.$$

For the total number $M^{(n)} = \alpha_1 + \cdots + \alpha_n$, it is proved that as $n \to \infty$

$$\mathbf{E}\, M^{(n)} = \log n + \gamma + o(1),$$

$$\sqrt{\operatorname{Var} M^{(n)}} = \sqrt{\log n} - (\pi^2/2 - \gamma/2)/\sqrt{\log n} + O(1/\sqrt{\log n}),$$

where γ is the Euler constant.
The random variable $(M^{(n)} - \log n)/\sqrt{\log n}$ has in limit, as $n \to \infty$, the Gaussian distribution with parameters $(0, 1)$.
It is also shown that for the random variable α_r

$$\mathbf{E}\, \alpha_r = 1/r, \quad \operatorname{Var} \alpha_r = 1/r$$

and the random variable α_r has in limit, as $n \to \infty$, the Poisson distribution with parameter $1/r$.
Denote $\mu^{(n)}$ the length of the maximum cycle of a random permutation from S_n. It is shown that

$$\mathbf{P}\{\mu^{(n)} < m\} = \sum_{h=0}^{\infty} \frac{(-1)^h}{h!} S_h(m, n),$$

where

$$S_0(m, n) = 1, \quad S_h(m, n) = \sum_{\substack{k_1+\cdots+k_h \le n, \\ k_1, \ldots, k_h \ge m}} \frac{1}{k_1 \cdots k_h}, \qquad h \ge 1.$$

Let

$$I_0(x, 1-x) = 1,$$

$$I_h(x, 1-x) = \int_{\substack{x_1+\cdots+x_h<1-x,\\ x_1,\ldots,x_h>x}} \frac{dx_1 \cdots dx_h}{x_1 \cdots x_h}, \qquad 0 < x < 1.$$

It is proved that, as $n \to \infty$, the random variable $\mu^{(n)}/n$ has the limit distribution with the density

$$\varphi(x) = \frac{1}{x} \sum_{h=0}^{\lambda-1} \frac{(-1)^h}{h!} I(x, 1-x), \qquad \frac{1}{1+\lambda} \le x \le \frac{1}{\lambda},$$

which, as is clear from the given formula, is defined by different analytic expressions on the sequential intervals of the form $[1/(1+\lambda), 1/\lambda]$, where λ is an integer. For example,

$$\varphi(x) = \frac{1}{x}, \qquad \frac{1}{2} \le x \le 1;$$

$$\varphi(x) = \frac{1}{x}\left(1 - \log\frac{1-x}{x}\right), \qquad \frac{1}{3} \le x \le \frac{1}{2}.$$

Goncharov's results on random sequences consisting of zeros and ones are less known than the results on permutations. Let us consider a Bernoulli sequence or a sequence of independent trials with two outcomes, 0 and 1, such that at each trial the outcome 1 appears with probability p and the probability of the outcome 0 is equal to $q = 1 - p$. In (Goncharov, 1943, 1944) Goncharov investigates the characteristics of runs of units in Bernoulli sequences. Let $M^{(n)}$ be the total number of runs in n Bernoulli trials with fixed probability p, $0 < p < 1$. It is shown that the characteristic function of the random variable

$$(M^{(n)} - \mathbf{E}\, M^{(n)})/\sqrt{\operatorname{Var} M^{(n)}}$$

tends, as $n \to \infty$, to the characteristic function of the Gaussian law with parameters $(0, 1)$.

Let $M^{(n)}(r)$ be the number of runs of length r in n trials. Goncharov proves that as $n \to \infty$ the limit distribution of the random variable

$$(M^{(n)}(r) - \mathbf{E}\, M^{(n)}(r))/\sqrt{\operatorname{Var} M^{(n)}(r)}$$

is Gaussian.

Of particular interest are the results on the limit distribution of the random variable $\Lambda^{(n)}$ which is the length of maximum run of units in n trials. In (Goncharov, 1943, 1944) Goncharov proved that the generating function $\varphi_m(t)$ of the probabilities $\Pi_m^{(n)} = \mathbf{P}\{\Lambda^{(n)} < m\}$ is rational and equal to

$$\varphi_m(t) = \sum_{n=0}^{\infty} \Pi_m^{(n)} t^n = \frac{1 - p^m t^m}{1 - t + p^m q t^{m+1}}.$$

Decomposing it into partial fractions and analyzing the roots of the denominator Goncharov obtains an asymptotic formula for $\Pi_m^{(n)}$. Let

$$m = [\log n/(-\log p)], \qquad \theta_n = \{\log n/(-\log p)\},$$

i.e., let m be the integer part and θ_n be the fractional part of $\log n/(-\log p)$. It is shown that for any fixed integer l, as $n \to \infty$,

$$\mathbf{P}\{\Lambda^{(n)} < m + l\} \sim e^{-qp^{l-\theta_n}}.$$

Thus, the random variable $\Lambda^{(n)} - m$ has no a limit distribution, but for any α, $0 \le \alpha \le 1$, there exists a sequence n_k such that $\theta_{n_k} \to \alpha$ and for any fixed integer l

$$\mathbf{P}\{\Lambda^{(n_k)} < m_k + l\} \to e^{-qp^{l-\alpha}},$$

where $m_k = [\log n_k/(-\log p)]$.
This phenomenon, described firstly by Goncharov, is rather general and later it was found for many other random variables in similar situations, for example, for the maximum occupancy number in the problems of random allocations (Kolchin et al., 1978) and for the maximum size of components in various sets of random graphs (Kolchin, 1986).
Concluding this short survey I would like to express the deep admiration for the papers of V. L. Goncharov which are ahead of his time by several decades.
For completeness we present here the list of other Goncharov's papers which we succeeded to find.

REFERENCES

Goncharov, V. L. (1937a): Interpolation processes and entire functions. *Uspekhi Math. Sci.* **3**, 113–143 (in Russian).

Goncharov, V. L. (1937b). On the interpolation of functions with finite number of singularities with the help of rational equations. *Soviet Math. Izv.* **2**, 171–184 (in Russian).

Goncharov, V. L. (1941). On correcting factors of interpolation processes. *Soviet Math. Dokl.* **32**, No. 7, 471–473 (in Russian).

Goncharov, V. L. (1942). On the distribution of cycles in permutations. *Soviet Math. Dokl.* **35**, No. 9, 299–301 (in Russian).

Goncharov, V. L. (1943). On the alternation of events in a sequence of Bernoulli trials. *Soviet Math. Dokl.* **36**, No. 9, 295–297 (in Russian).

Goncharov, V. L. (1944). On the field of combinatorics. *Soviet Math. Izv., Ser. Math.* **8**, 3–48 (in Russian).

Goncharov, V. L. (1948). On an interpolation scheme of a general form. *Soviet Math. Dokl.* **59**, No. 9, 1529–1532 (in Russian).

Goncharov, V. L. (1950). Sergei Natanovich Bernstein. The seventieth anniversary. *Uspekhi Math. Sci.* **3**, 172–183 (in Russian).

Goncharov, V. L. (1962). On the field of combinatory analysis. *Trans. Amer. Math. Soc.* **19**, 1–46.

Goncharov, V. L. and Kolmogorov, A. N. (1940). The sixtieth anniversary of Sergei Natanovich Bernstein. *Soviet Math. Izv.* 4, No. 3, 249–260 (in Russian).
Kolchin, V. F., Sevastyanov, B. A., and Chistyakov, V. P. (1978). *Random Allocations*. John Wiley, New York.
Kolchin, V. F. (1986). *Random Mappings*. Springer, New York.

Probabilistic Methods in Discrete Mathematics, pp. 28 – 39
V. F. Kolchin et al. (Eds.)
1993 TVP/VSP

PROBABILITY DISTRIBUTIONS IN PRIME NUMBER THEORY

S. W. GOLOMB

Communication Sciences Institute, EEB-504A, University of Southern California, Los Angeles, CA 90089-2565, U.S.A.

ABSTRACT

For any probability distribution $D = \{\alpha(n)\}$ on $\mathbf{Z}^+$, we define $\beta(m) = \sum_{j=1}^{\infty} \alpha(jm)$, the probability in D that a "random" integer is a multiple of m; and $\gamma(k) = \sum_{d|k} \mu(d) \times \beta(d)$, the probability in D that a "random" integer is relatively prime to k. We specialize this general situation to three important *families* of distributions: $D_s = \{\alpha_s(n)\} = \{n^{-s}/\zeta(s)\}$ for $s > 1$ (the *Dirichlet* family); $L_z = \{\alpha_z(n)\} = \{(1-z)z^{n-1}\}$ for $0 < z < 1$ (the *Lambert* family); and $U_N = \{\alpha_N(n)\} = \{1/N, 1 \le n \le N; 0, n > N\}$ for $N \in \mathbf{Z}^+$ (the *finite uniform* family). Several basic results and concepts from analytic prime number theory are revisited from the perspective of these families of probability distributions, and the Shannon entropy for each of these families is determined. Other families of distributions are also considered.

1. INTRODUCTION

When analytic number theory was being formulated in the nineteenth century, probability theory was not yet a reputable branch of mathematics. Moreover, one is tempted to make probabilistic statements as if there were a uniform probability distribution on the positive integers, which of course does not exist.

We will see that much of classical number theory deals with three families of (genuine) probability distributions on the positive integers, and each family is primarily studied as its critical parameter tends to a value at which the fictitious uniform distribution would be attained. Abelian and Tauberian theorems play an important role in interpreting the significance and validity of relationships obtained when the critical parameter is allowed to go to its limiting value. Euler products, formulas for the $(n+1)$st prime, averages of arithmetic functions, and various identities are readily seen to have direct probabilistic interpretations. Also, the Shannon entropy can be calculated in closed form for each of the three families of distributions as a function of the critical parameter of that family.

This paper incorporates and adds to results contained in (Golomb, 1960; 1966; 1970 a, b; 1974; 1976).

2. GENERAL PROBABILISTIC RELATIONSHIPS

Definition 1. Let $\{\alpha(n)\}$ be a probability distribution D on $\mathbf{Z}^+$, so that $0 \le$

$\alpha(n) \leq 1$ for all $n \in \mathbf{Z}^+$, with $\sum_{n=1}^{\infty} \alpha(n) = 1$.

Definition 2. Let $\beta(m)$ be the probability that a positive integer selected randomly from the distribution D is a multiple of m, for each $m \in \mathbf{Z}^+$.

Clearly

$$\beta(m) = \sum_{j=1}^{\infty} \alpha(jm). \tag{1}$$

Definition 3. Let $\gamma(k)$ be the probability that a positive integer selected randomly from the distribution D is relatively prime to k, for each $k \in \mathbf{Z}^+$.

Clearly

$$\gamma(k) = 1 - \sum_{p|k} \beta(p) + \sum_{p_1 p_2|k} \beta(p_1 p_2) - \cdots = \sum_{d|k} \mu(d)\beta(d), \tag{2}$$

where $\mu(d)$ is the Möbius function, and $p_1, p_2, \ldots$ are distinct primes.

Suppose that $f(n)$ and $F(n)$ are two arithmetic functions related by

$$F(n) = \sum_{d|n} f(d). \tag{3}$$

Then we readily derive the formal identity

$$\sum_{n=1}^{\infty} F(n)\alpha(n) = \sum_{n=1}^{\infty} \alpha(n) \sum_{m|n} f(m) = \sum_{m=1}^{\infty} f(m) \sum_{\substack{j=1 \\ (n=jm)}}^{\infty} \alpha(jm) = \sum_{m=1}^{\infty} f(m)\beta(m), \tag{4}$$

which is valid analytically if the left-most and right-most sums both converge absolutely.

Regarding (2) as a special case of (3), and substituting into (4), we get

$$\sum_{n=1}^{\infty} \alpha(n)\gamma(n) = \sum_{m=1}^{\infty} \mu(m)\beta^2(m), \tag{5}$$

where the left side clearly converges absolutely, since $|\gamma(n)| \leq 1$ for all n. Each side of (5) can be interpreted directly as the probability that if two integers are selected randomly from the distribution D, they will be relatively prime: the left-side viewpoint is that any integer n can be selected first, with probability $\alpha(n)$, and then the second integer has probability $\gamma(n)$ of being relatively prime to it; the right-side viewpoint is that for each prime p, the first integer is divisible by p with probability $\beta(p)$, as is the second integer, independently of the first, so that the two integers have the common prime factor p with probability $\beta^2(p)$, and by the usual inclusion-exclusion argument, the probability that they have no prime factor p in common is $\sum_{m=1}^{\infty} \mu(m)\beta^2(m)$.

Another instance of (4), when $f(m) = 1$ for all m, is

$$\sum_{n=1}^{\infty} \tau(n)\alpha(n) = \sum_{m=1}^{\infty} \beta(m), \tag{6}$$

where $\tau(n)$ is the "divisor function" (the number of divisors of n). Since it is known (Titchmarsh, 1951, p. 262) that the average order of $\tau(n)$ is $\log n$, the sum in (6) may diverge in some cases where $\sum_{n=1}^{\infty} \alpha(n)$ converges very slowly.

Definition 4. Let $\delta(n)$ be the probability that a positive integer selected randomly from the distribution D is a divisor of n.
Clearly

$$\delta(n) = \sum\nolimits_{d|n} \alpha(d). \tag{7}$$

From (7), with (3) and (4), we find the initially surprising identity

$$\sum_{n=1}^{\infty} \alpha(n)\delta(n) = \sum_{m=1}^{\infty} \alpha(m)\beta(m), \tag{8}$$

which expresses the truism that if two integers are selected "at random", the probability that the first is a divisor of the second is the same as the probability that the second is a multiple of the first.

Definition 5. Let $\beta(m, a)$ be the probability that a positive integer selected randomly from the distribution D is congruent to a modulo m, for each a, $1 \leq a \leq m$.
Clearly,

$$\beta(m, a) = \sum_{j=0}^{\infty} \alpha(jm + a), \tag{9}$$

and $\beta(m, m) = \beta(m)$ as per Definition 2.
Each of the next three sections explores the effects of specializing $\alpha(n)$, $\beta(m)$, and $\gamma(k)$ from an arbitrary distribution D to a *specific family* of distributions on $\mathbf{Z}^+$.

3. THE DIRICHLET FAMILY OF DISTRIBUTIONS

For each real $s > 1$, wé define (Golomb, 1970 a) the *Dirichlet distribution* $D_s = \{\alpha_s(n)\}$ on $\mathbf{Z}^+$ by

Definition 6. Let $\alpha_s(n) = \frac{1}{\zeta(s)} n^{-s}$, where $\zeta(s) = \sum_{n=1}^{\infty} n^{-s}$ is the Riemann Zeta Function.
Clearly $\sum_{n=1}^{\infty} \alpha_s(n) = \frac{1}{\zeta(s)} \sum_{n=1}^{\infty} n^{-s} = 1$.
For the distribution D_s, we have

$$\beta_s(m) = \sum_{j=1}^{\infty} \alpha_s(jm) = \frac{1}{\zeta(s)} \sum_{j=1}^{\infty} \frac{1}{(jm)^s} = \frac{1}{\zeta(s)} m^{-s} \sum_{j=1}^{\infty} j^{-s} = m^{-s}. \tag{10}$$

That is, in D_s the probability that a "random" integer is divisible by m is m^{-s}, and therefore the probability that a "random" integer is *not* divisible by m is $1 - m^{-s}$.
If $(a, b) = 1$, the probability (in D_s) that a "random" integer is divisible by both a and b is the probability that it is divisible by ab; hence

$$\beta_s(ab) = (ab)^{-s} = a^{-s}b^{-s} = \beta_s(a)\beta_s(b). \tag{11}$$

Thus, if $(a,b)=1$, divisibility by a and divisibility by b are statistically independent. We thus have:

$$\gamma_s(k)=\sum_{d|k}\mu(d)d^{-s}=\prod_{p|k}(1-p^{-s}). \tag{12}$$

Note that although $\{\alpha_s(n)\}$ is no longer a probability distribution at $s=1$ (since $\zeta(s)$ has a pole at $s=1$), both $\beta_s(m)$ and $\gamma_s(k)$ behave reasonably at $s=1$. Specifically, at $s=1$, the probability that a "random" positive integer is a multiple of m is $\beta_1(m)=1/m$, and the probability that it is *not* divisible by m is $1-\beta_1(m)=1-1/m$. Also, the probability that a "random" positive integer will be relatively prime to k is $\gamma_1(k)=\varphi(k)/k=\sum_{d|k}\mu(d)/d=\prod_{p|k}(1-1/p)$.
Going back to the case $s>1$, we ask: "What is the probability, in D_s, that a "random" positive integer is divisible by *none* of the primes?" Since the only $n\in\mathbf{Z}^+$ with no prime factors is $n=1$, the answer must be $\alpha_s(1)=1/\zeta(s)$. On the other hand, "n has no prime factors" is equivalent to "n is relatively prime to 0". Hence

$$\alpha_s(1)=\frac{1}{\zeta(s)}=\gamma_s(0)=\sum_{d=1}^{\infty}\frac{\mu(d)}{d^s}=\prod_{\text{all } p}\left(1-\frac{1}{p^s}\right), \tag{13}$$

which yields Euler's product identity for $\zeta(s)$.
What is the probability that if *two* positive integers, a and b, are selected at random from D_s, they will be relatively prime? From (5), this probability is given by

$$\sum_{n=1}^{\infty}\alpha_s(n)\gamma_s(n)=\sum_{m=1}^{\infty}\mu(m)\beta_s^2(m)=\sum_{m=1}^{\infty}\frac{\mu(m)}{m^{2s}}=\frac{1}{\zeta(2s)}. \tag{14}$$

This is also well-behaved at $s=1$, and justifies the assertion that if two integers are chosen "at random" in $\mathbf{Z}^+$ (i.e., in the fictitious but intuitive uniform distribution), the probability that they will be relatively prime is $1/\zeta(2)=6/\pi^2$.
If $F(n)=\sum_{d|n}f(d)$, then by (4) we have

$$\frac{1}{\zeta(s)}\sum_{n=1}^{\infty}\frac{F(n)}{n^s}=\sum_{m=1}^{\infty}\frac{f(m)}{m^s}, \tag{15}$$

a well-known formal identity (Titchmarsh, 1951, p. 4).
We also have, for $1\le a\le m$, and $m\in\mathbf{Z}^+$,

$$\beta_s(m,a)=\sum_{j=0}^{\infty}\alpha_s(jm+a)=\frac{1}{\zeta(s)}\sum_{j=0}^{\infty}\frac{1}{(jm+a)^s}=\frac{\zeta(s,r)}{\zeta(s)}m^{-s}, \tag{16}$$

where $r=a/m$, and the generalized zeta-function

$$\zeta(s,r)=\sum_{n=0}^{\infty}\frac{1}{(n+r)^s},\qquad 0<r\le 1,\quad \mathrm{Rl}\,(s)>1 \tag{17}$$

is as defined in (Titchmarsh, 1951, p. 36), with $\zeta(s,1)=\zeta(s)$.

As we have seen, the critical limits for the Dirichlet family of distributions are the ones taken as $s \to 1^+$.

4. THE LAMBERT FAMILY OF DISTRIBUTIONS

For each real $z \in (0,1)$, we define the *Lambert distribution* $L_z = \{\alpha_z(n)\}$ on $\mathbf{Z}^+$ by

Definition 7. Let $\alpha_z(n) = (1-z)z^{n-1}$ for each $n \in \mathbf{Z}^+$. (This is a *geometric distribution* for each $z \in (0,1)$.)
In L_z, we have

$$\begin{aligned}\beta_z(m) &= \sum_{j=1}^{\infty} \alpha_z(jm) = (1-z)\sum_{j=1}^{\infty} z^{jm-1} = \frac{1-z}{z}\sum_{j=1}^{\infty}(z^m)^j \\ &= \frac{1-z}{z}\cdot\frac{z^m}{1-z^m} = \frac{z^{m-1}}{1+z+z^2+\cdots+z^{m-1}}.\end{aligned} \tag{18}$$

We also have

$$\gamma_z(k) = \sum_{d|k}\mu(d)\beta(d) = \frac{1-z}{z}\sum_{d|k}\frac{\mu(d)z^d}{1-z^d}. \tag{19}$$

If $F(n) = \sum_{d|n} f(d)$, then by (4) we have, formally,

$$(1-z)\sum_{n=1}^{\infty} F(n)z^{n-1} = (1-z)\sum_{m=1}^{\infty}\frac{f(m)z^{m-1}}{1-z^m}, \tag{20}$$

which is obviously equivalent to

$$\sum_{n=1}^{\infty} F(n)z^n = \sum_{m=1}^{\infty}\frac{f(m)z^m}{1-z^m}, \tag{21}$$

the usual form of Lambert's identity.
Let $P_r = p_1p_2\dots p_r$ be the product of the first r primes. Then, from (19), the probability that an integer from L_z is divisible by none of the first r primes is

$$\gamma_z(P_r) = \frac{1-z}{z}\sum_{d|P_r}\frac{\mu(d)z^d}{1-z^d} = \alpha_z(1) + \alpha_z(p_{r+1}) + \cdots . \tag{22}$$

At $z = 1/2$, where $(1-z)/z = 1$, and $\alpha_z(n) = 2^{-n}$, this becomes

$$\gamma_{1/2}(P_r) = \sum_{d|P_r}\frac{\mu(d)}{2^d-1} = \frac{1}{2} + \frac{1}{2^{p_{r+1}}} + \cdots . \tag{23}$$

From this, we immediately get J. N. Gandhi's formula (Gandhi, 1971) for the $(r+1)$st prime, as the unique integer t such that

$$1 < 2^t\left(\sum_{d|P_r}\frac{\mu(d)}{2^d-1} - \frac{1}{2}\right) < 2. \tag{24}$$

(For an equivalent form of this proof, see (Golomb, 1974)).
Results in the fictitious "uniform distribution" on $\mathbf{Z}^+$ are those obtained by taking the limit as $z \to 1^-$ in the Lambert distribution. For example, in this limit, each residue class modulo m has a probability of $1/m$, as one would expect. Specifically, we have

$$\beta_z(m,a) = \frac{1-z}{z}\sum_{j=0}^{\infty} z^{jm+a} = \frac{1-z}{z}\cdot\frac{z^a}{1-z^m} = \frac{z^{a-1}}{1+z+z^2+\cdots+z^{m-1}}, \quad (25)$$

and $\beta_1(m,a) = 1/m$ for all a, $1 \le a \le m$.

5. THE FINITE UNIFORM FAMILY OF DISTRIBUTIONS

For each positive integer N, we define the "finite uniform distribution" $U_N = \{\alpha_N(n)\}$ on $\mathbf{Z}^+$ by

Definition 8. Let

$$\alpha_N(n) = \begin{cases} 1/N & \text{if } 1 \le n \le N, \\ 0 & \text{if } N < n. \end{cases}$$

For the distribution U_N we have

$$\beta_N(m) = \sum_{j=1}^{\infty} \alpha_N(jm) = \frac{1}{N}\left[\frac{N}{m}\right], \quad (26)$$

$$\gamma_N(k) = \sum_{d|k} \mu(d)\beta_N(d) = \frac{1}{N}\sum_{d|k}\mu(d)\left[\frac{N}{d}\right]. \quad (27)$$

Since $\gamma(0) = \alpha(1)$ for all distributions on $\mathbf{Z}^+$, we have

$$\frac{1}{N} = \alpha_N(1) = \gamma_N(0) = \frac{1}{N}\sum_{d=1}^{\infty}\mu(d)\left[\frac{N}{d}\right] = \frac{1}{N}\sum_{d=1}^{N}\mu(d)\left[\frac{N}{d}\right], \quad (28)$$

from which we obtain the identity

$$\sum_{d=1}^{N}\mu(d)\left[\frac{N}{d}\right] = 1 \qquad \text{for all} \quad N \in \mathbf{Z}^+. \quad (29)$$

If $F(n) = \sum_{d|n} f(d)$, then from (4) we obtain

$$\frac{1}{N}\sum_{m=1}^{N} F(n) = \frac{1}{N}\sum_{m=1}^{N} f(m)\left[\frac{N}{m}\right] \quad (30)$$

or, in its more familiar form,

$$\sum_{n=1}^{N} F(n) = \sum_{m=1}^{N} f(m)\left[\frac{N}{m}\right]. \quad (31)$$

Actually, it is the limit as $N \to \infty$ in (30) that is used to estimate the average values of certain arithmetic functions. Also, if $K(x) = \sum_{1 \le m \le x} f(m)$, we get the multiple identity

$$\sum_{n=1}^{N} F(n) = \sum_{n=1}^{N} \sum_{m|n} f(m) = \sum_{m=1}^{N} f(m) \left[\frac{N}{m}\right] = \sum_{m=1}^{N} K\left(\frac{N}{m}\right). \tag{32}$$

When $\Lambda(n)$ is von Mangoldt's function, we have $\sum_{d|n} \Lambda(d) = \log n$ and $\sum_{1 \le m \le x} \Lambda(m) = \Psi(x)$. These expressions can be used to define $\Lambda(n)$ and Chebyshev's function $\Psi(x)$. More directly, $\Lambda(n) = \{\log p \text{ if } n = p^k; 0 \text{ otherwise}\}$, and $\Psi(x) = \log \text{L.C.M.}\ (1, 2, 3, \ldots, [x])$. For these functions, (32) yields

$$\log N! = \sum_{n=1}^{N} \log n = \sum_{n=1}^{N} \sum_{m|n} \Lambda(m) = \sum_{m=1}^{N} \Lambda(m) \left[\frac{N}{m}\right] = \sum_{m=1}^{N} \Psi\left(\frac{N}{m}\right). \tag{33}$$

6. PROOFS OF THE PRIME NUMBER THEOREM

The "prime number theorem" is the assertion that $\pi(x) \sim x/\log x$ as $x \to \infty$, where $\pi(x)$ is the number of primes $\le x$. It is almost always *proved* in the equivalent form

$$\frac{\Psi(N)}{N} = \frac{1}{N} \sum_{n=1}^{N} \Lambda(n) \sim 1 \quad \text{as} \quad N \to \infty. \tag{34}$$

The original proof, outlined by Riemann (Riemann, 1859) and ultimately completed (independently) by J. Hadamard and by C. de la Vallée–Poussin, is based on first calculating the average value of $\Lambda(n)$ in the Dirichlet distributions D_s:

$$\sum_{n=1}^{\infty} \alpha_s(n) \Lambda(n) = \frac{1}{\zeta(s)} \sum_{n=1}^{\infty} \frac{\Lambda(n)}{n^s} = -\frac{\zeta'(s)}{\zeta^2(s)} = \frac{d}{ds} \frac{1}{\zeta(s)}. \tag{35}$$

Here we have used (15) and the fact that $\Lambda(n) = -\sum_{d|n} \mu(d) \log d$, so that

$$\frac{1}{\zeta(s)} \sum_{n=1}^{\infty} \frac{\Lambda(n)}{n^s} = -\sum_{m=1}^{\infty} \frac{\mu(m) \log m}{m^s} = \frac{d}{ds} \left(\sum_{n=1}^{\infty} \frac{\mu(m)}{m^s} \right). \tag{36}$$

In the vicinity of $s = 1$, it is well-known (Titchmarsh, 1951) that $\zeta(s) = (s-1)^{-1} + c_0 + c_1(s-1) + \cdots$, from which

$$\lim_{s \to 1^+} \left(-\zeta'(s)/\zeta^2(s) \right) = 1. \tag{37}$$

Referring back to (35), this says that the average value of $\Lambda(n)$ in the distribution D_s tends to 1 as $s \to 1^+$. However, a nontrivial argument is needed to show that the middle sum in (36) actually converges at $s = 1$. (This was done by Hadamard and de la Vallée–Poussin by showing that $\zeta(s)$ is zero-free, and thus $1/\zeta(s)$ is non-singular, on and near the line $\mathrm{Rl}(s) = 1$ in the complex plane.) Finally, a theorem is needed to justify that this Dirichlet mean is the same as the "ordinary" mean in (34).

N. Wiener (Wiener, 1933) gave a proof of the Prime Number Theorem by calculating the "Lambert mean" of $\Lambda(n)$. By (20),

$$\frac{(1-z)}{z}\sum_{n=1}^{\infty}\Lambda(n)z^n = -\frac{1}{z}\sum_{m=1}^{\infty}\frac{\mu(m)\log m\cdot z^m}{1+z+\cdots+z^{m-1}}, \tag{38}$$

and with the appropriate Abelian theorem (that the Lambert mean is *regular* in the sense of Schur–Toeplitz (Hardy, 1949), one obtains

$$\lim_{z\to 1^-}\frac{(1-z)}{z}\sum_{n=1}^{\infty}\Lambda(n)z^n = -\sum_{m=1}^{\infty}\frac{\mu(m)\log m}{m} = -\frac{\zeta'(s)}{\zeta^2(s)}\bigg|_{s=1} = 1. \tag{39}$$

Wiener then applies his Tauberian theorem to go from "The Lambert mean of $\Lambda(n)$ is 1" to "The 'ordinary' mean of $\Lambda(n)$ is 1", i.e., that (34) is true.

The so-called "elementary" proof of the Prime Number Theorem, by A. Selberg and/or P. Erdös, works with the Finite Uniform Distribution U_N, in which the average value of $\Lambda(n)$ is

$$\frac{1}{N}\sum_{n=1}^{N}\Lambda(n) = \frac{\Psi(N)}{N}, \tag{40}$$

and considers the limit as $n \to \infty$. That is, (34) is proved "directly", without recourse to the theory of functions of a complex variable. Even so, a careful reading of the "elementary" proof reveals that an Abelian theorem and a Tauberian theorem still play a role.

As we will also see in § 8, the Abelian theorems are usually more critical than the Tauberian theorems. Hardy (Hardy, 1949) even asserts that the *reqularity* of the Lambert mean is "equivalent" to the Prime Number Theorem, so that in his view of Wiener's proof, it is the Abelian theorem, and not Wiener's Tauberian theorem, which is crucial.

7. THE ENTROPY OF DISTRIBUTIONS ON $\mathbf{Z}^+$

For any probability distribution $\{\alpha(n)\}$ on $\mathbf{Z}^+$, Shannon's *entropy* $H(\alpha(n))$ is defined by

$$H(\alpha(n)) = -\sum_{n=1}^{\infty}\alpha(n)\log\alpha(n). \tag{41}$$

For the finite uniform distribution $U_N = \{\alpha_N(n)\}$, from Definition 8, we have

$$H(\alpha_N(n)) = \frac{1}{N}\sum_{n=1}^{N}\log N = \log N. \tag{42}$$

For the Lambert distribution $L_z = \{\alpha_z(n)\}$, where $\alpha_z(n) = (1-z)z^{n-1} = qp^{n-1}$ with $p = z, q = 1-z$ (the familiar *geometric distribution*), we have

$$H(\alpha_z(n)) = \frac{1}{q}H(p,q) = \frac{-1}{1-z}\big(z\log z + (1-z)\log(1-z)\big). \tag{43}$$

Finally, for the Dirichlet distribution $D_s = \{\alpha_s(n)\}$, where $\alpha_s(n) = n^{-s}/\zeta(s)$, we have

$$\begin{aligned} H(\alpha_s(n)) &= -\frac{1}{\zeta(s)}\sum_{n=1}^{\infty} n^{-s}\big(-\log\zeta(s) - s\log n\big) \\ &= \log\zeta(s) + \frac{s}{\zeta(s)}\sum_{n=1}^{s}\frac{\log n}{n^s} = \log\zeta(s) - s\frac{\zeta'(s)}{\zeta(s)}. \end{aligned} \tag{44}$$

It is shown in (Golomb, 1970 a) that the question "How much information is contained in the prime factorization of an integer chosen 'at random' from the distribution D_s?" has the same answer, $H(\alpha_s(n))$, given in (44).
A generating function method for obtaining the entropy $H(\alpha(n))$ in closed form, where $\alpha(n)$ is a probability distribution on $\mathbf{Z}^+$, is described in (Golomb, 1966).

8. THE "LAMBDA METHOD"

Let $a_1, a_2, \ldots, a_k$ be positive integers which all exceed 1 and are pairwise relatively prime. Then, as shown in (Golomb, 1970 b) (and originally in (Golomb, 1956))

$$\prod_{i=1}^{k}\Lambda(a_i) = \frac{(-1)^k}{k!}\sum_{d|a_1a_2\cdots a_k}\mu(d)\log^k d. \tag{45}$$

The case $k = 1$ is $\Lambda(a) = -\sum_{d|a}\mu(d)\log d$, mentioned earlier, and usually obtained by Möbius Inversion from $\log n = \sum_{d|n}\Lambda(d)$. With $k = 2$, this gives

$$\Lambda(a_1)\Lambda(a_2) = \frac{1}{2}\sum_{d|a_1a_2}\mu(d)\log^2 d \tag{46}$$

provided that $a_1 > 1,\ a_2 > 1$, and $(a_1, a_2) = 1$.
The average value of $\Lambda(2a-1)\Lambda(2a+1)$ is directly related to the density of the *twin primes*. In (Golomb, 1956) this average is explored relative to both the Dirichlet and the Lambert distributions, making use of (46). For example, the Lambert approach yields

$$\begin{aligned} &\frac{2(1-z)}{z}\sum_{\text{even } a=2}^{\infty}\Lambda(a-1)\Lambda(a+1)z^a - z(z-1)\log^2 3 \\ &= \frac{(1-z)}{z}\sum_{\text{even } a=2}^{\infty} z^a \sum_{d|(a^2-1)}\mu(d)\log^2 d = \frac{(1-z)}{z}\sum_{\text{odd } d=1}^{\infty}\frac{\mu(d)\log^2 d}{1-z^{2d}}\sum_{i=1}^{2^{\nu(d)}} z^{a_i} \\ &= \frac{1}{z}\sum_{\text{odd } d=1}^{\infty}\frac{\mu(d)\log^2 d}{1+z+\cdots+z^{2d-1}}\sum_{i=1}^{2^{\nu(d)}} z^{a_i}, \end{aligned} \tag{47}$$

where the a_i are the $2^{\nu(d)}$ even roots of $a^2 \equiv 1 \pmod{d}$ between 1 and $2d$, and $\nu(d)$ is the number of distinct prime factors of d. If the term-by-term limit, as $z \to 1^-$, could be justified, this would give

$$\lim_{z \to 1^-} \frac{2(1-z)}{z} \sum_{\text{even } a=2}^{\infty} \Lambda(a-1)\Lambda(a+1)z^a$$
$$= \sum_{\text{odd } d=1}^{\infty} \frac{\mu(d)2^{\nu(d)} \log^2 d}{2d} = 4 \prod_{p>2} \left(1 - \frac{1}{(p-1)^2}\right). \tag{48}$$

(The evaluation of this sum is performed in (Golomb, 1956).) The appropriate Tauberian theorem is available, and gives

$$T(x) \sim C_2 x / \log^2 x \quad \text{as} \quad x \to \infty, \tag{49}$$

where $T(x)$ is the number of twin primes $\leq x$, and C_2 is given by

$$C_2 = 2 \prod_{p>2} \left(1 - \frac{1}{(p-1)^2}\right) = 1.320\ldots, \tag{50}$$

the "correct" constant as determined by other methods, including a probabilistic argument in (Golomb, 1960).

Only the *Abelian theorem needed to justify* the term-by-term limit in (48) is missing! Unfortunately, as shown in (Golomb, 1956), the summation method embodied in the rightmost sum of (47) is *not* "regular" in the sense of Schur–Toeplitz (cf. (Hardy, 1949)). It is still possible that a justification for (48) may be found using a weaker Abelian result than "regularity".

In terms of a general distribution D on $\mathbf{Z}^+$, (47) takes the form

$$2 \sum_{\text{even } a=2}^{\infty} \Lambda(a-1)\Lambda(a+1)\alpha(a) - \alpha(2)\log^2 3 = \sum_{\text{odd } d=1}^{\infty} \mu(d) \log^2 d \sum_{i=1}^{2^{\nu(d)}} \beta(2d, a_i), \tag{51}$$

where the a_i are the $2^{\nu(d)}$ even roots of $a^2 \equiv 1 \pmod{d}$ between 1 and $2d$.

As we shall see in § 10, there are many other families of probability distributions on $\mathbf{Z}^+$ which one might consider, and perhaps in one of these other families the term-by-term limit

$$\sum_{\text{odd } d=1}^{\infty} \mu(d) \log^2 d \sum_{i=1}^{2^{\nu(d)}} \beta(2d, a_i) \to \sum_{\text{odd } d=1}^{\infty} \frac{\mu(d)2^{\nu(d)} \log^2 d}{2d} = 2C_2 \tag{52}$$

can be more easily justified.

The "Lambda method" is equally applicable to other problems involving sets of two or more primes.

9. MORE FORMULAS FOR THE NEXT PRIME

Gandhi's formula (24) for the $(n+1)$st prime, p_{n+1}, given the first n primes, $p_1, p_2, \ldots, p_n$, comes from the underlying relationship

$$\gamma(P_n) - \alpha(1) = \alpha(p_{n+1}) + \cdots, \tag{53}$$

where $P_n = p_1 p_2 \cdots p_n$, and the right side of (53) is the sum of $\alpha(m)$ for all $m > 1$ with $(m, P_n) = 1$. Instead of interpreting (53) in the Lambert distribution to obtain (24), we may interpret it in the Dirichlet distribution, as in (Golomb, 1976), to yield

$$\prod_{i=1}^{n}\left(1-\frac{1}{p_i^s}\right)-\frac{1}{\zeta(s)}=\frac{1}{p_{n+1}^s}+\cdots. \tag{54}$$

This in turn leads (Golomb, 1976) to four expressions for p_{n+1}:

$$p_{n+1} = \lim_{s\to\infty}\{P_n(s)\zeta(s)-1\}^{-1/s}, \tag{55}$$

$$p_{n+1} = \lim_{s\to\infty}\{P_n(s)-\zeta^{-1}(s)\}^{-1/s}, \tag{56}$$

$$p_{n+1} = \lim_{s\to\infty}\{\zeta(s)-Q_n(s)\}^{-1/s}, \tag{57}$$

$$p_{n+1} = \lim_{s\to\infty}\{1-\zeta^{-1}(s)Q_n(s)\}^{-1/s}, \tag{58}$$

where $P_n(s) = \prod_{i=1}^{n}(1-p_i^{-s})$ and $Q_n(s) = \{P_n(s)\}^{-1} = \sum_{n=1}^{\prime\infty} n^{-s}$, where the *prime* indicates that summation is extended over those values of n having no prime factors exceeding p_n.

10. OTHER FAMILIES OF DISTRIBUTIONS

Since $\sum_{n=1}^{\infty} nt^n = t/(1-t)^2$ for $|t| < 1$, there is a family G_t of distributions on $\mathbf{Z}^+$ given by

$$\alpha_t(n) = (1-t)^2 nt^{n-1}, \qquad 0 < t < 1, \quad \text{all} \quad n \in \mathbf{Z}^+. \tag{59}$$

In G_t, the probability that a "random" integer is divisible by m is given by

$$\beta_t(m) = \frac{(1-t)^2}{t}\sum_{j=1}^{\infty}(jm)t^{jm} = \frac{mt^{m-1}}{(1+t+t^2+\cdots+t^{m-1})^2}. \tag{60}$$

More generally, the probability that a "random" integer from G_t is congruent to a modulo m, $1 \le a \le m$, is given by

$$\begin{aligned}\beta_t(m,a) &= \frac{(1-t)^2}{t}\sum_{j=0}^{\infty}(jm+a)t^{jm+a} \\ &= \frac{mt^{m+a-1}}{(1+t+t^2+\cdots+t^{m-1})^2}+\frac{a(1-t)t^{a-1}}{(1+t+t^2+\cdots+t^{m-1})^2}.\end{aligned} \tag{61}$$

Reassuringly,

$$\lim_{t\to 1^-}\beta_t(m,a) = \frac{1}{m}. \tag{62}$$

For $|z| < 1$, the series

$$\sum_{n=1}^{\infty} n^{-s}z^n = f(z,s) \tag{63}$$

converges for each complex value of s, and, at $z = 1$, it converges for $\mathrm{Rl}(s) > 1$. Keeping z and s real and within the region of convergence of the series (63), we have the two-parameter family of probability distributions

$$\alpha_{z,s}(n) = n^{-s} z^n / f(z, s), \tag{64}$$

which specializes to the Dirichlet family D_s when $z = 1$, to the Lambert family L_z when $s = 0$, and to the family G_t just considered when $s = -1$. Obviously there are innumerable other ways to specialize z alone, or s alone or z and s jointly, to obtain a one-parameter family of distributions from (64). The simultaneous limit as $s \to 1^+$, $z \to 1^-$ approaches the mythical uniform distribution on $\mathbf{Z}^+$.

As general as (64) appears to be, it is far from all-inclusive. For example, the Poisson family of distributions, P_λ, given by

$$\alpha_\lambda(n) = e^{-\lambda} \lambda^{n-1} / (n-1)!, \qquad \text{all} \quad n \in \mathbf{Z}^+, \tag{65}$$

is not a specialization of (64). (Also, there is no limiting value of λ toward which $\alpha_\lambda(n)$ approaches the mythical uniform distribution on $\mathbf{Z}^+$.) For that matter, neither is $\alpha(n) = 1/(n(n+1))$, although this distribution exhibits behavior similar to D_2, which has $\alpha(n) = c/n^2$, with $c = 1/\zeta(2) = 6/\pi^2$.

The opportunities to tailor a probability distribution, or a family of probability distributions, on $\mathbf{Z}^+$, to fit a particular application are virtually unlimited.

REFERENCES

Gandhi, J. N. (1971). Formulae for the Nth prime. In: *Proceedings of the Washington State University Conference on Number Theory*, pp. 96–107.

Golomb, S. (1956). *Problems in the Distribution of the Prime Numbers*. Doctoral Dissertation, Harvard University, May.

Golomb, S (1960). The twin prime constant. *Amer. Math. Monthly* **67**, 767–769.

Golomb, S. (1966). The information generating function of a probability distribution. *IEEE Trans. Inform. Theory* **IT-12**, 75–77.

Golomb, S. (1970 a). A class of probability distributions on the integers. *J. Number Theory* **2**, 189–192.

Golomb, S. (1970 b). The lambda method in prime number theory. *J. Number Theory* **2**, 193–198.

Golomb, S. (1974). A direct interpretation of Gandhi's formula. *Amer. Math. Monthly* **81**, 752–754.

Golomb, S. (1976). Formulas for the next prime. *Pacific J. Math.* **63**, 401–404.

Hardy, G. H. (1949). *Divergent Series*. Oxford Univ. Press, Oxford.

Riemann, G. F. B. (1858). Über die Anzahl der Primzahlen unter einer gegebenen Grösse. *Monatsberichte der Berliner Akademie*, November.

Titchmarsh, E. C. (1951). *The Theory of the Riemann Zeta Function*. Oxford Univ. Press, Oxford.

Wiener, N. (1933). *The Fourier Integral and Certain of Its Applications*. Cambridge Univ. Press, Cambridge.

Probabilistic Methods in Discrete Mathematics, pp. 40 – 59
V. F. Kolchin et al. (Eds.)
1993 TVP/VSP

POLYNOMIAL AND POLYNOMIAL-LIKE ALLOCATIONS: RECENT DEVELOPMENTS

V. G. MIKHAILOV
Steklov Mathematical Institute, Vavilov Street 42, 117966 Moscow GSP-1, Russia

ABSTRACT

In this paper we give a survey of the results on polynomial and polynomial-like allocations obtained in the last 5 years since the preceding Petrozavodsk Conference.

During the period of more then 50 years passed from its beginning random allocation theory had gone a difficult and interesting way. The origins of the theory were connected with the classical occupancy problem described in detail by Feller in his famous book (Feller, 1970). Then the equiprobable and polynomial allocations were investigated. The main results of that period are concerned with the limit distributions of spectral statistics μ_r. These results were described in (Kolchin et al., 1978). Further developments of the theory went in a way of generalizations and complications of both the allocations schemes and statistics in question. Every complication was excused by some special application. The class of statistics under consideration was extended to decomposable statistics, then to randomized decomposable statistics, statistics defined on several allocation schemes, vector statistics and so on. A very extensive survey of the history and the terminology one can find in (Ivanov et al., 1984).

And once an impression arose that the theory may be collapsed by boundless extensions. But its way was different. Kolchin (1968) and Holst (1972) had presented the distributions of frequencies as conditional distributions of independent Poisson random variables. This representation led to the so-called generalized allocation scheme where the crucial point is an investigation of the conditional distribution of a sum of independent random variables

$$f_1(\xi_1) + \cdots + f_N(\xi_N) \tag{0.1}$$

under the condition

$$\xi_1 + \cdots + \xi_N = n. \tag{0.2}$$

Now the generalized allocation scheme plays one of the main roles in allocation theory and its applications, for example, in random mapping theory (see (Kolchin, 1986)).

In such a way the allocation theory has gone from the classical occupancy problem, combinatorial in its essence, to the central problems of probability theory about properties of sums of independent random variables under additional condition (0.2) in this case. The last condition distinguishes allocation theory from the classical theory of summation. The allocation theory had arrived to such a result with a rich experience of solving problems on random allocations. We describe the achievements obtained during the period passed from the preceding Petrozavodsk's meeting.

In the centre of our survey we have posed the results of (Bykov, 1991; Dyakonova and Mikhailov, 1992; Mikhailov, 1989 b). Bykov obtained the whole range of limit distributions for the so-called indicator randomized decomposable statistics, and for famous statistics μ_r among them. Bykov proved that the unbounded growth of variance is a necessary and sufficient condition for asymptotic normality of these statistics. This is one of the most remarkable results obtained for the polynomial allocations in recent years.

Two other papers investigate properties of decomposable statistics in polynomial-like allocation schemes: nonhomogeneous polynomial allocations and tuple allocations. In these papers a new method of asymptotic analysis of decomposable statistics is used. Sufficient conditions of asymptotic normality obtained in these papers include some standard restrictions and lower bounds for the rate of growing of the variance.

The survey consists of two parts. In the first part we describe polynomial allocations with finite or countable number of cells. In the second part we describe schemes with nonidentical laws of allocations of particles and particularly tuple allocations and conditionally independent allocations.

We state some of the results in a form different from original one but equivalent to it. We formulate also several new results.

1. POLYNOMIAL ALLOCATIONS

1.1. *Definitions*

In the polynomial allocation scheme n particles are distributed into N cells independently of each other, each particle occupies the cell with outcome probabilities $p_1, \ldots, p_N$, $p_1 + \cdots + p_N = 1$. An important question of the polynomial allocation is the problem of limit behaviour of decomposable and randomized decomposable statistics.

A *decomposable* statistic (D-statistics) L is defined by the formula

$$L = \sum_{k=1}^{N} f_k(\nu_k), \tag{1.1.1}$$

where $\nu_1, \ldots, \nu_N$ are the numbers of particles in the cells and the kernels $f_1, \ldots, f_N$ are functions of integer valued argument.

It is convenient to impose the condition $f_k(0) = 0$ which eliminates the problem of convergence in (1.1.1) as $N \to \infty$.

In the case where all kernels are the same, i. e., $f_1 = \cdots = f_N = f$, formula (1.1.1) gives a *symmetric* D-statistic. Among them we have the spectral statistics μ_r. The random variable μ_r is defined as the number of cells containing exactly r particles each. Any symmetric D-statistic can be decomposed into a linear combination of the random variables μ_r:

$$L = \sum_{k=1}^{N} f(\nu_k) = \sum_{r=0}^{\infty} f(r)\mu_r.$$

The well-known Pearson statistic

$$\chi^2 = \sum_{k=1}^{N} \frac{(\nu_k - np_k)^2}{np_k}$$

is a nonsymmetric D-statistic.
In some applications it is convenient to suppose that the kernels of D-statistic are random. More precisely, let $\xi_1(\nu), \ldots, \xi_N(\nu)$ be independent random functions of integer valued argument independent of the random variables $\nu_1, \ldots, \nu_N$. Then the formula

$$L = \sum_{k=1}^{N} \xi_k(\nu_k) \tag{1.1.2}$$

defines a *randomized* decomposable statistic (RD-statistic).
Note that in the case $N = \infty$ it is necessary to suppose that the sum in (1.1.2) converges almost surely (a. s.).
The Pearson statistic with statistical estimates substituted instead of the probabilities p_k, $k = 1, \ldots, N$, is a RD-statistic.

1.2. *Limit theorems for RD-statistics*

The study of RD-statistics began in 1980th. Basically, sufficient conditions of asymptotic normality have been studied. The first versions of such conditions included considerable restrictions on outcome probabilities (see (Ivanov et al., 1984)). In the late 1980th the conditions suitable for almost all polynomial schemes with an increasing but finite number of cells were obtained in (Bykov and Ivanov, 1991).
The following results give sufficient conditions of asymptotic normality for RD-statistic in the general case where N can be infinite.
Let x_m be independent random variables having Poisson distributions with means $\alpha_m = np_m$, $m = 1, 2, \ldots$, i. e., $\mathcal{L}(x_m) = \Pi(np_m)$. Set

$$\zeta_k(\nu) = \xi_k(\nu) - \mathbf{E}\,\xi_k(x_k) - \frac{\nu - np_k}{n} \sum_{m=1}^{\infty} \mathbf{E}\{\xi_m(x_m)(x_m - np_m)\},$$

$$A = \sum_{k=1}^{N} \mathbf{E}\,\xi_k(x_k), \quad \sigma^2 = \sum_{k=1}^{N} \mathbf{E}\,\zeta_k^2(x_k), \quad \hat{\zeta}_k(\nu) = \sigma^{-1}\zeta_k(\nu).$$

THEOREM 1.2.1 (Bykov, 1991). *Let $n \to \infty$ and the probabilities $p_1, p_2, \dots, p_N$ and the distributions of kernels $\xi_1(\nu), \dots, \xi_N(\nu)$ vary in such a way that:*

a) $p_k \le p < 1$, $k = 1, 2, \dots, N$, *where p is a constant,*

b) *for some $\delta > 0$*

$$\sum_{k=1}^{N} \mathbf{E} \left| \hat{\zeta}_k(x_k) \right|^{2+\delta} \to 0.$$

Then

$$\mathcal{L}(\sigma^{-1}(L - A)) \to \mathcal{N}(0, 1).$$

Remarks. 1. Bykov (1991) proved this theorem for $\mathbf{R}^m$-valued *RD*-statistics defined for several allocation schemes. In that paper different versions of condition a) are given also.

2. In case $N < \infty$ this result was proved in (Bykov and Ivanov, 1991).

3. Other sufficient conditions were obtained in (Mirakhmedov, 1989; 1990) for generalized allocation schemes and (as corollaries) for countable polynomial allocations.

Theorem 1.2.1 gives a simple criterion of asymptotic normality for the so-called *indicator RD*-statistics. *RD*-statistic is the indicator one if its kernels take the values 0 and 1 only.

The spectral statistics μ_r are indicator *RD*-statistics, for example.

THEOREM 1.2.2 (Bykov, 1991). *Let L be an indicator RD-statistic and $n \to \infty$. Then the following conditions are equivalent:*

a) $\operatorname{Var} L \to \infty$,

b) $\mathcal{L}(\widetilde{L}) \to \mathcal{N}(0, 1)$.

Here and below we use the notation

$$\widetilde{L} = (L - \mathbf{E}\, L)(\operatorname{Var} L)^{-1/2}.$$

Remark. Bykov (1991) proved this theorem for random vectors formed by indicator *RD*-statistics. In that case condition a) has a more complicated form. Bykov (1991) has studied also limit distributions in the cases where the variance of *RD*-statistic is bounded and the variances of summands of the statistic are negligible. In these cases the limit distributions are convolutions of one, two, or three Poisson distributions with different supports. Bykov's results give an exhaustive information in these cases.

1.3. *Limit theorems for spectral statistics*

The results cited above help to make an important step in the study of the spectral statistics μ_r in polynomial and countable polynomial allocations. First we present the results for μ_0 (the number of empty cells) in the case $N < \infty$. Put

$$A_0 = \sum_{k=1}^{N} e^{-np_k}, \quad \mathcal{P}_0 = \sum_{k=1}^{N} p_k \frac{e^{-np_k}}{1+np_k},$$

$$D_0^2 = \sum_{k=1}^{N} e^{-np_k}(1-e^{-np_k}) - n\left(\sum_{k=1}^{N} p_k e^{-np_k}\right)^2 .$$

The values A_0 and D_0^2 are the main terms of $\mathbf{E}\,\mu_0$ and $\operatorname{Var}\mu_0$ as $n \to \infty$ correspondingly.

THEOREM 1.3.1 (Bykov, 1991). *Let $N < \infty$, $n, N \to \infty$.*
a) *If $n\mathcal{P}_0(1-\mathcal{P}_0) \to \infty$, then*

$$\mathcal{L}((\mu_0 - A_0)D_0^{-1}) \to \mathcal{N}(0,1).$$

b) *If*

$$\max_k np_k \to 0, \quad n^2 \sum_{k=1}^{N} p_k^2 \to 2\lambda < \infty,$$

then

$$\mathcal{L}(\mu_0 - N + n) \to \Pi(\lambda).$$

c) *If*

$$\max_k np_k e^{-np_k} \to 0, \tag{1.3.1}$$

$$\sum_{k=1}^{N} e^{-np_k}(1-e^{-np_k})^2 \to \lambda_1 < \infty,$$

$$\sum_{k=1}^{N} e^{-2np_k}(1-e^{-np_k}) \to \lambda_2 < \infty,$$

then

$$\mathcal{L}(\mu_0 - M) \to \mathcal{L}(\xi_1 - \xi_2).$$

Here $M = \operatorname{card}\{k\colon np_k < 1\}$, the random variables ξ_1 and ξ_2 are independent and $\mathcal{L}(\xi_i) = \Pi(\lambda_i)$, $i = 1, 2$.

Remarks. 1. Condition a) of Theorem 1.3.1 is equivalent to the condition $\operatorname{Var}\mu_0 \to \infty$. Therefore, it is a necessary condition of normality.
2. It may seem that there exists an arbitrariness in the definition of M. It does not exist. It follows from condition (1.3.1) that for large n the value M is independent of the constant in the definition of M.
3. Point b) of Theorem 1.3.1 is the well-known result (see (Kolchin et al., 1978)). The other points are more general than the previous results. Theorem 1.3.1 covers

all limit distributions in the cases where contribution of a single cell into the variance of μ_0 is negligible.
In case $N = \infty$, one studies the number of occupied cells μ instead of μ_0. Dutko (1989) examined the countable polynomial allocation with fixed outcome probabilities as $n \to \infty$. Using Karlin's method he proved the following assertion.

THEOREM 1.3.2 (Dutko, 1989). *Let the outcome probabilities of the countable polynomial allocation be fixed and fulfil the condition*

$$\operatorname{Var}\mu \to \infty \qquad \text{as} \quad n \to \infty.$$

Then

$$\mathcal{L}(\tilde{\mu}) \to \mathcal{N}(0,1) \qquad \text{as} \quad n \to \infty.$$

Recall that

$$\tilde{\mu} = (\mu - \mathbf{E}\,\mu)(\operatorname{Var}\mu)^{-1/2}.$$

The last theorem does not use conditions of regular varying of outcome probabilities in contrary to the classical result of (Karlin, 1967).
Let us formulate limit theorems for the statistics μ_r. Now N can be infinite. Let α be the random variable with the distribution

$$\mathbf{P}\{\alpha = np_k\} = p_k, \qquad k = 1, \dots, N,$$

and

$$\pi_r(x) = \frac{x^r}{r!}e^{-x}, \quad A_r = \sum_k \pi_r(np_k),$$

$$D_r^2 = \sum_k \pi_r(np_k)\big(1 - \pi_r(np_k)\big) - n\big(\mathbf{E}\,(\pi_{r-1}(\alpha) - \pi_r(\alpha))\big)^2.$$

As above the values A_r and D_r^2 are the main terms of $\mathbf{E}\,\mu_r$ and $\operatorname{Var}\mu_r$ correspondingly as $n \to \infty$.

THEOREM 1.3.3 (Bykov, 1991). *Let $n \to \infty$.*
a) *If*

$$n\mathbf{E}\,e^{-\alpha}(1 - \mathbf{E}\,e^{-\alpha}) \to \infty,$$

then

$$\mathcal{L}(\tilde{\mu}_1) \to \mathcal{N}(0,1).$$

b) *If*

$$\max_k np_k \to 0, \quad n^2 \sum_k p_k^2 \to 2\lambda < \infty,$$

then

$$\mathcal{L}((n-\mu_1)/2) \to \Pi(\lambda).$$

c) *If*

$$\max_k \pi_1(np_k) \to 0, \quad A_1 \to \lambda < \infty,$$

then

$$\mathcal{L}(\mu_1) \to \Pi(\lambda).$$

THEOREM 1.3.4 (Bykov, 1991). *Let $n \to \infty$ and the outcome probabilities and the parameter r vary in such a way that $A_r \to \infty$, $r \geq 2$. Then*

$$\mathcal{L}(\tilde{\mu}_r) \to \mathcal{N}(0,1).$$

THEOREM 1.3.5 (Bykov, 1991). *Let $n \to \infty$ and the outcome probabilities and the parameter r vary in such a way that $n - r \to \infty$, $\max_k \pi_r(np_k) \to 0$, $A_r \to \lambda < \infty$. Then*

$$\mathcal{L}(\mu_r) \to \Pi(\lambda).$$

Remarks. 1. Conditions a) of Theorem 1.3.3 and of Theorem 1.3.4 are equivalent to the condition $\operatorname{Var}\mu_r \to \infty$. Therefore, they are necessary conditions of normality.
2. Theorems 1.3.3–1.3.5 cover all limit distributions of μ_r in the cases where the contributions of a single cell into the variance of μ_r are negligible. It is important to note that Theorems 1.3.4 and 1.3.5 cover the case $r \to \infty$.
3. Bykov (1991) also proved the convergence of the distribution of the vector ($\tilde{\mu}_r$, $r = 0, \ldots, d-1$) to a d-dimensional normal distribution under the condition

$$\operatorname{Var}\mu_r \to \infty, \qquad r = 0, \ldots, d-1.$$

4. Theorems analogous to Theorems 1.3.3–1.3.5 can be also proved for many other indicator statistics. Bykov in (Bykov, 1991) made that for the statistics

$$\mu_r^+ = \sum_{m=r}^{n} \mu_m, \qquad r = 1, 2, \ldots.$$

1.4. *Refinements of* CLT *for RD-statistics*

Here we consider the estimates of closeness of distributions of *RD*-statistics to the normal distribution and to their Edgeworth's expansions. We use the notation of Section 1.2. Let

$$P_m = \sum_k p_k^m, \quad l_s = \sum_k \mathbf{E}\left|\hat{\zeta}_k(x_k)\right|^s + (n^{-1} + P_2)^{(s-2)/2},$$

$\Phi(x)$ be the standard normal distribution function,

$$\rho(L) = \sup_y \left|\mathbf{P}\{L - A < \sigma y\} - \Phi(y)\right|.$$

THEOREM 1.4.1 (Mirakhmedov, 1990). *Let*

$$P_2 \leq (1 - e^{-1})/(20 \log n). \tag{1.4.1}$$

Then there exists a constant $C < \infty$ such that

$$\rho(L) \leq C l_3.$$

Remark. In (Mirakhmedov, 1990) this result has a different but equivalent form. Its proof is based on an analogous result for the generalized allocation scheme. (see (Mirakhmedov, 1989)).
For the spectral statistics μ_r Theorem 1.4.1 implies the following assertions.

THEOREM 1.4.2 (Mirakhmedov, 1990). *Let $r \geq 2$ be fixed, $n \to \infty$ and the outcome probabilities of the countable polynomial allocation vary in such a way that $nP_2A_r^{-1} \to 0$. Then*

$$\rho(\mu_r) = O\big((A_r^{-1} + P_2)^{1/2}\big).$$

THEOREM 1.4.3 (Mirakhmedov, 1990). *Let $n \to \infty$ and the outcome probabilities vary in such a way that for some $m \geq 2$*

$$\liminf n^{-1}\mathbf{E}\,\mu_m > 0. \tag{1.4.2}$$

Then, for any fixed $r \geq 1$,

$$\rho(\mu_r) = O\big((n^{-1} + P_2)^{1/2}\big).$$

Remark. The last estimate was proved by Mirakhmedov under the condition that for any $m \geq 0$

$$\liminf n^{-1}\mathbf{E}\,\mu_m > 0. \tag{1.4.3}$$

As was remarked in (Mikhailov, 1981) conditions (1.4.2) and (1.4.3) are equivalent. It is clear that condition (1.4.2) is more convenient for a check.
Now about asymptotic expansions. In the general case these expansions are very complicated. Therefore, we confine ourselves by a listing of these results only.

Mirakhmedov (Mirakhmedov, 1990) estimated the accuracy of approximation for $\sigma\mathbf{P}\{L = m\}$ in the uniform metric under condition (1.4.1). For the general allocation scheme, an analogous estimate in a nonuniform metric was obtained in (Ivchenko and Mirakhmedov, 1992). They obtained also estimates in the uniform metric for the approximation of the distribution function of RD-statistic by its Edgeworth's expansion.

These results can be used in the study of some interesting statistics, specifically

$$\mu_r, \quad X^2 = \frac{N}{n}\sum_{k=1}^{n} \nu_k^2, \quad \Lambda = \sum_{k=1}^{N} \nu_k \log \nu_k$$

in the polynomial allocation. We present such a result for the statistic X^2. The others are similar in form.

Let

$$\alpha = nN^{-1}, \quad A = nNP_2 + N, \quad \sigma^2 = 2N^2P_2 + nN^2(P_3 - P_2^2),$$

$$a_0 = \sigma^{-3}\left\{4n^2P_2 + 2n^3(16P_3 - 9P_2^2) + 8n^4(4P_4 - 9P_2P_3 + 5P_2^3)\right\},$$

$$a_1 = 2nNP_2, \quad S(u) = [u] - u + \frac{1}{2},$$

$[u]$ is the integer part of u. Let $\varphi(x)$ be the standard normal density function.

THEOREM 1.4.4 (Ivchenko and Mirakhmedov, 1992). *Let $N < \infty$, $n, N \to \infty$ and the outcome probabilities $p_1, \ldots, p_N$ vary in such a way that $0 < C_1 \le Np_k \le C_2 < \infty$, $k = 1, \ldots, N$, where C_1 and C_2 are constant. Then for any positive number l there exists a positive number c_l such that*

$$\left|\mathbf{P}\{X^2 - A < \sigma y\} - \Phi(y) - \varphi(y)\left(\frac{1-y^2}{2}a_0 + \frac{a_1}{2} + \frac{2}{\alpha\sigma}S\left(\frac{y\alpha\sigma}{2} + \frac{\alpha A}{2}\right)\right)\right|$$

$$= O\left(1/(n\alpha) + 1/N + nr(\alpha)\exp\left\{-c_l r_l(\alpha)e^{-2\alpha}\right\}\right),$$

where $r(\alpha)$ and $r_l(\alpha)$ have the forms:

a) *if $\alpha \to \infty$, then*

$$r(\alpha) = \alpha^{-1/2}, \quad r_l(\alpha) = \alpha^l;$$

b) *if* $\limsup \alpha < \infty$, *then*

$$r(\alpha) = \alpha^{1/2}, \quad r_l(\alpha) = \alpha^3.$$

Finally note that for the equiprobable polynomial allocation with $p_k = 1/N$ one can prove a simple but useful estimate of closeness of the distribution function of the statistic μ_0 to its Edgeworth's expansion. Using the property that μ_0 is equal

in distribution to a sum of independent random indicators (Vatutin and Mikhailov, 1982) and an estimate from (Mikhailov, 1991 b) we obtain the following result.
Put

$$a = \mathbf{E}\,\mu_0, \quad \sigma^2 = \mathrm{Var}\,\mu_0, \quad \beta = \mathbf{E}\,(\mu_0 - a)^3,$$

$$\Gamma(y) = \Phi(y) + \varphi(y)(1 - y^2)\beta/(6\sigma^3).$$

THEOREM 1.4.5. *For* $\sigma \geq 10$

$$\left|\mathbf{P}\{\mu_0 \leq m\} - \Gamma\left(\sigma^{-1}(m - a + 1/2)\right)\right| \leq (\sigma + 3)/(4\sigma^3).$$

2. POLYNOMIAL-LIKE ALLOCATIONS

2.1. *Definitions*

In this section we present some results for nonhomogeneous allocations where outcome probabilities depend on particle's numbers. The simplest version of such allocation is as follows.
Let n particles fall independently into N cells ($N \leq \infty$) and every particle has its own outcome probabilities:

$$p_1^{(i)}, \ldots, p_N^{(i)}, \qquad p_1^{(i)} + \cdots + p_N^{(i)} = 1, \quad i = 1, \ldots, n.$$

We call this scheme a nonhomogeneous polynomial allocation or simple nonhomogeneous allocation. A particular case of nonhomogeneous allocation is the allocation of particles of several different types. We can say that in a nonhomogeneous allocation every particle has its type. Of course, the polynomial allocation is a particular case of the nonhomogeneous one. Such an allocation scheme arises in mathematical statistics in cases where the properties of observable processes can vary in time, for example, as in the disorder problem.
A nonhomogeneous scheme is related to dependent allocations. In the latter case the information about preceding disposals has an influence on the posterior probabilities of the following allocations, making them dependent on the numbers of particles.
The range of problems on nonhomogeneous allocations is adopted from the polynomial scheme. But the methods which are usual for polynomial allocations cannot be mostly fitted for nonhomogeneous schemes.

2.2. *Asymptotic normality of D-statistics*

Let kernels $f_k(\nu)$ of the D-statistics L determined by (1.1.1) on the frequencies of outcomes of a nonhomogeneous scheme fulfil the conditions

$$f_k(0) = 0, \quad |f_k(\nu)| \leq e^{d\nu}, \qquad \nu = 0, 1, \ldots, \quad k = 1, \cdots, N. \tag{2.2.1}$$

Let X_i be the number of the cell into which the ith particle falls. Let ξ_r be the number of r-fold repetitions in the sequence $X_1, \dots, X_n$, i. e.,

$$\xi_r = \sum_{1 \le i_1 < \cdots < i_r \le n} \mathbf{I}\{X_{i_1} = \cdots = X_{i_r}\}.$$

Here $\mathbf{I}\{A\}$ denotes the indicator of the event A.

THEOREM 2.2.1 (Dyakonova and Mikhailov, 1992). *Let $n \to \infty$ and the distributions of the random variables $X_1, \dots, X_n$ and the functions f_k vary in such a way that condition (2.2.1) is fulfilled uniformly in n,*

$$a(n) = \sup_k \sum_{i=1}^{n} \mathbf{P}\{X_i = k\} \le a < \infty, \tag{2.2.2}$$

where a is a constant,

$$\log n = O(\log \mathbf{E}\,\xi_2),$$

$$\mathbf{E}\,\xi_2 = O(\mathrm{Var}\, L). \tag{2.2.3}$$

Then

$$\mathcal{L}(\widetilde{L}) \to \mathcal{N}(0,1).$$

It is possible to weaken condition (2.2.3) restricting the class of statistics in question by the condition

$$f_1(1) = f_2(1) = \cdots = f_N(1). \tag{2.2.4}$$

THEOREM 2.2.2 (Dyakonova and Mikhailov, 1992). *Let the conditions of Theorem 2.2.1 be fulfilled with condition (2.2.4) and the condition*

$$\mathbf{E}\,\xi_2 \to \infty, \quad \mathbf{E}\,\xi_2 = O(\mathrm{Var}\, L) \quad \text{as} \quad n \to \infty \tag{2.2.5}$$

instead of condition (2.2.3). Then

$$\mathcal{L}(\widetilde{L}) \to \mathcal{N}(0,1) \quad \text{as} \quad n \to \infty.$$

If

$$f_k(\nu) = 0, \qquad \nu = 0, \dots, r-1, \quad k = 1, \dots, N, \tag{2.2.6}$$

and $r \ge 3$, it is convenient to formulate the conditions in terms of $\mathbf{E}\,\xi_r$. Note that there exist cases where $\mathbf{E}\,\xi_r = o(\mathbf{E}\,\xi_2)$ for $r \ge 3$ as $n \to \infty$.

THEOREM 2.2.3 (Dyakonova and Mikhailov, 1992). *Let $n \to \infty$ and the distributions of the r.v.'s $X_1, \dots, X_n$ and the functions f_k vary in such a way that* *conditions (2.2.1), (2.2.2) and (2.2.6) are fulfilled (the first of them is fulfilled uniformly in n) and*

$$\mathbf{E}\,\xi_r \to \infty, \quad \mathbf{E}\,\xi_r = O(\operatorname{Var} L). \tag{2.2.7}$$

Then

$$\mathcal{L}(\widetilde{L}) \to \mathcal{N}(0,1).$$

In the case $a(n) \to 0$ this result can be formulated in such a way that the second part of condition (2.2.7) follows from the other conditions.

THEOREM 2.2.4 (Dyakonova and Mikhailov, 1992). *Let the functions f_k satisfy condition (2.2.1) uniformly in n, and let the value*

$$r = \min\{\nu\colon\, \exists\, k \in \{1,\ldots,N\},\; f_k(\nu) \neq 0\}$$

be fixed and $r \geq 2$. If $n \to \infty$, the distributions of the random variables $X_1,\ldots,X_n$ and the functions f_k vary in such a way that

$$|f_k(r)| \geq f > 0, \quad a(n) \to 0, \quad \mathbf{E}\,\xi_r \to \infty, \tag{2.2.8}$$

where f is a constant, then

$$\mathcal{L}(\widetilde{L}) \to \mathcal{N}(0,1).$$

Remarks. 1. In the important particular case where $\operatorname{Var} L \asymp n$ as $n \to \infty$, i. e., if the variance of L is equal to n in order, condition (2.2.5) of Theorem 2.2.2 and condition (2.2.7) of Theorem 2.2.3 can be omitted. Such a rate of growth of the variance of L is typical for the so-called central parameter domain in the polynomial allocation (see (Kolchin et al., 1978)).
2. Let

$$\xi_r(C_n) = \sum_{1 \leq i_1 < \cdots < i_r \leq n} \mathbf{I}\{X_{i_1} = \cdots = X_{i_r} \in C_n\},$$

where $C_n \subseteq \{1,\ldots,n\}$ is a given sequence of sets. The random variables $\xi_r(C_n)$ can be substituted into the statements of Theorems 2.2.1–2.2.4 for the random variables ξ_r, if the condition

$$f_k(\nu) \equiv 0, \qquad k \notin C_n,$$

is fulfilled. In this case it is naturally to suppose that

$$a(n) = a(n, C_n) = \sup_{k \in C_n} \sum_{i=1}^{n} \mathbf{P}\{X_i = k\}.$$

The limit theorems 2.2.1–2.2.3 have analogues for vector D-statistics. The assersions of Remarks 1 and 2 are applicable to these analogues.

Proofs of the theorems above are based on an approximation of decomposable statistics by sums of random variables with a small number of dependencies and on the results from (Janson, 1988; Mikhailov, 1991 a) about conditions of asymptotic normality for such sums.

The approximation uses the expansion

$$\mathbf{I}\{\nu_k = r\} = \sum_{l=r}^{n} (-1)^{l-r} \binom{l}{r} \sum_{1 \le i_1 < \cdots < i_r \le n} \mathbf{I}\{X_{i_1} = \cdots = X_{i_r} = k\},$$

which gives an opportunity to represent the statistic L in the form

$$L = \sum_{r=1}^{n} \zeta_r, \quad \zeta_r = \sum_{\alpha \in W_r} Z_\alpha,$$

$$Z_\alpha = \sum_{k} C_k(r) \mathbf{I}\{X_{i_1} = \cdots = X_{i_r} = k\},$$

where

$$W_r = \{\alpha = (i_1, \ldots, i_r) \colon 1 \le i_1 < \cdots < i_r \le n\}.$$

The coefficients $C_k(r)$ are determined by the kernels of statistic L. Let

$$L_t = \zeta_1 + \cdots + \zeta_t, \quad \varepsilon_t = \zeta_{t+1} + \cdots + \zeta_n.$$

The random variable L_t is a sum of the random variables Z_α with a small number of dependencies. Its behaviour is described by Theorem 2 from (Mikhailov, 1991 a). It is not difficult to check that the asymptotic normality of L follows from the asymptotic normality of L_t if

$$\operatorname{Var} \varepsilon_t = o(\operatorname{Var} L_t)$$

as $n \to \infty$ and $t \to \infty$ arbitrary slowly. All we need to check is the last property. Different versions of this method were used in (Mikhailov, 1989 b; Dyakonova and Mikhailov, 1992). This method is also suitable for dependent allocations (see (Gorchakov, 1993) in this volume).

2.3. *Tuple allocation*

Let

$$X_1, \ldots, X_{n+m-1} \tag{2.3.1}$$

be a sequence of independent random variables taking only the values $1, 2, \ldots$ with the probabilities

$$p_k^{(i)} = \mathbf{P}\{X_i = k\}, \quad \sum_{k=1}^{\infty} p_k^{(i)} = 1, \qquad i = 1, \ldots, n+m-1.$$

We say that the ith tuple $Y_i = (X_i, \ldots, X_{i+m-1})$ falls into the cell $k = (k_1, \ldots, k_m)$ if $Y_i = k$. A D-statistic is determined by the formula

$$L = \sum_{k \in \mathbf{N}^m} f_k(\nu_k),$$

where $\mathbf{N} = \{1, 2, \ldots\}$ and ν_k is the number of tuples that are equal to k in sequence (2.3.1).

For $m = 1$ this allocation scheme coincides with a nonhomogeneous allocation (see subsection 2.1). For $m \geq 2$ the allocations form a sequence of $(m-1)$-dependent random trials and, on the other hand, they form a Markov chain on the set $\mathbf{N}^m$. Tuple allocations were studied by many authors (see bibliography in (Ivanov et al., 1984) and (Mikhailov, 1984; Karlin and Ost, 1987, 1988). But these investigations concerned with some special families of D-statistics. Now there exists an opportunity to study rather general classes of D-statistics in the tuple allocation by the method described in subsection 2.2.

Let

$$p = \max_i \sup_{k \in \mathbf{N}} \mathbf{P}\{X_i = k\}.$$

THEOREM 2.3.1 (Mikhailov, 1989 b). *Let $n \to \infty$ and the probabilities $\mathbf{P}\{X_i = k\}$ vary in such a way that*

a) *for some C, $d < \infty$ and for every n and k*

$$|f_k(\nu)| \leq C e^{d\nu}, \qquad \nu = 0, 1, \ldots;$$

b) $np^m \leq c$ *for some* $c < \infty$;

c) $\operatorname{Var} L \asymp n$.

Then

$$\mathcal{L}(\widetilde{L}) \to \mathcal{N}(0, 1).$$

This result allows an extension to vector D-statistics. Let $L = (L^1, \ldots, L^r)$ be the vector formed by the decomposable statistics with families of kernels $\{f_k^1\}, \ldots, \{f_k^r\}$ correspondingly.

THEOREM 2.3.2 (Mikhailov, 1989 b). *Let $n \to \infty$ and the probabilities $\mathbf{P}\{X_i = k\}$ and the functions f_k^l vary in such a way that for the statistics $L^1, \ldots, L^r$ conditions* a) *and* b) *of Theorem 2.3.1 are fulfilled and*

d) *the covariance matrix of the random vector*

$$L' = n^{-1/2}(L - \mathbf{E}\, L)$$

converges to a nondegenerate matrix Σ.

Then

$$\mathcal{L}(L') \to \mathcal{N}(0, \Sigma).$$

Remarks. 1. These theorems are close in their forms to the results of subsection 2.2. However, the conditions of the results for the tuple allocation are more limiting due to the restrictions on the value np^m. The following example for the case $m = 1$ shows that these conditions restrict essentially a class of schemes in question. Let $n = n_1 s$ and $N = N_1 s$. The first n_1 particles fall equiprobably into the first N_1 cells, the next n_1 particles fall equiprobably into the next N_1 cells and so on. Consider an arbitrary symmetric D-statistic L. It is a sum of i.i.d. random variables $L_1, \ldots, L_s$, where L_j is the statistic L restricted on the jth group of cells. It is clear that for n_1 and N_1 fixed and a fixed kernel the statistic L is asymptotically normal as $s \to \infty$. But condition b) of Theorem 2.3.1 is not fulfilled:

$$np = nN_1^{-1} = (n_1 N_1^{-1}) s \to \infty \qquad \text{as} \quad s \to \infty.$$

This example and the results of subsection 2.2 lead us to the idea that the expression

$$a_m(n) = \sup_{k \in \mathbf{N}^m} \sum_{i=1}^{n} \mathbf{P}\{Y_i = k\} = \sup_{k \in \mathbf{N}^m} \mathbf{E}\, \nu_k$$

should be used in the theorem instead of the expression np^m. For $m = 1$ we have $a_1(n) = a(n)$ with $a(n)$ from subsection 2.2. Note that $a(n)$ is a constant in our example.

2. The theorems above answer to the question about normality conditions in general. But the check of the conditions of these theorems in typical cases can be a rather difficult problem. Thus, Tikhomirova in (Tikhomirova, 1992) proves an asymptotic normality of the number of nonoccurring tuples in polynomial trials by this method. Her proof exposes difficulties in checking condition c) of Theorem 2.3.1 in this case.

We have presented above the forward problems of the random allocation theory. There exists a different range of problems called the inverse problems. They study the times of the first occurrence of some events in the sequential allocations. Connections between the forward and inverse problems are well known (see (Kolchin et al., 1978)) and we will not present them.

Consider the equiprobable polynomial allocation with N outcomes and a sequence of sets of m-tuples $H_m \subseteq \{1, \ldots, N\}^m$. Let $X_1, X_2, \ldots$ be the sequence of our polynomial trials,

$$T(A) = \inf\{n\colon (X_{n-m+1}, \ldots, X_n) = A\},$$

$$W(H_m) = \max\{T(A)\colon A \in H_m\}.$$

Here A denotes a m-tuple. Denote by $Y(H_m)$ the normalized maximum waiting time, i. e.,

$$Y(H_m) = N^{-m} W(H_m) - \log |H_m|,$$

where $|H_m|$ is the cardinality of the set H_m. Let

$$F(y) = \exp(-\exp(-y)).$$

THEOREM 2.3.3 (Mori, 1990). *Let N be fixed, $m \to \infty$, $|H_m| \to \infty$. Then uniformly in H_m*

$$|H_m|^{\frac{1}{6}\frac{N-1}{N+1}} \sup_y \left|\mathbf{P}\{Y(H_m) \le y\} - F(y)\right| \to 0.$$

THEOREM 2.3.4 (Mori, 1990). *Let N be fixed, $m \to \infty$,*

$$\liminf_{m\to\infty} \frac{\log |H_m|}{\log m} > 6\,\frac{N+1}{N-1}.$$

Then

$$\liminf_{m\to\infty} \frac{Y(H_m)}{\log\log m} = -1$$

and

$$\limsup_{m\to\infty} \frac{Y(H_m)}{\log\log m} = 1$$

with probability one.

2.4. *Conditionally independent allocations*

Nonhomogeneous allocation schemes are related in natural way with some schemes of dependent allocations. For example, consider an allocation scheme controlled by a Markov chain $\{Y_i\}$. In this scheme the allocation of the ith particle, under the condition that Y_i is fixed, depends on Y_i only. For fixed $Y_1, \dots, Y_n$ the allocations of all particles are independent and nonidentically distributed. Thus fixing $Y_1, \dots, Y_n$ we transform this scheme into a nonhomogeneous allocation. This connection gives us a way of using the results for nonhomogeneous allocations in the study of dependent schemes.

Let a probability space $(\Omega, \mathfrak{A}, \mu)$ and a family of functions measurable on $(\Omega, \mathfrak{A}, \mu)$

$$0 \le p_{ki}(\omega) \le 1, \quad \sum_{k=1}^{\infty} p_{ki}(\omega) = 1, \qquad i = 1, \dots n,$$

be given. For every $\omega \in \Omega$ this family determines the nonhomogeneous allocation scheme S^ω with the probabilities $p_{ki}(\omega)$, $k \in \mathbf{N}$.

This scheme is called a *conditionally independent* allocation (CI- allocation). The allocation controlled by a Markov chain considered above is a particular case of the CI-allocations.

There exists another reason to study CI-allocations. While a polynomial allocation scheme works as a mathematical model of some natural process statistical estimates are used instead of the theoretical probabilities of outcomes. If these

estimates are obtained from an independent sample such a change transforms the polynomial allocation into the CI-allocation.
Let L be a decomposable statistic in a CI-allocation scheme. Denote by L^{ω} the same statistic in the scheme S^{ω}. Let

$$a(\omega) = \sup_k \sum_{i=1}^{n} p_{ki}(\omega), \quad b^2(\omega) = \operatorname{Var} L^{\omega}/\operatorname{Var} L,$$

$$m(\omega) = (\mathbf{E}\, L^{\omega} - \mathbf{E}\, L)(\operatorname{Var} L)^{-1/2}.$$

THEOREM 2.4.1. *Let $n \to \infty$ and the functions f_k determining the statistic L and the functions $p_{ki}(\omega)$ vary in such a way that*
a) *condition* (2.2.1) *is fulfilled uniformly in n;*
b) *for some function $h(x)$ such that $h(x) \to 0$ as $x \to 0$ the inequality*

$$\limsup_{n\to\infty} \mu\{\omega\colon a(\omega) > x\} \le h(x)$$

holds;
c) $\operatorname{Var} L \asymp n$ *as* $n \to \infty$;
d) $m(\omega) \to 0,\ b(\omega) \to 1$ *in measure μ as $n \to \infty$.*
Then

$$\mathcal{L}(\widetilde{L}) \to \mathcal{N}(0,1).$$

Remark. This result was proved in (Mikhailov, 1989 a) under a more restrictive than b) condition in which the role of $a(n)$ is played by

$$\max_i \sup_k np_{ki}(\omega)$$

and for symmetric D-statistics only. Theorem 2.4.1 is proved analogously to Theorem 6 in (Mikhailov, 1989 a) with the help of the results given in Remark 1 of subsection 2.2.

3. COMMENTARY

The results presented above do not cover the whole range of problems in polynomial and polynomial-like allocations. For example, investigations of inverse problems for polynomial allocations in (Ivchenko, 1991) and a study of large deviations for RD-statistics for polynomial allocations in (Sirazhdinov et al., 1989; Suyunov, 1989 a, b) are omitted. The results for RD-statistics in the generalized allocation scheme such as the results of (Ronzhin, 1988; Ivchenko and Mirakhmedov, 1988, 1992; Mirakhmedov, 1989) are beyond the scope of this paper.
It is not difficult to guess that in the near future different generalizations and refinements of the reviewed results will be obtained. Theorem 1.2.2 calls for a special attention. It shows that the introduction of indicator RD-statistics was

very successful. It is natural to make an attempt to extend this class of statistics in such a way that the property $\operatorname{Var} L \to \infty$ should be preserved as a sufficient condition. We illustrate this idea by the following result simply derived from the vector version of Theorem 1.2.2.

THEOREM 3.1. *Let the kernels of RD-statistic L take integer values only and be bounded by some constant. The following conditions are equivalent as $n \to \infty$:*

a) $\operatorname{Var} L \to \infty$;

b) $\mathcal{L}(\widetilde{L}) \to \mathcal{N}(0,1)$.

It is clear that in the near future the investigations of the polynomial and generalized allocation schemes will go on in analogous ways. Therefore, the progress in study of local limit theorems for the triangular arrays in the classical probability theory will play a principal role. Note here that these problems were studied in (Mukhin, 1991; Mirakhmedov, 1988, 1992 b).

The study of polynomial-like allocation schemes is in its beginning. We hope that the results of subsection 2.2 will be supplemented to a rather complete range of limit theorems. An extension of the class of statistics in question is also possible. The methods described in subsection 2.2 permit to make this. For the tuple allocation a natural way is the application of Theorems 2.3.1 and 2.3.2 to specific families of statistics which was started in (Tikhomirova, 1992).

The main problem in the CI-allocations is the search of conditions under which the change of the probabilities of a polynomial scheme by their statistical estimates has no influence on the limit behaviour of D-statistics. This problem is connected with the method of bootstrap estimation in mathematical statistics (see, for example, (Do and Hall, 1991)).

REFERENCES

Bykov, S. I. (1991). *Limit Theorems for Randomized Decomposable Statistics and Their Applications.* Thesis. Institute of Automatics, Moscow (in Russian).

Bykov, S. I. and Ivanov, V. A. (1991). On the conditions of asymptotic normality of multidimensional randomized decomposable statistics. *Discrete Math. Appl.* **1**, 219–227.

Do, K.-A. and Hall, P. (1991). On importance of resampling for the bootstrap. *Biometrika* **78**, 161–167.

Dutko, M. (1989). Central limit theorems for infinite urn models. *Ann. Probab.* **17**, 1255–1263.

Dyakonova, Ye. Ye. and Mikhailov, V. G. (1992). On sufficient conditions of asymptotic normality for decomposable statistics in a nonhomogeneous allocation scheme. *Discrete Math. Appl.* **2**, 325–336.

Feller, W. (1970). *An Introduction to Probability Theory and Its Applications.* Vol. 1. Wiley, New York.

Gortchakov, A. B. (1993). Limit theorems for decomposable statistics on dependent random variables. In: *Probabilistic Methods in Discrete Mathematics.* TVP/VSP, Moscow/Utrecht, pp. 187–199.

Holst, L. (1972). Asymptotic normality in a generalized occupancy problem. *Z. Wahrscheinlichkeitstheor. verw. Geb.* **21**, 109–120.

Ivanov, V. A., Ivchenko, G. I., and Medvedev, Yu. I. (1984). Discrete problems of probability theory. *J. Soviet Math.* **35**, 5–61.

Ivchenko, G. I. (1991). Decomposable statistics in an inverse allocation problem. *Discrete Math. Appl.* **1**, 81–96.

Ivchenko, G. I. and Mirakhmedov, Sh. A. (1988). On the asymptotic normality of randomized separable statistics in a generalized allocation scheme. *Theory Probab. Appl.* **33**, 749–755.

Ivchenko, G. I. and Mirakhmedov, Sh. A. (1992). On limit theorems for decomposable statistics and the efficiency of corresponding statistical tests. *Discrete Math. Appl.* **2**, 533–548.

Janson, S. (1988). Normal convergence by higher semiinvariants with applications to sums of dependent random variables and random graphs. *Ann. Probab.* **16**, 305–312.

Karlin, S. (1967). Central limit theorems for certain infinite urn schemes. *J. Math. Mech.* **17**, 373–401.

Karlin, S. and Ost, F. (1987). Count of long aligned word matches among random letter sequences. *Adv. Appl. Probab.* **19**, 293–351.

Karlin, S. and Ost, F. (1988). Maximal length of common words among random letter sequences. *Ann. Probab.* **16**, 535–563.

Kolchin, V. F. (1968). A class of limit theorems for conditional distributions. *Lithuanian Matem. Sb.* **8**, 53–63 (in Russian).

Kolchin, V. F. (1986). *Random Mappings.* Optimization Software, New York.

Kolchin, V. F., Sevastyanov, B. A., and Chistyakov, V. P. (1978). *Random Allocations.* Wiley, New York.

Kudlaev, E. M. (1988). Decomposable statistics. In: *Itogi Nauki i Tekhniki (Ser. Probab. Theory. Math. Statist. Theor. Cybern.).* V. 26. VINITI, Moscow, pp. 97–150 (in Russian).

Mikhailov, V. G. (1981). A central limit theorem for the independent allocation scheme. *Proc. Steklov Inst. Math.* **157**, 138–152 (in Russian).

Mikhailov, V. G. (1984). Asymptotic normality in a scheme of finitely dependent distribution of particles in cells. *Math. USSR Sbornik* **47**, 499–512.

Mikhailov, V. G. (1989 a). On the asymptotic normality of symmetric decomposable statistics in a nonhomogeneous scheme. *Diskretnaya matematika* 1, 2, 15–27 (in Russian).

Mikhailov, V. G. (1989 b). Asymptotic normality of decomposable statistics on the frequencies of m-tuples. *Discrete Math. Appl.* **1**, 335–347.

Mikhailov, V. G. (1991 a). On a theorem of Janson. *Theory Probab. Appl.* **36**, 173–176.

Mikhailov, V. G. (1991 b). On the approximation for a sum of random indicators. *Theory Probab. Appl.* **36**, 798.

Mirakhmedov, Sh. A. (1988). On a local limit theorem for lattice random variables in a triangular array. In: *Probabilistic Problems of Discrete Mathematics*, Moscow Inst. Electron. Eng., Moscow, pp. 114–119 (in Russian).

Mirakhmedov, Sh. A. (1989). Randomized decomposable statistics in a generalized allocation scheme with a countable set of cells. *Diskretnaya matematika* **1**, 4, 46–62 (in Russian).

Mirakhmedov, Sh. A. (1990). Randomized decomposable statistics in the scheme of independent allocating particles into boxes. *Diskrete Math. Appl.* **2**, 91–108.

Mirakhmedov, Sh. A. (1992). Probabilities of large deviations in limit theorems for conditional distributions of sums of independent random variables. *Diskretnaya matematika* **4**, 1, 22–32 (in Russian).

Mori, T. F. (1990). More on the waiting time till each of some given patterns occurs as a run. *Canadian J. Math.* **42**, 915–932.

Mukhin, A. B. (1991). Local limit theorems for lattice random variables. *Theory Probab. Appl.* **36**, 698–713.

Ronzhin, A. F. (1988). Goodness-of-fit tests for generalized urn schemes based on separable statistics. *Theory Probab. Appl.* **33**, 86–96.

Sirazhdinov, S. Kh., Mirakhmedov, Sh. A., and Ismatullaev, Sh. A. (1989). Probabilities of large deviations for randomized divisible statistics in the multinomial scheme. *Theory Probab. Appl.* **34**, 645–657.

Suyunov, A. M. (1989 a). On the probability of large deviations for the number of empty cells in the polynomial allocation scheme. In: *Functionals on Random Processes and Statistical Inferences*, Tashkent, pp. 85–96 (in Russian).

Suyunov, A. M. (1989 b). On the probability of large deviations for the number of cells containing exactly r particles in the polynomial scheme. *Izv. AN UzSSR, Ser. Phiz.-Mat.* **5**, 26–28 (in Russian).

Tikhomirova, M. I. (1992). On the asymptotic normality for the number of nonoccurring s-tuples. *Diskretnaya matematika* **4**, 2, 122–129 (in Russian).

Vatutin, V. A. and Mikhailov, V. G. (1982). Limit theorems for the number of empty cells in the equiprobable allocation scheme of series of particles. *Theory Probab. Appl.* **27**, 684–692.

Probabilistic Methods in Discrete Mathematics, pp. 60 – 63
V. F. Kolchin et al. (Eds.)
1993 TVP/VSP

PROPERTIES OF RANDOM PERMUTATIONS WITH CONSTRAINTS ON THE MAXIMUM CYCLE LENGTH

A. A. GRUSHO
TVP Laboratories, Krasikov Street 32, 117418 Moscow, Russia

ABSTRACT

Let $\mathcal{C}_n^l$ be the class of all permutations of a finite set X with cycle lengths not greater than l, and let n be the number of elements in X. On the set $\mathcal{C}_n^l$ we define the uniform probability measure and choose at random a permutation from this set. Independently of the choice of the permutation, we choose a point from X according to the uniform measure on X. Let the random variable ζ be equal to the length of the cycle of the random permutation which contains the random point chosen from X.
In the present paper we investigate the probability that the random point is on the cycle of length l.

Let $\mathcal{C}_n^l$ be the class of all permutations of a finite set $X = \{1, \dots, n\}$ with cycle lengths not greater than l, $C_n^l = |\mathcal{C}_n^l|$ be the number of elements in $\mathcal{C}_n^l$. On the set $\mathcal{C}_n^l$ we define the uniform probability measure and choose at random a permutation from this set. Independently of the choice of the permutation we choose a point from X according to the uniform measure on X.
Let the random variable (r. v.) ζ be equal to the length of the cycle of the random permutation which contains the random point chosen from X. If g is a random permutation from $\mathcal{C}_n^n$, then as was shown in (Goncharov, 1944) the length of the maximal cycle of g is close to n. Thus the probability of the event $\{\zeta \geq \varepsilon n\}$ for some ε, $0 < \varepsilon < 1$, converges to a positive number. The main goal of our article is to investigate the probability that a random point in a random permutation g from $\mathcal{C}_n^l$ is on the cycle of length l.
Let $\lambda(g, l)$ denote the number of l-cycles in g. It is clear that

$$\mathbf{P}\{\zeta = l\} = \frac{l}{n}\mathbf{P}\{\lambda(g,l) = 1\} + \frac{2l}{n}\mathbf{P}\{\lambda(g,l) = 2\} + \cdots + \frac{kl}{n}\mathbf{P}\{\lambda(g,l) = k\} + \cdots$$

Therefore

$$\mathbf{P}\{\zeta = l\} = \frac{l}{n}\mathbf{E}\,\lambda(g,l). \tag{1}$$

According to (Sachkov, 1977)

$$\sum_{n=0}^{\infty} C_n^l \frac{t^n}{n!} = \exp\left\{\sum_{j=1}^{l} \frac{t^j}{j}\right\}. \tag{2}$$

If C^l_{nk} is the number of permutations in C^l_n with k cycles of length l, then

$$C^l_n(x) = \sum_{k=0}^{[n/l]} C^l_{nk} x^k, \tag{3}$$

$$\sum_{n=0}^{\infty} C^l_n(x)\frac{t^n}{n!} = \exp\left\{(x-1)\frac{t^l}{l} + \sum_{j=1}^{l} \frac{t^j}{j}\right\}. \tag{4}$$

Since we consider the uniform distribution on C^l_n,

$$\sum_{k=0}^{[n/l]} \mathbf{P}\{\lambda(g,l) = k\}x^k = \frac{C^l_n(x)}{C^l_n}, \tag{5}$$

$$\mathbf{E}\,\lambda(g,l) = \frac{1}{C^l_n}\frac{dC^l_n(x)}{dx}\bigg|_{x=1}, \tag{6}$$

$$\sum_{n=0}^{\infty} \frac{dC^l_n(x)}{dx}\bigg|_{x=1}\frac{t^n}{n!} = \frac{t^l}{l}\exp\left\{\sum_{j=1}^{l}\frac{t^j}{j}\right\}. \tag{7}$$

From (1), (6) and (7) we obtain

$$\mathbf{P}\{\zeta = l\} = \frac{1}{n}\frac{\operatorname{coef}_{t^n} t^l \exp\left\{\sum_{j=1}^{l} t^j/j\right\}}{\operatorname{coef}_{t^n} \exp\left\{\sum_{j=1}^{l} t^j/j\right\}}. \tag{8}$$

THEOREM 1. *Let $n \to \infty$, $l = [n^\gamma]$, $1/2 < \gamma < 1$. Then*

$$\mathbf{P}\{\zeta = l\} = \frac{(1-\gamma)\log n}{n^\gamma}(1 + o(1)).$$

Proof. Let $n \to \infty$, $l = [n^\gamma]$, $1/2 < \gamma < 1$. We use the saddle-point method. It is clear that

$$\frac{C^l_n}{n!} = \frac{1}{2\pi i}\oint \exp\left\{\sum_{j=1}^{l}\frac{t^j}{j}\right\}\frac{dt}{t^{n+1}}, \tag{9}$$

where the integral is taken around any contour containing zero. Denote

$$h(t) = \sum_{j=1}^{l}\frac{t^j}{j} - n\log t.$$

The equation for the saddle point has the form

$$\sum_{j=0}^{l-1} t^j = \frac{n}{t}. \tag{10}$$

Let $1+\alpha(n,l)$ be the positive solution of this equation. It is easy to show that $\alpha(n,l)>0$ and equation (10) takes the form

$$(1+\alpha(n,l))^l = 1 + \frac{n\alpha(n,l)}{1+\alpha(n,l)}. \tag{11}$$

By iteration we obtain that

$$\alpha(n,l) = \frac{\log(n/l)+\log\log(n/l)}{l} + O\left(\frac{\log\log\log(n/l)}{l}\right). \tag{12}$$

The integral in (9) may be taken around the circle of radius $1+\alpha$ (we omit the arguments n and l in $\alpha(n,l)$):

$$\frac{C_n^l}{n!} = \frac{1}{2\pi}\int_{-\varphi_0}^{\varphi_0} \exp\left\{h((1+\alpha)e^{i\varphi})\right\} d\varphi + \frac{1}{2\pi}\int_{|\varphi|>\varphi_0} \exp\left\{h((1+\alpha)e^{i\varphi})\right\} d\varphi. \tag{13}$$

We denote by I_1 and I_2, correspondingly, the two integrals in (13). If

$$|\varphi_0| = \frac{f_n}{\sqrt{nl}} = O\left(\frac{n^\varepsilon}{\sqrt{nl}}\right), \qquad 0<\varepsilon<n^{\gamma/6}, \tag{14}$$

we can expand $h((1+\alpha)e^{i\varphi})$ by the Taylor formula and obtain

$$h((1+\alpha)e^{i\varphi}) = h(1+\alpha) - \frac{h''(1+\alpha)}{2}\varphi^2 + O(nl\log(n/l)\varphi^3), \tag{15}$$

where

$$h''(1+\alpha) = O(nl). \tag{16}$$

Introducing new variables $u=\varphi/\sqrt{nl}$ and taking into account (16), we can rewrite the integral I_1 in the form

$$I_1 = \frac{\exp\{h(1+\alpha)\}}{2\pi\sqrt{nl}}\left\{\int_{-A}^{A}\exp\left\{-\frac{u^2}{2}\right\}du(1+o(1)) + O(e^{-\delta A^2})\right\},$$

where $\delta>0$, $A=o(\sqrt{\log n})$. Therefore

$$I_1 = \frac{\exp\{h(1+\alpha)\}}{\sqrt{2\pi nl}}(1+o(1)). \tag{17}$$

Since on the circle of radius $1+\alpha$ the absolute value of the function $\exp\left\{\sum_{j=1}^{l} t^j/j\right\}$ attains the maximum on the positive semi-axis, in view of (14) we can obtain a bound for $|I_2|$ estimating the absolute value of the integrand function at the point φ_0. Then

$$\begin{aligned}|I_2| &\le (1+\alpha)^{-n}\exp\left\{\sum_{j=1}^{l}\frac{(1+\alpha)^j}{j}\right\}\exp\left\{-\frac{h^n(1+\alpha)}{2nl}f_n^2\right\}(1+o(1)) \\ &= o(I_1).\end{aligned} \tag{18}$$

From (17) and (18) we obtain the asymptotic formula for the number of permutations with the cycle lengths not greater than l:

$$\frac{C_n^l}{n!} = \frac{\exp\left\{\sum_{j=1}^{l} j^{-1}(1+\alpha(n,l))^j\right\}(1+o(1))}{\sqrt{2\pi nl}\,(1+\alpha(n,l))^n}. \tag{19}$$

The estimation of the numerator in (8) replicates the above reasoning, and we obtain that

$$\begin{aligned} \operatorname{coef}_{t^{n-l}} \exp\left\{\sum_{j=1}^{l} \frac{t^j}{j}\right\} &= \exp\left\{\sum_{j=1}^{l} j^{-1}(1+\alpha(n-l,l))^j\right\} \\ &\quad \times \big(1+\alpha(n-l,l)\big)^{-n+l}\big(2\pi(n-l)l\big)^{-1/2}\big(1+o(1)\big). \end{aligned} \tag{20}$$

From (19) and (20)

$$\begin{aligned} \mathbf{P}\{\zeta = l\} &= \exp\left\{h\big(1+\alpha(n-l,l)\big) - h\big(1+\alpha(n,l)\big)\right\} \\ &\quad \times \big(1+\alpha(n-l,l)\big)^l\big(n\sqrt{(n-l)/n}\big)^{-1}\big(1+o(1)\big). \end{aligned} \tag{21}$$

Making use of (11), (15), (16) and the relation

$$\alpha(n-l,l) - \alpha(n,l) \le \frac{\alpha(n-l,l)}{cn}, \qquad c > 0,$$

we can easily show that $h(1+\alpha(n-l,l)) - h(1+\alpha(n,l)) = o(1)$. Therefore

$$\big(1+\alpha(n-l,l)\big)^l = \frac{n-l}{l}\left(\log\frac{n-l}{l} + \log\log\frac{n-l}{l}\right)\big(1+o(1)\big).$$

Substituting these estimates into (21), we obtain the asymptotic formula

$$\mathbf{P}\{\zeta = l\} = \frac{\log\left((n-l)/l\right)}{l}\left(1+o(1)\right),$$

which can be transformed into the formula pointed out in the theorem.

REFERENCES

Goncharov, V. L. (1944). Du domaine d'analyse combinatoire. *Bull. Acad. Sci. URSS, Ser. Math.* **8**, 3–48.

Sachkov, V. N. (1977). *Combinatorial Methods of Discrete Mathematics*. Nauka, Moscow (in Russian).

Stepanov, V. E. (1969). Limit distributions of certain characteristics of random mappings. *Theory Probab. Appl.* **14**, 612–626.

Probabilistic Methods in Discrete Mathematics, pp. 64 – 70
V. F. Kolchin et al. (Eds.)
1993 TVP/VSP

STOCHASTIC PROPERTIES OF SYSTEMS OF RANDOM LINEAR EQUATIONS OVER FINITE ALGEBRAIC STRUCTURES

I. N. KOVALENKO and A. A. LEVITSKAYA

V. M. Glushkov Institute of Cybernetics of the Academy of Sciences of Ukraine, Glushkova pr. 20, 252207 Kiev, Ukraine

ABSTRACT

The paper presents a survey of results on systems of random linear equations over finite algebraic structures.

The investigations of systems of linear equations over finite algebraic structures is a branch of the general theory of systems of equations. The development of this area of discrete mathematics was initiated by a lot of applied problems in such fields as design of large-scale systems, theory of coding, theory of impulse system testing, theory of random graphs. The result of (Slepian, 1955) which concerned the coding theory seems to be the first one belonging to the subject. This result consists of the following statement. For the rank $r_{\mathbf{A}}$ of a random Boolean $N \times n$ matrix $\mathbf{A}$ with independent symmetrically distributed elements the equality

$$\mathbf{P}\{r_{\mathbf{A}} = N\} = \prod_{i=n-N+1}^{n} (1 - 2^{-i}) \tag{1}$$

holds. Intensive and thorough investigations of stochastic matrices over finite fields have been originated by the pioneering paper (Erdős and Rényi, 1963): they have studied the limit behaviour of the permanent rank and the linear rank of a random Boolean $n \times n$ matrix. The investigations were continued in (Balakin, 1967, 1968; Kovalenko, 1972; Kolchin, 1988; Levitskaya, 1991 a). The paper (Balakin, 1968) deals with a random $N \times n$ matrix $\mathbf{A} = (a_{ij})$ over the field $\mathbf{GF}(q)$, the elements a_{ij} being independent random variables with the following distribution:

$$\mathbf{P}\{a_{ij} = 0\} = 1 - \frac{\log n + x}{n}, \quad \mathbf{P}\{a_{ij} = z \in \mathbf{GF}(q) \setminus 0\} = \frac{\log n + x}{(q-1)n}$$

(x is a constant). The author obtained the relation between N and n sufficient for the following effect. The distributions of differences between $\min(n, N)$ and the determinant, permanent and linear ranks of $\mathbf{A}$, respectively, asymptotically

coincide and asymptotically have either the Poisson distribution or the degenerate in zero distribution.
Consider, for example, the case

$$N/n = \alpha + O(n^{-1}),$$

where α is a constant. If $\alpha > 1$, then the three differences mentioned above are equal to zero with probabilities approaching 1 as $n \to \infty$. If $\alpha \in (0,1)$, then the differences asymptotically have the Poisson distribution with parameter αe^{-x}. The differences are asymptotically Poissonian in the case $N = n + \beta n/\log n + o(n/\log n)$, β is a constant as well; the parameter of the Poisson distribution equals e^{-x} if $\beta < 0$ and $e^{-x-\beta}$ if $\beta > 0$. The results of (Balakin, 1968) concerning the limit behaviour of the linear rank of the random matrix **A** were extended in (Kovalenko, 1972) to the case $q = 2$, $\mathbf{P}\{a_{ij} = 1\} = n^{-1}(\log n + x_{ij})$, $|x_{ij}| < T$, $i = 1, \dots, N$, $j = 1, \dots, n$, where $N/n = O(1)$.
Results from (Kolchin, 1988; Levitskaya, 1991a) supplement considerably the results of (Balakin, 1968). In the lecture notes (Kolchin, 1988) the effectiveness of a new approach was illustrated . A new proof of the well-known result of (Balakin, 1968) concerning the distribution of the determinant rank of the matrix **A** provided $q = 2$, $\alpha \in (0,1)$, is proposed and in this case the limit probability of consistency of a nonhomogeneous system of equations $\mathbf{AX} = \mathbf{B}$, where $\mathbf{B} = (b_1, \dots, b_N)$ is a random Boolean vector independent of the matrix **A**, is obtained. In the paper (Levitskaya, 1991a) expressions for the limit distributions of the determinant rank and the linear rank of a random $(n + \beta n/\log n + o(n/\log n)) \times n$ Boolean matrix are found, where β is the same as in the paper (Balakin, 1968). The number of units in each row of the matrix is assumed to belong to the interval $[r_0, \log n + s_0]$ or, alternatively, to $[n - \log n + s_0, r_1]$, where s_0 is a constant, $0 \le r_0 \le \log n + s_0$, $n - \log n + s_0 \le r_1 \le n$. The results of (Levitskaya, 1991a) agree with the results of (Balakin, 1968).
Of a practical interest is to solve the following problem. In many cases the desired stochastic properties of a linear transformation **A** can be achieved by means of the simplest procedure, namely, by the uniform independent generation of the elements a_{ij} of **A**. But such a procedure is too complicated. The question arises: what is the minimal filling of the matrix **A** which is enough to guarantee the desired properties? The problem has been treated, as a rule, in various limit cases, see (Kovalenko, 1965, 1967, 1975; Kozlov, 1966; Kovalenko and Levitskaya, 1975, 1977; Kovalenko et al., 1986; Levitskaya, 1978, 1982, 1983, 1984, 1985, 1988, 1991b; Masol, 1980). Thus, a kind of "invariance theory" has been developed. The essence of the theory is that, for certain classes of systems of random linear equations, admissible regions for the distributions of the system coefficients can be pointed out within which the limit distribution of the number of solutions (or some other limit characteristics) is invariant and coincides for example with the distribution for the "independent uniform" case. Thus, for any fixed number s under some additional conditions the limit distribution of the rank of a $(n - s) \times n$ random matrix **A** is invariant provided each row of the matrix contains at least $\alpha \log n$ elements, where α depends on the corresponding algebraic structure.
In an early publication (Kovalenko, 1965) a random $n \times n$ Boolean matrix $\mathbf{A} = (a_{ij})$

with independent elements was considered under the condition

$$0 < \delta \leq \mathbf{P}\{a_{ij} = z\}, \qquad i,j = 1,\ldots,n, \quad z = 0,1, \tag{2}$$

where δ is a constant. It was found that the $\lim_{n\to\infty} \mathbf{P}\{r_{\mathbf{A}} = n\}$ is of the form (1) if $n = N$. In (Kozlov, 1966) a more general result was obtained that the invariance property holds under the condition

$$\frac{(\log n)^{1+\varepsilon}}{n} \leq \mathbf{P}\{a_{ij} = z\}, \qquad i,j = 1,\ldots,n, \quad z = 0,1.$$

A further development of the invariance theory for systems of random linear equations was made by Kovalenko in the paper (Kovalenko, 1967) devoted to the detailed analysis of the system

$$\mathbf{A}\mathbf{x}' = \mathbf{0}'. \tag{3}$$

Let the matrix $\mathbf{A} = (a_{ij})$ be a $(n-s) \times n$ random matrix which satisfies the conditions introduced in (Kovalenko, 1965) and let s be a fixed integer of any sign. Let us recall the main theoretical result of (Kovalenko, 1967). The limit distribution of the number ν_n of solutions of system (3) exists under conditions (2), does not depend on the distributions of the elements of $\mathbf{A}$, and has the form

$$\lim_{n\to\infty} \mathbf{P}\{\nu_n = 2^k\} = 0 \text{ if } k < s, \tag{4}$$

$$\lim_{n\to\infty} \mathbf{P}\{\nu_n = 2^k\} = 2^{-k(k-s)}\left\{\prod_{i=1}^{k-s}(1-2^{-i})\right\}^{-1} \prod_{i=k+1}^{\infty}(1-2^{-i}) \text{ if } k \geq \max\{0,s\}.$$

In addition, the geometric structure of any full set of nontrivial solutions of system (3) is described.

The results of (Kovalenko, 1967) were generalized in (Kovalenko, 1975; Kovalenko and Levitskaya, 1977) for the case of an arbitrary finite field $\mathbf{GF}(q)$. The limit distribution of ν_n is still given by the equation (4), where the number 2 should be replaced by q. The further generalization in (Kovalenko, 1975) of the original method of (Kovalenko, 1965, 1967) made it possible to find the following natural bounds of the invariance region within which the limit distribution is of the same form as (4):

$$\delta_i = \delta_i(n) \leq \mathbf{P}\{a_{ij} = z\}, \qquad i = 1,\ldots,n-s, \quad j = 1,\ldots,n, \quad z \in \mathbf{GF}(2),$$

where $\sum_{i=1}^{n-s} \exp\{-n\delta_i\} \to 0$ as $n \to \infty$.

In (Kovalenko, 1975) a similar problem is solved for random Boolean systems of type (3) with a given number $r_i = r_i(n)$ of unit elements in the ith row of the matrix $\mathbf{A}$ provided $r_i \equiv 1 \pmod 2$ for at least one of the indices $i = 1,\ldots,n-s$. (It is assumed that the places of the r_i unit elements are being chosen at random.) Then the condition

$$\sum_{i=1}^{n-s} \exp\{-r_i\} \to 0, \qquad n \to \infty,$$

describes the boundaries of the invariance region. This result was then generalized and complemented in (Levitskaya, 1991b). In this paper the case of the fixed number of units in each row of a Boolean matrix is considered, the number varies within the interval $[r_0, r_1]$, where $0 \le r_0 \le n - \log c_n n$, $\log c_n n \le r_1 \le n$, $r_0 \le r_1$, $c_n \to \infty$. The case

$$r_i = r_i(n) \equiv 0 \pmod 2, \qquad i = 1, \dots, n - s,$$

was discussed in terms of the paper (Kovalenko, 1975).

In (Masol, 1980) restrictions on the distribution of coefficients of system (3) are determined which imply the invariance of the limit distribution of the rank of $\mathbf{A}$. These restrictions lead to the condition that the matrix $\mathbf{A}$ contains a fixed number of zero rows and columns.

In the important lecture notes (Kolchin, 1988) the author proposes new approaches to the investigation of various types of random linear Boolean systems based on the relations of the matrices of systems with the classical allocation models and random graphs.

It should be noted that in all the publications mentioned above the investigation of (3) means a study of the rank of the matrix $\mathbf{A}$ or the number of linearly independent solutions of the corresponding system under an appropriate condition on the distributions of the elements of $\mathbf{A}$. If system (3) is considered over an arbitrary associative ring, the rank and the linear independence cannot be defined. But the problems having been solved previously for finite fields are still of interest. The methods which make it possible to treat a wide class of problems for system (3) over arbitrary finite fields or rings from a unified viewpoint were elaborated by Levitskaya in (Levitskaya, 1978, 1982, 1983, 1984, 1985; Kovalenko et al., 1986). For the most detailed presentation of the subject see the monograph (Kovalenko et al., 1986). Let us present main results of the monograph.

Let $\mathbf{R}$ be a finite associative ring with $|\mathbf{R}| = m$ elements. Consider the system of equations

$$\sum_{j=1}^{n} a_{ij} x_j = b_i, \qquad i = 1, \dots, n - s, \tag{5}$$

where s is a fixed integer of any sign, a_{ij} are independent random variables, b_i are assumed to be fixed. Denote by ν_n the number of solutions of (5).

Under these conditions the bounds of the invariance regions are established for $\lim_{n\to\infty} \mathrm{M}\nu_n^r$ $(r \in \mathbf{Z}^+)$ and for the limit distribution of ν_n as $n \to \infty$.

It is interesting to note that in the case of a ring containing the left unit the invariance regions for the $\lim_{n\to\infty} \mathrm{M}\nu_n^r$ $(r \in \mathbf{Z}^+)$ and for the limit distribution of ν_n are the same and are described by the following inequalities:

$$\begin{gathered} \frac{l_0}{m(l_0-1)} \frac{\log c_n n}{n} \le \mathbf{P}\{a_{ij} = z\}, \\ i = 1, \dots, n-s, \quad j = 1, \dots, n, \quad z \in \mathbf{R}, \end{gathered} \tag{6}$$

if the distributions of the coefficients of the ith equation are independent of i, and otherwise

$$\delta_{in} \le \mathbf{P}\{a_{ij} = z\}, \qquad i = 1, \dots, n - s, \quad j = 1, \dots, n, \quad z \in \mathbf{R}, \tag{7}$$

where

$$\sum_{i=1}^{n-s} \exp\left\{-\frac{m(l_0-1)}{l_0} n\delta_{in}\right\} \to 0, \qquad n \to \infty.$$

Here l_0 is the minimal number of elements of the proper left ideals of $\mathbf{R}$, c_n is an arbitrary sequence tending to infinity.
Within the invariance regions (6) or (7),

$$\lim_{n\to\infty} \mathbf{M}\nu_n^r = \sum_{M_r \subseteq \mathbf{R}^r;\, \mathbf{b}^i \in M_r,\, i=1,\ldots,n-s} |M_r|^s, \tag{8}$$

where the summation is over all the left submodules $M_r \subseteq \mathbf{R}^r$ containing all the r-vectors $\mathbf{b}^i = (b_i, \ldots, b_i)$. With the help of (8) the geometric structure of the set of the solutions of (5) is described as well.
For homogeneous systems (5) in the invariance region (6) (or (7)) we suggested an approach for obtaining the limit distribution of ν_n. The simplest case is the case of the ring $\mathbf{Z}_{p_1\cdots p_t}$ of residues modulo $p_1p_2\cdots p_t$, where p_i, $i = 1, \ldots, t$, are primes. System (5) can be decomposed into t independent systems over fields $\mathbf{GF}(p_i)$, thus

$$\nu_n = \prod_{i=1}^{t} \nu_n^i,$$

where ν_n^i is the number of solutions of (5) over $\mathbf{GF}(p_i)$, and the limit distribution of ν_n can be written in the following form:

$$\lim_{n\to\infty} \mathbf{P}\left\{\nu_n = \prod_{i=1}^{t} p_i^{k_i}\right\} = 0$$

if $\min_{1\le i\le t}(k_i) < s$;

$$\lim_{n\to\infty} \mathbf{P}\left\{\nu_n = \prod_{i=1}^{t} p_i^{k_i}\right\} = \prod_{i=1}^{t} p_i^{-k_i(k_i-s)} \left\{\prod_{j=1}^{k_i-s} (1-p_i^{-j})\right\}^{-1} \prod_{j=k_i+1}^{\infty} (1-p_i^{-j}),$$

if $k_i \ge \max\{0, s\}$, $i = 1, \ldots, t$.
As to nonhomogeneous systems (5), within the invariance region an expression for the $\lim_{n\to\infty} \mathbf{P}\{\nu_n \ge 1\}$ is obtained.
For homogeneous systems (5) over an arbitrary finite ring with dependent coefficients but with independent rows of the matrix some sufficient conditions were derived under which the limit distribution of ν_n coincides with the limit distribution for the independent symmetrical case.
It should be noted that the probabilistic analysis of system (5) presented in the monograph (Kovalenko et al., 1986) is based on the theory of summation of random variables on a finite Abelian group. For example, while constructing the invariance theory we use an extreme property of the distribution of the sum of independent random variables defined on some subgroups of an additive Abelian group. It turns out that the theory of summation of random variables on finite Abelian groups is a

base of the probabilistic analysis of another type of linear systems (see (Kovalenko et al., 1986)).
Let $\mathbf{X}$, $\mathbf{G}$ be finite Abelian groups, and let $f(a, x)$ be a mapping $\{0, 1, \dots, r-1\} \times \mathbf{X} \to \mathbf{G}$ which is reduced to a homeomorphism for any fixed a,

$$f(a, x+y) = f(a, x) + f(a, y), \quad f(0, x) = 0, \qquad x \in \mathbf{X}.$$

Consider the system of equations

$$L_i \mathbf{x} \triangleq \sum_{j=1}^{n} f(a_{ij}, x_j) = b_i, \qquad i = 1, \dots, n-s, \tag{9}$$

where s is a fixed integer of any sign, b_i are fixed, a_{ij} are independent random variables on $\{0, 1, \dots, r-1\}$, $\mathbf{x} = (x_1 \dots x_n) \in \mathbf{X}^n$. It is proved that the condition

$$\begin{gathered} \frac{c \log n}{n} \leq \mathbf{P}\{a_{ij} = a\}, \\ a = 0, 1, \dots, r-1; \quad i = 1, \dots, n-s, \quad j = 1, \dots, n, \end{gathered} \tag{10}$$

implies the invariance of the limit distribution of ν_n. The constant c is evaluated in (Levitskaya, 1988).

REFERENCES

Balakin, G. V. (1967). On random matrices. *Theory Probab. Appl.* **12**, 301–338.

Balakin, G. V. (1968). The distribution of the rank of random matrices over a finite field. *Theory Probab. Appl.* **13**, 594–614.

Erdös, P. and Rényi, A. (1963). On random matrices. *Magyar Tud. Akad. Kutató Int. Kőzl.* **8**, 455–461.

Kolchin, V. F. (1988). *Systems of Random Equations*. Moscow Inst. Electron. Eng., Moscow (in Russian).

Kovalenko, I. N. (1965). On a limit theorem for determinants in the class of Boolean functions. *Soviet Math. Dokl.* **161**, 517–519 (in Russian).

Kovalenko, I. N. (1967). On the limit distribution of the number of solutions of a random system of linear equations in the class of Boolean functions. *Theory Probab. Appl.* **12**, 47–57.

Kovalenko, I. N. (1972). Distribution of the linear rank of a random matrix. *Theory Probab. Appl.* **17**, 487.

Kovalenko, I. N. (1975). On invariance theorems for random Boolean matrices. *Cybernetics* **5**, 138–152 (in Russian).

Kovalenko, I. N. and Levitskaya, A. A. (1975). The limit behaviour of the number of solutions of a system of linear equations over a finite field or a finite ring. *Soviet Math. Dokl.* **221**, 778–781 (in Russian).

Kovalenko, I. N. and Levitskaya, A. A. (1977). The limit behaviour of the number of solutions of a system of linear equations over a finite field or a finite ring. *Theory Probab. Math. Statist.* **13**, 70–83 (in Russian).

Kovalenko, I. N., Levitskaya, A. A., and Savchuk, M. N. (1986). *Selected Problems of the Probability Combinatorics*. Naukova Dumka, Kiev (in Russian).

Kozlov, M. V. (1966). On the rank of matrices with random Boolean elements. *Soviet Math. Dokl.* **169**, 1013–1016 (in Russian).

Levitskaya, A. A. (1978). Invariance theorems for the number of solutions of a system of random linear equations over a finite ring. *Cybernetics* **2**, 140–141 (in Russian).

Levitskaya, A. A. (1982). Invariance theorems for the number of solutions of a system of random linear equations over a finite ring. *Soviet Math. Dokl.* **263**, 279–281 (in Russian).

Levitskaya, A. A. (1983). Limit theorems for a system of random linear equations over a finite ring. In: *First Petrozavodsk Conference on Probabilistic Methods in Discrete Mathematics. Abstracts of Communications*, Karelian Branch Acad. Sci. USSR, Petrozavodsk, pp. 43–44 (in Russian).

Levitskaya, A. A. (1984). Invariance theorems for a system of random linear equations over a finite ring without a left unity. *Soviet Math. Dokl.* **9**, 71–72 (in Russian).

Levitskaya, A. A. (1985). The probability of consistency of a system of random linear equations over an arbitrary finite ring. *Theory Probab. Appl.* **30**, 364–385.

Levitskaya, A. A. (1988). Invariance theorems for a system of random linear equations over a finite Abelian group. In: *Second Petrozavodsk Conference on Probabilistic Methods in Discrete Mathematics. Abstracts of Communications*, Karelian Branch Acad. Sci. USSR, Petrozavodsk, pp. 56–57 (in Russian).

Levitskaya, A. A. (1991 a). Boundary distribution of the rank of a Boolean random matrix with a fixed number of units in the rows. *Soviet Math. Dokl.* **8**, 49–52 (in Russian).

Levitskaya, A. A. (1991 b). Invariance theorems for the rank of a Boolean random matrix with a fixed number of units in the rows. *Soviet Math. Dokl.* **9**, 53–56 (in Russian).

Masol, V. (1980). An expansion of the invariance domain for random Boolean matrices. *Cybernetics* **3**, 125–128 (in Russian).

Slepian, D. (1955). A class of binary signalling alphabets. In: *Proceedings of the Symposium on Information Theory*, Royal Institution, London.

Probabilistic Methods in Discrete Mathematics, pp. 71 – 98
V. F. Kolchin et al. (Eds.)
1993 TVP/VSP

ON THE NUMBER OF SOLUTIONS OF SYSTEMS OF PSEUDO-BOOLEAN RANDOM EQUATIONS

G. V. BALAKIN
TVP Laboratories, Krasikov Street 32, 117418 Moscow, Russia

ABSTRACT

We consider systems of equations of the form

$$f_i(x_1,\dots,x_n)=a_i, \qquad i=1,\dots,t,$$

over the field of real numbers, the unknowns $x_1,\dots,x_n$ taking values 0 and 1 only. Such systems are called pseudo-Boolean. The functions $f_1,\dots,f_n$ are chosen at random from a set Φ of pseudo-Boolean functions, and values $a_1,\dots,a_t$ either are chosen at random and independently of the set of all values of the functions $f_1,\dots,f_t$ accordingly, or $a_i=f_i(\bar{x}^0)$, $i=1,\dots,t$, where $\bar{x}^0=(x_1^0,\dots,x_n^0)$ is a fixed vector. In the first case the system is called random, and in the second case it is called a priori solvable random system.
For some random and a priori solvable random systems of equations the mathematical expectation of the number of solutions and the distribution of the number of solutions are obtained. We consider methods for investigation of random and a priori solvable random systems, namely, geometric and combinatorial methods.

1. LINEAR PSEUDO-BOOLEAN SYSTEMS WITH INDEPENDENT COEFFICIENTS

Let us consider a system of equations over the field of real numbers

$$f_i(x_1,\dots,x_n)=b_i, \qquad i=1,\dots,t, \tag{1.1}$$

where the unknowns $x_1,\dots,x_n$ take values 0 and 1 only. Systems (1.1) and the functions $f_1,\dots,f_t$ are called pseudo-Boolean (see (Ivanescu and Rudeanu, 1966)). The functions $f_1,\dots,f_t$ are chosen from a set Φ of pseudo-Boolean functions.
If $b_i=f_i(\bar{x}^0)$, $i=1,\dots,t$, for some fixed vector $\bar{x}^0=(x_1^0,\dots,x_n^0)$, then such a system is called *a priori solvable*, and the vector $\bar{x}^0$ is called the true solution (see (Balakin, 1983)). If $b_1,\dots,b_t$ are random variables which are independent of the functions $f_1,\dots,f_t$, then such a system is called *random*.
In Sections 1 and 2, linear systems (1.1) for which $f_i(\bar{x})=(\bar{a}_i,\bar{x})$, where $\bar{a}_i=(a_{i1},\dots,a_{in})$ are independent integer-valued random vectors with the same distribution, are investigated.

1.1. The average number of solutions for systems with (1,−1) coefficients

Suppose that a_{ij} ($i = 1,\dots,t,\ j = 1,\dots,n$) are independent random variables which take values 1 and −1 only with equal probabilities. Let us derive asymptotic formulas for the average number of solutions of a priori solvable and random systems.
Let system (1.1) be a priori solvable with the true solution

$$\bar{x}^0 = (x_1^0,\dots,x_n^0),\quad x_1^0 = \dots = x_m^0 = 0,\quad x_1^0 + \dots + x_n^0 = n - m,\quad m = 1,\dots,n.$$

Denote $\eta(t)$ the number of solutions of a priori solvable system (1.1).

THEOREM 1.1. *If*

$$\lim_{n\to\infty} t\frac{\log n}{n} > 2\log 2,$$

then $\mathbf{E}\,\eta(t) \to 1$ *and* $\mathbf{P}\{\eta(t) = 1\} \to 1$. *If*

$$\lim_{n\to\infty} t\frac{\log n}{n} < 2\log 2,$$

then $\mathbf{E}\,\eta(t) \to \infty$.

Proof. Consider a vector $\bar{x}^k = (x_1^k,\dots,x_n^k)$ which differs from $\bar{x}^0$ in k coordinates. For this vector

$$\mathbf{P}\{a_{i1}x_1^k + \dots + a_{in}x_n^k = a_{i1}x_1^0 + \dots + a_{in}x_n^0\} = \mathbf{P}\{a_{i1} + \dots + a_{ik} = 0\}.$$

Hence we obtain

$$\begin{aligned}\mathbf{E}\,\eta(t) &= \sum_{k=0}^{[n/2]} C_n^{2k}\left(\frac{C_{2k}^k}{2^{2k}}\right)^t \\ &= 1 + \sum_{0<k<n/2}\sqrt{\frac{n}{2\pi k(n-2k)}}\exp\left\{-nH\left(\frac{2k}{n}\right) - t\left(\frac{1}{2}\log\pi k + O\left(\frac{1}{k}\right)\right)\right\} \\ &\quad + o(1), \end{aligned} \tag{1.2}$$

where $H(p) = p\log p + (1-p)\log(1-p)$; here and below we use the notation C_a^b for $\binom{a}{b}$.
Let $t = \alpha n/\log n$, where α is a positive constant. Then the exponent can be written in the form

$$n\left(\beta_k\log\frac{1}{\beta_k} + (1-\beta_k)\log\frac{1}{1-\beta_k} - \frac{\alpha}{2} + O\left(\frac{1}{\log n}\right)\right),$$

where $\beta_k = 2k/n$. The maximum of this expression is equal to

$$n\big(\log 2 - (\alpha/2) + O(1/\log n)\big)$$

and is attained at $\beta_k \sim 1/2$, which proves the assertion of the theorem.

Let us consider now random system (1.1) with the right-hand sides $b_1, \ldots, b_t$ such that $\mathbf{P}\{b_i = l\} = p_l$, $i = 1, \ldots, t$, $l = -n, \ldots, n$. Denote $\xi(t)$ the number of solutions of system (1.1) in this case. Obviously,

$$\mathbf{E}\,\xi(t) = \sum_{\bar{x}} \prod_{i=1}^{t} \mathbf{P}\{(\bar{a}_i, \bar{x}) = b_i\},$$

where

$$\mathbf{P}\{(\bar{a}_i, \bar{x}) = b_i\} = \sum_{l} p_l \mathbf{P}\{(\bar{a}_i, \bar{x}) = l\}.$$

THEOREM 1.2. *Let*

$$p_l = 1/(2n+1), \qquad l = -n, \ldots, n; \qquad t = \big(n \log 2 + C_n\big)/\big(\log(2n+1)\big).$$

Then $\mathbf{E}\,\xi(t) = e^{-C_n}$ *and therefore* $\mathbf{E}\,\xi(t) \to \infty$ *if* $C_n \to -\infty$ *and* $\mathbf{E}\,\xi(t) \to 0$ *if* $C_n \to \infty$ as $n \to \infty$.

Proof. It is easy to verify that

$$\mathbf{P}\{(\bar{a}_i, \bar{x}) = b_i\} = 1/(2n+1), \qquad i = 1, \ldots, t,$$

$$\begin{aligned} \mathbf{E}\,\xi(t) &= 2^n (2n+1)^{-t} = \exp\{n \log 2 - t \log(2n+1)\} \qquad (1.3) \\ &= \exp\{n \log 2 - (n \log 2 + C_n)\} = \exp\{-C_n\}. \end{aligned}$$

The results of the theorem follow from formula (1.3).
Note that the critical number of equations for a priori solvable system is equal to $t_1 \sim 2n \log 2/\log n$, and for random systems it is equal to $t_2 \sim n \log 2/\log n$ so that $t_1/t_2 \to 2$ as $n \to \infty$.

COROLLARY 1.1. *Under the assumptions of Theorem 1.2*

$$\mathbf{P}\{\xi(t) = 0\} \to 1 \quad \textit{if} \quad C_n \to \infty.$$

With the help of (1.2), (1.3) one can find $\operatorname{Var} \xi(t)$. Indeed, we have

$$\xi(t) = \sum_{j=1}^{2^n} \xi_j,$$

where $\xi_j = 1$, $j = 1, \ldots, 2^n$, if the vector

$$\bar{x} = (x_1(j), \ldots, x_n(j)), \qquad 2^{n-1} x_n(j) + \cdots + 2x_2(j) + x_1(j) = j$$

is a solution of random system (1.1), and $\xi_j = 0$ otherwise. It is clear that

$$\mathbf{E}\,\xi^2(t) = \mathbf{E}\left(\sum_{j=1}^{2^n} \xi_j\right) = \sum_{k \neq j} \mathbf{E}\,\xi_k \xi_j + \mathbf{E}\,\xi(t).$$

Now, taking into account that

$$\mathbf{E}\,\xi_k\xi_j = \mathbf{P}\{\xi_k = 1\}\mathbf{P}\{\xi_j = 1 \mid \xi_k = 1\},$$

$$\mathbf{P}\{\xi_k = 1\} = \prod_{i=1}^{t} \mathbf{P}\{(\bar{a}_i, \bar{x}(k)) = b_i\} = 1/(2n+1)^t,$$

$$\mathbf{P}\{\xi_j = 1 \mid \xi_k = 1\} = \prod_{i=1}^{t} \mathbf{P}\{(\bar{a}_i, \bar{x}(j)) = (\bar{a}_i, \bar{x}(k))\},$$

$$\sum_{j\neq k} \mathbf{P}\{\xi_j = 1 \mid \xi_k = 1\} = \mathbf{E}\,\eta(t) - 1,$$

$$\sum_{k=1}^{2^n}\sum_{j\neq k} \mathbf{P}\{\xi_j = 1 \mid \xi_k = 1\} = 2^n(\mathbf{E}\,\eta(t) - 1),$$

$$\sum_{j\neq k} \mathbf{E}\,\xi_k\xi_j = \left(\frac{1}{2n+1}\right)^t 2^n(\mathbf{E}\,\eta(t) - 1) = \mathbf{E}\,\xi(t)\mathbf{E}\,\eta(t) - \mathbf{E}\,\xi(t),$$

we get

$$\mathbf{E}\,\xi^2(t) = \mathbf{E}\,\xi(t)\mathbf{E}\,\eta(t), \quad \operatorname{Var}\xi(t) = \mathbf{E}\,\xi(t)(\mathbf{E}\,\eta(t) - \mathbf{E}\,\xi(t)). \tag{1.4}$$

Applying (1.2), (1.3) to (1.4), we obtain a formula for $\operatorname{Var}\xi(t)$. Notice that (1.4) follows from the main result of the paper (Balakin, 1988 a).

1.2. The average number of solutions for systems with (0,1) coefficients

Let system (1.1) be a priori solvable, its coefficients take values 0 and 1 with equal probabilities, and the true solution $\bar{x}^0 = (x_1^0, \dots, x_n^0)$ has m zero components, $x_1^0 = \dots = x_m^0 = 0$,

$$0 \le \theta = \lim_{n\to\infty} \frac{m}{n} < 1.$$

THEOREM 1.3. *If*

$$\lim_{n\to\infty} t\frac{\log n}{n} > 2\left(\theta\log\frac{1}{\theta} + (1-\theta)\log\frac{1}{1-\theta}\right),$$

then $\mathbf{E}\,\eta(t) \to 1$ *and* $\mathbf{P}\{\eta(t) = 1\} \to 1$. *If*

$$0 < \alpha = \lim_{n\to\infty} t\frac{\log n}{n} < 2\left(\theta\log\frac{1}{\theta} + (1-\theta)\log\frac{1}{1-\theta}\right),$$

then $\mathbf{E}\,\eta(t) \to \infty$.

Proof. Consider a vector $\bar{x} = (x_1, \dots, x_n)$ which satisfies the relations

$$x_1 + \dots + x_m = p, \quad x_{m+1} + \dots + x_n = n - m - q.$$

It is clear that for such a vector

$$\mathbf{P}\{a_{i1}x_1 + \cdots + a_{in}x_n = a_{i1}x_1^0 + \cdots + a_{in}x_n^0\}$$

$$= \mathbf{P}\{a_{i1} + \cdots + a_{ip} = a_{i,m+1} + \cdots + a_{i,m+q}\} = \sum_{j\geq 0} C_p^j 2^{-p} C_q^j 2^{-(p+q)}$$

$$= 2^{-(p+q)} \operatorname{Coef}_{y^0}(1+1/y)^q(1+y)^p = C_{p+q}^p 2^{-(p+q)}.$$

Therefore, the average number of solutions is equal to

$$\mathbf{E}\,\eta(t) = \sum_{p,q} C_q^p C_{n-m}^q (C_{p+q}^p/2^{p+q})^t.$$

Put $p = \beta_p n$ and $q = \lambda_q n$. Making use of Stirling's formula, one can obtain

$$I = \sum_{\varepsilon m < p < (1-\varepsilon)m} C_m^p \sum_{\varepsilon(n-m) < q < (1-\varepsilon)(n-m)} C_{n-m}^q (C_{p+q}^p/2^{p+q})^t$$

$$= \sum_{\substack{\varepsilon\theta < \beta_p < (1-\varepsilon)\theta \\ \varepsilon(1-\varepsilon) < \lambda_q < (1-\varepsilon)(1-\theta)}} \exp\Big\{n\big(H(\theta) - \beta_p \log \beta_p - (\theta - \beta_p)\log(\theta - \beta_p)$$

$$-\lambda_q \log \lambda_q - (1-\theta-\lambda_q)\log(1-\theta-\lambda_q)\big)$$

$$-\frac{\alpha n^2}{\log n}\left((\beta_p + \lambda_q)\log\frac{1}{\beta_p + \lambda_q} + \beta_p \log \beta_p + \lambda_q \log \lambda_q + (\beta_p + \lambda_q)\log 2\right)$$

$$-\frac{\alpha n}{2} + O\left(\frac{n}{\log n}\right)\Big\}.$$

It is easy to see that if $\alpha = \lim_{n\to\infty} t \log n/n > 0$, the maximum value of the exponent is attained if the expression in the second brackets is equal to zero, i. e. if $\beta_p = \lambda_q$. Under this condition the maximum for the first brackets is attained at $\beta = \theta(1-\theta)$.

Therefore, the maximum value of the exponent is

$$n\left(\theta \log\frac{1}{\theta} + (1-\theta)\log\frac{1}{1-\theta} - \frac{\alpha}{2} + O\left(\frac{1}{\log n}\right)\right).$$

To complete the proof, note that if we choose an appropriate $\varepsilon > 0$ for the domain of summation for I, then the condition $\beta = \lambda = \theta(1-\theta)$ are not fulfilled outside this domain and $\mathbf{E}\,\eta(t) \sim 1 + I$.

Theorem 1.3 was presented in (Balakin, 1988 b) without proof.

Consider random system (1.1) and assume that

$$\mathbf{P}\{b_i = l\} = \frac{1}{n+1}, \qquad l = 0, \ldots, n.$$

As in the proof of Theorem 1.2, we obtain that

$$\mathbf{E}\,\xi(t) = 2^n(n+1)^{-t} = e^{-w_n}, \tag{1.5}$$

if $t = (n \log 2 + w_n)/\log(n+1)$. Therefore $\mathbf{E}\,\xi(t) \to \infty$ as $w_n \to -\infty$, and $\mathbf{E}\,\xi(t) \to 0$, as $w_n \to \infty$, as $n \to \infty$. Formula (1.4) cannot be used here since according to Theorem 1.3 the expected value $\mathbf{E}\,\eta(t)$ depends on the number of zero coordinates in the true solution (see also formula (1.5)).

1.3. The structure and the number of solutions of the systems for $t = 3$ and $t = 2$

Consider the structure and the number of solutions of a priori solvable system with $t = 3$ under assumptions of Subsection 1.1:

$$\begin{aligned} a_{11}x_1 + \cdots + a_{1n}x_n &= b_1; \\ a_{21}x_1 + \cdots + a_{2n}x_n &= b_2; \\ a_{31}x_1 + \cdots + a_{3n}x_n &= b_3. \end{aligned} \tag{1.6}$$

Let us make a change of variables in (1.6) according to the following rule

$$z_j = \begin{cases} x_j, & \text{if } a_{1j} + a_{2j} + a_{3j} > 0; \\ 1 - x_j, & \text{if } a_{1j} + a_{2j} + a_{3j} < 0. \end{cases}$$

Remind that a_{ij} take values 1 and -1. Divide the set of the new variables into two groups in such a way that the first group consists of z_j with $|a_{1j} + a_{2j} + a_{3j}| = 3$ and the second group consists of the remaining z_j with $|a_{1j} + a_{2j} + a_{3j}| = 1$. System (1.6) can be rewritten in the form

$$\begin{aligned} L_1 + L_2^* + L_3 + L_4 &= a_1, \\ L_1 + L_2 + L_3^* + L_4 &= a_2, \\ L_1 + L_2 + L_3 + L_4^* &= a_3, \end{aligned} \tag{1.7}$$

where

$$\begin{aligned} L_1 &= y_1 + \cdots + y_{n_1}, \\ L_2 &= y_{n_1+1} + \cdots + y_{n_1+n_2}, \\ L_3 &= y_{n_1+n_2+1} + \cdots + y_{n_1+n_2+n_3}, \\ L_4 &= y_{n_1+n_2+n_3+1} + \cdots + y_n, \\ & \quad y_i = z_{c_i}, \quad \{c_1, \dots, c_n\} = \{1, \dots, n\}, \\ L_2^* &= \bar{y}_{n_1+1} + \cdots + \bar{y}_{n_1+n_2}, \qquad \bar{y}_i = 1 - y_i, \end{aligned}$$

and L_3^*, L_4^* are defined similarly.

Suppose that

$$L_1 = k_1, \quad L_2 = k_2, \quad L_3 = k_3, \quad L_4 = k_4.$$

Then k_1, k_2, k_3, k_4 satisfy the following a priori solvable system of equations:

$$\begin{aligned} k_1 + (n_2 - k_2) + k_3 + k_4 &= a_1, \\ k_1 + k_2 + (n_3 - k_3) + k_4 &= a_2, \\ k_1 + k_2 + k_3 + (n_4 - k_4) &= a_3, \end{aligned}$$

and the system of inequalities

$$0 \le k_1 \le n_1, \quad 0 \le k_2 \le n_2, \quad 0 \le k_3 \le n_3, \quad 0 \le k_4 \le n_4.$$

From the system of equations we obtain

$$k_2 = B_1 - k_1, \quad k_3 = B_2 - k_1, \quad k_4 = B_3 - k_1,$$

where

$$\begin{aligned} B_1 &= \frac{1}{2}((a_2 + a_3) - (n_3 + n_4)), \\ B_2 &= \frac{1}{2}((a_1 + a_3) - (n_2 + n_4)), \\ B_3 &= \frac{1}{2}((a_1 + a_2) - (n_2 + n_3)), \end{aligned} \tag{1.8}$$

and from the system of inequalities we obtain the following bounds for k_1

$$m_1 = \max\{0, B_1 - n_2, B_2 - n_3, B_3 - n_4\} \le k_1 \le \min\{n_1, B_1, B_2, B_3\} = m_2 \tag{1.9}$$

Therefore, the number of solutions of the system under consideration is equal to

$$\eta = \sum_{m_1 \le k \le m_2} C_{n_1}^k C_{n_2}^{B_1 - k} C_{n_3}^{B_2 - k} C_{n_4}^{B_3 - k}.$$

Let us formulate the corresponding assertion.

THEOREM 1.4. *The number of solutions of system* (1.7) *is equal to*

$$\eta = \sum_{m_1 \le k \le m_2} C_{n_1}^k C_{n_2}^{B_1 - k} C_{n_3}^{B_2 - k} C_{n_4}^{B_3 - k},$$

and the set of solutions is

$$Y = \bigcup_{m_1 \le k \le m_2} Y_k,$$

where

$$Y_k = \{\bar{y}: L_1 = k, L_2 = B_1 - k, L_3 = B_2 - k, L_4 = B_3 - k\};$$

$n_1, \ldots, n_4$ are the numbers of the unknowns in the linear forms $L_1, \ldots, L_4$ respectively; the quantities B_1, B_2, B_3, m_1, m_2 are defined by formulas (1.8), (1.9).

COROLLARY 1.2. *The number of solutions of system* (1.7) *satisfying the condition $z_1 + \cdots + z_n = S$ is equal to*

$$C_{n_1}^r C_{n_2}^{B_1 - r} C_{n_3}^{B_2 - r} C_{n_4}^{B_3 - r},$$

where

$$r = (a_1 + a_2 + a_3 - n_2 - n_3 - n_4 - S)/2.$$

COROLLARY 1.3. *Random system (1.7) with linear forms $L_1, \dots, L_4$ having no common variables has a solution if and only if the following conditions are fulfilled:*

1) B_1, B_2, B_3 *are non-negative integers;*

2) $\min\{n_1, B_1, B_2, B_3\} \geq \max\{0, B_1 - n_2, B_2 - n_3, B_3 - n_4\}$.

If we are looking for solutions among vectors with S nonzero components, then the following two additional conditions should be satisfied:

3) $r = (a_1 + a_2 + a_3 - n_2 - n_3 - n_4 - S)/2$ *is an integer;*

4) $m_1 \leq r \leq m_2$.

Consider an example of how a priori solvable system (1.6) can be solved for $n = 10$ and

$$\bar{x}^0 = (0, 0, 0, 0, 0, 1, 1, 1, 1, 1).$$

Let the matrix and the right-hand side of the system have the following forms:

$$\begin{array}{cccccccccc} 1 & 2 & 3 & 4 & 5 & 6 & 7 & 8 & 9 & 10 \end{array}$$
$$\begin{pmatrix} 1 & 1 & -1 & -1 & 1 & 1 & 1 & -1 & -1 & 1 \\ 1 & -1 & 1 & -1 & -1 & -1 & -1 & 1 & 1 & 1 \\ 1 & -1 & 1 & -1 & 1 & 1 & -1 & -1 & 1 & -1 \end{pmatrix} \begin{pmatrix} 1 \\ 1 \\ -1 \end{pmatrix}.$$

The system can be transformed into (1.7), where

$$L_1 = x_1 + (1 - x_4), \quad L_2 = (1 - x_2) + x_3 + (1 - x_7) + x_9, \quad L_3 = x_5 + x_6 + (1 - x_4),$$

$$L_4 = x_{10}, \quad a_1 = 5, \quad a_2 = 6, \quad a_3 = 4.$$

The numbers $B_1 = 3$, $B_2 = 2$, $B_3 = 2$, $m_1 = 1$, $m_2 = 2$ define the number of solutions of this system

$$\eta = C_2^1 C_4^2 C_3^1 C_1^1 + C_2^2 C_4^1 C_3^0 C_1^0 = 40,$$

and the solutions themselves can be easily obtained with the help of Theorem 1.4. The similar result can be obtained for a priori solvable systems from Subsection 1.2 for $t = 2$. Indeed, such a system can be transformed into the form:

$$L_1 + 0 + L_3 + 0 = a_1, \quad L_1 + L_2 + 0 + 0 = a_2,$$

where

$$L_1 = y_1 + \dots + y_{n_1}, \quad L_2 = y_{n_1+1} + \dots + y_{n_1+n_2},$$

$$L_3 = y_{n_1+n_2+1} + \dots + y_{n_1+n_2+n_3}, \quad L_4 = 0 \cdot y_{n-n_4+1} + \dots + 0 \cdot y_n = 0.$$

The number of solutions of this system is equal to

$$\eta = 2^{n_4} \sum_{m_1 \leq k \leq m_2} C_{n_1}^k C_{n_2}^{a_2 - k} C_{n_3}^{a_1 - k},$$

$$m_1 = \max\{0, a_2 - n_2, a_1 - n_3\}, \quad m_2 = \min\{n_1, a_2, a_1\}.$$

In conclusion we call your attention to the fact that the systems of linear equations considered in Subsections 1.1 and 1.2 become equivalent if we join the equation

$$x_1 + \cdots + x_n = n - m. \tag{1.10}$$

to both the systems.

Indeed, if we add equation (1.10) to each equation of the system from Subsection 1.1 and divide the obtained equation by 2 we will have a system of equation from Subsection 1.2. To obtain the system from Subsection 1.1 we can subtract equations of Subsection 1.2 multiplied by 2 from equation (1.10).

2. LINEAR PSEUDO-BOOLEAN SYSTEMS OF EQUATIONS WITH THE SAME FIXED NUMBER OF UNKNOWNS

2.1. The average number of solutions of random systems

Consider system (1.1) with a fixed number of unknowns in each equation

$$x_{s_{i1}} + \cdots + x_{s_{ir}} = b_i, \qquad i = 1, \ldots, t. \tag{2.1}$$

Suppose that the numbers $s_{i1}, \ldots, s_{ir}$ are chosen from the set $\{1, 2, \ldots, n\}$ by sampling without replacement, the probability of any sample is $(C_n^r)^{-1}$ and the samples for different $i = 1, \ldots, t$ are independent. Let b_i $(i = 1, \ldots, t)$ be independent random variables taking values $0, 1, \ldots, r$ with probabilities $p_0, p_1, \ldots, p_r$. Let us find the average number of solutions of system (2.1) under these assumptions. Consider a vector $\bar{x} = (x_1, \ldots, x_n)$ with k coordinates which are equal to 1. This vector satisfies all t equations of system (2.1) with probability

$$\left(\sum_{i=0}^{r} C_k^i C_{n-k}^{r-i} p_i / C_n^r \right)^t .$$

Therefore the average number of such solutions is equal to

$$a(k) = C_n^k \left(\sum_{i=0}^{r} C_k^i C_{n-k}^{r-i} p_i / C_n^r \right)^t ,$$

and the average number of all solutions is

$$\mathbf{E}\,\xi(t) = \sum_{k=0}^{n} a(k) = \sum_{k=0}^{n} C_n^k \left(\sum_{i=0}^{r} C_k^i C_{n-k}^{r-i} p_i / C_n^r \right)^t . \tag{2.2}$$

Consider expression (2.2) in the three interesting cases.

Case 1. $\mathbf{P}\{b_i = d\} = 1$. In this case

$$\mathbf{E}\,\xi(t) = \sum_{k=0}^{n} C_n^k \left(C_k^d C_{n-k}^{r-d} / C_n^r \right)^t = \sum_{0}^{\varepsilon n} a(k) + \sum_{\varepsilon n+1}^{(1-\varepsilon)n} a(k) + \sum_{(1-\varepsilon)n+1}^{n} a(k)$$

$$= I_1 + I_2 + I_3.$$

a) If $d=0$, then

$$\mathbf{E}\,\xi(t)=\sum_{k=0}^{n}C_n^k\left(\frac{(n-k)_r}{(n)_r}\right)^t.$$

The system is a priori solvable and $\bar{x}^0=(0,\dots,0)$. If $t\sim r^{-1}\log n$, the system has a unique zero solution with probability tending to 1 as $n\to\infty$. The main contribution into $\mathbf{E}\,\xi(t)$ is made by the first sum I_1.
b) If $d=r$, then we have a similar result for $\bar{x}^0=(1,\dots,1)$. The main contribution into $\mathbf{E}\,\xi(t)$ is made by the third sum I_3.
c) If $0<d<r$, we consider the middle sum I_2. As $n\to\infty$ the numbers $n-k$ and k also converge to infinity. Therefore the hypergeometric probability

$$C_k^d C_{n-k}^{r-d}/C_n^r$$

converges to the corresponding binomial probability

$$C_r^d(\beta_k)^d(1-\beta_k)^{r-d}\left(1-\frac{1}{n}\left(\frac{C_d^2}{\beta_k}+\frac{C_{r-d}^2}{1-\beta_k}-C_r^2\right)+O\left(\frac{1}{n^2}\right)\right),$$

where $\beta_k=k/n$.
Putting $t=\alpha n$, where α is a constant,

$$f_{r,d}(\beta)=\log(C_r^d\beta^d(1-\beta)^{r-d}),\quad g(\beta)=\beta\log\beta+(1-\beta)\log(1-\beta),$$

we obtain that

$$I_2\sim\sum_{\epsilon n<k<(1-\epsilon)}\left(2\pi\beta_k(1-\beta_k)n\right)^{-1/2}$$
$$\times\exp\left\{-\alpha\left(\frac{C_d^2}{\beta_k}+\frac{C_{r-d}^2}{1-\beta_k}-C_r^2\right)+n\left(\alpha f_{r,d}(\beta_k)-g(\beta_k)\right)\right\}.$$

Let us find the maximum value of the function

$$\varphi(\beta)=\alpha f_{r,d}(\beta)-g(\beta).$$

We have

$$\varphi'(\beta)=\alpha\frac{d-r\beta}{\beta(1-\beta)}-\log\frac{\beta}{1-\beta},\quad \varphi''(\beta)=\frac{1}{\beta(1-\beta)}-\frac{r\beta^2-2d\beta+d}{\beta(1-\beta)}<0.$$

Hence, there is a unique root $\beta_0(\alpha)$ of the equation

$$\alpha r\frac{d/r-\beta}{\beta(1-\beta)}-\log\frac{\beta}{1-\beta}=0,$$

and the function $\varphi(\beta)$ takes its maximum value at the point $\beta=\beta_0(\alpha)$. Notice that $\beta_0(\alpha)$ is a monotone function of α, and as α runs from 0 to ∞ the function $\beta_0(\alpha)$ varies from $1/2$ to d/r .

The critical value is equal to

$$\alpha_0 = \max_{0<\beta<1} \frac{\beta \log \beta + (1-\beta)\log(1-\beta)}{\log(C_r^d \beta^d (1-\beta)^{r-d})}. \tag{2.3}$$

If $\alpha > \alpha_0$, then $\alpha f_{r,d}(\beta) - g(\beta) < 0$ for all $\beta \in (0,1)$ and hence $I_2 \to 0$. But if $\alpha < \alpha_0$, then for some $\beta \in (0,1)$

$$\alpha f_{r,d}(\beta) - g(\beta) > 0$$

and I_2 tends to infinity.
Let us estimate I_1 and I_3 for $\alpha > 0$. For some $\varepsilon > 0$

$$I_1 = \sum_{k \le \varepsilon n} C_n^k (C_k^d C_{n-k}^{r-d}/C_n^r)^t \le n C_n^{\varepsilon n}(C_r^d \varepsilon^d)^t = O(\sqrt{n}) \left(\frac{(C_r^d \varepsilon^d)^d}{\varepsilon^\varepsilon (1-\varepsilon)^{1-\varepsilon}} \right)^t = o(1).$$

The sum I_3 is similarly estimated.
Let us formulate the assertion proved above.

THEOREM 2.1. *Let $b_i = d$, $0 < d < r$, $i = 1, \ldots, t$, and let α_0 be given by formula (2.3). Then* $\mathbf{E}\,\xi(t) \to 0$ *if $\alpha > \alpha_0$ and* $\mathbf{E}\,\xi(t) \to \infty$ *if $\alpha < \alpha_0$ as $n \to \infty$.*

Case 2. $\mathbf{P}\{b_i = d\} = 1/(r+1)$, $i = 1, \ldots, t$, $d = 0, \ldots, r$. In this case

$$\mathbf{E}\,\xi(t) = 2^n \left(\frac{1}{r+1} \right)^t \to \infty,$$

if $t \log(r+1) - n \log n \to -\infty$ and $\mathbf{E}\,\xi(t) \to 0$ if $t \log(r+1) - n \log n \to \infty$.
Case 3. $\mathbf{P}\{b_i = d\} = C_r^d (1/2)^r$, $i = 1, \ldots, t$, $d = 0, \ldots, r$. Here

$$\mathbf{E}\,\xi(t) = \sum_{k=0}^n C_n^k \left(\sum_{i \ge 0} C_r^i C_k^i C_{n-k}^{r-i} / (2^r C_n^r) \right)^t = I_1 + I_2 + I_3.$$

If $\alpha > 0$ then the sums I_1 and I_3 can be made arbitrarily small by choice of sufficiently small ε. Indeed,

$$\begin{aligned} I_1 = I_3 &\le \varepsilon n \left(\frac{1}{\varepsilon^\varepsilon (1-\varepsilon)^{1-\varepsilon}} \right)^n \max_{k<\varepsilon n} \left(\sum_{i \ge 0} C_r^i C_k^i C_{n-k}^{r-i} / (2^r C_n^r) \right)^{\alpha n} \\ &\le \varepsilon n \big((C_r^{r/2})^\alpha / (2^{\alpha r} \varepsilon^\varepsilon (1-\varepsilon)^{1-\varepsilon}) \big)^n = o(1). \end{aligned}$$

Consider the sum

$$I_2 = \sum_{\varepsilon n < k < (1-\varepsilon)n} a(k).$$

By standard transformations we obtain that

$$I_2 = \sum_{\varepsilon n < k < (1-\varepsilon)n} \frac{\exp\{-\alpha h(\beta_k) + O(1/n)\}}{\sqrt{2\pi n \beta_k (1-\beta_k)}} \exp\big\{ n(\alpha f_r(\beta_k) - g(\beta_k)) \big\}, \tag{2.4}$$

where $\beta_k = k/n$, $t = \alpha n$,

$$h(\beta) = \sum_{i=0}^{r} 2^{-r}(C_r^i)^2\beta^i(1-\beta)^{r-i}\left(\frac{C_i^2}{\beta} + \frac{C_{r-i}^2}{1-\beta} - C_r^2\right),$$

$$f_r(\beta) = \log\left(\sum_{i=0}^{r} 2^{-r}(C_r^i)^2\beta^i(1-\beta)^{r-i}\right),$$

$$g(\beta) = \beta\log\beta + (1-\beta)\log(1-\beta).$$

Let us find the maximum value of the function

$$\varphi(\beta) = \sum_{i=0}^{r} 2^{-r}(C_r^i)^2\beta^i(1-\beta)^{r-i}$$

for $\beta \in (0,1)$. The function $\varphi(\beta)$ has a simple interpretation:

$$\varphi(\beta) = \mathbf{P}\{\xi_0 = \eta_0\},$$

where ξ_0 and η_0 are independent random variables with the binomial distribution with probability of the "success" 1/2 and β correspondingly. It is natural to expect that $\varphi(\beta)$ takes its maximum value when ξ_0 and η_0 have the same distribution, i. e., at $\beta = 1/2$. Indeed,

$$\varphi'(\beta) = 2^{-r}\sum_{i=0}^{r}(C_r^i)^2\beta^{i-1}(1-\beta)^{r-i-1}\left(i(1-\beta) - (r-i)\beta\right), \quad \varphi'\left(\frac{1}{2}\right) = 0;$$

$$\varphi''(\beta) = 2^{-r}\sum_{i=0}^{r}(C_r^i)^2\beta^{i-2}(1-\beta)^{r-i-2}\left((i-r\beta)^2 - i(1-\beta)^2 - (r-i)\beta^2\right);$$

$$\varphi''\left(\frac{1}{2}\right) = 2^{-2r+4}C_{2r}^r\sum_{i=0}^{r}(i-r/2)^2\left((C_r^i)^2/C_{2r}^r - C_r^i/2^r\right) < 0.$$

Therefore, the maximum value of the exponent

$$\alpha f_r(\beta) - g(\beta)$$

in (2.4) is attained at $\beta = 1/2$ and it is equal to

$$\alpha\log(C_{2r}^r/2^{2r}) + \log 2.$$

Therefore the critical value α_0 is given by the expressions

$$\max_{0<\beta<1}\frac{g(\beta)}{f_r(\beta)} = \frac{g(1/2)}{f_r(1/2)} = \frac{\log 2}{2r\log 2^r - \log C_{2r}^r}. \tag{2.5}$$

THEOREM 2.2. *Let* $\mathbf{P}\{b_i = d\} = C_r^d(1/2)^r$, $i = 1,\dots,t$, $d = 0,\dots,r$, *and let* α_0 *be given by* (2.5) *and* $n \to \infty$. *Then* $\mathbf{E}\,\xi(t) \to \infty$, *if* $\lim_{n\to\infty} t/n < \alpha_0$,

$\mathbf{E}\,\xi(t) \to 0$, *if* $\lim_{n\to\infty} t/n > \alpha_0$, *where* $\xi(t)$ *is the number of solutions of system* (2.1).

2.2. The distribution of the number of solutions of a priory solvable random system

Let system (2.1) be a priori solvable. Let us find the distribution of the number of solutions $\eta(t)$ of this system.

THEOREM 2.3. *If* $t = n(\log n + x)/r$, $\limsup_{n\to\infty} r/\log n < 1$, *then as* $n \to \infty$ *for any fixed* $k \geq 0$

$$\mathbf{P}\{\eta(t) = 2^k\} = \frac{(e^{-x})^k}{k!}\exp\{-e^{-x}\} + o(1).$$

Proof. Notice that

$$2^\zeta \leq \eta(t) \leq \eta^*(t),$$

where ζ is the number of absent variables in system (2.1) and $\eta^*(t)$ is the number of solutions of system (2.1) considered over the field $\mathbf{GF}(2)$. Under the assumptions of the theorem $\eta^*(t)$ converges in probability to 2^ζ as $n \to \infty$, and for the random variable ζ we have

$$\mathbf{P}\{\zeta = k\} = \frac{(e^{-x})^k}{k!}\exp\{-e^{-x}\} + o(1)$$

for any fixed $k \geq 0$. Thus the theorem is proved.

COROLLARY 2.1. *If* $n \to \infty$, *then* $\mathbf{E}\,\eta(t) \to \infty$ *as* $x \to -\infty$, *and* $\mathbf{E}\,\eta(t) \to 1$, $\mathbf{P}\{\eta(t) = 1\} \to 1$ *as* $x \to \infty$.

2.3. The average number of solutions of systems with restrictions on the values of unknowns

Consider system (1.1) provided that their coefficients and unknowns satisfy some constraints. Assume that

$$n = rn_1, \qquad a_{i1} + \cdots + a_{in} = r, \quad i = 1, \ldots, t, \qquad a_{i1} + \cdots + a_{in_1} = 1,$$

$$a_{i\,n_1+1} + \cdots + a_{i\,2n_1} = 1, \ldots, a_{i\,(r-1)n_1+1} + \cdots + a_{i\,rn} = 1$$

Consider new variables

$$x_{ij} = x_{(i-1)n_1+j}, \qquad i = 1, \ldots, r, \quad j = 1, \ldots, n_1,$$

and assume that

$$\sum_{j=1}^{n_1} x_{ij} = 1, \qquad i = 1, \ldots, r.$$

We obtain a system of t linear pseudo-Boolean equations with respect to the new variables

$$x_{1\,d_{i1}} + \cdots + x_{r\,d_{ir}} = b_i, \qquad i = 1, \ldots, t, \tag{2.6}$$

and r restrictions on the values of variables

$$x_{j1} + \cdots + x_{jn_1} = 1, \qquad j = 1, \ldots, r. \tag{2.7}$$

Assume that d_{ij} ($i = 1, \ldots, t$, $j = 1, \ldots, r$) are independent random variables uniformly distributed on the set $\{1, 2, \ldots, n_1\}$.
At first consider the case where the system is a priori solvable and a vector

$$\bar{x}^0 = (x_{11}^0, \ldots, x_{1n_1}^0, \ldots, x_{r1}^0, \ldots, x_{rn_1}^0)$$

is its true solution. Let us obtain an expression for the average number of solutions. It is clear that the number of possible solutions is equal to n_1^r, $C_r^l(n_1-1)^l$ vectors have $2l$ components which differ with the corresponding components of the true solution $\bar{x}^0$. For the vector

$$\bar{x}^l = (x_{11}^l, \ldots, x_{1n_1}^l, \ldots, x_{r1}^l, \ldots, x_{rn_1}^l),$$

which differs from the true solution $\bar{x}^0$ in $2l$ coordinates, the probability is

$$p(2l) = \mathbf{P}\{(x_{1\,d_{i1}}^l, \ldots, x_{r\,d_{ir}}^l) = (x_{1\,d_{i1}}^0, \ldots, x_{r\,d_{ir}}^0)\} = \sum_{j\ge 0} C_l^j C_{l-j}^j \frac{(n_1-2)^{l-2j}}{n_1^l}.$$

Therefore, the average number of solutions for the a priori solvable system is given by

$$\mathbf{E}\,\eta(t) = 1 + \sum_{l=1}^{r} C_r^l(n_1-1)^l \left(\sum_{j\ge 0} \frac{(l)_{2j}}{(j!)^2} \frac{(n_1-2)^{l-2j}}{n_1^l} \right)^t. \tag{2.8}$$

Let us analyze the asymptotic behaviour of expression (2.8) if $n_1 \to \infty$, $t = (n_1 \log n_1 + x n_1)/2$, $r = o(\sqrt{n_1/\log n_1})$.
We have

$$p(2l) = \sum_{j\ge 0} \frac{(l)_{2j}}{(j!)^2} \frac{(n_1-2)^{l-2j}}{n_1^l} = ((n_1-2)/n_1)^l (1 + O(r^2/n_1^2)),$$

$$(p(l))^t = \exp\left\{ \frac{n_1 \log n_1 + x n_1}{2} \big(-2l/n_1 + O(r^2/n_1^2) \big) \right\} = \frac{(e^{-x})^l}{n_1^l}(1 + o(1)),$$

$$\mathbf{E}\,\eta(t) \sim \sum_{l=0}^{r} C_r^l (e^{-x})^l = (1 + e^{-x})^r.$$

For random system (2.6)–(2.7), assume that $b_1, \ldots, b_t$ are independent identically distributed random variables,

$$\mathbf{P}\{b_i = j\} = C_r^j \frac{(n_1-1)^{r-j}}{n_1^r}, \qquad i = 1, \ldots, t, \quad j = 0, \ldots, r.$$

In this case expression (2.8) is changed in the following way:

$$\mathbf{E}\,\xi(t) = n_1^r \left(\sum_{j=0}^{r} (C_r^j)^2 \frac{(n_1-1)^{2r-2j}}{n_1^{2r}}\right)^t = n_1^r \left(1 - \frac{1}{n_1}\right)^{2rt} \left(\sum_{j=0}^{r} \frac{(C_r^j)^2}{(n_1-1)^{2j}}\right)$$

Under the same assumptions which have been made for a priori solvable system

$$\mathbf{E}\,\xi(t) \sim n_1^r e^{-2rt/n_1} = (e^{-x})^r.$$

We can formulate the obtained assertion as follows.

THEOREM 2.4. *Let $n_1 \to \infty$, $t = (n_1 \log n_1 + x n_1)/2$, $x = o(\log n_1)$, $r = o(\sqrt{n_1/\log n_1})$. Then the average number of solutions for a priori solvable system* (2.6)–(2.7)

$$\mathbf{E}\,\eta(t) \sim (1 + e^{-x})^r,$$

and the average number of solutions for the corresponding random system with independent and equally distributed right-hand sides

$$\mathbf{E}\,\xi(t) \sim n_1^r e^{-2rt/n_1} = (e^{-x})^r.$$

Consider the following r systems of equations

$$x_{j\,a_{ij}} = b_{ij}, \qquad j = 1, \dots, r, \quad i = 1, \dots, t, \tag{2.9}$$

which can be obtained from system (2.6). Let $\eta(j)$ be the number of solutions of the jth a priori solvable system of equations (2.9). Then b_i from (2.6) is equal to

$$b_i = b_{i1} + b_{i2} + \cdots + b_{ir}, \qquad i = 1, \dots, t,$$

and for the number of solutions for a priori solvable system (2.6) we have $\eta \geq \prod_{j=1}^r \eta(j)$. From system (2.6) we can obtain $N = \prod_{i=1}^t C_r^{b_i}$ systems of equations (2.9). It may happen that some of these systems of equations have no solutions. That is why the number of solutions for system (2.6) does not exceed the total number N of solutions of systems (2.9).

Suppose that the number of nonzero columns in the matrix of system (2.9) is equal to θ. Then taking into account constraints (2.7) for the values of variables, we have

$$\eta(j) = \begin{cases} 1 & \text{if } \sum_{i=1}^{t} b_{ij} > 0, \\ \theta & \text{if } b_{1j} = b_{2j} = \cdots = b_{tj} = 0, \end{cases}$$

and therefore $\eta \geq \prod_{j \in I} \theta_j$, where $I = \{k\colon \sum_{i=1}^t b_{ik} = 0,\ k = 1, \dots, r\}$. Notice that if $b_{1j} = b_{2j} = \cdots = b_{tj} = 0$, then $\theta_j \geq 1$, $j = 1, \dots, r$. If the unknowns of system (2.9) satisfy the inequalities

$$\sum_{j=1}^{n_1} x_{ij} \leq 1, \qquad i = 1, \dots, r, \tag{2.10}$$

then $\eta(j) = \theta_j + 1$, $j = 1, \dots, r$, and the number of solutions of system (2.6)

$$\eta \geq \prod_{j=1}^{r} (\theta_j + 1).$$

THEOREM 2.5. *For the system of equations* (2.6)–(2.7)

$$\mathbf{P}\{\eta = 1\} \leq \mathbf{P}\Big\{ \prod_{j \in I} \theta_j = 1 \Big\}$$

$$= \prod_{j=1} \Big(\mathbf{P}\Big\{ \sum_{i=1}^{t} b_{ij} \geq 1 \Big\} + \mathbf{P}\Big\{ \theta = 1, \ \sum_{i=1}^{t} b_{ij} = 0 \Big\} \Big)$$

$$= \left(1 - \sum_{k=1}^{n_1 - 1} C_{n_1 - 1}^{k} \left(\frac{n_1 - 1 - k}{n_1} \right)^{t} \right)^{r},$$

and for the system of equations (2.6)–(2.10)

$$\mathbf{P}\{\eta = 1\} \leq \mathbf{P}\{\theta_1 = 0, \dots, \theta_r = 0\} = \left(\sum_{k=0}^{n_1} (-1)^k C_{n_1}^{k} \left(\frac{n_1 - k}{n_1} \right)^{t} \right)^{r}.$$

The proof of the theorem is obvious.

3. A GEOMETRICAL METHOD FOR INVESTIGATION OF SYSTEMS OF EQUATIONS

3.1. The matrix of solutions. Dependent and independent elements

Let ξ be the number of solutions of system (1.1) and let X_i be the set of solutions of the ith equation. Then

$$\mathbf{P}\{\xi = k\} = \mathbf{P}\{ |X_1 \cap \dots \cap X_t| = k \}.$$

Let us equip all the possible solutions $\bar{x}$ of system (1.1) with indices $1, 2, \dots, 2^n$ and construct the matrix of solutions $A = \| a_{ij} \|$ of size $t \times 2^n$ in the following way:

$$a_{ij} = \begin{cases} 1 & \text{if } \bar{x}_j \in X_i, \\ 0 & \text{if } \bar{x}_j \notin X_i. \end{cases}$$

It is clear that the vector $\bar{x}_k$ is a solution of the system of equation if and only if the kth column of the matrix A contains no zeros. Thus, the number ξ of solutions of system (1.1) is equal to the number of unity columns of the matrix A.

System (1.1) is constructed in the following way. At first we choose the function f_i from Φ in the ith equation, i. e., on the set of all vectors $\bar{x} = (x_1, \dots, x_n)$ with $(0,1)$ coordinates we determine the partition

$$X = \bigcup_{a} X_{i\,a},$$

where

$$X_{ia} = \{\bar{x}: f_i(\bar{x}) = a\}.$$

Then according to a certain probabilistic rule we choose the right-hand side a_i. Thus the set of solutions of the ith equation $X_i = X_{ia}$ is determined.
The probability measure determined on the set of couples (f_i, a_i) induces the probability measure on the subsets X_i of the set X:

$$\mathbf{P}\{X_i\} = \sum_{\{(f_i,a_i):\, X_{i a_i} = X_i\}} p(f_i, a_i).$$

Therefore we can turn our attention to random $(0,1)$ matrices A. These matrices have independent rows and the ith row defines the set X_i.
Let $\xi_i = 1$, $i = 1, \ldots, 2^n$, if the ith column of the matrix A does not contain zeros and $\xi_i = 0$ otherwise. Then

$$\xi = \xi_1 + \xi_2 + \cdots + \xi_{2^n}$$

and

$$\mathbf{P}\{\xi_i = 1\} = \prod_{j=1}^{t} \mathbf{P}\{\bar{x}_i \in X_j\}.$$

For a priori solvable systems one of the indicators, for example ξ_{2^n}, is identically equal to unity.
In some cases the indicator representation of the random variable ξ can be useful. For example, if $\xi_1, \ldots, \xi_{2^n}$ are independent random identically distributed variables, then

$$\mathbf{P}\{\xi = k\} = C_{2^n}^k p^k (1-p)^{2^n - k},$$

where $p = \mathbf{P}\{\xi_i = 1\}$. If Φ is the set of all Boolean functions of n variables over the field $\mathbf{GF}(2)$, $|\Phi| = 2^{2^n}$, and every function takes values from Φ with equal probabilities 2^{-2^n}, then for any right-hand sides $a_1, \ldots, a_t$ the events $\{\bar{x}_1 \in X_i\}, \ldots, \{\bar{x}_{2^n} \in X_i\}$ are independent and have the probabilities that are equal to $1/2$. Thus,

$$\mathbf{P}\{\xi = k\} = C_{2^n}^k (2^{-t})^k (1 - 2^{-t})^{2^n - k}.$$

If $n \to \infty$, $n - t = O(1)$, then for any $k \geq 0$

$$\lim_{n\to\infty} \left(\mathbf{P}\{\xi = k\} - \frac{(2^{n-t})^k}{k!} \exp\{-2^{n-t}\} \right) = 0. \tag{3.1}$$

Let the events $\{\bar{x}_1 \in X_i\}, \ldots, \{\bar{x}_{2^n} \in X_i\}$ be independent. In this case it may be useful to apply the inclusion–exclusion principle. For example, let Φ be the set of the functions in n variables having the weight $m = 2^{n-1}$, $|\Phi| = 2^n!(2^{n-1}!)^{-2}$, and let $f_1, \ldots, f_t$ be uniformly distributed on Φ. The choice of a function and a value of a_i in the equation $f_i(\bar{x}) = a_i$ is equivalent to the choice of a set X_i from $C_{2^n}^{2^{n-1}}$ possible choices or, what is the same, to a replacement of 2^{n-1} units in the ith row of the matrix A.

Therefore

$$\mathbf{P}\{\xi = k\} = \sum_{i\geq 0}(-1)^i C_{k+i}^i C_{2^n}^{k+i}(C_{2^n-k-i}^{2^{n-1}-k-i}/C_{2^n}^{2^{n-1}})^t.$$

If $n \to \infty$, $n - t = O(1)$, then for any $k \geq 0$

$$\mathbf{P}\{\xi = k\} = \frac{(2^{n-t})^k}{k!}\exp\{-2^{n-t}\} + o(1) \quad \text{as} \quad n \to \infty$$

what is equivalent to (3.1).

3.2. Examples of systems analyzed by the geometrical method

Consider first a pseudo-Boolean system of equations. Remind that a Boolean and pseudo-Boolean function in n variables maps Boolean n-dimension vectors into the field $\mathbf{GF}(2)$ and into the field of real numbers respectively. In particular, a pseudo-Boolean system of equations can be written in the following form:

$$\varphi_i(\bar{x}) = f_{i1}(\bar{x}) + \cdots + f_{ir}(\bar{x}) = a_i, \qquad i = 1, \ldots, t,$$

where all functions are Boolean and the summation is carried in the field of real numbers.

From geometrical point of view the pseudo-Boolean function $\varphi_i(\bar{x})$ defines the decomposition of the set of all Boolean vectors

$$X = \bigcup_{j=0}^{r} X_{ij},$$

where

$$X_{ij} = \{\bar{x} \mid \varphi_i(\bar{x}) = j\}.$$

The right-hand side a_i shows which of the sets X_{ij}, $j = 0, \ldots, r$, is the set of solutions X_{ij} of the equation $\varphi_i(\bar{x}) = a_i$. Obviously, $X_i = X_{i\,a_i}$.

Let the functions $f_{ij}, i = 1, \ldots, t, j = 0, \ldots, r$, be chosen from the set of all Boolean functions in n variables independently and with equal probabilities. Then the events

$$\{\bar{x}_i \in X_i\}, \ldots, \{\bar{x}_{2^n} \in X_i\}$$

are independent and for any $k \in \{1, \ldots, 2^n\}$

$$\mathbf{P}\{\bar{x}_k \in X_i\} = \sum_{l=0}^{r}\mathbf{P}\{a_i = l\}\mathbf{P}\{f_{i1}(\bar{x}_k) + \cdots + f_{ir}(\bar{x}_k) = l\}$$

$$= \sum_{l=0}^{r} p_l C_r^l/2^r = p. \tag{3.2}$$

Therefore

$$\mathbf{P}\{\xi = k\} = C_{2^n}^k(p^t)^k(1 - p^t)^{2^n - k}.$$

If $p_l = 1/(r+1), l = 0,\dots,r$, then $p = 1/(r+1)$. For $p_l = C_k^l/2^r$, $l = 0,\dots,r$, we have $p = C_{2^r}^r/2^{2^r}$. In the first case for $t = (n\log 2 - \log\lambda + o(1))/\log(r+1)$

$$\lim_{n\to\infty} \mathbf{P}\{\xi = k\} = \frac{\lambda^k}{k!}e^{-\lambda}, \qquad k \geq 0,$$

and in the second case this fact takes place, if

$$t = \big(n\log 2 - \log\lambda + o(1)\big)/(2r\log 2 - \log C_{2^n}^r).$$

Consider a more complicated case. Let Φ be the set of Boolean functions of weight $m = \theta 2^n$, $0 < \theta < 1$, $\theta \neq 1/2$, $|\Phi| = C_{2^n}^m$, $\mathbf{P}\{a_i = 1\} = p$. In this case m or $2^n - m$ units are placed in the ith row of the matrix A depending on the value of a_i. Let $\theta > 1/2$. If $\theta < 1/2$ then in all formulas we must substitute p for $1-p$. Let us find the distribution of the number of solutions ξ for the random system of equations.

THEOREM 3.1. *Let* $\theta > 1/2$, $0 < p < 1$,

$$t = -\frac{n\log 2}{p\log\theta + (1-p)\log(1-\theta)} + \sqrt{n}\frac{t_\alpha(\log\theta - \log(1-\theta))\sqrt{p(1-p)\log 2}}{(-p\log\theta - (1-p)\log(1-\theta))^{3/2}} + O(1).$$

Then

$$\lim_{n\to\infty} \mathbf{P}\{\xi = 0\} = \frac{1}{\sqrt{2\pi}}\int_{-\infty}^{t_\alpha} e^{-x^2/2}\,dx = 1-\alpha, \tag{3.3}$$

for any fixed $k = 1,2,\dots$

$$\lim_{n\to\infty} \mathbf{P}\{\xi = k\} = 0$$

and $\mathbf{E}\,\xi \to \infty$ *as* $n \to \infty$.

Proof. Obviously, in our case

$$\mathbf{P}\{\xi = k\} = \sum_{t_1\geq 0} C_t^{t_1} p^{t_1}(1-p)^{t-t_1} \sum_{i\geq 0}(-1)^i C_{k+i}^i C_{2^n}^{k+i} \frac{(C_{2^n}^{m-k-i})^{t_1}(C_{2^n-k-i}^m)^{t-t_1}}{(C_{2^n}^m)^t}.$$

For any fixed $k \geq 0$ as $n \to \infty$

$$\mathbf{P}\{\xi = k \mid a_1 + \dots + a_t = t_1\} = \sum_{i\geq 0}(-1)^i \frac{(2^n\theta^{t_1}(1-\theta)^{t-t_1})^{k+i}}{k!\,i!} + o(1)$$

$$= \frac{(2^n\theta^{t_1}(1-\theta)^{t-t_1})^k}{k!}\exp\{-2^n\theta^{t_1}(1-\theta)^{t-t_1}\} + o(1),$$

i.e., provided $a_1 + \dots + a_t = t_1$, the random variable ξ has in limit the Poisson distribution with parameter

$$\lambda(t_1) = 2^n(1-\theta)^t\big(\theta/(1-\theta)\big)^{t_1}.$$

Therefore

$$\mathbf{P}\{\xi = k\} = \frac{1}{k!\sqrt{2\pi p(1-p)^t}} \sum_{|j-tp|<t^{3/5}} \exp\left\{-\frac{(j-tp)^2}{2tp(1-p)} - \lambda(j) + k\log\lambda(j)\right\} + o(1).$$

Consider the expression

$$\log\lambda(j) = n\log 2 + j\log\theta + (t-j)\log(1-\theta)$$

for $j = tp + t_\alpha\sqrt{tp(1-p)} + l_j$, $l_j = O(t^{3/5})$, $t_\alpha = O(1)$,

$$t = -\frac{n\log 2}{p\log\theta + (1-p)\log(1-\theta)} + \frac{\sqrt{n}t_\alpha \log(\theta/(1-\theta))\sqrt{p(1-p)\log 2}}{\left(-p\log\theta - (1-p)\log(1-\theta)\right)^{3/2}} + O(1).$$

We have

$$\log\lambda(j) = n\log 2 + t\big(p\log\theta + (1-p)\log(1-\theta)\big)$$
$$+\sqrt{t}\,t_\alpha\sqrt{p(1-p)}\log\frac{\theta}{1-\theta} + l_j\log\frac{\theta}{1-\theta}$$
$$= L + l_j\log\frac{\theta}{1-\theta}, \qquad L = O(1).$$

Hence we obtain that

$$\lambda(j) = e^L\left(\frac{\theta}{1-\theta}\right)^{l_j},$$

and for fixed $\theta > 1/2$

$$\lambda(j) \to 0 \qquad \text{if} \quad l_j = j - tp - t_\alpha\sqrt{tp(1-p)} \to -\infty;$$

$$\lambda(j) \to \infty \qquad \text{if} \quad l_j \to \infty.$$

Therefore for fixed $k = 1, 2, \ldots$

$$\lim_{n\to\infty} \mathbf{P}\{\xi = k \mid a_1 + \cdots + a_t = j\} > 0$$

only if

$$l_j = j - tp - t_\alpha\sqrt{tp(1-p)} = O(1).$$

Hence for any $k = 1, 2, \ldots$

$$\lim_{n\to\infty} \mathbf{P}\{\xi = k\} = 0;$$

and

$$\lim_{n\to\infty} \mathbf{P}\{\xi = 0\} = \frac{1}{\sqrt{2\pi}}\int_{-\infty}^{t_\alpha} e^{-x^2/2}\,dx = 1 - \alpha.$$

To complete the proof of (3.3), we have to show that $\mathbf{E}\,\xi \to \infty$ as $n \to \infty$. Indeed,

$$\mathbf{E}\,\xi = 2^n\big(p\theta + (1-p)(1-\theta)\big)^t$$

$$= \exp\left\{n \log 2 \left(\frac{\log((p\theta + (1-p)(1-\theta))/(\theta^p(1-\theta)^{1-p}))}{-p\log\theta - (1-p)\log(1-\theta)} + O(n^{-1/2})\right)\right\}.$$

Since the arithmetic mean is not less then the geometric mean,

$$p\theta + (1-p)(1-\theta) \geq \theta^p(1-\theta)^{1-p}.$$

The equality is possible only if $\theta = 1 - \theta = 1/2$. Therefore

$$(\log \mathbf{E}\,\xi)/n \to C, \qquad C > 0,$$

and $\mathbf{E}\,\xi \to \infty$ as $n \to \infty$. Thus the theorem is proved.
It is interesting to note a fact following from the proof of Theorem 3.1. If the number of units in the right-hand side of system (3.2) satisfies the condition

$$a_1 + \cdots + a_t \leq tp + t_\alpha\sqrt{tp(1-p)},$$

then the system have none solution with probability tending to one as $n \to \infty$. If the inverse inequality is fulfilled, then $\mathbf{P}\{\xi > 0\} \to 1$ as $n \to \infty$. Therefore as $n \to \infty$ we can make the decision with the probability approaching one whether the system is consistent or not judging by the right-hand side.

4. A COMBINATORIAL METHOD FOR INVESTIGATION OF SYSTEMS OF EQUATIONS

4.1. Conditions of the convergence of the number of solutions in probability

Consider the system of equations

$$f_i(x_{s_{i1}}, \ldots, x_{s_{ir}}) = a_i, \qquad i = 1, \ldots, t. \tag{4.1}$$

Let $s_i = (s_{i1}, \ldots, s_{ir})$ be a hypergeometric sample from $\{1, \ldots, n\}$, $s_1, \ldots, s_t$ be independent, and let $f_1, \ldots, f_t$ be independent identically distributed elements of a class of pseudo-Boolean functions Φ, and $a_1, \ldots, a_t$ be independent random variables.
The matrix $B = |\,b_{ij}\,|$ with elements

$$b_{ij} = \begin{cases} 1, & \text{if } j \in \{s_{i1}, \ldots, s_{ir}\}, \\ 0, & \text{if } j \notin \{s_{i1}, \ldots, s_{ir}\}, \end{cases}$$

is called the incidence matrix of the variables and the equations of system (4.1). For linear systems in the field $\mathbf{GF}\,(2)$ the incidence matrix B coincides with the matrix A of the system. For nonlinear systems the matrix B indicates the variables which are not present in the system. In any case the number ξ of solutions of a priori solvable systems satisfies the inequality $\xi \geq 2^\eta$, where η is the number of absent variables in the system or, what is the same, the number of zero columns in the matrix B.
As $n \to \infty$

$$\mathbf{E}\,\xi - \mathbf{E}\,2^\eta = o(1),$$

therefore

$$\xi \to 2^\eta \quad \text{in probability.}$$

Indeed,

$$\mathbf{P}\{\xi \neq 2^\eta\} = \mathbf{P}\{\xi > 2^\eta\} \leq \mathbf{E}\,\xi - \mathbf{E}\,2^\eta = o(1).$$

In this case the system has a unique solution with respect to the variables which are really contained in the system.
It is useful to consider the incidence matrix for some a priori solvable nonlinear systems of equations as well. Remind that for these systems

$$a_i = f_i(x^0_{s_{i1}}, \ldots, x^0_{s_{ir}}), \qquad i = 1, \ldots, t,$$

where $\bar{x}^0 = (x^0_1, \ldots, x^0_n)$ is a vector.
For example, consider system (4.1) which is constructed in the following way. We choose r from n possible variables and assign $m + 1$ values to this set of variables taking into account that one of this values corresponds to the r-subset of the solution $\bar{x}^0$ and the other m values correspond to m different r-sets from the $2^r - 1$ possible subsets. For this case

$$|\Phi| = C^r_n C^m_{2^r-1}.$$

In practice such a situation can arise if we transmit parts of the true vector through a communication channel with a noise.

THEOREM 4.1. *Let $n \to \infty$, $r \to \infty$, $\lim_{n\to\infty} r/\log n < 1$, $m = o(2^r/\log n)$, $t = (n \log n + x)/r$, $x = O(1)$. Then the number of solutions ξ of system (4.1) converges in the probability to 2^η and*

$$\mathbf{P}\{\xi = 2^k\} = \frac{(e^{-x})^k}{k!} \exp\{-e^{-x}\} + o(1).$$

Proof. Consider a mathematical expectation of the number of solutions

$$\begin{aligned}
\mathbf{E}\,\xi &= \sum_{k=0}^{n} C^k_n \{(C^r_{n-k} C^m_{2^r-1} + (C^r_n - C^r_{n-k}) C^{m-1}_{2^r-2})/(C^r_n C^m_{2^r-1})\}^t \\
&= \sum_{0 \leq k \leq \log n} b_k + \sum_{\log n < k < n \log 3/\log n} b_k + \sum_{n \log 3/\log n \leq k \leq n} b_k.
\end{aligned}$$

Let us estimate the first sum:

$$\begin{aligned}
I_1 &= \sum_{0 \leq k \leq \log n} C^k_n \left(1 - \left(1 - \frac{m}{2^r - 1}\right)\left(1 - \frac{(n-k)_r}{(n)_r}\right)\right)^t \\
&= \sum_{0 \leq k \leq \log n} \exp\left\{-k\left(x - (\log n + x)\frac{m}{2^r - 1}\right) + O\left(\frac{\log^5 n}{n}\right)\right\} \\
&= \exp\{e^{-x} + o(1)\}.
\end{aligned}$$

Let us prove that I_2 and I_3 tend to zero. Indeed,

$$\begin{aligned}
I_3 &= \sum_{n\log 3/\log n \le k \le n} C_n^k \left(\frac{m}{2^r-1} + \frac{(n-k)_r}{(n)_r}\left(1-\frac{m}{2^r-1}\right)\right)^t \\
&\le 2^n \left(\frac{m}{2^r-1} + \left(1-\frac{\log 3}{\log n}\right)^r\right)^t = o(1); \\
I_2 &= \sum_{\log n < k < n\log 3/\log n} b_k \\
&\le \sum_{\log n < k < n\log 3/\log n} e^k \left(\frac{n}{k}\right)^k \left(1-\left(1-\frac{m}{2^r-1}\right)\left(1-\left(1-\frac{k}{n}\right)^r\right)\right)^t \\
&\le \sum_{\log n < k < n\log 3/\log n} \exp\left\{-k\left(\log k - k\frac{r\log n}{2n} + O(1)\right)\right\}.
\end{aligned}$$

The function

$$\varphi(k) = \log k - \frac{kr\log n}{2n}$$

is convex and it attains the maximum at $k = 2n/(r\log n)$. Therefore for sufficiently large n

$$\min_{\log n < k < n\log 3/\log n} \varphi(k) \ge \min(\varphi(\log n), \varphi(n\log 3/\log n)) \ge \frac{1}{2}\log\log n$$

and hence $I_2 = o(1)$.
Let us find now the distribution of the number of zero columns of the incidence matrix:

$$\mathbf{P}\{\eta = k\} = \sum_{i\ge 0}(-1)^i C_n^{k+i} C_{k+i}^i (C_{n-k-i}^r / C_n^r)^t = \frac{(e^{-x})^k}{k!}\exp\{-e^{-x}\} + o(1).$$

To complete the proof of the theorem, notice that

$$\lim_{n\to\infty} \mathbf{E}\, 2^\eta \ge \exp\{e^{-x}\}$$

and

$$\mathbf{E}\,\xi - \mathbf{E}\, 2^\eta = o(1).$$

4.2. Analysis of stair systems of equations

Consider the incidence matrix B of system (4.1). The matrix B belongs to the set M of $t \times n$ matrices with r units and $n-r$ zeros at each row. The number N of such matrices is equal to $|M| = (C_n^r)^t$. Assign the probability $\mathbf{P}\{B\} = 1/N$ to any matrix $B \in M$. Let us analyze the structure of these random matrices.

LEMMA 4.1. *Let $r = o(n^{1/2})$, $t = \sqrt{2\lambda n}/r$, $\lambda = O(1)$. Then the matrix B can be transformed into a step-matrix, each step being equal to r, by a transposition of columns with probability $e^{-\lambda} + o(1)$.*

Proof. First we place r units in the first row of the matrix B, then r units at the second one and so on. After the first replacement we occupy $\eta_1 = r$ columns then another $\eta_2 \le r$ columns are so on. After t replacements we have $\eta_1 + \cdots + \eta_t$ occupied columns. Remaining

$$n - \sum_{i=1}^{t} \eta_i$$

columns will consist of zeros only.
The probability that we obtain a step-matrix is equal to

$$\begin{aligned} \mathbf{P}\{\eta_1 = r, \ldots, \eta_t = r\} &= \prod_{i=0}^{t-1} (C_{n-ir}^r / C_n^r) \\ &= \exp\left\{ \sum_{i=0}^{t-1} \sum_{j=0}^{r-1} \left(\log\left(1 - \frac{ir+j}{n}\right) - \log\left(1 - \frac{j}{n}\right)\right)\right\} \\ &= \exp\{-\lambda + o(1)\}. \end{aligned}$$

The lemma is proved.
Let A_k $(k = 1, \ldots, t)$ be the set of indices of those columns which have units in the kth row. Consider the event $B_{i_1 \ldots i_k}$ which means that

1. $\eta_{i_1} = \cdots = \eta_{i_k} = r - 1$, $\eta_1 + \cdots + \eta_t = rt - k$.
2. $\sum_{j \ne i} |A_i \cap A_j| \le 1$, $i = 1, \ldots, t$.

It follows from point 2 that $i_1, \ldots, i_k$ should satisfy the inequalities $i_l \ge 2l$, $l = 1, \ldots, k$.
The event $B_{i_1 \ldots i_k}$ means that there are exactly k couples of rows (i, j) in the matrix B such that $|A_i \cap A_j| = 1$. All the other couples have zero intersections.

LEMMA 4.2. *Under the conditions of Lemma 4.1 for any fixed $k \ge 1$*

$$\mathbf{P}\{B_{i_1 \ldots i_k}\} \left(\prod_{l=1}^{k} (i_l - 2l + 1) \right)^{-1} = \frac{r^{2k}}{n^k} \exp\{-\lambda + o(1)\} \tag{4.2}$$

uniformly in $2 \le i_1 < i_2 < \cdots < i_k \le t$, $i_l \ge 2l$, $l = 1, \ldots, k$.

Proof. It is easy to see that

$$\mathbf{P}\{B_{i_1 \ldots i_k}\} = \prod_{l=0}^{k} \prod_{j=0}^{i_{l+1}-i_l-2} \frac{C_{n-r(i_l+j)+l}^r}{C_n^r} \prod_{l=1}^{k} \frac{r(i_l - 2l + 1) C_{n-r(i_l-1)+(l-1)}^{r-1}}{C_n^r} = C_1 C_2,$$

where $i_0 = 0$, $i_{k+1} = t+1$. For the first factor

$$C_1 = \prod_{l=0}^{k} \prod_{j=0}^{i_{l+1}-i_l-2} \frac{C^r_{n-r(i_l+j)+l}}{C^r_n}$$
$$= \exp\left\{-\sum_{l=0}^{k} \sum_{j=0}^{i_{l+1}-i_l-2} \frac{r(i_l+j)-l}{n} + O\left(\frac{r^2t}{n}\right)\right\}$$
$$= \exp\{-\lambda + o(1)\},$$

uniformly in $2 \le i_1 < i_2 < \cdots < i_k \le t$.
For the second factor

$$C_2 = r^{2k}\left(1 + O\left(\frac{r^2t}{n}\right)\right) \prod_{l=1}^{k} \frac{(i_l - 2l + 1)}{n}.$$

Therefore

$$\mathbf{P}\{B_{i_1\cdots i_k}\}\left(\prod_{l=1}^{k}(i_l - 2l + 1)\right)^{-1} = \frac{r^{2^k}}{n^k}\exp\{-\lambda + o(1)\}.$$

The lemma is proved.

LEMMA 4.3. *Under the conditions of Lemma 4.1 for any fixed* $k \geq 1$

$$\mathbf{P}\left\{\bigcup_{i_1,\ldots,i_k} B_{i_1\cdots i_k}\right\} = \frac{\lambda^k}{k!}e^{-\lambda} + o(1).$$

Proof. Since estimate (4.2) is uniform in $i_1, \ldots, i_k$ and the events $B_{i_1\cdots i_k}$ are independent

$$\mathbf{P}\left\{\bigcup_{i_1\ldots i_k} B_{i_1\cdots i_k}\right\} = r^{2k}\exp\{-\lambda + o(1)\} \sum_{2\le i_1 < i_2 < \cdots < i_k \le t} \prod_{l=1}^{k} \frac{(i_l - 2l + 1)}{n}.$$

Let us estimate the value

$$S = \frac{1}{t^{2k}} \sum_{2\le i_1 < i_2 < \cdots < i_k \le t} \prod_{l=1}^{k}(i_l - 2l + 1).$$

After the substitution $i_l = tx_l$, $l = 1, \ldots, k$, we see that

$$S = \sum_{2/t \le x_1 < \cdots < x_k \le 1} \frac{1}{t^k} \prod_{l=1}^{k}\left(x_l - \frac{2l-1}{t}\right)$$
$$\sim \int_{2/t \le x_1 < \cdots < x_k \le 1} \cdots \int x_1 \cdots x_k \, dx_1 \cdots dx_k = \frac{2^{-k}}{k!}.$$

Therefore

$$\mathbf{P}\left\{\bigcup_{i_1,\dots,i_k} B_{i_1\cdots i_k}\right\} = \frac{1}{k!}\left(\frac{r^2t^2}{2n}\right)e^{-\lambda} + o(1).$$

The assertion is proved. Thus we obtained the following result.

THEOREM 4.2. *Let $r = o(n^{1/2}), \sqrt{2\lambda n}/r$, $\lambda = O(1)$. Then by a transposition of columns the matrix B can be transformed into a step-matrix with k steps of length $r-1$ and $t-k$ steps of length r with probability*

$$\frac{\lambda^k}{k!}e^{-\lambda} + o(1).$$

Every step may have an intersection (a common column with units) with no more than one step.

Let us consider possible applications of this assertion. Let X_i be the set of solutions of the ith equation of system (4.1). Set

$$\tilde{p}_l = \mathbf{P}\{|X_i \cap X_j| = lq^{n-2r+1} \,|\, |A_i \cap A_j| = 1\};$$

$$q_l = \mathbf{P}\{|X_i| = lq^{n-r}\}, \qquad l \geq 0.$$

If the event

$$B_k = \bigcup_{i_1,\dots,i_k} B_{i_1\cdots i_k}$$

takes place then the number of solutions of the system is equal to

$$\xi = q^{n-tr+k}\prod_{i=1}^{k}\eta_i\prod_{j=1}^{t-2k}\xi_j, \tag{4.3}$$

where

$$\mathbf{P}\{\eta_i = k\} = \tilde{p}_k, \quad \mathbf{P}\{\xi_i = k\} = q_k, \qquad k = 0, 1, \dots, \quad i = -1, \dots, t,$$

and the random variables are independent.
Taking into account that

$$\mathbf{P}\{B_k\} = \frac{\lambda^k}{k!}e^{-\lambda} + o(1),$$

we can find the distribution of the number of solutions ξ with the help of (4.3). In particular,

$$\mathbf{P}\{\xi > 0\} = \sum_{k\geq 0}\mathbf{P}\{B_k\}(1-\tilde{p}_0)^k(1-q_0)^{t-2k} + o(1)$$

$$= (1-q_0)^t \exp\left\{-\lambda\frac{\tilde{p}_0 - 2q_0 + q_0^2}{(1-q_0)^2}\right\} + o(1). \tag{4.4}$$

For $q_0 = 0$ formula (4.4) can be written as

$$\mathbf{P}\{\xi > 0\} \sim \exp\{-\tilde{p}_0\lambda\}$$

and as a corollary we obtain the following assertion: if $q_0 = 0$ and $t = o(\sqrt{n}/r)$, then

$$\lim_{n\to\infty} \mathbf{P}\{\xi > 0\} = 1.$$

Consider several examples of calculating $\mathbf{P}\{\xi > 0\}$ by formula (4.4). Assume that r is fixed.

a) Φ is the set of functions of r variables over the field $\mathbf{GF}(2)$, $|\Phi| = C_n^r 2^{2^r}$, the functions f_i, $i = 1, \ldots, t$, are uniformly distributed on Φ. In this case for any right-hand sides of system (4.1).

$$q_0 = \mathbf{P}\{|X_i| = 0\} = 2^{-2^r},$$

$$\tilde{p}_0 = \mathbf{P}\{|X_i \cap X_j| = 0 \,|\, |A_i \cap A_j| = 1\} = 2^{-2^r}(4 - 2^{-2^r})$$

b) Consider a pseudo-Boolean system of equations

$$f_{i1}(x_{s_{i1}}, \ldots, x_{s_{ir}}) + \cdots + f_{il}(x_{s_{i1}}, \ldots, x_{s_{ir}}) = a_i, \qquad i = 1, \ldots, t.$$

The functions f_{ij} ($i = 1, \ldots, t$, $j = 1, \ldots, l$) are chosen from the set of all Boolean functions of r variables independently and with equal probabilities. Here events $\{\bar{x}_1 \in X_i\}, \ldots, \{\bar{x}_{2^r} \in X_i\}$ are independent as in the previous case, $|\Phi| = C_n^r (2^{2^r})^l$. Therefore,

$$q_0 = \mathbf{P}\{|X_i| = 0\} = \left(1 - \sum_{i=0}^{l} p_i \frac{C_l^i}{2^l}\right)^{2^r};$$

$$\tilde{p}_0 = \mathbf{P}\{|X_i \cap X_j| = 0 \,|\, |A_i \cap A_j| = 1\}$$

$$= 1 - (1 - q_0)^2 + 2\left(1 - \sum_{i=0}^{l} p_i \frac{C_l^i}{2^l}\right)^{2^{r-1}} \left(1 - \left(1 - \sum_{i=0}^{l} p_i \frac{C_l^i}{2^l}\right)^{2^{r-1}}\right)^2.$$

c) Φ is the set of Boolean functions in r variables of the weight m, $1 \le m \le 2^r - 1$. In this case

$$q_0 = \mathbf{P}\{|X_i| = 0\} = 0;$$

$$\tilde{p}_0 = \mathbf{P}\{|X_i \cap X_j| = 0 \,|\, |A_i \cap A_j| = 1\} = 2(p_1 C_{2^r-1}^m / C_{2^r}^m + p_0 C_{2^r-1}^{2^r-m} / C_{2^r}^m)^2.$$

If $m = 2^{r-1}$, then

$$\tilde{p}_0 = 2/(C_{2^r}^{2^{r-1}})^2.$$

If $m > 2^{r-1}$, then

$$\tilde{p}_0 = 2p_0^2 (C_{2^r-1}^{m-2^{r-1}} / C_{2^r}^{2^r-m})^2.$$

d) Consider the pseudo-Boolean linear system of equations from Section 2

$$x_{s_{i1}} + \cdots + x_{s_{ir}} = a_i, \qquad i = 1, \ldots, t.$$

Here

$$|\Phi| = C_n^r, \qquad q_0 = \mathbf{P}\{|X_i| = 0\} = 0,$$
$$\tilde{p}_0 = \mathbf{P}\{|X_i \cap X_j| = 0 \,|\, |A_i \cap A_j| = 1\} = 2p_0 p_r.$$

Therefore

$$\mathbf{P}\{\xi > 0\} \sim \exp\{-2p_0 p_r \lambda\}.$$

REFERENCES

Balakin, G. V. (1973). On the distribution of the number of solutions of random Boolean systems of equations. *Theory Probab. Appl.* **18**, 627–632.

Balakin, G. V. (1983). A priori solvable systems of random equations over a finite field. In: *First Petrozavodsk Conference on Probabilistic Methods in Discrete Mathematics. Abstracts of Communications.* Karelian Branch Acad. Sci. USSR, Petrozavodsk, pp. 8–10.

Balakin, G. V. (1988 a). On a possibility to distinguish a random case from a law-governed one. In: *Second Petrozavodsk Conference on Probabilistic Methods in Discrete Mathematics. Abstracts of Communications.*, Karelian Branch Acad. Sci. USSR, Petrozavodsk, pp. 10–11.

Balakin, G. V. (1988 b). Systems of random linear equations. In: *Abstracts of the Eighth Conference on Problems of Theoretical Cybernetics*, Gorkii, pp. 35–36.

Ivanescu, P. L. and Rudeanu, S. (1966). Pseudo-Boolean methods for bivalent programming. *Lecture Notes Math.* **23**.

Probabilistic Methods in Discrete Mathematics, pp. 99 - 127
V. F. Kolchin et al. (Eds.)
1993 TVP/VSP

PROBLEMS OF SECURITY IN INFORMATION PROCESSING SYSTEMS

A. F. RONZHIN
SCB "Progress", Mironovskaya Street 10, Box 3, 105058 Moscow, Russia

ABSTRACT

Up till now in our contry the problems of information protection in computers were practically not discussed in publications. The purpose of the present report is to make the participants of the conference acquainted with some examples of practical and mathematical problems which must be solved for the organization of a reliable secret communication system.
In this survey we consider methods of data protection in computers, the main ideas of organization, constructing and usage of cryptographical means. We also include into this survey a short description of cryptographical possibilities for data authentication, signature verification, user authentication.

1. INTRODUCTION

The wide spread of computers and extending the range of their applications make the protection of information in computers extremely necessary and generate a considerable interest throughout the world to methods of the protection and to the problems of its standardization especially nowadays when information assumes a status of goods with specified value. The necessity of data protection in computers stems from the requirements to keep the safety, reliability and privacy of the storing information in spite of actions of unexperienced users, breakages or malfunctions of hardware or penetration of imposters. Here and in what follows by an imposter is meant both the persons who wants deliberately obtain, destruct, or change the storing information and the nonprofessionals who construct computer viruses whose penetration into computers lead to loss of part or all the storing information. The data security is achieved by combining some measures which can be approximately divided into four groups, namely organizational, technical, program, and cryptological methods.
The program methods of information protection are intended to provide a differentiated access to computers, i.e., such a system of software which guarantees that each user has an access only to the data specified for the user. The protection methods realized by computer programs are well known to all users, these are passwords for restricted access and systems of privileges for access to some resources and devices of the computer. The system of passwords fulfils authentication functions with the aim to identify the user or the component of the equipment and specify and control, by the access control function, the resources and devices of

the computer which may be submitted to the user. Usually an audit control is used for registration and analysis of all computer events which can be useful for the system protection. This control is intended to detect attempts of unauthorized actions and to distinguish the situations where the total actions of many users can lead to unfavorable consequences.

The protection of computer networks is a complex problem due to the controversy between the requirement to the system to be open to easy general use and to have a restricted access for the protection.

Unfortunately the methods of protection realized by software appeared to be very vulnerable because a slight mistake can generate the changes which reduce the effectiveness of the protection system to zero. In addition, in view of the organizational principles of information storage, high skilled specialists have a possibility to reach any data which are not protected by special hardware or organizational measures. Hence there exists a necessity to provide an illegibility of the data that can be accidentally called by a unauthorized user. This is attained by cryptographic means which provide the illegibility and are used practically in all information protection methods from enciphering the information itself to the organization of password systems, access control and so on. By the opinion of the specialists from the National Bureau of Standards of the USA encryption is the unique reliable method of data protection during their transmission and is very useful one during their storage on various data media (Data Encryption, 1978; Konheim et al., 1980; With data encryption, 1980). This is the reason of great interest in enciphering of information. Cryptographic methods do not exclude completely technical and organizational measures but reduce them to a minimum. In the last years the technical protection measures are often based on the cryptographic ideas.

For data enciphering and deciphering corresponding methods and devices are needed. Moreover, they must depend on time, i.e., the method of enciphering must depend on certain variable which is called a cryptographic key. These keys should be generated, distributed, and controlled. All these constitute an enciphering system or a cryptosystem. The cryptosystem is not isolated from the computer but exists and works in the computer media and consequently the protocols which describe and control the usage of cryptographic means are needed. However it should be noted that no cryptographic protection measures that solve problems such as the initial installation of keys in terminal and central devices of the information network, and protection from direct reading of the keys. These problems must be solved by hardware and organizational measures.

Up till now in our contry the problems of information protection in computers were practically not discussed in the publications. The purpose of the present report is to make the participants of the conference acquainted with some examples of practical and mathematical problems which must be solved while organizing of a reliable secret communication system.

In this survey we consider the methods of data protection in computers, the main ideas of organization, design and usage of cryptographical means, and their dependence on the available technical means and technical devices which are in possession of imposters. We also include into this survey a description of cryptographical possibilities for data authentication, signature verification, user authentication.

A more detail discussion of the computer information security can be found in (Denning, 1983; Gasser, 1987; Special issue on security technology, 1983).

2. CRYPTOGRAPHIC ALGORITHMS

2.1. Areas of applications and main types of algorithms

Computers are widely used in modern cryptology and its development keeps pace with the development of information processing technology. The wide spread of cryptographic information protection has become possible owing to elaboration of enciphering algorithms which can be realized by modern computer systems. The cryptographic protection is used in three areas while processing information by computers: during data transmission, during information storage, and for authentication of users and information. The most-used is encryption for data transmission through communication networks and for computer data processing in distributed systems and computer networks. The other important area of application of cryptographic protection is enciphering during information storage, in this case a possibility of the encrypted data processing must be preserved. Compatibility of enciphering and the principles of authentication and access control pose additional requirements to cryptographic transformations. The authentication possibility is especially important if distant computer terminals are used. The authentication procedures by cryptographic means assume an additional importance in connection with the wide spread of electronic mail (transmission of messages and documents from one computer to another by communication channels) and contract making with the help of computers. The knowledge of properties of cryptosystems adapted for the computer use is a necessary condition for applications of cryptographic methods of information protection in computers and computer networks.

The cryptographic methods are used to provide security of information by data transformation (ciphering) such that the transformed data (ciphertext) give no possibility for unauthorized reconstruction of the plaintext. A cryptosystem consists of an invertible cryptographic transformation (which provides a possibility of encryption and decryption) and a set of keys which are the parameters of the transformation. The knowledge of the used key gives a possibility to decrypt the ciphertext. The scheme of cryptographic protection can be described as follows. Let plaintexts are written in some alphabet $A = \{A_1, \ldots, A_m\}$, where m is the alphabet cardinality. The set of plaintexts X is a set of words $X_1, \ldots, X_n$ over the alphabet A:

$$X = \{X_1, \ldots, X_n\}$$

and the set of ciphertexts Y is a set of words $Y_1, \ldots, Y_l$ over the same alphabet:

$$Y = \{Y_1, \ldots, Y_l\}.$$

The symbols of ASCII, letters of national alphabets, binary symbols can be taken as the alphabet. Enciphering as a cryptographic transformation is a function E which to each plaintext $X_i \in X$ assigns a single element $Y_i \in Y$, $Y_i = E(X_i)$, $i = 1, \ldots, n$. Deciphering is the function D which is inverse to E. To each ciphertext $Y_j \in Y$ the function D assigns a single element $X_j \in X$, $X_j = D(Y_j)$,

$j = 1, \ldots, l$. Thus the function E, called a cipher, with the set of plaintexts X as the domain of definition and the set of ciphertexts Y as the range of values, for each element of X determines a single element of Y. The total number of such functions is equal to $l!/(l-n)!$. Let us label these functions in an arbitrary way with the elements of a set K. The elements $k \in K$ are called keys and can be considered as values of a cryptographic variable.

The process of encryption consists of a choice of $k \in K$ and mapping of a plaintext by the function E_k into an element of the set Y. The selected key k is secret, but the set of functions and the correspondence between K and this set is nonsecret. There are straight requirements to the choice of the set of the functions and the set of the keys K (Introduction to FFT security, 1976; Shannon, 1951) which guarantee that imposters cannot reconstruct a plaintext and find the used key k on the base of ciphertext, but here we will not discuss these conditions.

Let us consider cryptographic functions and their strength and weakness in applications for information protection in computers in more detail. All the cryptographic functions can be partitioned into classes of block encryption functions and sequential encryption functions.

2.2. Block encryption algorithms or ciphers

Block ciphers transfer input binary vectors of length n into output blocks of length m (encrypted sequences) in such a way that each bit of the output block depends on each bit of the input block (Don, 1985). In what follows for the sake of simplicity we assume that $n = m$. The scheme of block encryption can be described as follows. Let a key k be selected, then each input block X of length n is transformed into a block Y of length n of the ciphertext by a function E_k,

$$Y = E_k(X), \quad X = D_k(Y),$$

where D_k is the function inverse to E_k. Let $\{E_k, k \in K\} = \{D_k, k \in K\}$, what is usually fulfilled, then $D_k = E_{k'}$ and

$$Y = E(k, X), \quad X = E(k', E(k, X)).$$

The following two situations are possible: for given k the key k' and $E_{k'}$ are easily calculated or the calculation of k' is as difficult as the reconstruction of the corresponding plaintext X on the base of a given ciphertext Y. In the latter case the key k together with the ciphertext Y can be transmitted to the recipient through a nonsecret channel. This does not destroy the cryptosystem. With respect to this property all block encryption systems are divided into the class of conventional encryption procedures and the class of public key encryption procedures. A typical representative of the class of conventional algorithms is the DES algorithm which we discuss later. Now we turn to the second class of algorithms.

2.3. Public key algorithms

The main feature of these algorithms (Diffie and Hellman, 1976) is the use of a nonsecret key E_k for encryption and a secret key D_k for decryption. Such algorithms have the following properties:

- users have a possibility to calculate the keys E_k and D_k effectively;
- the knowledge of E_k gives no possibility to calculate D_k effectively;
- the decryption which follows an encryption gives an initial information $X = D_k(E_k(X))$.

Public key algorithms are constructed on the basis of well-known difficult mathematical problems of high complexity, and conventional algorithms realize specific transformations which are difficult to analyze. While the use of public key algorithms the transmission of the enciphering key E_k in encrypted form is not needed.

In spite of essential differences between the classes of algorithms, the conventional algorithms can be adapted in such a way that they provide many useful properties of public key procedures (Ehrsam et al., 1978; Zennon and Matyas, 1979; Zennon et al., 1980, 1981, 1982).

Let us consider two examples of public key algorithms.

2.4. RSA algorithms

The Rivest, Shamir and Adleman (RSA) public key encryption algorithms are based on the fact that factoring of a large number is a very laborious computation problem (Ore, 1948). To determine an RSA algorithm, it is needed to select carefully two large primes p and q which are secret. Their product $r = pq$ is not secret, but the number $f(r) = (p-1)(q-1)$ is also secret. By the Euler formula

$$a^{f(r)} = 1 \bmod r \tag{1}$$

where a is relatively prime to r and $f(r)$ is the Euler function.

Let $E_k D_k = 1 \bmod f(r)$, then as it follows from (1) for any numerical record of a plaintext X

$$X^{E_k D_k} = X \bmod r.$$

Therefore it is possible to encrypt and decrypt the information by the formulas

$$Y = E(E_k, X) = X^{E_k} \bmod r,$$

$$X = D(D_k, Y) = Y^{D_k} \bmod r.$$

Thus the procedure of design and use of an RSA algorithm is the following.

1. Two secret primes p and q are randomly selected.
2. The public number $r = pq$ and a secret number $f(r) = (p-1)(q-1)$ are calculated.
3. A number E_k relatively prime to $f(r)$ is selected and a secret D_k such that $E_k D_k = 1 \bmod f(r)$ is calculated.
4. A plaintext X and a ciphertext Y are numbers (maybe in the binary representation) from the set $\{0, 1, \ldots, r-1\}$,

$$Y = X^{E_k} \bmod r \in \{0, 1, \ldots, r-1\},$$

$$X = Y^{D_k} \bmod r \in \{0, 1, \ldots, r-1\}.$$

Let us consider some mathematical problems related to the construction of an RSA algorithm.
The first problem concerns the choice of E_k and D_k meeting the condition of item 3. By a number theory theorem from (Ore, 1948) the congruence

$$D_k x = 1 \bmod f(r)$$

has a solution if and only if D_k and $f(r)$ are relatively prime. Therefore the Euclidean algorithm for calculation of the greatest common divisor can be used for the selection of D_k and E_k.
The second problem concerns the amount of prime numbers. It is well known that

$$\frac{\pi(x)\log x}{x} \to 1$$

as $x \to \infty$ where $\pi(x)$ is the number of primes which are not greater than x. Thus if p is a 200-bit binary number then the number of such primes is about 10^{58}.
Finding large primes is the third problem. Such a choice requires testing of primality. As the tests of primality we can use the Wilson theorem (Ore, 1948) which states that if p is prime then $(p-1)! = -1 \bmod p$ and the simple fact that p is prime if it is not divisible by all integers not greater than $\sqrt{p}$. For realization of these tests much calculations are needed, in return they certainly determine primality of p. There exist some methods which confirm primality with sufficiently large probability and are easy in calculations (Solovay and Strassen, 1977). The other tests of primality with effective calculation are described in (Miller, 1976; Polland, 1974; Rabin, 1976).
In recent years the problems related to public key procedures have become a subject of considerable interest of mathematicians and cryptologists (Blackley and Blackley, 1978; 1979; Data Encryption Standard, 1977; Davies et al., 1979; Hellman et al., 1976; Herlestam, 1978; Meyer, 1978; Rivest, 1978 a, 1978 b; Simmons and Norris, 1977; Williams and Schmid, 1978).
From the results of (Rivest et al., 1978) it follows that for a RSA system with blocks of length 338 bits the time needed for factoring r is the same as the time of the exhaustive checking of all 2^{56} keys of the DES algorithm.

2.5: Knapsack algorithms

These algorithms are based on a classical problem of the number theory (Merkle and Hellman, 1978). Let $X = (x(1), \ldots, x(n))$ be a secret binary vector which is a plaintext and $A = (a(1), \ldots, a(n))$ be a nonsecret vector with integer components. Then the encrypted information is the number

$$Y = (A, X) = a(1)x(1) + \cdots + a(n)x(n).$$

Simplicity of encryption is obvious. A possibility to reconstruct X with the help of Y can be obtained by appropriate choice of A. The only problem is an introduction of a secret value which could allow an easy decryption of the received hidden information. This can be done in the following way. One choose a secret vector $A' = (a'(1), \ldots, a'(n))$ such that

$$a'(l+1) > a'(1) + \cdots + a'(l), \qquad l = 1, \ldots, n-1.$$

For such a vector the problem of determination of X from $Y' = (A', X) = a'(1)x(1) + \cdots + a'(n)x(n)$ is easily solved since $x(n) = 1$ if $Y' \geq a'(1) + \cdots + a' \times (n-1)$, $x(n-1) = 1$, if $Y' - a'(n) \geq a'(1) + \cdots + a'(n-2)$ and so on. Let us introduce secret relatively prime numbers r and t and find s such that $st = 1 \bmod r$. Now we can organize the encryption in the following way. Define a nonsecret A with coordinates $a(i) = a'(i)t \bmod r$ and choose $Y = (A, X)$ as the ciphertext leaving all difficulties of decryption on A. For the decryption we transform Y into $Y' = Ys \bmod r$ and taking into account that $Y' = (A', X)$ we decrease the complexity of X reconstruction. Estimates of the corresponding complexities can be found in (Shamir, 1982).

2.6. Sequential ciphering

Let a plaintext be a binary sequence and there exist a set of autonomous automata $A(k)$, $k \in K$, with the output alphabet $\{0, 1\}$. Then for a plaintext $X = (x(1), \ldots, x(n))$ and an output sequence $R(k) = (r(1, k), \ldots, r(n, k))$ of the automaton $A(k)$, the ciphertext $Y = (y(1), \ldots, y(n))$ is defined as

$$y(i) = x(i) + r(i, k), \qquad i = 1, \ldots, n.$$

It is usual but not necessary that the key k is an initial state of an automaton. Since

$$x(i) = y(i) - r(i, k), \qquad i = 1, \ldots, n,$$

the encryption and decryption are realized by the same scheme. This encryption-decryption scheme has the following essential features: if a key k is used for encryption of several plaintexts (for example, if a plaintext is divided into blocks and each block is encrypted separately by the same sequence) and one of these plaintexts becomes known to an imposter, then he can easily obtain the sequence $R(k) = X + Y$ and then, without of knowing the key k, he can decrypt all other records adding the corresponding ciphertexts and $R(k)$. This happens because the automaton $A(k)$ with the same initial states each time gives the same output. This disadvantage can be avoided if each time before encryption performance the automaton is forced to change its initial state. This can be reached by introduction of some parameter Z which control the initial state of the automaton. The only purpose of this introduction is to transfer the automaton with the key k into different initial states before each new encryption performance, but not to provide a cryptographic reliability. Therefore the requirements to Z are not so strong as the requirement to the keys K.

2.7. DES algorithm

Let us describe the algorithm which is known as the Data Encryption Standard algorithm. Before discussing the technical side of the procedure we shall make a journey into the legislation of the United States on the questions of secrecy and into history of elaboration of the Data Encryption Standard. It is clear that such an information can be useful for performing organizational measures of data protection.

To keep a secret in information processing systems it is needed to follow some rules which regulate the user possibilities to select, store and change the information at their own will. In addition users must have the copyright on their information. In the present time in the USA these rules are regulated by the act of 1974 (Privacy Act, 1974). As concerns the cryptographic measures of data protection, their applications are regulated by the act of 1977 (Public Law 95-213). More detailed description of the USA legislation on security of computer systems can be found in (Hoffman, 1977).

In 1973 the National Bureau of Standards of the USA published a note (Cryptographic Algorithms, 1973) where it was suggested to design a unified cryptographic algorithm for computer data processing in the commercial and also in the defence fields. The following requirements were formulated:

1. the algorithm must be completely definite without any uncertainty,
2. it must provide the corresponding security level which can be naturally estimated in time or amount of operations needed to break the system,
3. its security must be based on the key security only,
4. it can be used without any restrictions on the users.

On August 6, 1974, IBM suggested an algorithm worked out by a scientific group under the supervision of W. L. Tuchman (Ehrsan et al., 1976) as a candidate to satisfy these requirements. The algorithm was based on the Lucifer cryptosystem (IBM Research Report, 1971) which had been elaborated by Feistel (Feistel, 1974). The USA expert in cryptology and communication security is the National Security Agency (NSA). The National Bureau of Standards together with NSA estimated the security of the candidate (Report Workshop on Cryptograply, 1977) and this algorithm provided the basis for the Data Encryption Standard (DES) which was published by NSA on March 17 of 1975, and on July 15 of 1975 the algorithm was officially accepted as the federal standard. DES is permitted for use by federal agencies and the State department for encryption of data not classified by the Act on National Security of 1974 and the Act on Nuclear Energy of 1954. Moreover, it was recommended for application by industrial corporations. DES was adapted by the American National Standards Institute (ANSI) to the requirements CCIP(X3) and was accepted as a standard industrial algorithm DEC X3. 92 (American National Standard, 1981). The requirements to security and physical and electronic protection of the equipment of DES were approved in (Proposed Federal Standard 1026, 1980; Proposed Federal Standard 1027, 1980). A more wide information about the history of elaboration of the DES algorithm can be found there.

In (Unclassified Summary, 1978) the following official characteristics of the DES algorithm was given:

1. DES is more than acceptable in its area of applications.
2. IBM is the inventor and constructor of DES.
3. NSA took no part in the construction.
4. NSA testified the security of DES against the attacks by all known statistical methods.
5. NSA recommended to the Federal Bureau of Investigation to use DES for data transmission in electronic devices.

The DES algorithm is a typical representative of the class of block ciphers. Let us give its description following (Data Encryption Standard, 1977). The algorithm is

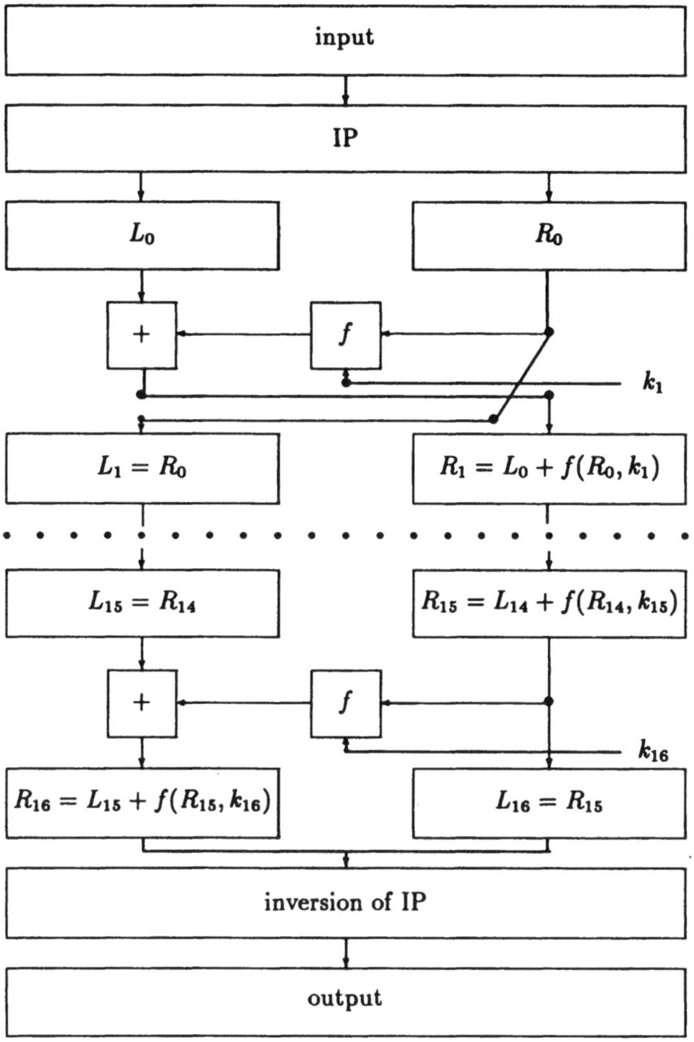

Fig. 1

used for encryption and decryption of data blocks of 64 bits length and is controlled by a key of 64 bits length.

As is seen from Fig. 1, the enciphering is fulfilled in the following way. The 64 bits of the input block are permuted according to the initial permutation (IP). This permutation places on the first position the 58th bit, on the second position the 50th bit and so on. The output of this permutation goes to the input of a processor which depends on a key. The output of the processor is called preoutput and is permuted according to the substitution inverse to IP. The processor fulfils 16 iterations with a function f which depends on two 32-blocks of the previous iteration and on the 48 bits of the key. The function f is called the encryption

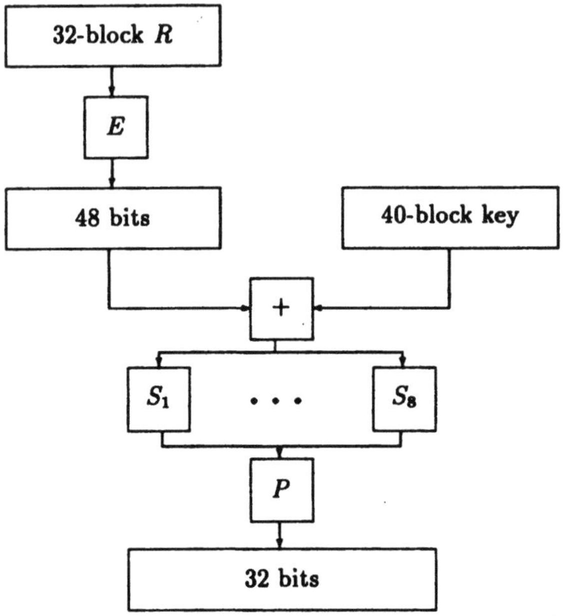

Fig. 2

function and takes values from the set of 32-blocks. For each iteration 64 bits of the input block are separated into the 32-block L and the 32-block R. Let k be a 48-block selected from 64-block of the key, then the output of the iteration consists of the 32-block L' and 32-block R' such that $L' = R$, $R' = L + f(R, k)$. In each iteration the key k is selected from a key sequence K which is a 64-block. The selection is performed by a function K_S which depends on the number of the iteration $n = 1, \ldots, 16$, and the key sequence K. The key k_n used in the nth iteration is

$$k_n = K_S(n, K).$$

Deciphering realized by the same algorithm with the only difference that the keys $k_1, \ldots, k_{16}$ are used in backward order.
The encryption function f can be represented by the scheme in Fig. 2.
The function E in the scheme allocates the 32-block R into 48 places by a prescribed rule. Each of the eight groups $B_1, \ldots, B_8$ of 6 sequential bits is the input of the corresponding selection function S_i, $i = 1, \ldots, 8$, with 4-bit outputs. For each i the function S_i is defined by a 16×4 matrix A_i with elements $a_{rs}^{(i)}$ taking values $0, 1, \ldots, 15$. The first and the last bits of the input considered as a binary number determine the number r, and the 4 middle bits considered as

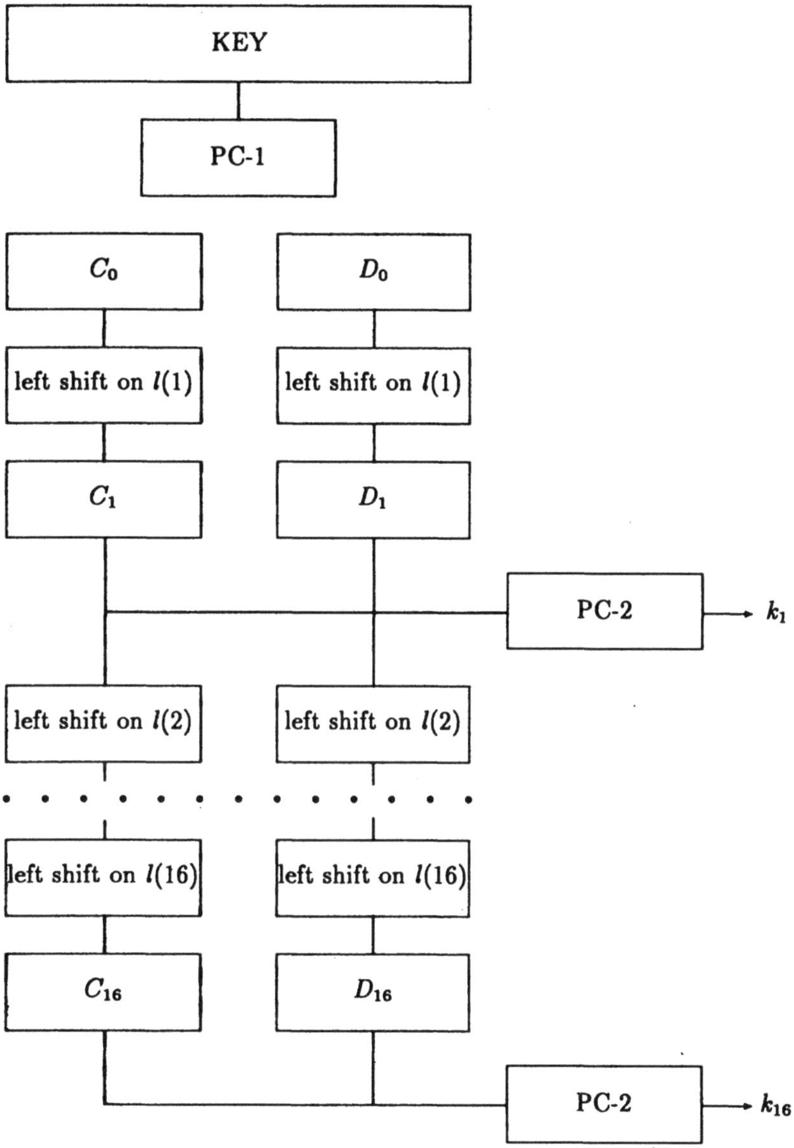
KEY
PC-1
C_0
D_0
left shift on $l(1)$
left shift on $l(1)$
C_1
D_1
PC-2
k_1
left shift on $l(2)$
left shift on $l(2)$
left shift on $l(16)$
left shift on $l(16)$
C_{16}
D_{16}
PC-2
k_{16}

Fig. 3

a binary number too determine the number s. Then the output of S_i is equal to $a_{rs}^{(i)}$ represented in binary notation.

Table 1. Initial permutation IP

58	50	42	34	26	18	10	2
60	52	44	36	28	20	12	4
62	54	46	38	30	22	14	6
64	56	48	40	32	24	16	8
57	49	41	33	25	17	9	1
59	51	43	35	27	19	11	3
61	53	45	37	29	21	13	5
63	55	47	39	31	23	15	7

Table 2. Function E

32	1	2	3	4	5	—B_1
4	5	6	7	8	9	—B_2
8	9	10	11	12	13	—B_3
12	13	14	15	16	17	—B_4
16	17	18	19	20	21	—B_5
20	21	22	23	24	25	—B_6
24	25	26	27	28	29	—B_7
28	29	30	31	32	1	—B_8

Table 3. PC-1

57	49	41	33	25	17	9
1	58	50	42	34	26	18
10	2	59	51	43	35	27
19	11	3	60	52	44	36
63	55	47	39	31	23	15
7	62	54	46	38	30	22
14	6	61	53	45	37	29
21	13	5	28	20	12	4

$$L = (1,1,2,2,2,2,2,2,1,2,2,2,2,2,2,1)$$

The output of S_i is transformed by a permutation P. Thus if the inputs of the function f are R and k, then

$$f(R,k) = P(S_1(B_1),\ldots,S_8(B_8)),$$

where

$$B_1 B_2 \cdots B_8 = k + E(R).$$

Note that we consider g-blocks A and h-block B written together as a single block AB of length $g + h$. The process of construction of the key k from the 64-block KEY is shown in Fig. 3. From 64 bits of KEY the function PC-1 selects 56 bits and divides them into 28-blocks C and D. In each iteration i, $i = 1, \ldots, 16$, the cyclic shift of the contents of the registers C_i and D_i on $l(i)$ bits to the left are performed and with the help of the function PC-2 the 48-block k_i is selected from the obtained block $C_i D_i$. To complete the description of the algorithm it remains to give the values of IP, E, $S_1, \ldots, S_8$, PC-1, PC-2 and $L = (l(1), \ldots, l(16))$.
The functions $S_1, \ldots, S_8$ and PC-2 can be found in (Data Encryption Standard, 1977). All they have no definite structure and we do not give their description here.
Thus we see that DES performs 16 iterations and depends on 16 keys of 48 bits each. These keys, as follows from definition of PC-1, are selected from one installed key. Besides, the contents of registers C_i, D_i are shifted so that $C_{16} = C_0$ and $D_{16} = D_0$. During the decryption the key selection functions act in the same manner but the shifts in registers are performed to the right.
We conclude the description of the DES algorithm by the matrix for the selection of the keys $k_1, \ldots, k_{16}$

k_1	=	10	51	...	41	22	28	...	31
k_2	=	2	43	...	33	14	20	...	23
...	...	...	...	...	...	...	...	...	...
k_{16}	=	18	59	...	49	30	5	...	39

Among these numbers there are no bits from the stages $8, 16, \ldots, 64$ which are used for control on the initial step of the key construction.
Note some features of the DES algorithm.

1. *Vulnerable keys* (Meyer and Matyas, 1982). There are the keys for which $k_1 = \cdots = k_{16}$, or the keys whose contents of the registers C and D consist of zeros only (or units only) or of sequences of the form 0101...01; 1010...10; 0011...0011; 0110...0110; 1001...1001; 1100...1100. Although the number of such keys is very small in comparison with the total number of keys that is equal to $7 \cdot 10^{17}$ and the probability that such a key occurs under the random generation of keys is very small it is necessary to exclude such keys by a special control.

2. *Criteria of choosing of mathematical transformations.* The first step in investigations of the DES algorithm continued 5 years. As a result some criteria were suggested, some of these criteria are rigorously formulated and can be used in cryptanalysis. In particular, for selection of permutations the principle was accepted that each output bit must really depend on each bit of the key and plaintext, and the transformation must include the minimum of iterations (Meyer, 1978). The other important principle is the requirement that the functions in DES use for their realization the minimum number of logic operations (Hong et al., 1974).

3. SECURITY OF CRYPTOSYSTEMS

3.1. Two approaches to definition of cryptosystem security

Consider the conditions that a cryptosystem must satisfy for reliable information protection. The security of enciphered information (cryptographic security or simply security) depends on the possibility of an unauthorized reading of the data. There exist two kinds of security: theoretical (or mathematical) security and practical security. These concepts were introduced in the classical paper (Shannon, 1949). The term practical security does not mean that the definition is not mathematically rigorous. The essence of the both kinds of security is the following.

The theoretical security is based on the fact that a cryptosystem is modelled by some formal object and for this model certain conditions of impossibility for an outsider to break the cryptosystem are formulated. Usually it is assumed that the information available to the imposter must be insufficient to determine the plaintext even if the information about the cryptosystem is not secret.

As a measure of practical security we take the work, i.e., the number of operations or time complexity of determination of the plain information by an outside or an average values of such characteristics over the set of all plaintexts. In this case the goal is to obtain the maximum complexity of the problem of an unauthorized deciphering.

It should be noted that there exist two approaches to construction of practically secure ciphers. In the first case the cryptosystem is constructed and then it is demonstrated that its breaking is a difficult problem. In the second case some difficult mathematical problem is selected and then the corresponding cryptosystem is constructed whose breaking is equivalent to the solving of that problem. In (Diffie and Hellman, 1976) the authors for the first time suggested to use a class of difficult problems for constructing cryptosystems. But it should be noted that in principle there is no great difference between the use of public key systems and, for example, DES algorithms.

The problems of theoretical security were investigated mainly in papers devoted to cryptology. The problems of practical security were investigated in pure mathematical papers on the computation complexity.

Finally we note that theoretical security is determined under the condition that there is no time restriction on unauthorized deciphering and consequently it answers the question what cryptosystems cannot be broken in principle. Such perfectly secure systems really exist (Wilkes, 1972). They can be constructed with the help of a random equiprobable encryption key whose length is not less than the length of plaintext. The perfectly secure systems are extremely expensive in realization. Therefore the systems which can be broken in principle but in unacceptable time are used in practice.

Note that the NSA experts (see (Matyas et al., 1980; Unclassified summary, 1978)) did not find a possibility to break the keys of DES by statistical methods, and an ICST working group (Wilkes, 1972) declared the impossibility to design up to 1990 a specialized device for exhausting search the DES keys with success probability 0.1–0.2. The NSA experts prolonged the guarantee of security of DES algorithms till 1995.

The cost and the time needed for breaking a cryptosystem also pose some restrictions on security of cryptosystems. For any cryptosystem there exists a correspondence between its cost and the time of its breaking. Enciphered data have also the value decreasing usually in time. This gives some correspondence between the value and the storage time of the information. These correspondences between the cost and breaking time and the value and storage time must be taken into account while we design and estimate the practical security of cryptosystems.

3.2. Mathematical security

Let a source of plaintexts X generate information messages in correspondence with some stochastic process $g(t)$ with a range of values $\{X\}$. Let a key $k \in K$ be selected on the basis of another stochastic process and probably $g(t)$. As a result the ciphertext is also a stochastic process, say $r(t)$. The main condition ensuring the security is the requirement that the probability of restoring a plaintext on the basis of the ciphertext $Y = r(t)$ calculated by an outsider is equal to the initial probability of X, i.e.,

$$\mathbf{P}\{g(t) = X\} = \mathbf{P}\{g(t) = X \mid r(t) = Y\}. \tag{2}$$

It is reasonable to consider the cryptosystems such that there exists only one message $Y = E(k, X)$ for any plaintext X and any key k, and for distinct pairs of (X, k) the corresponding ciphertexts are different. Then $|\{X\}| \leq |\{Y\}|$. We restrict ourselves by the case $|\{X\}| = |\{Y\}|$. In this case $E(k)$ is one-to-one correspondence and $D(k) = E^{-1}(k)$ exists. If the keys are selected with equal probabilities $1/|K|$ from the set K of all keys and the number of plaintexts which are preimages of a ciphertext is the same for any ciphertext then (2) is valid. The other sufficient conditions for (2) are given in (Shannon, 1949). Whereas the equiprobable selection of keys is natural, the question about the distribution on the set of plaintexts is not so clear. Shannon (see (Shannon, 1949, 1951)) proposed an approach based on the redundancy of natural languages. The set $\{X\}$ of plaintexts in the alphabet of a natural language is divided into the following two classes: the total probability of the first class is close to zero, and the probability of each message from the second class is equal approximately to $p = \exp(-NH)$ where N is the length of the message and H is the entropy of the language. Usually we can assume that

$$H = -p(1)\log p(1) - \cdots - p(z)\log p(z),$$

where $p(1), \ldots, p(z)$ are the relative frequencies of letters of the natural language. On the basis of the conditions discussed above we can obtain the following formulas for parameters of cryptosystems (Meyer and Matyas, 1982).

The probability to guess the key under a given ciphertext is equal approximately to $(1 - \exp(-h))/h$, where $h = |K|\exp(NH)/|\{Y\}|$. Here $|K|$ is the number of keys and $|\{Y\}|$ is the number of ciphertexts.

If in contrast to the above mentioned condition the number of keys translating a given ciphertext into a meaningful plaintext is not the same for all ciphertexts and is a random variable M, then

$$\mathbf{P}\{M = m\} = \exp(-h')(h')^{m-1}/(m-1)!,$$

where $h' = h(|K| - 1)/|K|$.
From this formula we can determine such an important parameter as the uniqueness length (Shannon, 1949), i.e., the minimum length N of a plaintext such that there exists a single key corresponding to a given ciphertext.
The probability to guess the plaintext from a ciphertext is equal to

$$\left(1 - \exp(-h)/h\right)\left(1 + h'/(2\exp(NH))\right).$$

Note that these formulas correspond to natural languages and include the number of meaningful texts such that the probability of the set of the meaningful messages is equal to one.
The second step of security evaluation considers enciphering as a transmission of a plaintext through a channel with a noise (the role of the noise is played by the cryptosystem). Then the problem of restoring a plaintext by a ciphertext coincides with the problem of recovering a signal transmitted through a noisy channel. In communication one wish to reduce the noise to minimum, in cryptography a channel with maximum noise is preferable. Therefore it is clear that the messages have a redundancy measured by $1 - H$ where H is the entropy of the language.
Now we list some results obtained with the help of this approach (Meyer and Matyas, 1982; Shannon, 1949). The uniqueness length is the solution of the equation

$$H(X) + H(K) - H(Y) = 0$$

with respect to the length of Y. In the case where both the plaintext and ciphertext are given the uniqueness length for the keys is the solution of the equation

$$H(K) - H(Y \mid X) = 0$$

with respect to the length of keys from K. In these formulas $H(\xi)$ is the entropy of the random variable ξ and

$$H(\xi) = \sum_{k=0}^{\infty} p_k \log p_k,$$

where $p_k = \mathbf{P}\{\xi = k\}$, i.e., $H(X), H(Y), H(K)$ are measures of uncertainty of the choice of plaintext, ciphertext, and key respectively, and $H(Y|X)$ is the conditional entropy of Y provided X is given.
Under the assumption that the keys in the DES algorithm are selected with equal probabilities and for a given plaintext all 2^{56} ciphertexts are selected with equal probabilities from the set of 2^{8N} possible ciphertexts (N is the length of plaintext) we obtain that the uniqueness length is equal to the length of one block for the keys and to twice length of a block for plaintexts. Thus the DES algorithm is not a perfectly secure system.
Examples of evaluation of the uniqueness length for permutation, substitution and other cryptosystems are given in (Deavours, 1977; Matyas, 1974; Meyer and Matyas, 1982).

3.3. Algorithmic security and possibilities of modern hardware

The algorithmic or computer complexity (Denning, 1983) determines the requirements to computer resources which are necessary for the analysis of cryptographic algorithms and is used for investigation of complexity of cipher breaking. The security of cryptosystems is measured by the computer complexity of algorithms used for its breaking. The computer complexity consists of time complexity T and space complexity S (memory capacity) used by the algorithm. The parameters T and S are functions of the length N of input sequences. Usually estimates of a function $f(n)$ can be presented in the form $f(n) = O(g(n))$, where $g(n)$ is a simply computable function. For example, if $g(n)$ is a polynomial of degree t then $f(n) = O(n^t)$. Under such a representation a complexity is rather independent of computer facility. In addition the real running time of the algorithm depends on characteristics of the keys and plaintext. A similar representation is possible for the average capacity of memory used in the algorithm.

The following classification of algorithms in accordance with their complexity is known.

Polynomial algorithms are the algorithms with polynomial time complexity $T = O(n^t)$, where t is a constant. If $t = 1$, the algorithm is called linear.

Exponential algorithms are the algorithms with exponential time complexity $T = O(t^{h(n)})$, where $h(n)$ is a polynomial and t is a constant.

It can be assumed that for large n the polynomial algorithms are realizable in general in multiprocessor computers or in specialized computers. The exponential algorithms cannot be realized effectively.

The algorithmic complexity is closely related to the mathematical NP-complexity. In the complexity theory the problems are classified in accordance with the minimal time and memory needed for their solution in Turing's machine or in some other abstract computer model (Aho et al., 1974; Garey and Johnson, 1979; Minsky, 1967). It is shown that the problems within a class are equivalent. All public key systems are based on the algorithms with polynomial complexity of encryption and decryption and with exponential complexity of their breaking (Konheim, 1981).

Many cryptosystems have polynomial complexity and their practical security is determined by the power of computers and, primarily, by the speed of computers and microprocessors. The speed of microprocessors determines the power of both the personal computers which are the most accessible to imposters and specialized multiprocessor systems used by professionals. Consider the state and trends in development of microprocessors and high-speed computers.

Table 4.

Type, producer	Working frequency, MHz	Speed, MIPS	Number of operations per command
80386, Intel	24	5.3	6.0
80486, Intel	40	18.0	1.0
68020, Motorola	25	4.0	7.0
68030, Motorola	25	6.8	—
68040, Motorola	40	18.0	—
VAX 8550, DEC	22.2	3.0	7.0
VAX 8800, DEC	22	11.0	—
R2000, MIPS	16.7	12.0	1.4
R3000, MIPS	25	20.0	1.25
Am2900, AMD	25	12.5	2.0
SPARC, Sun Microsystems	16.7	8.0	2.1

Table 5.

Type of computer	Number of processors	Speed, MIPS
Cray Y-MP/8	8	2000
SX-3	4	22000
TX-3	1024	20000
n-Cube 2	8192	27000

During the last two decades the stable tendency of considerable progress in microprocessors is observed. In view of the progress in microelectronics and in other areas the demand for microprocessors steadily grows. Now 1-, 4-, 8-, 16-, 32-, and even 64-bit microprocessors are available. We have no possibility to give here a full review of the current state of the microprocessor and computer industry. We present only two tables (Table 4 and Table 5) which show some characteristics of produced microprocessors and high-speed computers.

4. PROTOCOLS OF CRYPTOSYSTEMS

Organization of secure communications must satisfy certain rigorous rules which are regulated by protocols. For different types of secure communications different protocols are needed. We have no possibility to list here all types of secure communications and give even short descriptions of corresponding protocols. We restrict ourselves to some examples of authentication protocols and protocols of digital signatures.

4.1. Authentication problems

Authentication is a process which is used to confirm that somebody or something is authentic. Cryptographic methods provide the highest level of authentication of messages, data, keys, and devices. All authentication schemes have a common part in which authenticity of one or several parameters is being examined.

The Federal Standard of the USA (Federal Standard 1026, 1978) defines authentication as a process guaranteeing invariance of given parts or positions or confirming authenticity of a given information unit. It is necessary to distinguish authentication parameters and authentication process itself from digital signatures. In the authentication process it is not needed to prove to a third person that the information was sent just by this sender and was not created by the recipient. But it is needed that recipient must be sure that the information was sent just by the given sender.

The cryptographic algorithms used for authentication must satisfy the following conditions.

1. They must be reliable in the sense that the key used for encryption cannot be found on the basis of corresponding plaintext and ciphertext.
2. The keys of the cryptosystem must be protected.

It is clear that these conditions can be met by the use of classical algorithms, public key algorithms and procedures based on one-way functions (Evans et al., 1974; Purdy, 1974; Westin, 1968).

Let us point out three cases where authentication plays an important role. The first case is the confirmation that a connection is established between two authorized users of the system. The procedure is called the handshaking. The second case is the authentication of a message which must to confirm to a recipient authenticity of the message sender, authenticity of the message contents, authenticity of the order of messages (i.e., the recipient must be sure that he received the messages in the same order as they were sent), authenticity of the message address (i.e., the recipient must be sure that the message is transmitted just in the direction it was sent). The third case is the checking or authenticity examination of those system data that are constant for some period of time (for example, keys, user identifiers and so on).

As an example we describe a handshaking protocol.

4.2. Handshaking protocols

Consider a procedure allowing a user A to verify authenticity of a user B. It is clear that the inverse procedure allows the user B to verify authenticity of the user A.

The user A generates a pseudo-random number N which is enciphered with the help of a session key k_S and sent to the user B. The user B deciphers it and computes some public (nonsecret) function $y(N)$. Then the value $y(N)$ is ciphered and sent to the user A. The user A deciphers the received value and compares it with $y(N)$. The equality confirms authenticity of the user B.

This method makes impossible the "night attack" of an imposter because at the next session the terminal generates another pseudo-random number N and the imposter has no possibility to give the right answer to the system request since he knows only the value $E(k_S, N)$. Sometimes the inversion of some bits in the binary notation of the number N is used as the function Y. It is important that N would be generated by a terminal saved in protected memory of A and, moreover, the operator would not have a possibility to entry N. Otherwise the imposter could proceed as follows.

He installs $y(N')$ instead of N' in the terminal, receives $E(k_S, y(N'))$ and enciphers it and sends to the user B instead of $E(k_S, N')$. So, in the memory of the user there is now $E(k_S, y(N'))$ instead of the true record $E(k_S, y(N))$. Further, the imposter sends N' to the user B and succeeds in checking of authenticity.

To prevent such a kind of attacks one uses some number R_N stored in a protected memory instead of randomly generated N. To check authenticity of communication, A enciphers the number R_N by the key k_S, i.e., he takes $E(k_S, R_N)$. Then an imposter can input N' and obtain $E(k_S, N')$. Since y is a public function it is possible to obtain $y(E(k_S, N'))$. It is also possible to obtain $y(E(k_S, y(k_S, N')))$. He can replace $E(k_S, y(E(k_S, E(k_S, R_N))))$ with $E(k_S, y(E(k_S, N')))$, but he cannot replace N with N'.

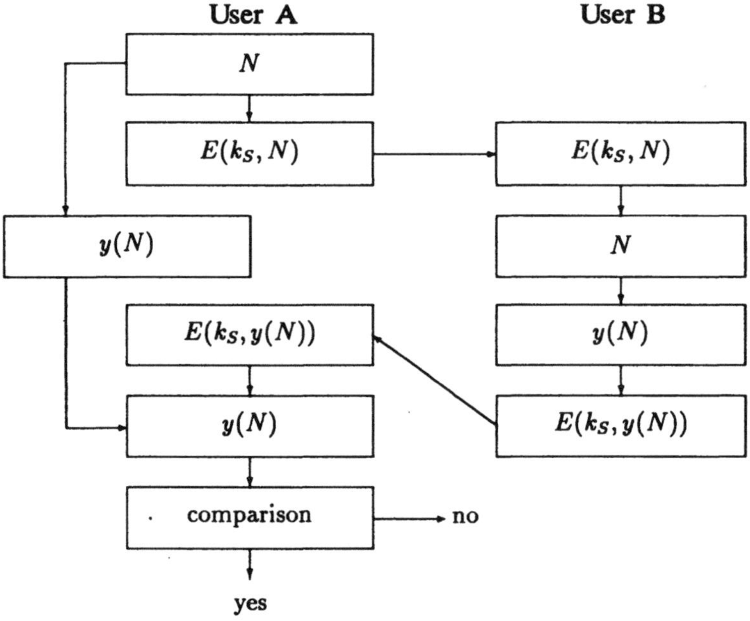

Fig. 4

4.3. Digital (electronic) signatures

According to the Uniform Commercial Code (Uniform Commercial Code, 1972) and the laws of the USA such as the Signature Law and the Agencies Law (Corley and Robert, 1971) a signature on a document has to guarantee that the document stipulating parties obligations is really signed by the given person at the given time. To this aim the third side (an arbiter, a notary and so on) is involved in the process. In (Lipton and Matyas, 1978) the legal regulations are adapted to

the case of electronic mail. A digital signature should be such that a recipient could prove the authenticity of the message to the third side and the sender. The digital signature is a function of the message, the protocol or the document that are signed, and a confidential information known to the sender only, and possibly a public information.
The transmission of a signed message from A to B can take place if A and B have the corresponding possibilities. A should have information to create a signature to each message, B should have information to establish the authenticity of the message and the evidence of the authorship of A. The sender A can demand a certification of the message, i.e., acknowledging the receipt of the message.
There exist several cryptosystems to realize digital signatures on the basis of conventional and public key algorithms. The digital signatures used in the systems where each person admitted to the system service can establish the authenticity of a sender and a message are called universal. They are easily constructed on the basis of public key algorithms. The universal signatures can be constructed on the basis of conventional algorithms also. In this case a transmission of an additional confidential or public information is needed for each message. Methods of constructing digital signatures based on the conventional algorithms providing properties of the public key algorithms are also of interest. Another approach to constructing digital signatures is to draw in the process of communication the third side which plays the role of inspector.

4.4. Protocols of universal signatures

Usage of universal signatures allows a recipient to establish the authenticity of a signature and a message independently of any other person. The third side can be involved in the process only to resolve moot points. Let us consider some protocols in detail.

4.5. Use of public key algorithm

This class of algorithms is described in (Cryptographic Algorithms, 1973; Hellman et al., 1976; Shamir, 1978). Let K_{E_i} be the key for enciphering the messages of the ith user and let K_{D_j} be the key for deciphering the messages obtained by the jth user. Then depending on the privacy of the keys the following types of communication systems are possible:

a) K_{E_i} is public and K_{D_j} is secret. In this case the communication system ensures the secrecy of a transmitted information. Any user applying the public key may transmit a message and the appropriate recipient only can decipher it.

b) K_{E_i} is secret and K_{D_j} is public. In this case the analogous communication system does not provide the privacy of the transmitted information. Any user can decipher the message and therefore to verify that the information is sent by the ith user since the ith user only could encrypt by K_{E_i}.

c) K_{E_i} and K_{E_j} are secret and K_{D_i} and K_{D_j} are public. The algorithm satisfies the condition

$$D(K_D, E(K_E, X)) = E(K_E, D(K_D, X)) = X \tag{3}$$

for all X or a subset of all messages (Hellman et al., 1976). The system provides the privacy of transmitted messages and the proof of signature authenticity.

The general scheme of transmission of a signed message is constructed on the basis of a file available to any person and containing the identifiers of users (the numbers $1, 2, \ldots$) and their encryption keys $K_{E_1}, K_{E_2}, \ldots$ Each user has a secret deciphering key $K_{D_1}, K_{D_2}, \ldots$ Besides, condition (3) holds for each pair of keys K_{E_i}, K_{D_i} and is not valid for almost all X if $K_E = K_{E_j}$, $K_D = K_{D_i}$. In spite of the fact that the public file does not contain any secret information it should be protected from destroying by other systems.
Let all messages have the form

$$M = \langle\text{sender identifier}\rangle\langle\text{recipient identifier}\rangle\langle\text{sequence of numbers}\rangle\langle\text{data}\rangle.$$

To transmit a message $M_n(i, j, n, X)$ from i to j, it is necessary to execute the following procedures:

1. The user i decipher the message M_n by the secret key K_{D_i}.
2. The user i enciphers it by the public key K_{E_j} and after these operations transmits the message of the form $E(K_{E_j}, D(K_{D_i}, X))$ to j.
3. The user j deciphering the message by the secret key K_{D_j} and enciphering it by the public key K_{E_i} obtains the message M_n.

The quantity $D(K_{D_i}, X)$ can be used, after the checking the message M_n, as an evidence of the fact that the message M_n was sent by the user i. Since the quantity K_{D_i} is known to the user i only nobody can create such a message.
Let a part of the message M_n consisting of

$$Q_n = \langle\text{sender identifier}\rangle\langle\text{recipient identifier}\rangle\langle\text{sequence of numbers}\rangle$$

contain r bits. Then the probability of correct deciphering of $D(K_{D_i}, M_n)$ by an arbitrary key different from K_{E_i} is 2^{-r}. This fact allows to accomplish the authentication of M_n by checking the value of Q_n. In the case under consideration $D(K_{D_i}, M_n)$ depends both on the message and on the secret key of the sender and can be calculated by the sender only. Besides, one can consider $D(K_{D_i}, M_n)$ as the signature since the authenticity of a message can be determined by the parameters of Q_n which are available to everybody.
In spite of the fact that the conventional block algorithms differ essentially from the public key algorithms from the point of view of encryption and decryption transformations, the cryptosystems based on them, including key control, can be constructed in such a way that the use of the conventional algorithms does not change the protocol described above. A detailed decryption of the communication scheme for the case of two users and a user and a system can be found in (Zennon et al., 1980, 1982).
We present the idea in short. Let K be a secret key and let the following keys be defined:

$$K_E = E(K_{M1}, K), \quad K_D = E(K_{M2}, K).$$

Here K_{Mi}, $i = 1, 2$, are some variants of the master key. Besides, the two cryptographic operations, encoding of data only and decoding of data only, are available:

$$ENCO\{K_E, X\} \rightarrow E(D_{K_{M1}}, K_E, X)$$

and

$$DECO\{K_D, Y\} \rightarrow D(D_{K_{M2}}, K_D, Y)$$

respectively. In this case for all X we have

$$DECO\{K_D, ENCO\{K_E, X\}\} = X$$

and $K_E \neq K_D$ for all K. It should be noted that the knowledge of K_E gives no possibility to restore K_D by virtue of security of the block algorithm itself. If we put K_E to be public and K_D to be secret we obtain all the properties of public key algorithms. The security of the system is provided while the security of the muster key of the system and the key K is ensured.

To create the keys K_E and K_D, the operation $GENKEY$ of key generation is used which has no inputs (it takes a pseudorandom numbers RN from its internal generator) and has two outputs:

$$GENKEY \rightarrow (E(K_{M1}, RN), E(K_{M2}, RN)).$$

4.6. The method of Diffie and Lumport

This algorithm is based on a conventional block algorithm (Chaum, 1979). Let signatures of n bit message must be confirmed. A sender generates in advance n pairs of keys of the cryptographic algorithm:

$$(K_{0i}, K_{1i}), \qquad i = 1, \ldots, n,$$

which are kept in secret, and provides a recipient by two sequences to verify authenticity:

$$(U_{0i}, U_{1i}), \quad (V_{0i}, V_{1i}), \qquad i = 1, \ldots, n,$$

where

$$V_{1i} = E(K_{li}, U_{li}), \quad l = 1, 2, \qquad i = 1, \ldots, , n.$$

The sequences are recorded by the sender and the recipient stores them in a non-secret file having an additional protection against destruction.

After the process described above is finished the following procedure of transmitting the signed message is used. Together with the message $M = m(1) \ldots m(n)$ the sender transmits the keys $K_{m(i)i}$ and, comparing the result with $V_{m(i)i}$, $i = 1, \ldots, n$, establishes the authenticity of the message and the sender, since nobody knows the keys K_{0i} and K_{1i}, $i = 1, \ldots, n$. Since the keys are unknown to the recipient also, he has no possibility to produce a false message.

The procedure has an essential shortcoming. It is necessary to transmit and to store a great amount of information proportional to the length of a transmitted message. One of the possibilities to diminish the shortcoming is to use compression methods.

Compression encoding (C_E) is a function which maps zero-one sequences of varying lengths into a bit sequences of a fixed length m. The compression encoding and the quantity m are chosen in such a way that the calculation of M and M' with the property $C_E(M) = C_E(M')$ is highly complex.

One of the ways to construct a function C_E by a conventional algorithm looks as follows. One chooses two nonsecret keys K_1 and K_2. For a message X consisting of blocks $X_1, \ldots, X_n$ two values U_1 and U_2 are calculated according to the rules

$$Y_i = E(K_r, X_i + Y_{i-1}), \quad Y_0 = 0, \qquad i = 1, \ldots, n,$$

$$U_r = E(K_r, X_1 + \cdots + X_n + Y_n), \qquad r = 1, 2,$$

and then we put

$$C_E(M) = (U_1, U_2).$$

4.7. Rabin's method

The method is described in (Rabin, 1978). A sender generates $2r$ cryptographic keys $K_1, \ldots, K_{2r}$ which are secret. The keys are used to generate parameters $U_1, \ldots, U_{2r}$ and $E(K_1, U_1), \ldots, E(K_{2r}, U_{2r})$ allowing to establish the authenticity. These parameters are sent to a recipient and are kept by the sender and recipient in files having a protection against destruction.

A digital signature is formed by a message M in the following way. The message is compressed to the form $C_E(M)$ and then $C_E(M)$ is enciphered by all the keys of the message. The resulting signature

$$E(K_1, C_E(M)), \ldots, E(K_{2r}, C_E(M)) \tag{4}$$

is transmitted to the recipient. The checking authenticity procedure looks as follows. The sender opens to the recipient a half of the keys and keeps secret the remaining half to prevent a possibility of a message falsification by the recipient. The r keys needed are chosen with participation of the recipient. He takes either a random sample of size r without replacement from the set of $2r$ elements or generates a random binary vector of length $2r$ containing exactly r units. The unit at the ith place means that the key K_i is chosen. The chosen keys are transmitted to the recipient and he verifies whether the values of the parameters U_i and $C_E(M)$ enciphered by the keys coincide with $E(K_i, U_i)$ and $E(K_i, C_E(M))$ respectively.

In the case that the recipient tries to ascribe a false message to the sender the last shows to the third side all the keys $K_1, \ldots, K_{2r}$ and it verifies the correspondence of all the parameters. If more than r elements of signature (4) coincide, the message is attributed to the sender. Otherwise it is considered as a falsification. To fabricate a false message the sender must specify exactly r parameters. Since the keys are inquired by the recipient the sender guesses right the places of the parameters with probability $(r!)^2/(2r)!$.

4.8. The method of Matyas and Meyer

This method (Matyas and Meyer, 1981 a, 1981 b) requires a construction of initial tables of parameters. The first table of size 31 × 31 consists of secret keys $K(i,j)$ and the second table of size 30 × 31 consists of code words $U(i,j)$ which are non-secret public parameters for checking authenticity.

These tables can be generated, for example, as follows. Suppose we are given a secret key K and a public parameter U. Then for the nth message, we construct the first table with the elements

$$U(i,j) = E(31^2(n-1) + 31(i-1) + j),$$
$$K(1,j) = E(K, 31(n-1) + j),$$
$$K(i+1,j) = E(K(i,j), U(i,j)), \qquad i = 1,\dots,30, \quad j = 1,\dots,31.$$

The keys from the 31th row of the key table are called final and are sent to a recipient in advance. The recipient can require to provide the proofs of their authenticity up to their inclusion into a public register. Besides, it is necessary to have 31 nonsecret keys $K_1,\dots,K_{31}$ which are included in a public register. Further for each message M its compression code $C_E(M)$ and the 31 code words

$$C_1 = E(K_1, C_E(M)), \dots, C_{31} = E(K_{31}, C_E(M))$$

are calculated. The code words $C_1,\dots,C_{31}$ are arranged in the increasing order in the row $C_{i(1)},\dots,C_{i(31)}$ and the numbers of keys of a digital signature are formed by the rows as the keys with the numbers $(i(1),1),\dots,(i(31),31)$ respectively. The keys located at these places are sent to the recipient who performs, excluding one step, the same steps as the sender.

The recipient calculates the compression code of the message M. Then he calculates and sorts out the code words and checks the numbers of the keys obtained and using them and the table of code words, restores the 31th row of the key matrix. At the final step he verifies the correspondence of the restored keys and the matrix of code words.

One can see that the protocols based on conventional algorithms are one-shot and require to change at least one parameter of authenticity checking while passing from one message to another. At the same time public key algorithms allow to use a single parameter, a public key of a sender, for all the messages signed.

In conclusion we have to note that the survey is mainly devoted to cryptography. The time and space restrictions did not permit us to present the results on the second part of cryptology, cryptanalysis. We hope that the information presented in the survey can help to receive a preliminary knowledge on this fascinating field of applied mathematics which poses a lot of interesting pure mathematical problems.

REFERENCES

Aho, A., Hopcroft, J., and Ullman, J. (1974). *The Design and Analysis of Computer Algorithms*. Addison-Wesley, Reading, Massachusetts.

American National Standard for PIN Management and Security. (1980). Draft Standard ANSI, Technical Committee, New York, X9, A3.

American National Standard X4.16-1980. (1980). *Magnetic Stripe Encoding for Financial Transaction Cards*. ANSI, New York, X4.

American National Standard X9.1-1980. (1980). *Magnetic Stripe Data Content Track* 3. ANSI, New York, X9.

American National Standard X3.92-1981. (1981). *Data Encryption Algorithm*. ANSI, New York, X3.

Andreassen A. L., Leighton, J., and Schrelber, D. F. (1988). Information security: an overview. *AT&T Techn. J.* **67**, No. 3, 2–8.

Barau, P. (1964). *On Distribution Communications*: IX *Security, Secrecy and Tamper-free Considerations Memo*. RRM-3765-PR. Rand Corporation, Santa Monica, CA.

Blackley, B. and Blackley, G. R. (1978). Security of number-theoretical public key cryptosystems against random attack, I. *Cryptologia* **2**, 305–321.

Blackley, B. and Blackley, G. R. (1979 a). Security of number-theoretical public key cryptosystems against random attack, II. *Cryptologia* **3**, 29–42.

Blackley, B. and Blackley, G. R. (1979 b). Security of number-theoretical public key cryptosystems against random attack, III. *Cryptologia* **3**, 105–118.

Campbell, C. M. (1979). A microprocessor-based module to provide security in electronic funds transfer systems. In: *Proceedings of COMCON'79*, pp. 148–153.

Chaum, D. Z. (1979). Untraceable electronic mail, return address and digital pseudonyms, Memo. In: *UCB/ERLM79/9*, Electronic Research Lab., Univ. of California, Berkeley.

Corley, R. N. and Robert, W. J. (1971). *Dillavou and Howard's Principles of Business Law*. Prentice-Hall, New York.

Cryptographic Algorithms for Protection of Computer Data During Transmission and Dormant Storage. (1973). Federal register 38, No. 93.

Cypser, R. J. (1978). *Communications Architecture for Distributed Systems*. Addison-Wesley, Reading, Massachusetts.

Data Encryption: Is for You? (1978). *EDP Analyzer* **16**, 1–13.

Data Encryption Standard. (1977). *FIPS Publication* **46**, NBS, U.S. Department of Commerce, Washington.

Davies, D. W., Price, W. L., and Parkin, G. I. (1979). An evaluation of public key cryptosystems. In: *NPL Report* CTU1. National Physical Laboratory, Teddington.

Deavours, G. A. (1977). Unicity points in cryptanalysis. *Cryptologia* **1**, 46–68.

Denning, D. E. (1983). *Cryptography and Data Security*. Addison-Wesley, Reading, Massachusetts.

DES Modes of Operation. (1981). *FIPS Publication* **81**, NBS, U.S. Department of Commerce, Washington.

Diffie, W. and Hellman, M. (1976). New directions in cryptography. *IEEE Transactions of Information Theory* **22**, 644–645.

Don, J. (1985). *Principles of Communication Systems*. Artech House.

Ehrsam, W. F., Matyas, S. M., Meyer, C. H., and Tuchman, W. L. (1978). A cryptographic key management scheme for implementing the DES. *IBM Systems J.* **17**, 106–125.

Ehrsam, W. F., Meyer, C. H., Powers, R. L., Smith, J. L., and Tuchman W. L. (1976). *Product Block Cipher System for Data Security*. U.S. Patent No. 3, 962, 539.

Evans, A. J., Kantrowitz, W., and Weiss, E. (1974). An user authentication system not requiring secrecy in the computer. *Communication of the ACM* **17**, 437–442.

Federal Standard 1026. (1978). *Telecommunications: Compatibility Requirements for Use of the DES General Services Administration*. Washington.

Feistel, H. (1974). *Block Cipher Cryptographic System.* U.S. Patent No. 3, 798, 359.

Garey, M. R. and Johnson, D. S. (1979). *Computers and Intractability. A Guide to the Theory of NP-Completeness.* Freeman, San Francisco.

Gasser, M. (1987). *Building a Secure Computer System.* Van Nostrand, Princeton.

Green, P. E. (1979). An introduction to network architectures and protocols. *IBM Systems J.* **18**, 202–222.

Hellman, M., Merkle, R., Schroeppel, R., Washington L., Diffie, W., Pohling, S., and Schweitzer. (1976). Results of an initial attempt to cryptanalyze the NBS DES. In: *Information Systems Laboratory Report*, Stanford Univ. Press, Stanford.

Herbst, N. M. and Liu, C. N. (1977). Automatic signature verification based on accelerometry. *IBM J. Research and Development* **21**, 245–253.

Herlestam, T. (1978). Critical remarks on some public key cryptosystems. *BIT* **18**, 493–496.

Hirsch, P. (1980). Freuch bring smart credit card to U.S. *Computerworld* **16**, 1–8.

Hoffman, L. J. (1977). *Modern Methods for Computer Security and Privacy.* Prentice-Hall, New York.

Hong, S. J., Cain, R. G., and Ostapko, D. L. (1974). MINI: A heuristic approach to logic minimization. *IBM J. Research and Development* **18**, 445–458.

IBM Research Reports **7**. (1971).

IBM Cryptographic Subsystem Concepts and Facilities. (1977). IBM Systems Library Order Number GC22-9063. IBM Corporation, Data Processing Division, White Plains, New York.

Introduction to FFT Security. (1976). Division of Management Systems and Economic Analysis. Federal Deposit Insurance Corporation, Washington.

Kahn, D. (1972). *The Codebreakers.* McMillan, New York.

Kaufman, D. and Averbach K. (1976). A secure national system for electronic funds transfer. In: *AFIPS Conference Proceedings* **46**, NCC, pp. 129–138.

Kent, S. T. (1979). Comparison of some aspects of PK and conventional cryptosystems. In: *Conference Record of 1979, ICC*, **1**, pp. 04.3.1–5.

Konheim, A. G. (1981). *Cryptography.* John Wiley, New York.

Konheim, A. G., Mack, M. N., McNeill, R. K., Tuckermann, L. B., and Waldbaum, G. (1980). The IPS cryptographic programs. *IBM Systems J.* **19**, 253–283.

Lipton, S. M. and Matyas, S. M. (1978). Making the digital signature legal- and safeguarded. *Data Communications* **7**, 41–52.

Martin, J. T. (1973). *Security, Accuracy and Privacy in Computer Systems.* Prentice–Hall, New York.

Matyas, S. M. (1974). *A Computer-Oriented Cryptanalytic Solution for Multiple Substitution Enciphering Systems.* Ph. D. Thesis, University of Iowa.

Matyas, S. M. (1979). Digital signatures—an overview. *Computer Network* **3**, 87–94.

Matyas, S. M. and Meyer, C. H. (1978). Generation distribution and installation of cryptographic keys. *IBM Systems J.* **17**, 126–137.

Matyas, S. M. and Meyer, C. H. (1981 a). Electronic signature for DES. *IBM Techn. Disclosure Bull.* **24**, 2335–2336.

Matyas, S. M. and Meyer, C. H. (1981 b). Electronic signature for DES. *IBM Techn. Disclosure Bull.* **24**, 2332–2334.

Matyas, S.M. and Meyer, C.H. (1981 c). Cryptographic authentication techniques in electronic funds transfer system. In: *Proceedings of National Electronic Conference* **35**, pp. 305–314.

Matyas, S. M., Meyer, C. H., and Tuckerman, L.B. (1980). *Method and Apparatus for Enciphering Blocks which Succeed Shot Blocks in a Key-Controlled Block-Cipher Cryptographic System.*

U.S. Patent No. 4, 229, 818.

McFayden, J. H. (1976). Systems network architectures and protocols. *IBM Systems J.* **15**, 4–23.

Merkle, R. and Hellman, M. (1978). Hiding information and receipts in trap-door knapsacks. *IEEE Transactions on Information Theory* **24**, 525–530.

Meyer, C. H. (1978). Ciphertext/plaintext and ciphertext/key dependence is number of rounds for DES. In: *AFIPS Conference Proceedings* **47**, pp. 1119–1126.

Meyer, C. H. and Matyas, S. M. (1982). *Cryptography: A New Dimension In Computer Data Security.* Jonh Wiley, New York.

Meyer, C. H., Matyas, S. M. and Zennon, R. E. (1981). Required cryptographic authentication criteria for EFT. In: *Proceedings of* 1981 *Symposium on Security and Privacy*, IEEE Computer Society, Oakland, CA, pp. 89–98.

Miller, G. L. (1976). Riemann's hypothesis and test for primality. In: *Proceedings of the 7th Annual ACM Symposium on the Theory of Computing*, New Mexico, pp. 234–239.

Minsky, M. (1967). *Computation: Finite and Infinite Machines.* Prentice-Hall, Englewood Cliffs, New York.

Needham, R. M. and Schröder, M. D. (1978). Using encryption for authentication in large networks of computers. *Communications of the ACM* **21**, 993–999.

Ore, O. (1948). *Number Theory and Its History.* McGraw Hill, New York.

Orr, W. (1980). The chip card is here, but where is it going? *ABA Banking J.* **72**, 93–95.

Personal Privacy in an Information Society. (1977). In: *The Report of the Privacy Protection Study Commission*, p. 527.

PIN Manual: A Guide to the Use of Personal Identification Numbers in Interchange. (1980). Interbank Card Association, New York.

Polland, J. M. (1974). Theorems on factorization and primality testing. *Cambridge Philosophical Soc. Proceedings* **76**, 521–528.

Privacy Act of 1974. (1974). Public Law 93-579, 5 U.S.C. 552 a(e)(10).

Proposed Federal Standard 1026. (1980). *Telecommunication: Interoperability and Security Requirements for Use of the DES in Data Communication Systems.* NCS, GSA, Washington.

Proposed Federal Standard 1027. (1980). *Telecommunication: General Security Requirements for Equipment Using DES.* CSO, NSA, Fort George.

Public Law 95-213. Title IS102, S1 Stat, 1494.

Purdy, G. B. (1974). A high security login procedure. *Communications of the ASM* **17**, 442–445.

Rabin, M. O. (1976). Probabilistic algorithms. In: *Algorithms and Complexityeightrm* , Academic Press, New York.

Rabin, M. O. (1978). *Digitized Signature in Foundations of Secure Computation.* (R.A. DeMillo et al., Eds.) Academic Press, New York.

Report Workshop on Cryptography in Support Computer Security. (1977). In: *NBSIR*, pp. 77-1291, Held at NBS, September 21–22. 1976. NBS, U.S. Department of Commerce, Washington.

Rivest, R. L. (1978 a). Remarks on a proposed cryptanalytic attack on the MIT public key cryptosystem. *Cryptologia* **2**, 62–65.

Rivest, R. L. (1978 b). Critical remarks on some public key cryptosystems by Tore Herlestam. *BIT* **19**, 1–3.

Rivest, R. L., Shamir, A. and Adleman, L. (1978). A method for obtaining digital signatures and public key cryptosystems. *Communications of the ACM* **21**, 120–126.

Schwartz, M. (1977). *Computer-Communication Network Design and Analysis.* Prentice–Hall, Englewood Cliffs, New York.

Shamir, A. (1978). *A Fast Signature Scheme.* MIT/LCM/TM-107.

Shamir, A. (1982). A polynomial time algorithm for breaking Merkle–Hellman cryptosystems. In: *Proceedings of the Weizmann Institute, Israel Appl. Math.* **17**, pp. 267–281.

Shannon, C. E. (1949). Communication theory of secrecy systems. *Bell System Techn. J.* **28**, 656–715.

Shannon, C. E. (1951). Predictions of entropy in printed English. *Bell System Techn. J.* **30**, 50–64.

Simmons, G. J. and Norris, J. N. (1977). Preliminary comments on the MIT public key cryptosystem. *Cryptologia* **1**, 406–414.

Smid, M. E. (1979). *A Key Notarization System For Computer Networks.* NBS Special Publication 500-54, Washington.

Solovay, R. and Strassen, V. (1977). A fast Monte-Carlo test to primality. *SIAM J. Computing* **6**, 84–85.

Special issue on security technology. (1983). *Computer* **16**, No. 7.

Thomas, O. T. (1980). *Funds Transfer. Research Department Working Paper.* U.S. League of Saving Associations, Chicago.

Transaction Security Products Security Module. (1980). *Darpo Reports on Banking* B-61-854-101, New York.

Unclassified Summary. (1978). *Involvement of NSA in the Development of the DES.* U.S. Senate Select Committee of Intelligence, Washington.

Uniform Commercial Code. (1972). *Official Text with Comments.* American Law Institute and National Conference of Commission on Uniform State Law.

Westin, A. F. (1968). *Privacy and Freedom.* Atheneum, New York.

Wilkes, M. V. (1972). *Time-Sharing Computer Systems.* Elsevier, Amsterdam.

Williams, H. C. and Schmid, B. (1978). *Some Remarks Concerning the MIT Public Key Cryptosystem.* Science Report No. 91, Manitoba Univ.

With Data Encryption, Scents Are Safe at IFF, DP Dialogue, Data Processing Division. (1980). *Computerworld* **14**, 95.

Zennon, R. E. (1978). Cryptography architecture for information security. *IBM Systems J.* **17**, 138–150.

Zennon, R. E. and Matyas, S. M. (1978). Cryptographic key distribution using composite keys. In: *NTC'78, Conference Record,* **2**, pp. 26.1.1–26.1.6.

Zennon, R. E. and Matyas, S. M. (1979). Unidirectional cryptographic functions using master key variants. In: *National Telecommunications Conference Record* **3**, pp. 43.4.1–43.4.5.

Zennon, R. E., Matyas, S. M., and Meyer, C. H. (1980). Cryptographic authentication method for emerging data processing applications. In: *Proceedings of INTLCOM'80*, pp. 337–341.

Zennon, R. E., Matyas, S. M., and Meyer, C. H. (1981). Cryptographic authentication of time-invariant quantities. *IEEE Transactions on Communications,* **COM-29**, 773–777.

Zennon, R. E., Matyas, S. M., and Meyer, C. H. (1982). Public-key enciphering/deciphering transformations using a conventional algorithm. *IBM Techn. Disclosure Bull.* **25**, 1241–1249. 100. 5 U.S.C. 552 (a), Sec. 3(E)(10).

Probabilistic Methods in Discrete Mathematics, pp. 128–132
V. F. Kolchin et al. (Eds.)
1993 TVP/VSP

DECOMPOSABLE STATISTICS BASED ON SPACINGS AND SPACING-FREQUENCIES OF HIGHER ORDERS

A. A. ASHURKOV
Laboratory of Mathematical Statistics, Department of Mechanics and Mathematics, Moscow State University, Vorobjevy Gory, 119899 Moscow, Russia

ABSTRACT

The decomposable statistics (DS) based on the spacings and spacing-frequencies of higher orders are considered. The DS are constructed from independent random variables with the uniform distribution on $(0,1)$. We show that it is possible to construct independent random vectors such that the distribution of the decomposable statistic is identical to the conditional distribution of the statistic based on these special random vectors.

1. INTRODUCTION

Some construction of dependent random variables are often used in the probability theory and its applications. Let us consider the sum

$$\xi_n = \sum_{k=1}^{k_n} f_{nk}(w_{n\,k}, w_{n\,k+1}, \dots, w_{n\,k+m}), \tag{1}$$

where f_{nk} are real Borel functions, the indices are considered modulo k_n and

$$w_{n1}, \dots, w_{nk_n}$$

are random variables.
Let $\theta_{n1}, \dots, \theta_{nk_n}$ be independent random variables such that the joint conditional distribution given

$$\theta_{n1} + \theta_{n2} + \dots + \theta_{nk_n} = a_n$$

is identical to the joint distribution

$$\mathcal{L}(w_{n1}, \dots, w_{nk_n}) = \mathcal{L}\Big((\theta_{n1}, \dots, \theta_{nk_n}) \mid \sum_{k=1}^{k_n} \theta_{nk} = a_n\Big).$$

In such a case, quantity (1) is called m-decomposable statistic (m-DS) if $m > 0$ and decomposable statistic (DS) if $m = 0$. The term DS was introduced by Yu. I. Medvedev (see (Medvedev, 1970)) who studied additive statistics in the

polynomial scheme with n independent trials and k_n outcomes. In that case $m = 0$, w_{nk} is the frequency ν_{nk} of outcomes k (the probability of the outcome k is $p_{nk} > 0$ in each trial), the random variables θ_{nk} has the Poisson distribution with parameter np_{nk}, and $a_n = n$, $k = 1, \ldots, k_n$.
In this scheme the χ^2-statistic may be considered as a decomposable statistic.
In the present paper we study additive statistics based on the spacings and spacing-frequencies of higher orders constructed from independent random variables with the uniform distribution on $(0,1)$.

2. MAIN RESULT

Let

$$U_1, \ldots, U_{N-1};\ V_1^1, \ldots, V_{n_1}^1; \ldots;\ V_1^t, \ldots, V_{n_t}^t$$

be $t+1$ independent samples consisting of independent random variables with uniform distribution on $(0,1)$. Let integer positive numbers r_s be such that

$$0 = r_0 < 1 \le r_1 < \cdots < r_m \le N - 1 < r_{m+1} = N.$$

Let us denote by $U_N(r)$ the rth order statistic constructed from the first sample and by $\varkappa_s^i$ the number of points of the sample $V_1^i, \ldots, V_{n_i}^i$ in the half-open interval

$$\big[U_N(r_{s-1});\ U_N(r_s)\big], \qquad s = 1, \ldots, m+1; \quad i = 1, \ldots, t;$$

$$U_N(0) = 0, \quad U_N(N) = 1.$$

Let

$$(X_s, Y_s^1, \ldots, Y_s^t), \qquad s = 1, \ldots, m+1, \tag{2}$$

be independent random vectors with $t+1$ components, the distributions of components being negative binomial:

$$\mathbf{P}\{Y_s^1 = k_s^1\} = \binom{k_s^1 + \Delta_s - 1}{k_s^1} p^{k_s^1} q^{\Delta_s},$$

$$\mathbf{P}\{Y_s^i = k_s^i \mid Y_s^j = k_s^j,\ j = 1, \ldots, i-1\}$$
$$= \binom{k_s^1 + \cdots + k_s^i + \Delta_s - 1}{k_s^i} p^{k_s^i} q^{k_s^1 + \cdots + k_s^{i-1} + \Delta_s}, \tag{3}$$

where $\Delta_s = r_s - r_{s-1}$, k_s^i ($i = 1, \ldots, t$, $s = 1, \ldots, m+1$) are non-negative numbers, and the conditional (under the conditions $Y_s^i = k_s^i$, $i = 1, \ldots, t$) distribution density of the first component is equal to the gamma-distribution density

$$f_{X_s \mid Y_s^1 \cdots Y_s^t}(x_s \mid k_s^1, \ldots, k_s^t) = \frac{\theta^{-(k_s^1 + \cdots + k_s^t + \Delta_s)}}{(k_s^1 + \cdots + k_s^t + \Delta_s - 1)!}$$

$$\times x_s^{k_s^1 + \cdots + k_s^t + \Delta_s - 1} \exp\{-x_s/\theta\}, \qquad \theta > 0. \tag{4}$$

THEOREM. *The joint distribution of the random vectors*

$$\left(N\left(U_N(r_s)-U_N(r_{s-1})\right);\ \varkappa_s^1,\dots,\varkappa_s^t\right),\qquad s=1,\dots,m+1,$$

coincide with the joint conditional distribution of the random vectors

$$(X_s;Y_s^1,\dots,Y_s^t),\qquad s=1,\dots,m+1,$$

under the condition

$$\sum_{s=1}^{m+1}X_s=N,\quad \sum_{s=1}^{m+1}Y_s^i=n_i,\qquad i=1,\dots,t. \tag{5}$$

Proof. The joint distribution density of the random variables $N(U_N(r_s)-U_N(r_{s-1}))$ is given by

$$\frac{(N-1)!}{N^{N-1}}\prod_{s=1}^{m+1}\frac{v_s^{\Delta_s-1}}{(\Delta_s-1)!},\qquad 0\le v_s\le N,\quad v_{m+1}=N-\sum_{s=1}^{m}v_s, \tag{6}$$

and

$$\mathbf{P}\left\{\varkappa_s^i=k_s^i,\ s=1,\dots,m+1\,\middle|\,N\left(U_N(r_j)-U_N(r_{j-1})\right)=v_j,\ j=1,\dots,m+1\right\}$$

$$=\frac{n_i!}{N^{n_i}}\prod_{j=1}^{m+1}v_j^{k_j^i}/k_j^i!. \tag{7}$$

On the other hand, the joint distribution of vectors (2) is

$$\prod_{s=1}^{m+1}\left[\frac{\theta^{-(k_s^1+\dots+k_s^t+\Delta_s)}}{(k_s^1+\dots+k_s^t+\Delta_s-1)!}x_s^{k_s^1+\dots+k_s^t+\Delta_s-1}e^{-x_s/\theta}p^{k_s^1}q^{\Delta_s}\frac{(k_s^1+\Delta_s-1)!}{k_s^1!(\Delta_s-1)!}\right.$$

$$\left.\times\dots\times q^{k_s^1+\dots+k_s^{t-1}+\Delta_s}p^{k_s^t}\frac{(k_s^1+\dots+k_s^t+\Delta_s-1)!}{k_s^t!(\Delta_s-1)!}\right]$$

$$=\theta^{-(n_1+\dots+n_t+N)}e^{-N/\theta}q^Np^{n_1}\dots q^{N+n_1+\dots+n_{t-1}}$$

$$p^{n_t}\prod_{s=1}^{m+1}\left(\frac{x_s^{k_s^1+\dots+k_s^t+\Delta_s-1}}{(k_s^1+\dots+k_s^t+\Delta_s-1)!(\Delta_s-1)!}\prod_{i=1}^{t}\frac{1}{k_s^i!}\right) \tag{8}$$

and

$$\mathbf{P}\left\{\sum_{s=1}^{m+1}Y_s^1=n_1\right\}\cdots\mathbf{P}\left\{\sum_{s=1}^{m+1}Y_s^t=n_t\,\middle|\,\sum_{s=1}^{m+1}Y_s^j=n_j,\ j=1,\dots,t-1\right\}$$

$$\times f_{\sum_{s=1}^{m+1}X_s\mid\sum_{s=1}^{m+1}Y_s^1\cdots\sum_{s=1}^{m+1}Y_s^t}(N\mid n_1,\dots,n_t,)$$

$$= \frac{\theta^{-(n_1+\cdots+n_t+N)}}{(n_1+\cdots+n_t+N-1)!} \exp^{-N/\theta} N^{n_1+\cdots+n_t+N-1}$$

$$\times \binom{n_1+N-1}{n_1} p^{n_1} q^N \cdots \binom{n_1+\cdots+n_t+N-1}{n_t} p^{n_t} q^{n_1+\cdots+n_{t-1}+N}. \quad (9)$$

Considering the ratio of the last parts of (8) and (9) and comparing it with the product of (6) and (7), one obtains the required result.
Note that the conditional distribution introduced in Theorem does not depend on the parameters p and θ.
The following terminology is used usually: in the case $\Delta_s = r_s - r_{s-1} > 1$ the differences

$$U_N(r_s) - U_N(r_{s-1}), \qquad s = 1, \ldots, m+1,$$

are called the uniform spacings of higher orders and $\varkappa_s^i$ are called the uniform spacing-frequencies of higher orders; in the case $\Delta_s = 1$ we have accordingly uniform spacings and spacing-frequencies.
Following the general difinition of decomposable statistic (1), we consider the quantity

$$\xi = \sum_{s=1}^{m+1} h_s\Big(N\big(U_N(r_s) - U_N(r_{s-1})\big);\ \varkappa_s^1, \ldots, \varkappa_s^t\Big), \quad (10)$$

where h_s are real Borel functions defined on the set

$$\big\{(x; y_1, \ldots, y_t) : x \geq 0,\ y_i = 0, 1, 2 \ldots;\ i = 1, \ldots, t\big\}.$$

Remark. According to Theorem, the distribution of statistic (10) is identical to the conditional distribution of the random variable

$$\xi^* = \sum_{s=1}^{m+1} h_s(X_s; Y_s^1, \ldots, Y_s^t)$$

under condition (5).
The case $t = 1$ was stadied in (Kudlaev, 1986). If $\Delta_s = 1$ for all $s = 1, \ldots, m + 1$, then statistic (10) is a statistic considered by E. M. Kudlaev (see (Kudlaev, 1978 a,b; 1985)).

COROLLARY 1 (Kudlaev, 1978; Holst, 1979). *If all functions h_s do not depend on the first argument, then ξ is a statistic for testing the homogeneity of two samples.*

Note that if $\Delta_s = 1$ for all s, then in the case $t = 1$ distribution (3) is replaced by the corresponding geometric distribution and distribution (4) is replaced by the standart exponential distributions, as was noted in (Le Cam, 1958).

COROLLARY 2 (Kudlaev, 1985). *If $\Delta_s = 1$ for all $s = 1, \ldots, m+1$, $t = 1$, and h_s does not depend on the second argument, then ξ is an additive goodness-of-fit test statistic based on the spacings.*

REFERENCES

Holst, L. (1979). Two conditional limit theorems with applications. *Ann. Statist.* **7**, 551–557.

Kudlaev, E. M. (1978 a). Asymptotic normality of statistics of some classes of goodness-of-fit and homogeneity test. *Dokl. Akad. Nauk SSSR* **240**, 534–537 (in Russian).

Kudlaev, E. M. (1978 b). Integral representations for a comparison of two samples. *Matem. Zametki* **23**, 809–817 (in Russian).

Kudlaev, E. M. (1985). Limit theorems for a class of separable statistics. *Theory Probab. Appl.* **30**, 187–191.

Kudlaev, E. M. (1986). Decomposable statistics in one-sample and two-sample schemes. In: *Statistical Methods of Estimation and Hypotheses Testing*, Perm. Gos. Univ., Perm, pp. 84–87 (in Russian).

Le Cam, L. (1958). Une theoreme sur la division d'un intervalle par des points pris au hasard. *Publs. Inst. Statist. Univ. Paris* **7**, 7–16.

Medvedev, Yu. I. (1970). Some theorems on the asymptotic distribution of the χ^2-statistic. *Dokl. Akad. Nauk SSSR* **192**, 987–989 (in Russian).

Probabilistic Methods in Discrete Mathematics, pp. 133 – 142
V. F. Kolchin et al. (Eds.)
1993 TVP/VSP

ON THE ALLOCATION PROCESSES WITH GIVEN DISTRIBUTIONS OF FREQUENCY VECTORS

S. I. CHECHOTA
TVP Laboratories, Krasikov Street 32, 117418 Moscow, Russia

ABSTRACT

Let particles be successively allocated into N labelled cells in some dependent way and let the random variable ξ_n be the number of the cell into which the nth particle falls. Let X_n be the vector of cell fillings after the allocation of the nth particle. Thus the random process $\xi = \{\xi_n, n \geq 1\}$ generates a sequence of distributions of the random vectors X_n. In the present paper an inverse problem is considered and necessary and sufficient conditions are obtained for the existence of a random allocation process ξ with distributions of the vectors X_n given in advance.
The proofs of theorems are based on the well-known result from the theory of flows in networks. As an application generalized allocation schemes are considered.

Let $(\Omega, \mathcal{F}, \mathbf{P})$ be a probability space, $\xi = \{\xi_n, \mathcal{F}_n; n \geq 1\}$ be a random process with $\mathcal{F}_n = \sigma(\xi_1, \dots, \xi_n)$, and let random variables (r.v.'s) ξ_n take values from the finite set $\mathcal{U} = \{1, \dots, N\}$, $N \geq 2$. We can interpret $\mathcal{U}$ as a set of labelled cells in which particles are thrown successively in some dependent way, and the r.v. ξ_n as the number of the cell into which a particle falls on the nth step; therefore ξ is called the *allocation process* (a.p.).
Let $e(j)$ denote the N-dimensional vector whose jth component is unit and all the others are zeros. Let

$$X_0 = (0, \dots, 0), \quad X_n = \sum_{k=1}^{n} e(\xi_k), \qquad n \geq 1.$$

The random vector X_n is the frequency vector of the elements of $\mathcal{U}$ in the sequence $\xi_1, \dots, \xi_n$, or, what is the same, the vector of the cell fillings after the allocation of the nth particle. Evidently, the sum of all components of X_n is n.
Notice that there exists a one-to-one correspondence between the sequences $\{\xi_n,\ n \geq 1\}$ and $\{X_n,\ n \geq 0\}$, and the σ-algebras $\mathcal{F}_n = \sigma(\xi_1, \dots, \xi_n)$ and $\sigma(X_0, \dots, X_n)$ coincide. Therefore, instead of the original a.p. ξ we can study the equivalent N-dimensional process $X = \{X_n, \mathcal{F}_n;\ n \geq 0\}$, where $\mathcal{F}_0 = \{\varnothing, \Omega\}$.
Let $m = (m_1, \dots, m_N)$ be a vector with non-negative integer-valued components, and $\bar{m} = m_1 + \dots + m_N$. Denote by $S_{N,n}$ the discrete simplex of vectors m satisfying $\bar{m} = n$. It is well-known (see, e.g., (Feller, 1968)) that $|S_{N,n}| = \binom{n+N-1}{N-1}$.

We set

$$q(m) = \mathbf{P}\{X_{\bar{m}} = m\}. \tag{1}$$

Evidently, for every $n \geq 0$

$$\sum_{m \in S_{N,n}} q(m) = 1. \tag{2}$$

Thus the a.p. ξ determines a non-negative function $q = q_\xi$ on the set $\mathbf{N}_0^N$. Let us consider the inverse problem. Let the function

$$q\colon \mathbf{N}_0^N \to [0,1],$$

satisfying (2) be given in advance. We would like to determine necessary and sufficient conditions on q implying the existence of an a.p. ξ such that $q_\xi = q$.

Definition. An a.p. ξ is said to be *symmetrically predictable* (s.p.) if for every $n \geq 1$, $1 \leq j, j_1, \ldots, j_n \leq N$ the conditional probability $\mathbf{P}\{\xi_{n+1} = j \mid \xi_1 = j_1, \ldots, \xi_n = j_n\}$ does not depend on permutations of the elements $j_1, \ldots, j_n$.

It is easily seen that the class of all the symmetrically predictable allocation processes (s.p.a.p.) contains as a proper subset the class of sequences of symmetrically dependent r.v.'s (Neveu, 1964) taking values from the set $\mathcal{U}$.

PROPOSITION 1. *The following conditions are equivalent:*

(1) *the a.p. ξ is s.p.,*

(2) *the process X is a simple Markov chain.*

The straightforward proof is omitted.
For every $m \in S_{N,n}$ let us define the set $F(m) \subseteq S_{N,n+1}$:

$$F(m) = \{m + e(1), \ldots, m + e(N)\};$$

and for every $M \subseteq S_{N,n}$ let

$$F(M) = \bigcup_{m \in M} F(m), \quad q(M) = \sum_{m \in M} q(m).$$

THEOREM 1.

(1) *Let ξ be an a.p. with a function $q_\xi = q$. Then the inequality*

$$q(F(M)) \geq q(M) \tag{3}$$

holds for every $n \geq 0$, $M \subseteq S_{N,n}$.

(2) *Let the function $q\colon \mathbf{N}_0^N \to \mathbf{R}_+$ satisfy (2), (3). Then there exists a s.p.a.p. ξ such that $q_\xi = q$.*

For the proof of Theorem 1 we recall some notions from the theory of flows in networks (Ford and Fulkerson, 1962) and establish an auxiliary lemma.
A network is a directed graph $G = (\mathcal{N}, \mathcal{A})$ with a finite set of vertices $\mathcal{N}$. Each directed arc $(x, y) \in \mathcal{A}$ is associated with a non-negative real number $c(x, y)$ called

the capacity of the arc (x, y); if $(x, y) \notin \mathcal{A}$, we set $c(x, y) = 0$. The set of vertices $\mathcal{N}$ is partitioned into three subsets: sources S, intermediate nodes R and sinks T, and we have two non-negative real functions $a(x)$, $x \in S$, and $b(y)$, $y \in T$. The value a(x) is called the supply at the source x, and the value $b(y)$ is called the demand at the sink y.

For any subsets V and W of $\mathcal{N}$, let (V, W) denote the set of all arcs that lead from $x \in V$ to $y \in W$; and, for any function g from $\mathcal{A}$ to reals, let

$$g(V, W) = \sum_{(x,y)\in(V,W)} g(x, y).$$

Similarly, for any function h from $\mathcal{N}$ to reals, let

$$h(V) = \sum_{x\in V} h(x).$$

A non-negative real function f on $\mathcal{A}$ is called a flow in the network $G = (\mathcal{N}, \mathcal{A})$, if it satisfies the constrains

$$f(x, \mathcal{N}) - f(\mathcal{N}, x) \leq a(x), \qquad x \in S,$$

$$f(x, \mathcal{N}) - f(\mathcal{N}, x) = 0, \qquad x \in R, \tag{4}$$

$$f(\mathcal{N}, y) - f(y, \mathcal{N}) \geq b(y), \qquad y \in T,$$

$$0 \leq f(x, y) \leq c(x, y), \qquad (x, y) \in \mathcal{A}. \tag{5}$$

A problem of network flow is called feasible if there exists at least one flow f. The following theorem due to Gale (Gale, 1957) gives necessary and sufficient conditions for a network flow problem to be feasible.

THEOREM 2 (Ford and Fulkerson, 1962). *Constrains* (4), (5) *are feasible if and only if the inequality*

$$b(T \cap \bar{V}) - a(S \cap \bar{V}) \leq c(V, \bar{V}), \qquad \bar{V} = \mathcal{N} \backslash V, \tag{6}$$

holds for every $V \subseteq \mathcal{N}$.

Let us consider one particular case and show that in this case instead of (6) we can check some simpler condition.

Assume that in a network G there are no intermediate nodes, $G = (\mathcal{N}, \mathcal{A})$ is bipartite directed graph in which all the arcs have the origin in the part S and the end in the part T, and at least one arc emanates from any $x \in S$. We also set $a(S) = b(T) = 1$ and $c(x, y) = 1$, $(x, y) \in \mathcal{A}$.

Thus, we have the following network flow problem. We are given

$$G = (\mathcal{N}, \mathcal{A}), \quad \mathcal{N} = S \cup T, \quad S \cap T = \varnothing,$$
$$a(x) \geq 0, \quad x \in S, \quad b(y) \geq 0, \quad y \in T, \tag{7}$$
$$a(S) = b(T) = 1,$$

and have to determine feasibility conditions for constrains

$$f(x, T) = a(x), \quad x \in S, \qquad f(S, y) = b(y), \quad y \in T,$$
$$0 \leq f(x, y) \leq 1, \qquad (x, y) \in \mathcal{A}. \tag{8}$$

Let us introduce the following notations

$$F(x) = \{y \in T\colon\ (x, y) \in \mathcal{A}\}, \qquad x \in S,$$
$$B(y) = \{x \in S\colon\ (x, y) \in \mathcal{A}\}, \qquad y \in T,$$
$$F(S_1) = \bigcup_{x \in S_1} F(x), \qquad S_1 \subseteq S,$$
$$B(T_1) = \bigcup_{y \in T_1} B(y), \qquad T_1 \subseteq T.$$

LEMMA 1. *Flow problem* (8) *in network* (7) *is feasible if and only if the inequality*

$$b(T_1) \leq a(B(T_1)) \tag{9}$$

holds for every $T_1 \subseteq T$.

Proof. Let $V \subseteq \mathcal{N}$. Then V can be represented in the form $V = S_1 \cup T_1$, where $S_1 = S \cap V$, $T_1 = T \cap V$, and $\overline{V} = S_2 \cup T_2$, where $S_2 = S \backslash S_1$, $T_2 = T \backslash T_1$. Feasibility condition (6) takes the form

$$b(T_2) - a(S_2) \leq c(S_1, T_2) \tag{10}$$

for every $S_1 \subseteq S$, $T_2 \subseteq T$.
Let us show that (9) and (10) are equivalent.
Let (10) hold. Choose an arbitrary $T_2 \subseteq T$ and put $S_2 = B(T_2)$. Then from the definition of the set $B(T_2)$ it follows that the set of arcs (S_1, T_2) is empty, hence, $c(S_1, T_2) = 0$, and condition (10) takes form (9).
Conversely, let (9) hold. Let $S_2 \subseteq S$, $T_2 \subseteq T$. We put $R_0 = S_2 \backslash B(T_2)$, $R_1 = B(T_2) \backslash S_2$, and consider two cases.
(a) $R_1 = \varnothing$, i.e., $B(T_2) \subseteq S_2$. Then the set of arcs (S_1, T_2) is empty, and $c(S_1, T_2) = 0$. On the other hand,

$$\begin{aligned} b(T_2) - a(S_2) &= \{b(T_2) - a(B(T_2)\} - \{a(S_2) - a(B(T_2))\} \\ &= \{b(T_2) - a(B(T_2))\} - a(R_0) \leq 0. \end{aligned}$$

(b) Let $R_1 \neq \varnothing$. We have

$$\begin{aligned} b(T_2) - a(S_2) &= \{b(T_2) - a(B(T_2))\} + \{a(B(T_2)) - a(S_2)\} \\ &= \{b(T_2) - a(B(T_2))\} + \{a(R_1) - a(R_0)\} \\ &\leq a(R_1) \leq a(S) = 1. \end{aligned}$$

On the other hand,

$$c(S_1, T_2) = c(R_1, T_2) \geq |R_1| \geq 1.$$

Lemma 1 is proved.

Remark 1. Condition (9) is equivalent to the following one: $b(F(S_1)) \geq a(S_1)$ for every $S_1 \subseteq S$.

Proof of Theorem 1. (1) The first part of Theorem 1 follows directly from the inclusion $\{X_n \in M\} \subseteq \{X_{n+1} \in F(M)\}$.
(2) For $n \geq 0$, $N \geq 2$ let us consider the network $G_{N,n}$ with the set of sources $S_{N,n}$ and the set of sinks $S_{N,n+1}$. Two vertices $m \in S_{N,n}$ and $m' \in S_{N,n+1}$ are connected with the arc (m, m') if and only if $m' = m + e(j)$ for some $j = 1, \ldots, N$. Supply and demand values are given by the function q. The capacity of each arc in $G_{N,n}$ equals one. Thus, we have the flow problem of form (7), (8). In virtue of Lemma 1 and Remark 1 the existence of a flow f in the network $G_{N,n}$ follows from (3), therefore

$$\sum_{m' \in F(m)} f(m, m') = q(m), \qquad m \in S_{N,n}, \tag{11}$$

$$\sum_{m \in B(m')} f(m, m') = q(m'), \qquad m' \in S_{N,n+1}. \tag{12}$$

Let us consider a sequence of measurable spaces $\{(E_n, \mathcal{A}_n);\ n \geq 0\}$, where $E_n = S_{N,n}$, $\mathcal{A}_n = \{M : M \subseteq S_{N,n}\}$. Notice that E_0 consists of the only point $m^{(0)} = (0, \ldots, 0)$.
For every $n \geq 0$ we define a function

$$p_{n+1}^{0\cdots n} \colon \prod_{t=0}^{n} E_t \times \mathcal{A}_{n+1} \to [0,1]$$

by the equalities

$$\begin{aligned} p_{n+1}^{0\cdots n}(m^{(0)}, \ldots, m^{(n)}; m^{(n+1)}) &= \varphi(m^{(n)}, m^{(n+1)}), \\ p_{n+1}^{0\cdots n}(m^{(0)}, \ldots, m^{(n)}; M) &= \sum_{m^{(n+1)} \in M} \varphi(m^{(n)}, m^{(n+1)}), \end{aligned} \tag{13}$$

where $M \subseteq S_{N,n+1}$, and

$$\varphi(m, m') = \begin{cases} f(m, m')/q(m), & m' \in F(m),\ q(m) > 0, \\ 1/N, & m' \in F(m),\ q(m) = 0, \\ 0, & m' \notin F(m). \end{cases} \tag{14}$$

Notice that the conditions $m' \in F(m)$ and $m \in B(m')$ are equivalent.
From condition (11) and definitions (13), (14) it follows that the function $p_{n+1}^{0\cdots n}$ is a transition probability (Neveu, 1964) for the spaces $(\prod_{t=0}^{n} E_t, \bigotimes_{t=0}^{n} \mathcal{A}_t)$ and $(E_{n+1}, \mathcal{A}_{n+1})$. Then by the theorem of Tulcéa (Neveu, 1964) there exists a unique probability $\mathbf{P}$ on the space $(\Omega, \mathcal{F}) = \prod_{t=0}^{\infty}(E_t, \mathcal{A}_t)$ such that for every measurable rectangle $\prod_t M_t$

$$\begin{aligned} \mathbf{P}\left\{\prod_t M_t\right\} = \sum_{m^{(1)} \in M_1} \varphi(m^{(0)}, m^{(1)}) \sum_{m^{(2)} \in M_2} \varphi(m^{(1)}, m^{(2)}) \cdots \\ \times \sum_{m^{(k)} \in M_k} \varphi(m^{(k-1)}, m^{(k)}), \end{aligned} \tag{15}$$

where k is such that $M_t = E_t$ for $t > k$.
Let $m \in E_n$. For the rectangle $\prod_t M_t$, where

$$M_t = \begin{cases} E_t, & t \neq n, \\ \{m\}, & t = n, \end{cases}$$

formula (15) gives

$$\begin{aligned} \mathbf{P}\{\{m\}\} &= \sum_{m^{(1)} \in F(m^{(0)})} \varphi(m^{(0)}, m^{(1)}) \sum_{m^{(2)} \in F(m^{(1)})} \varphi(m^{(1)}, m^{(2)}) \cdots \\ &\times \sum_{m^{(n-1)} \in F(m^{(n-2)})} \varphi(m^{(n-2)}, m^{(n-1)}) \varphi(m^{(n-1)}, m) \\ &= \sum_{m^{(n-1)} \in B(m)} \varphi(m^{(n-1)}, m) \sum_{m^{(n-2)} \in B(m^{(n-1)})} \varphi(m^{(n-2)}, m^{(n-1)}) \cdots \\ &\times \sum_{m^{(0)} \in B(m^{(1)})} \varphi(m^{(0)}, m^{(1)}). \end{aligned} \tag{16}$$

By induction, with the help of (12) and (16) it is easy to prove that

$$\mathbf{P}\{\{m\}\} = q(m). \tag{17}$$

Thus, we have constructed the probability space $(\Omega, \mathcal{F}, \mathbf{P})$. Let X be a random function in this space, and X_n, $n \geq 0$, be corresponding coordinate maps. Equality (17) now can be written in the form

$$\mathbf{P}\{X_n = m\} = q(m), \qquad m \in S_{N,n}.$$

From definitions (13), (14) it follows that the probability

$$\mathbf{P}\{X_n = m, X_{n+1} = m'\}, \qquad m \in S_{N,n}, \quad m' \in S_{N,n+1},$$

may be not equal to zero only if $m' = m + e(j)$ for some $j = 1, \ldots, N$.
Let $\mathcal{F}_n = \sigma(X_0, \ldots, X_n)$. We define an a.p. $\xi = \{\xi_n, \mathcal{F}_n; n \geq 1\}$ by the condition $e(\xi_n) = X_n - X_{n-1}$, $n \geq 1$. Since by definition the transition probability $p_{n+1}^{0\ldots n}$ depends only on two latter variables, the process X is a Markov process. Thus in virtue of Proposition 1 the a.p. ξ is s.p. The proof of Theorem 1 is complete.

COROLLARY 1. *For every a.p. ξ there exists an s.p.a.p. ξ' such that $q_\xi = q_{\xi'}$.*

The proof of Corollary 1 consists of the application of items (1), (2) of Theorem 1.

Remark 2. Condition (3) may be verified not for all subsets $M \subseteq S_{N,n}$, but only for the connected ones, which are defined in the following way. Let us define a graph $L_{N,n}$ on the set of vertices $S_{N,n}$: two vertices $m, \tilde{m}$ are connected by an edge in $L_{N,n}$ if and only if the Hamming distance $d(m, \tilde{m})$ equals 2, which is equivalent to the condition $F(m) \cap F(\tilde{m}) \neq \varnothing$. A subset $M \subseteq S_{N,n}$ is said to be connected if the restriction $L_{N,n} \mid M$ is a connected graph. Suppose that (3) holds for every connected M. Let K be an arbitrary subset of $S_{N,n}$, and $M_1, \ldots, M_t$ are all the connected components of the graph $L_{N,n} \mid K$. Then the sets $F(M_s)$, $s = 1, \ldots, t$ are pairwise disjoint, and

$$q(F(K)) = \sum_{s=1}^{t} q(F(M_s)) \geq \sum_{s=1}^{t} q(M_s) = q(K),$$

which is what we had to show.

Remark 3. The s.p.a.p. ξ in item (2) of Theorem 1 is determined, generally speaking, not uniquely. Indeed, let us consider (11), (12) as a system of linear equations with respect to the variables $f(m, m')$, where $m' = m + e(j)$, $j = 1, \ldots, N$. This system consists of $l_{N,n} = |S_{N,n}| + |S_{N,n+1}|$ equations with $k_{N,n} = N|S_{N,n}|$ variables. It can be easily shown that the rank of this system is equal to $(l_{N,n} - 1)$, and the dimension of the set of solutions is

$$d_{N,n} = k_{N,n} - l_{N,n} + 1 = \binom{n+N-1}{N-1}\left(N - 1 - \frac{n+N}{n+1}\right) + 1 \geq 0,$$

and $d_{N,n} = 0$ only if $N = 2$ or $n = 0$. Therefore, the s.p.a.p. ξ with the property $q_\xi = q$ is uniquely determined only for $N = 2$.

We apply Theorem 1 to studying sequences of generalized allocation schemes (Kolchin et al., 1978).

Non-negative integer-valued random vector (n. i. r. v.) $X_n = (X_1^{(n)}, \ldots, X_N^{(n)})$, for which $\overline{X}_n = \sum_{j=1}^N X_j^{(n)} = n$, is said to be a generalized allocation scheme if there exists a n.i.r.v. $Y = (Y_1, \ldots, Y_N)$ with independent components, generally not identically distributed, such that the equality

$$\mathbf{P}\{X_n = m\} = \mathbf{P}\{Y = m \mid \overline{Y} = n\}. \tag{18}$$

holds for every $m \in S_{N,n}$.

Let

$$p_t^{(j)} = \mathbf{P}\{Y_j = t\}, \qquad t \geq 0, \quad j = 1, \ldots, N,$$

$$A_j(s) = \mathbf{E}s^{Y_j} = \sum_{t=0}^{\infty} p_t^{(j)} s^t, \tag{19}$$

$$b_{k,t} = \operatorname{coef}_{s^t} \prod_{j=N-k+1}^{N} A_j(s), \qquad t \geq 0, \quad k = 1, \ldots, N. \tag{20}$$

Taking into account (1), (19), (20), we can rewrite equality (18) in the form

$$q(m) = p_m / b_{N,\overline{m}} \tag{21}$$

where $p_m = p_{m_1}^{(1)} \cdots p_{m_N}^{(N)}$.

THEOREM 3. *Let a function $q: \mathbf{N}_0^N \to \mathbf{R}_+$ have the form (21). Then the following conditions are equivalent:*

(1) *There exists a s. p. a. p. ξ such that $q_\xi = q$.*

(2) *The inequality*

$$\frac{1}{b_{k+1,n}} \sum_{t=u}^{v} p_{n-t}^{(N-k)} b_{k,t} \leq \frac{1}{b_{k+1,n+1}} \sum_{t=u}^{v+1} p_{n-t+1}^{(N-k)} b_{k,t} \tag{22}$$

holds for every $n \geq 0$, $0 \leq u \leq v \leq n$, $1 \leq k < N$.

Proof. (a) Let $N = 2$. Consider the network $G_{2,n}$. In virtue of Remark 2 we can verify condition (3) only for the connected subsets $M \subseteq S_{2,n}$. An arbitrary connected subset $M \subseteq S_{2,n}$ has the form

$$M = \{(n-u,u),(n-u-1,u+1),\dots,(n-v,v)\}, \qquad 0 \le u \le v \le n,$$

and the corresponding subset $F(M) \subseteq S_{2,n+1}$ has the form

$$F(M) = \{(n-u+1,u),\dots,(n-v+1,v),(n-v,v+1)\}.$$

In this case condition (3) can be written as

$$\sum_{t=u}^{v} q(n-t,t) \le \sum_{t=u}^{v+1} q(n-t+1,t),$$

which coincides with (22) for $k = 1$.

(b) Let the assertion of Theorem 3 be true for $2,\dots,N-1$. Let us prove it for N. Consider the network $G_{N,n}$. To construct a flow in $G_{N,n}$, let us consider some auxiliary networks of lower dimension.

We partition the simplex $S_{N,n}$ into $n+1$ strata:

$$S_{N,n} = \bigcup_{t=0}^{n} S_{N,n}^{(t)},$$

where

$$S_{N,n}^{(t)} = \{m \in S_{N,n}\colon\ m_1 = n-t\}, \qquad 0 \le t \le n.$$

Let us write m in the form $m = (m_1, \tilde{m})$, where $\tilde{m} = (m_2,\dots,m_N)$. Evidently, if $m \in S_{N,n}^{(t)}$, then $\tilde{m} \in S_{N-1,t}$; thus, each stratum $S_{N,n}^{(t)}$ may be thought of as a simplex of lower dimension $S_{N-1,t}$ with the label $m_1 = n-t$, and we can write

$$S_{N,n} = (n, S_{N-1,0}) \cup (n-1, S_{N-1,1}) \cup \dots \cup (0, S_{N-1,n}).$$

Similarly,

$$S_{N,n+1} = (n+1, S_{N-1,0}) \cup (n, S_{N-1,1}) \cup \dots \cup (0, S_{N-1,n+1}).$$

Let us consider the network $G'_{2,n}$ with the set of sources $S = \{x_0,\dots,x_n\}$ and the set of sinks $T = \{y_0,\dots,y_{n+1}\}$, where

$$x_t = (n-t, S_{N-1,t}), \qquad 0 \le t \le n,$$

$$y_t = (n-t+1, S_{N-1,t}), \qquad 0 \le t \le n+1.$$

From the source x_t two arcs lead to the sinks y_t, y_{t+1}; all the arcs have capacity one. Supply and demand values are given by the function q and have the form

$$a(x_t) = p_{n-t}^{(1)} b_{N-1,t}/b_{N,n}, \qquad 0 \le t \le n,$$

$$b(y_t) = p^{(1)}_{n-t+1} b_{N-1,t}/b_{N,n+1}, \qquad 0 \le t \le n+1.$$

By the item (a) already proved, for the existence of a flow in the network $G'_{2,n}$ it is necessary and sufficient that the inequality

$$\sum_{t=u}^{v} a(x_t) \le \sum_{t=u}^{v+1} b(y_t)$$

holds for every $0 \le u \le v \le n$, which coincides with (22) for $k = N-1$. For the flow f in the network $G'_{2,n}$ let us introduce the notations $\omega_{2t} = f(x_t, y_t)$, $\omega_{2t+1} = f(x_t, y_{t+1})$, $0 \le t \le n$.

According to the induction hypothesis, the inequalities (22) for $k = 1, \dots, N-2$ are necessary and sufficient for the existence of a flow in every network $G_{N-1,t}$ with the set of sources $S_{N-1,t}$, the set of sinks $S_{N-1,t+1}$, and supply and demand values

$$a(\tilde{m}) = p_{\tilde{m}}/b_{N-1,t}, \qquad \tilde{m} \in S_{N-1,t},$$

$$b(\tilde{m}') = p_{\tilde{m}'}/b_{N-1,t+1}, \qquad \tilde{m}' \in S_{N-1,t+1},$$

where $p_{\tilde{m}} = p^{(2)}_{m_2} \cdots p^{(N)}_{m_N}$.

Let $f_t(\tilde{m}, \tilde{m} + e(j))$ $(j = 1, \dots, N-1)$ denote a flow in the network $G_{N-1,t}$, $t = 0, \dots, n$.

We define a non-negative function h on the arcs (m, m') of the network $G_{N,n}$ as follows. Let $m = (n-t, \tilde{m}) \in S_{N,n}$. Then

$$h(m, m+e(j)) = \begin{cases} \omega_{2t} a(\tilde{m}), & j = 1, \\ \omega_{2t+1} f_t(\tilde{m}, \tilde{m} + e(j-1)), & j = 2, \dots, N. \end{cases} \tag{23}$$

By simple calculations we obtain that

$$\sum_{m' \in F(m)} h(m, m') = q(m) = p_m/b_{N,n}, \qquad m \in S_{N,n},$$

$$\sum_{m \in B(m')} h(m, m') = q(m') = p_{m'}/b_{N,n+1}, \qquad m' \in S_{N,n+1},$$

hence the function h is a flow in the network $G_{N,n}$.

Thus, it is shown that inequalities (22) are necessary and sufficient for the existence of a flow in the network $G_{N,n}$. The remaining reasoning replicates the proof of Theorem 1. Theorem 3 is proved.

COROLLARY 2. *There exists a s.p.a.p. ξ with the function*

$$q(m) = \binom{\tilde{m} + N - 1}{N-1}^{-1}.$$

Proof. Notice that for a random vector X_n uniformly distributed on $S_{N,n}$, $n \ge 0$, representation (18) holds with independent identically distributed r.v.'s Y_j having the geometric distribution (Kolchin, 1986), and the inequalities (22) are readily verified.

Remark 4. Consider definition (23) of a flow for the s.p.a.p. ξ in Corollary 2. If the partitions of the simplexes $S_{k,t}$ on every step k are done by the first component of vectors, inductively we obtain that

$$\mathbf{P}\{\xi_{n+1} = j \mid X_n = m\} = (m_j + 1)/(n + N), \qquad j = 1, \dots, N. \tag{24}$$

COROLLARY 3. *The inequality*

$$|F(M)| \geq \left(1 + \frac{N-1}{n+1}\right)|M|$$

holds for every $n \geq 0$, $N \geq 2$, $M \subseteq S_{N,n}$, *and the equality takes place only for* $M = S_{N,n}$.

The proof of Corollary 3 consists of applying of Theorem 1 to the s.p.a.p. (24).

COROLLARY 4. *Let* $n \geq 0$, $N \geq 2$, $M \subseteq S_{N,n}$,

$$l_j = \min\{m_j\colon\ m \in M\}, \qquad j = 1, \dots, N, \quad l = l_1 + \dots + l_N.$$

Then

$$|F(M)| \geq \left(1 + \frac{N-1}{n-l+1}\right)|M|.$$

Proof. Let us consider the set $M' = M - \sum_{j=1}^{N} l_j e(j)$. Evidently, $|M'| = |M|$ and $|F(M')| = |F(M)|$. Now we can apply Corollary 3 to the set $M' \subseteq S_{N,n-l}$.

Acknowledgements
The author is thankful to A. A. Grusho for the statement of the problem, and his attention and help, and to A. M. Protasov for useful discussions.

REFERENCES

Feller, W. (1968). *An Introduction to Probability Theory and its Applications*. Wiley, New York.

Ford, L. R. and Fulkerson, D. R. (1962). *Flows in Networks*. Princeton Univ. Press, Princeton.

Gale, D. (1957). A theorem on flows in networks. *Pacific J. Math.* **7**, 1073–1082.

Kolchin, V. F. (1986). *Random Mappings*. Springer, New York.

Kolchin, V. F., Sevastyanov, B. A., and Chistyakov, V. P. (1978). *Random Allocations*. Wiley, New York.

Neveu, J. (1964). *Bases Mathématiques du Calcul des Probabilités*. Masson, Paris.

Probabilistic Methods in Discrete Mathematics, pp. 143 - 147
V. F. Kolchin et al. (Eds.)
1993 TVP/VSP

STATISTICAL PROBLEMS CONCERNED WITH ESTIMATION OF THE COMPOSITION OF β-RADIOACTIVE WASTES

P. V. CHISTYAKOV, V. P. CHISTYAKOV, V. I. DOROGOV, and V. I. HANDOGIN

Mechanical Institute, Moscow State University, 119899 Moscow, Russia;
Steklov Mathematical Institute, Vavilov Street 42, 117966 Moscow GSP-1, Russia;
NPO INTEK HOLDING, Zagorodnoe sh. 8(7), 117449 Moscow, Russia;
NPO INTEK HOLDING, Zagorodnoe sh. 8(7), 117449 Moscow, Russia

ABSTRACT

The problem of determination of the composition of β-radiators mixture in radioactive wastes can be reduced to the problem of parameter estimation in polynomial schemes. For mixtures of β-radioactive nuclei, which can contain all R types of nuclei, the accuracy of usual asymptotically effective estimators is not enough.

More precise estimators can be constructed when we have the following additional information: the mixture contains only r ($r < R$) nuclei and it can be any collection from all $\binom{R}{r}$ collections. The construction of such estimators is based on looking over and minimization of $\binom{R}{r}$ expression having the form of χ^2 distribution. We suggest that estimators are asymptotically normal.

For applied problems connected with the processing and the following burial of radioactive wastes, it is essential to know the composition and concentration of radiators of the mixture. In these cases summary spectrum and activity is often used. Usually it is more simple to reveal γ-radiating nuclei because in this case energy distributions of registered particles have (with a proper choice of detector) a small variance and can be closed to degenerate distributions. Different types of the degenerate distributions correspond to different types of nuclei. For the mixture of nuclei with pure β-radiation, energy distribution of particles is continuous and nondegenerate. Therefore for decomposition of the mixture containing β-radiators a more accurate solution is required. This problem leads to the following statistical problem of parameter estimation.

Let $\Pi = \{\pi_i = (\pi_{i1}, \dots, \pi_{iN}),\ i = 1, \dots, R\}$ denote the collection of probability distributions

$$\sum_{k=1}^{N} \pi_{ik} = 1, \qquad \pi_{ik} > 0, \quad i = 1, \ldots, R, \quad k = 1, \ldots, N.$$

Suppose that the vectors of Π are linearly independent. In practice, π_{ik} is a probability that the particle generated by a nucleus of ith type is recorded in the kth energy interval.

We define the collection $\mathcal{P} = \{(p_1, \ldots, p_N)\}$ of mixtures of these distributions:

$$p_k = \sum_{i=1}^{R} \delta_i \pi_{ik}, \qquad \delta_i \geq 0, \quad \sum_{i=1}^{R} \delta_i = 1. \tag{1}$$

Let m_k ($k = 1, \ldots, N$) be a number of events in kth energy interval, $n = m_1 + \cdots + m_N$.

We have to estimate the parameters $\delta_1, \ldots, \delta_R$ given by the values $m_1, \ldots, m_N$.

In (Vatutin, Televinova, and Chistyakov, 1985) it was shown that under the usual regularity conditions the estimators of unknown parameters obtained by minimization of the modified chi-squared expression

$$Q = \sum_{k=1}^{N} \frac{(m_k - np_k)^2}{m_k}, \tag{2}$$

are asymptotically normal as $n \to \infty$ and their covariation matrix coincides with the covariation matrix of maximum likelihood estimator. Note that in the problem under consideration the minimization of (2) instead of the usual chi-squared expression leads to the linear equations for estimators.

Define $(R-1) \times (R-1)$ matrix $\mathbf{B}^* = \|b^*_{ij}\|$ by the following way:

$$b^*_{ij} = \sum_{l=1}^{N} \frac{1}{p_l} (\pi_{il} - \pi_{Rl})(\pi_{jl} - \pi_{Rl}), \qquad i, j = 1, \ldots, R-1. \tag{3}$$

Theorem 1 is easily obtained, for example, from results in (Vatutin, Televinova, and Chistyakov, 1985).

THEOREM 1. *The estimator $(\tilde{\delta}_1, \ldots, \tilde{\delta}_{R-1})$ minimizing the expression (2) satisfies the system of equations*

$$\sum_{j=1}^{R-1} c_{ij} \delta_j = f_i, \qquad i = 1, \ldots, R-1,$$

where

$$c_{ij} = \sum_{l=1}^{N} \frac{1}{m_l} (\pi_{il} - \pi_{iR})(\pi_{jl} - \pi_{Rl}), \tag{4}$$

$$f_i = -\sum_{l=1}^{N} \frac{1}{m_l} (\pi_{il} - \pi_{Rl}) \pi_{Rl}. \tag{5}$$

If the determinant $|\mathbf{B}^*|$ is not equal to 0, then the vector $(\tilde{\delta}_1 - \delta_1, \ldots, \tilde{\delta}_{R-1} - \delta_{R-1})\sqrt{n}$ is asymptotically normal as $n \to \infty$ with the zero vector of expectations and the covariation matrix inversed to the matrix $\mathbf{B}^*$ from (3).

For large R the variances of estimators are too large for actual problems. To obtain the estimators with smaller variances one can use the additional information. In applied problems under consideration the mixture can usually contain no more than r ($r < R$) radioactive nuclei but the types of these nuclei are unknown.

Let $\mathcal{P}_r$ ($\mathcal{P}_r \subset \mathcal{P}$) be the mixtures of the distributions for which exactly r parameters among the whole set $\delta_1, \ldots, \delta_R$ are not equal to zero:

$$\delta_i > 0, \quad i \in I = (i_1, \ldots, i_r); \qquad \delta_i = 0, \quad i \notin I.$$

Define $\tilde{\delta}_i^{(r)}$ by the following way. For every fixed collection $I = (i_1, \ldots, i_r)$, we find estimators $\hat{\delta}_i^{(I)}$, $i \in I$, minimizing

$$Q_I = \sum_{k=1}^{N} \frac{(m_k - np_k^I)^2}{m_k}, \tag{6}$$

where

$$p_k^I = \sum_{i \in I} \delta_i \pi_{ik}.$$

The equations for these estimators have the form (4) and (5); $(r-1) \times (r-1)$ matrix $\mathbf{B}_I^*$ is defined similarly to (3). Assume that $|\mathbf{B}_I^*| \neq 0$ for each collection I. We next find $\min_I \widehat{Q}_I$, where $\widehat{Q}_I$ is determined by (6) with p_k^I replaced by

$$\hat{p}_k^I = \sum_{i \in I} \hat{\delta}_i^{(I)} \pi_{ik}.$$

Let

$$\widehat{Q}_{I_0} = \min_I \widehat{Q}_I, \qquad I_0 = (i_1^0, \ldots, i_r^0).$$

Then

$$\tilde{\delta}_i^{(r)} = \hat{\delta}_i^{(I_0)}, \quad i \in I_0; \qquad \tilde{\delta}_i^{(r)} = 0, \quad i \notin I_0.$$

THEOREM 2. *The probability that collection I chosen from $\binom{r}{R}$ collections is the true one converges to 1 as $n \to \infty$ and the vector*

$$(\tilde{\delta}_{i_1^0}^{(r)} - \delta_{i_1^0}, \ldots, \tilde{\delta}_{i_{r-1}^0}^{(r)} - \delta_{i_{r-1}^0})\sqrt{n}$$

is asymptotically normal as $n \to \infty$ with zero expection vector and a covariation matrix inversed to $\mathbf{B}_{I_0}^*$.

The construction of the estimators for the collection of mixtures $\bigcup_{s=1}^{r} \mathcal{P}_s$ is carried out similarly.

To illustrate the above method we consider a problem of determination of the structure of the mixture which can be formed by nuclei given in Table 1.

The probability π_{lk} was calculated by the data adopted from (Kolobashkin, 1978); it was assumed that response function (the density of measured energy distribution) was Gaussian with the variance equal 0.01 of measured energy.
Values $m_1, \dots, m_N$ were simulated with $n = 10^5$ for the mixture formed by the nuclei (1, 3, 8, 9, 15), $\delta_1 = \delta_3 = \delta_8 = \delta_9 = \delta_{15} = 0.2$, with the help of a generator of random numbers. Thus we have to look over $\binom{15}{5} = 3003$ collections.
The smallest values of $\widehat{Q}_I$ are given in Table 2. Consequently $\widehat{Q}_{I_0} = \min_I \widehat{Q}_I$ and $I_0 = (1, 3, 8, 9, 15)$.
The estimators $\tilde{\delta}_i^{(5)}$ are

$$\tilde{\delta}_1^{(5)} = 0.2000; \qquad \tilde{\delta}_3^{(5)} = 0.2020; \qquad \tilde{\delta}_8^{(5)} = 0.2009;$$

$$\tilde{\delta}_9^{(5)} = 0.2023; \qquad \tilde{\delta}_{15}^{(5)} = 0.1947.$$

If we don't use the additional information $(p_1, \dots, p_N) \in \mathcal{P}_5$ then the following estimators are obtained from (4) with $R = 15$:

$$\tilde{\delta}_1^{(5)} = 0.200; \qquad \tilde{\delta}_2^{(5)} = -0.103; \qquad \tilde{\delta}_3^{(5)} = 0.151; \qquad \tilde{\delta}_4^{(5)} = -0.027;$$

$$\tilde{\delta}_5^{(5)} = -0.127; \qquad \tilde{\delta}_6^{(5)} = -0.058; \qquad \tilde{\delta}_7^{(5)} = -0.047; \qquad \tilde{\delta}_8^{(5)} = 0.203;$$

$$\tilde{\delta}_9^{(5)} = 0.158; \qquad \tilde{\delta}_{10}^{(5)} = -0.005; \qquad \tilde{\delta}_{11}^{(5)} = -0.212; \qquad \tilde{\delta}_{12}^{(5)} = 0.070;$$

$$\tilde{\delta}_{13}^{(5)} = 0.125; \qquad \tilde{\delta}_{14}^{(5)} = 0.388; \qquad \tilde{\delta}_{15}^{(5)} = 0.191.$$

In practical measurements, a background is always present. When the structure of mixture is estimated by means of β-radiation, the radiation of radioactive nuclei which have been already determined by any other method may be also regarded as the background. If we investigate the mixture taking into account the background then probabilities $\tilde{p}_k$ $(k = 1, \dots, N)$ of a polynomial scheme must be determined by the formula which differs from (1):

$$\tilde{p}_k = \left(p_k + \frac{\omega_k}{\lambda}\right) \Big/ \left(1 + \frac{\omega}{\lambda}\right),$$

where p_k are determined by (1), $\omega = \sum_{k=1}^{N} \omega_k$, ω_k is the expectation of a number of the particles recorded during the unit of time in kth energy interval, λ is the expectation of a number of all particles recorded during the unit of time excluding the background.
The theorems which are similar to Theorems 1 and 2 can be formulated. Let us give the expressions for $\tilde{c}_{ij}$, $\tilde{f}_i$ similarly to c_{ij}, f_i of (4) and (5):

$$\tilde{c}_{ij} = c_{ij}, \qquad \tilde{f}_i = -\sum_{l=1}^{N} \frac{1}{m_l} (\pi_{il} - \pi_{Rl}) \left(\pi_{Rl} - \frac{\omega_l}{\lambda}\right).$$

Let us also give the formulas connecting mixture coefficients δ_i with physical parameters

$$\delta_i = \alpha_i \mathcal{M}_i \Big/ \left(\sum_{k=1}^{R} \alpha_k \mathcal{M}_k \right),$$

where α_i is the expectation of the number of the particles generated by a nucleus of ith type during a unit of time, $\mathcal{M}_i$ is the number of radioactive nuclei of ith type in investigating value.

Table 1. Some of β-radiators

1	2	3	4	5	6	7	8
^{3}H	^{14}C	^{10}Be	^{36}Cl	^{79}Se	^{87}Rl	^{90}Sr	^{93}Zr

9	10	11	12	13	14	15
^{99}Tc	^{113m}Cd	^{135}Cs	^{204}Te	^{45}Ca	^{35}S	^{89}Sr

Table 2. The smallest $\widehat{Q}_I$

$\widehat{Q}_I$	156.37	292.77	311.54	324.12
I	1,3,8,9,15	1,7,8,9,15	1,6,7,8,15	1,7,8,13,15

REFERENCES

Vatutin, V. A., Televinova, T. M., and Chistyakov, V. P. (1985). ***Probability Methods in Physical Researches.*** Nauka, Moscow (in Russian).

Kolobashkin, V. M. (1978). ***Beta-Radiation of Fission Products.*** Atomisdat, Moscow (in Russian).

Probabilistic Methods in Discrete Mathematics, pp. 148–154
V. F. Kolchin et al. (Eds.)
1993 TVP/VSP

ON TRANSIENT PHENOMENA FOR BRANCHING MIGRATION PROCESSES

E. E. DYAKONOVA
Steklov Mathematical Institute, Vavilov Street 42, 117966 Moscow GSP-1, Russia

ABSTRACT

The behavior of a Galton–Watson process with immigration and emigration is investigated. A limit theorem is obtained under the near critical conditions.

1. INTRODUCTION

We shall study the Galton–Watson process with homogeneous migration. The model of such process can be defined in the following way.
Let $\{\xi_i^{(n)}\}_{i=1,2,\dots}^{n=0,1,\dots}$, $\{\eta_n\}_{n=0,1,\dots}$, be two independent sets of random variables on the probability space $(\Omega,\ \mathcal{F},\ \mathbf{P})$.
We assume that the $\xi_i^{(n)}$ are non-negative independent identically distributed random variables with the generating function $f(z) = \mathbf{E}z^{\xi_i^{(n)}}$.
Let us suppose that the η_n are independent identically distributed random variables and

$$\mathbf{P}\{\eta_n = k\} = p_k, \qquad k = 0, 1, 2, \dots,$$

$$\mathbf{P}\{\eta_n = -1\} = q > 0, \qquad q + \sum_{k=0}^{\infty} p_k = 1.$$

We define a stochastic process $\{Z_n\}_{n=0,1,\dots}$, by the following relations

$$Z_0 = 0, \quad Z_{n+1} = \begin{cases} \xi_1^{(n)} + \dots + \xi_{Z_n+\eta_n}^{(n)}, & Z_n + \eta_n > 0, \\ 0, & Z_n + \eta_n \le 0. \end{cases}$$

It is easily seen that $\{Z_n\}_{n=0}^{\infty}$ is a Galton–Watson process with immigration and emigration. In other words, one can consider Z_n as a population of the particles which evolve according to the Galton–Watson process with the offspring generating function $f(z)$, and at each moment $n = 0, 1, 2, \dots$ either k $(k = 0, 1, \dots)$ particles immigrate into the population with probability p_k or one particle of those that are present at time n emigrates from the population with probability q.
We note that in our model, as in the paper (Vatutin, 1977), at each moment $n = 1, 2, \dots$ only one particle can leave the population. It seems our result can be

extended to the case where at each moment $n = 1, 2, \ldots$ more than one particle have possibility to emigrate.
Analogous processes with random number of emigrating particles were studied in (Zubkov, 1974; Mitov and Yanev, 1984; Nagaev and Han, 1980; Yanev and Yanev 1991; Kaverin, 1990). However, all these papers consider the migration branching processes with constant mean of offspring particles.
Our aim is to study the Galton–Watson migration process under the near critical conditions.

2. STATEMENT OF RESULTS

In what follows we will assume that the generating function $f(z)$ belongs to the class $K = \{f(z)\}$ of generating functions satisrfying the following conditions

$$f(1-z) = 1 - mz + bz^2 + \alpha(z), \tag{1}$$

$$f(0) \geq l > 0, \tag{2}$$

where $l = \text{const}$, $|\alpha(z)| \leq \gamma(z) = o(z^2)$, $0 < c_1 \leq m < 1$, $b \leq c_2$, $c_1 = \text{const}$, $c_2 = \text{const}$.
We will suppose also that

$$\sum_{k=1}^{\infty} kp_k - q = 0, \tag{3}$$

$$\sum_{k=1}^{\infty} k^2 p_k < \infty. \tag{4}$$

Let $g(z) = P(z) + q/z$, where $P(z) = \sum_{k=0}^{\infty} p_k z^k$. We note that conditions (3), (4) imply that for the function $g(z)$ the following relation is valid

$$g(z) = 1 + A(z-1)^2 + o((z-1)^2), \qquad z \to 1-. \tag{5}$$

It is known (Yanev and Yanev, 1991) that the process $\{Z_n\}_{n=0}^{\infty}$ has a stationary distribution. Let ν be a random variable having the distribution coinciding with the stationary distribution of $\{Z_n\}$.

THEOREM. *If $m \uparrow 1$ in such a way that conditions* (1)–(4) *are valid, then*

$$\mathbf{P}\Big\{\frac{\log \nu}{\log(1/(1-m))} \leq x\Big\} \to x, \qquad x \in [0,1].$$

Proof. Let $\Psi(z) = \mathbf{E}z^{\nu} = \sum_{i=0}^{\infty} \pi_i z^i$, where $\{\pi_i\}$ is the stationary distribution of the process $\{Z_n\}$. One can easily obtain that

$$\Psi(z) = \Psi(f(z))\Big(P(f(z)) + \frac{q}{f(z)}\Big) + \pi_0 q\Big(1 - \frac{1}{f(z)}\Big). \tag{6}$$

We introduce the sequence of iterations $f_k(z) = f(f_{k-1}(z))$, $f_1(z) = f(z)$.

It follows from (6) that

$$\Psi(z) = \pi_0 q \sum_{i=1}^{\infty} \left(1 - \frac{1}{f_i(z)}\right) \prod_{k=1}^{i-1} \left(P(f_k(z)) + \frac{q}{f_k(z)}\right)$$

$$+ \prod_{k=1}^{\infty} \left(P(f_k(z)) + \frac{q}{f_k(z)}\right). \tag{7}$$

Using the condition $\Psi(0) = \pi_0$ we obtain

$$\pi_0 = \pi_0 q \sum_{i=1}^{\infty} \left(1 - \frac{1}{f_i(0)}\right) \prod_{k=1}^{i-1} g(f_k(0)) + \prod_{k=1}^{\infty} g(f_k(0)). \tag{8}$$

From (7), (8) we deduce that

$$\Psi(z) = \prod_{k=1}^{\infty} g(f_k(z))$$

$$\times \left[1 - q \sum_{i=1}^{\infty} \left(\frac{1}{f_i(z)} - 1\right) \prod_{k=i}^{\infty} g^{-1}(f_k(z))\right.$$

$$\left. \times \left(\prod_{k=1}^{\infty} g^{-1}(f_k(0)) + q \sum_{i=1}^{\infty} \left(\frac{1}{f_i(0)} - 1\right) \prod_{k=i}^{\infty} g^{-1}(f_k(0))\right)^{-1}\right]. \tag{9}$$

To complete the proof we need auxiliary statements.

LEMMA 1. *For each $\varepsilon \in (0,1)$ there exists $k_0 = k_0(\varepsilon)$ such that for m sufficiently close to 1*

$$(1 - f_{k_0}(0)) < \varepsilon$$

uniformly in $f \in K$.

Proof. Since

$$1 - f(z) = m(1-z) - b(z)(1-z)^2, \tag{10}$$

condition (2) implies $0 < d < m - 1 + l < b(0) < 1$, $d = \text{const}$ for m sufficiently close to 1 (uniformly in $f \in K$). Using this fact and relation (10) we obtain

$$1 - f(z) \le m(1-z) - d(1-z)^2 < (1-z)\left[1 - d(1-z)\right]. \tag{11}$$

From (11) it follows that for $0 < \varepsilon < 1 - z$

$$1 - f(z) < (1-z)(1 - d\varepsilon), \tag{12}$$

i.e., $1 - f_n(z) \le \max(\varepsilon, (1-d\varepsilon)^n)$, where $d < 1$. The statement of Lemma follows now from the last estimate.

LEMMA 2. *For $c_1^2 > \varepsilon > 0$ there exists $\delta = \delta(\varepsilon)$ such that for all $z \in U_\delta(1) = (1-\delta, 1]$*

$$f^-(u) \stackrel{\text{def}}{=} \frac{mu}{1+(b+\varepsilon)u} \le 1 - f(z) \le \frac{mu}{1+(b-\varepsilon)u} \stackrel{\text{def}}{=} f^+(u), \tag{13}$$

where $u = 1 - z$.

Proof. Since $b - \varepsilon > 0$ for $c_1^2 > \varepsilon > 0$, condition (1) implies

$$mu - (b+\varepsilon)u + \beta^+(u^3) \le mu - bu^2 + \alpha(u^2) = 1 - f(z) \tag{14}$$

$$\le mu - (b-\varepsilon)u + \beta^-(u^3),$$

where

$$\beta^+(u^3) = mu\left(1 + \frac{(b+\varepsilon)u}{m}\right)^{-1} - mu + (b+\varepsilon)u^2,$$

$$\beta^-(u^3) = mu\left(1 + \frac{(b-\varepsilon)u}{m}\right)^{-1} - mu + (b-\varepsilon)u^2$$

and u is sufficiently small.

Remark. From Lemma 2 it follows that for $z \in U_\delta(1)$

$$f_n^-(u) \le 1 - f_n(u) \le f_n^+(u),$$

where $f_n^-(u)$, $f_n^+(u)$, $f_n(u)$ are the nth iterations of the functions $f^-(u)$, $f^+(u)$, $f(u)$ correspondingly.

LEMMA 3. *For $f(z) = 1 - mu/(1+bu)$, $u = 1 - z$,*

$$\sum_{i=1}^{\infty}\left(\frac{1}{f_i(z)} - 1\right) \underset{m\uparrow 1}{\sim} \frac{1}{b}\log\frac{1}{1-m} \tag{15}$$

(where z is fixed),

$$\sum_{i=1}^{\infty}\left(\frac{1}{f_i(1-(1-m)^x)} - 1\right) \underset{m\uparrow 1}{\sim} \frac{1}{b}\log\frac{1}{1-m}(1-x), \qquad x \in [0,1]. \tag{16}$$

Proof. It is easy to verify that

$$1 - f_i(z) = \frac{m^i u}{1 + bu(1-m^i)/(1-m)},$$

hence

$$\frac{1-f_i(z)}{f_i(z)} = m^i u\left[1 + \frac{bu}{1-m} - m^i u\left(1 + \frac{b}{1-m}\right)\right]^{-1}.$$

Put $a = 1 + bu/(1-m)$, $D = 1 + b/(1-m)$. Since $(1/f_i(z) - 1)$ decreases in i, we obtain

$$\sum_{i=1}^{\infty}\Big(\frac{1}{f_i(z)} - 1\Big) = \int_1^{\infty} \frac{m^x u}{a - m^x uD}dx + \theta\Big(\frac{1}{f(z)} - 1\Big), \qquad 0 \le \theta \le 1. \tag{17}$$

One can easily calculate the integral in (17):

$$\int_1^{\infty} \frac{m^x u \, du}{a - m^x uD} = \Big[1 + \frac{b}{1-m}\log\frac{1}{m}\Big]^{-1}$$

$$\times \log\Big(1 + \frac{mu(b+1-m)}{(1-m)}\frac{1}{(1+u(b-m))}\Big). \tag{18}$$

Now Lemma 2 follows from (18) by virtue of the boundness of the remainder term in (17). In view of (5) and the monotonic decreasing of $g(z)$ in z the following chain of relation is valid

$$0 < \sum_{i=1}^{\infty}\Big(\frac{1}{f_i(z)} - 1\Big)\Big(1 - \prod_{k=i}^{\infty} g^{-1}(f_k(z))\Big) \le \sum_{i=1}^{\infty}\Big(\frac{1}{f_i(z)} - 1\Big)\sum_{k=i}^{\infty}\log\Big(g(f_k(z))\Big)$$

$$= \sum_{i=1}^{\infty}\Big(\frac{1}{f_i(z)} - 1\Big)\sum_{k=i}^{\infty}(1 - f_k(z))^2(1 + o(1))$$

$$\le \sum_{i=1}^{\infty}\Big(\frac{1}{f_i(0)} - 1\Big)\sum_{k=i}^{\infty}(1 - f_i(0))^2. \tag{19}$$

From Lemmas 1, 2 it follows that for every $\varepsilon > 0$ there exists i_0 such that for all

$$i > i_0 \ \sum_{k=i}^{\infty}(1 - f_i(0))^2 < \varepsilon.$$

Combining (2) and (19), we obtain

$$(1-\varepsilon)\sum_{i=1}^{\infty}\Big(\frac{1}{f_i(z)} - 1\Big) + H(z) \le \sum_{i=1}^{\infty}\Big(\frac{1}{f_i(z)} - 1\Big)\prod_{k=i}^{\infty} g^{-1}(f_k(z))$$

$$\le \sum_{i=1}^{\infty}\Big(\frac{1}{f_i(z)} - 1\Big), \tag{20}$$

where $|H(z)| \le i_0(\varepsilon)/l$ and $\varepsilon > 0$ is arbitrary. Let ε_1 be fixed. Using Lemma 2, we can find $\delta_1 = \delta(\varepsilon_1)$ such that relation (13) will be valid. Then from (14) it follows that for $z \in U_{\delta_1}(1)$

$$\frac{f_i^-(u)}{1 - f_i^-(u)} \le \Big(\frac{1}{f_i(z)} - 1\Big) \le \frac{f_i^+(u)}{1 - f_i^+(u)}.$$

In view of Lemma 1 and the obvious inequality $f(z) \geq f(0)$ for all $z \in [0,1)$ there exists $k_1 = k_1(\delta_1)$ such that $1 - f_{k_1}(\dot{z}) < \delta_1$. Thus we have

$$(\mathrm{I}) = \sum_{i=k_1}^{\infty} \frac{f_i^-(u)}{1 - f_i^-(u)} \leq \sum_{i=1}^{\infty} \left(\frac{1}{f_i(z)} - 1\right)$$

$$\leq \sum_{i=k_1}^{\infty} \frac{f_i^+(u)}{1 - f_i^+ u} + G_2 = (\mathrm{II}) + G_2, \tag{21}$$

where G_2 is bounded while the sums (I), (II) are similar to those investigated in Lemma 3.
Thus for all $z \in U_{\delta_1}$ (1)

$$\sum_{i=1}^{\infty} \left(\frac{1}{f_i(z)} - 1\right)$$

$$\leq \left[\left(1 + \frac{b-\varepsilon}{1-m}\right) \log \frac{1}{m}\right]^{-1} \log\left(1 + \frac{mu(b-\varepsilon+1-m)}{1-m} \frac{1}{1+u(b-\varepsilon+1-m)}\right)$$

$$+\mathrm{const}, \qquad u = 1 - z, \tag{22}$$

$$\sum_{i=1}^{\infty} \left(\frac{1}{f_i(z)} - 1\right)$$

$$\geq \left[\left(1 + \frac{b+\varepsilon}{1-m}\right) \log \frac{1}{m}\right]^{-1} \log\left(1 + \frac{mu(b+\varepsilon+1-m)}{1-m} \frac{1}{1+u(b+\varepsilon-m+1)}\right). \tag{23}$$

From Lemma 1 it follows that

$$\sum_{i=1}^{\infty} \left(\frac{1}{f_i(z_0)} - 1\right) \leq \sum_{i=1}^{\infty} \left(\frac{1}{f_i(0)} - 1\right) \leq \sum_{i=1}^{\infty} \left(\frac{1}{f_i(z_0)} - 1\right) + \frac{N_0}{l}, \tag{24}$$

where N_0 is such that for all $N > N_0$

$$f_N(0) > f_{N_0}(0) = z_0 > 1 - \delta_1.$$

We introduce the notation

$$J(z) = \frac{q \sum_{i=1}^{\infty} \left(f_i(z)^{-1} - 1\right) \prod_{k=i}^{\infty} g^{-1}(f_k(z))}{\left(\prod_{k=1}^{\infty} g(f_k(0))\right)^{-1} + q \sum_{i=1}^{\infty} \left(f_i(0)^{-1} - 1\right) \prod_{k=1}^{\infty} g^{-1}(f_k(0))}.$$

From (22), (23), (24), (20) we deduce

$$\limsup_{m \uparrow 1} J\left(1 - (1-m)^x\right) \leq \frac{(1-x)(b+\varepsilon_1)}{b - \varepsilon_1},$$

$$\liminf_{m \uparrow 1} J\left(1 - (1-m)^x\right) \geq \frac{b-\varepsilon}{b+\varepsilon}(1-\varepsilon)(1-x).$$

Combining the estimates and letting ε_1 and ε tend to zero we obtain

$$\lim_{m\uparrow 1} J\big(1-(1-m)^x\big) = 1-x. \tag{25}$$

Since $\lim_{z\uparrow 1}\prod_{k=1}^{\infty} g\big(f_k(z)\big) = 1$ it follows from (25) that

$$\Psi\big(1-(1-m)^x\big) \underset{m\uparrow 1}{\rightarrow} x, \qquad x\in[0,1] \tag{26}$$

The statement of Theorem follows from (26).

REFERENCES

Kaverin, S. B. (1990). A refinement of limit theorems for critical branching processes with emigration. *Theory Probab. Appl.* **35**, 352–358.

Mitov, K. V. and Yanev, N. M. (1984). Critical Galton–Watson processes with decreasing state-dependent immigration. *J. Appl. Probab.* **21**, 226–239.

Nagaev, S. V. and Han, L. V. (1980). Limit theorems for a critical Galton–Watson branching process with migration. *Theory Probab. Appl.* **25**, 514–525.

Vatutin, V. A. (1977). A critical Galton–Watson branching process with emigration. *Theory Probab. Appl.* **22**, 465–481.

Yanev, G. P. and Yanev, N. M. (1991). On a new model of branching migration processes. *Comp. Rend. Acad. Sci. Bulgaria* **44**, 19–22.

Zubkov, A. M. (1974). Analogies between Galton–Watson processes and φ-branching processes. *Theory Probab. Appl.* **19**, 309–329.

Probabilistic Methods in Discrete Mathematics, pp. 155 – 185
V. F. Kolchin et al. (Eds.)
1993 TVP/VSP

ASYMPTOTIC BEHAVIOUR OF WAITING TIMES IN A SCHEME OF PARTICLE ALLOCATION

N. J. ENATSKAYA and E. R. KHAKIMULLIN
Moscow Institute of Electronic Engineering, B. Vuzovskii per. 3/12, 109028 Moscow, Russia

ABSTRACT

We consider N cells divided into $k+1$ groups. Let N_i be the number of cells in the ith group, $M = \sum_{i=1}^{k} N_i \leq N$, $i = 1, \dots, k$. The problem concerns a sequential allocation of particles into the cells, first we allocate m particles into the cells under assumption that all C_n^m allocations have the same probability. Then we continue the same process with the second set of m particles an so on. Among these $k+1$ groups we are interested in the behaviour of the first k groups of cells only. Let $t_1, \dots, t_k$ be non-negative integers (levels). The process of allocating particles lasts up to the time when the numbers of particles at least in r groups ($r = 1, \dots, k$) exceed the corresponding levels.
Let $\nu^{(r)} = \nu^{(r)}(t_1, \dots, t_k)$ be the mentioned random stopping time. The first part of the present paper contains a survey of results on the limit distributions and rates of convergence for the waiting time in the case $k = 1$, $N \to \infty$.
The results about asymptotic behaviour of the random vector $\nu_{1,k} = (\nu^{(1)}, \nu^{(k)})$ are given in the second part for the case $2 \leq k < \infty$, $N \to \infty$.

INTRODUCTION

We consider N cells divided into $k+1$ groups; let N_i be the number of cells in the ith group, $i = 1, \dots, k$,

$$M = \sum_{i=1}^{k} N_i \leq N.$$

The problem concerns a sequential allocation of particles into the cells. First, we allocate m particles into the cells under assumption that all C_n^m allocations have the same probability. Then we repeat the same process for the second set of m particles an so on.
Among these $k+1$ groups we are interested in the behaviour of the first k groups of cells only.
Let $t_1, \dots, t_k$ be non-negative integers (levels). The process of allocating particles lasts up to the time when the numbers of particles, at least in r groups exceed the corresponding levels. Let $\nu^{(r)} = \nu^{(r)}(t_1, \dots, t_k)$ be the mentioned random stopping time, $r = 1, \dots, k$.

We study the asymptotic behaviour of the random vector $\nu_{1,k} = (\nu^{(1)}, \nu^{(k)})$ as $N \to \infty$ under some conditions on the parameters of the scheme.
The investigation of the problem began in (Ivchenko and Medvedev, 1966) with the case $M = N$, $k = 1$. The limit distribution and the rate of convergence were considered in (Enatskaya and Khakimullin, 1989, 1990); asymptotics for $\nu_{1,k}$ were studied in (Enatskaya and Khakimullin, 1991). In the present paper we study the problem for $k > 1$ and $N \to \infty$ in the case where the numbers N_i, $i = 1, \dots, k$, have the same order as N.
We give also a brief survey of the previous results correcting some misprint in (Enatskaya and Khakimullin, 1989, 1990).

1. LIMIT DISTRIBUTIONS AND RATES OF CONVERGENCE FOR THE WAITING TIME IN THE CASE $k=1$

1.1. *Notation*

Let $[z]$ ($\{z\}$) be the integral (fractional) part of the number z, and let $\langle z \rangle$ be the minimal integer which is not less than z. Put

$$b(r,n,p) = C_n^r p^r (1-p)^{n-r}, \quad B(r,n,p) = \sum_{i=0}^{r} b(i,n,p), \quad 0 \le p \le 1, \ r = 0, \dots, n;$$

$$\pi_n(\lambda) = \lambda^n e^{-\lambda}/n!, \quad \Pi_r(\lambda) = \sum_{i=0}^{r} \pi_i(\lambda), \qquad \lambda > 0, \ n = 0, 1, 2, \dots;$$

$$\rho_\lambda(\xi) = \frac{1}{2} \sum_{j=0}^{\infty} \left| \mathbf{P}\{\xi = j\} - \pi_j(\lambda) \right|, \qquad \lambda > 0;$$

$$\varphi(z) = \frac{e^{-z^2/2}}{\sqrt{2\pi}}, \quad \Phi(x) = \int_{-\infty}^{x} \varphi(z)\, dz, \quad \overline{\Phi}(x) = 1 - \Phi(x);$$

$$l^+ = [\log N / (\log N - \log (N - m))] \,.$$

Let l_0 be an integer from $[l^+ - 1, l^+ + 1]$ and let $\mu_0(n, L)$ be the number of empty cells among L cells after allocation n sets of particles; put

$$w(t,n) = (M - t - \mathbf{E}\,\mu_0(n,M)) / \sqrt{\operatorname{Var} \mu_0(n,M)},$$

$$u(t,l,\beta) = (M - (1-\beta)^l - t)) / \sqrt{M},$$

where $0 < \beta < 1$ and l is a positive integer;

$$w = w(x) = \left[(x + \log M) / \log (N/(N-m)) \right]; \qquad \lambda_l = M(1 - m/N)^{w+l},$$

$l = 0, 1, 2, \ldots$; c, c_i are positive constants, $i = 1, 2, \ldots$; if $k = 1$, $\nu = \nu(t) = \nu^{(1)} = \nu^{(k)}$ where $t = t_1$;

$$A = \log(1 - t/M)/\log(1 - m/N);$$

$$B^2 = t(1 - t/M) - (M - t)^2 A/(N(N - m)), \qquad 1 \le m \le N - 1.$$

1.2. *Results*

If M is fixed ($m = o(N)$), then the limit distribution of the normalized random variable (r.v.) ν is continiuos. If the size of series is large, more precisely, if $m/N \to \gamma$, $0 < \gamma \le 1$, then the r.v. ν has a discrete limit distribution concentrated on a countable set of points. If $\gamma = 1$, the limit distribution degenerates in the point 1, i.e. $\mathbf{P}\{\nu(t) = 1\} \to 1$.

We make these assertions more precise in Theorem 1.

The case $M \to \infty$ is more complicated becáuse of much more variants of interrelations between the scheme parameters. The limit distributions for this case may be represented as combinations of Poisson ones (Theorem 2.1) as well as combinations of discrete distributions concentrated on countable sets (Theorem 2.2). In some cases the supports of the limit distribution contain no more than two points (Theorems 3–6). For the majority of rest situations, the limit distributions are normal.

THEOREM 1. *Let $N \to \infty$, M be a fixed integer and $m/N \to \gamma$. Then the following assertions are valid:*

1) *if $\gamma = 0$, then for any fixed $x > 0$*

$$\mathbf{P}\{(m/N)\nu(t) \le x\} - B(M - t, M, e^{-x}) = te^{-x} b(M - t, M, e^{-x})/(2N(1 - e^{-x}))$$

$$\times \big(m(x - 2\{xN/m\} + (x/(1 - e^{-x}))(M - t - e^{-x}(M - 1))\big) + O(m^2/N^2);$$

2) *if $0 < \gamma < 1$, then for any natural fixed l*

$$\mathbf{P}\{\nu(t) \le l\} - B(M - t, M, p) = \big(lmt/(2N(N - m))\big)(M - t - p(M - 1))$$

$$\times b(M - t, M, p)/(1 - p)^2 + O(1/N^2),$$

where $p = (1 - m/N)^l$;

3) *if $\gamma = 1$, $s = N - m$, then*

$$\mathbf{P}\{\nu(t) = 1\} = \begin{cases} 1 & \text{for } 1 \le s \le M - t, \\ 1 - \binom{M}{t-1} \frac{s!}{(s-M+t-1)!} \left(\frac{1}{N}\right)^{M-t+1} \\ \quad + O\left(N^{-(M-t+2)}\right) & \text{for fixed } s > M - t, \\ 1 - \binom{M}{t-1} \left(\frac{s}{N}\right)^{M-t+1} + O\left(\frac{1}{s}\left(\frac{s}{N}\right)^{M-t+1}\right) & \text{for } s \to \infty. \end{cases}$$

THEOREM 2. 1) *Let $N,\ M \to \infty,\ m/N \to \gamma,\ M - t = s,$ where s is a fixed number. Then the following assertions are valid.*

a) *If $\gamma = 0$, then for any fixed x*

$$\left|\mathbf{P}\{\nu(M-s)\log(N/(N-m)) - \log M \le x\} - \Pi_s(\lambda)\right|$$

$$\le 2\lambda_0 \min\left\{\frac{\lambda_0}{M} + \frac{\lambda_0 N w m}{(N-1)^2(N-m)},\ \left(1 - \frac{m(N-1-Mw)}{(N-1)^2}\right)e^{-\frac{m(w-1)}{N-1}}\right\};$$

b) *If $0 < \gamma < 1$, then for any fixed integer l*

$$\left|\mathbf{P}\{\nu(M-s) - w \le l\} - \Pi_s(\lambda_l)\right|$$

$$\le 2\lambda_l \min\Big\{\lambda_l/M + \lambda_l N(w+l)m/((N-1)^2(N-m)),$$

$$(1 - m(N-1-M(w+l))/(N-1)^2)\exp\big(-m(w+l-1)/(N-1)\big)\Big\}.$$

2) *Let $N \to \infty,\ M/N \to 1,\ m/\sqrt{N} \to 0,\ t/\sqrt{N} \to \alpha,\ (N-M)\sqrt{N} \to \beta.$ Then, for any fixed integer $a \ge 0$,*

$$\left|\mathbf{P}\{\nu(t) = \langle t/m\rangle + a\} - \Pi_{(a+1)m-s}(\Lambda) + \Pi_{am-s+1}(\Lambda)\right| \le 2\rho + R,$$

where s is defined by the relations $t \equiv s \pmod m,\ 1 \le s < m,$ and $\lambda = \alpha\beta + \alpha^2/2$,

$$\Lambda = M(1 - m/N)^{\langle t/m\rangle + a} + t + am - s - M = \lambda + o(1),$$

$$\rho = \rho_\Lambda\big(\mu_0(\langle t/m\rangle + a - 1, M) - M + t + am - 1\big),$$

$$R = \sum_{i=1}^{m} \pi_{am-s+1}(\Lambda)$$

$$\times\left(\left(1 - \binom{N-M+t+1}{m-j}\binom{M-t+1}{j}\binom{N-M+t+1}{m-j}\binom{N}{m}^{-1}\right)\right)^{-1}.$$

Put $c(m) = (\alpha+\beta)m\pi_{(a+1)m-s}(\lambda)$.

COROLLARY 1. *Under the conditions of Theorem 2.2*

$$\rho \le \mathbf{E}\mu_0((\langle t/m\rangle + a - 1), M) - M + t + am - s - \operatorname{Var}\mu_0((\langle t/m\rangle + a - 1), M)$$

$$= \alpha(\lambda + \beta/3)/\sqrt{N} + o(1/\sqrt{N}),$$

$$\rho \leq \Lambda^2\Big(1/M + N(t+am-s)/((N-m)(N-1)^2)\Big).$$

$R = c(m)/\sqrt{N} + o(m/\sqrt{N})$ *if* m *is fixed or if* $m \to \infty$ *but* $m-s$ *is fixed;* $R = o(1/\sqrt{N})$ *if* $m - s \to \infty$.

THEOREM 3. *Let* $N, M \to \infty$, $s = N - m$ *be a natural fixed number. Then*

$$\mathbf{P}\{\nu(t) = 1\} = 1 \quad if \quad 1 \leq t \leq M - s;$$

$$\mathbf{P}\{\nu(t) = 1\} = B(M-t, s, M/N) + O(1/N),$$

$$\mathbf{P}\{\nu(t) = 2\} = 1 - B(M-t, s, M/N) + O(1/N),$$

if $M - s < t \leq M$.

THEOREM 4. *Let* $N \to \infty$, $M/N \to p$, $0 < p < 1$, $N - m \to \infty$, $m/N \to 1$. *Then*

1) *for all* t, $1 \leq t \leq M$,

$$\mathbf{P}\{\nu(t) \leq l_0 + 2\} \to 1;$$

2) *if* $t/M < c_1 < 1$, *then there exists a constant* c *such that*

$$1 - \mathbf{P}\{\nu(t) = 1\} \leq \max\Big\{c/\sqrt{N-m},\ \overline{\Phi}(w(t,1))\Big\};$$

3) *if* $t/M \to 1$, $M - t \to \infty$, *then there exists an integer* n *such that*

a) *if* $w(t,n) \to -\infty$, $w(t,n+1) \to \infty$, *then there exists a constant* c *such that for* $1 \leq n \leq l_0 - 1$

$$1 - \mathbf{P}\{\nu(t) = n+1\} \leq \max\Big\{cN^{n/2}(N-m)^{-(n+1)/2}, \overline{\Phi}(w(t,n+1)), \Phi(w(t,n))\Big\},$$

b) *if* $w(t,n) \to u_0$, *where* u_0 *is a constant, then there exists a constant* c *such that for* $1 \leq n \leq l_0 - 1$

$$\Big|\mathbf{P}\{\nu(t) = n\} - \Phi(w(t,n))\Big| \leq cN^{(n-1)/2}(N-m)^{-n/2},$$

$$\Big|\mathbf{P}\{\nu(t) = n+1\} - \overline{\Phi}(w(t,n))\Big| \leq \max\{cN^{n/2}(N-m)^{-(n+1)/2}, \overline{\Phi}(w(t,n+1))\},$$

$$|\mathbf{P}\{\nu(t) = l_0\} - \Phi(w(t,l_0))| \leq cN^{(l_0-1)/2}(N-m)^{-l_0/2},$$

$$\Big|\mathbf{P}\{\nu(t) = l_0 + 1\} - \overline{\Phi}(w(t,l_0))\Big| \leq c(N-m)^{-1};$$

4) *if* $M - t = s, \; s > 0$, *then there exists a constant c such that*

$$|\mathbf{P}\{\nu(t) = l_0 + 1\} - \Pi_s(\lambda)| \le c/(N - m),$$

$$|\mathbf{P}\{\nu(t) = l_0 + 2\} - 1 + \Pi_s(\lambda)| \le c(N - m)/N;$$

here $\lambda = \mathbf{E}\,\mu_0(l_0 + 1, M), \; \lambda > 0$.

5) *if* $\mathbf{E}\,\mu_0(l_0 + 1, M) \to 0$, *then there exists a constant c such that*

$$\left|\mathbf{P}\{\nu(t) = l_0\} - \Phi(w(t, l_0))\right| \le cN^{(l_0-1)/2}/(N - m)^{l_0/2},$$

$$\left|\mathbf{P}\{\nu(t) = l_0 + 1\} - \overline{\Phi}(w(t, l_0))\right|$$

$$\le \max\{cN^{(l_0-1)/2}/(N - m)^{l_0/2}, \; (N - m)^{(l_0+1)/2}/N^{l_0/2}\}.$$

THEOREM 5. *Let* $N \to \infty, \; m/N \to \gamma, \; 0 < \gamma < 1, \; t/M \to \beta, \; 0 < \beta < 1, \; 0 < c_1 < M/N < c_2 < 1$. *Then the following assertions are valid.*

1) *If* $\beta < \gamma$, *then there exists a constant c such that*

$$1 - \mathbf{P}\{\nu(t) = 1\} = \max\left\{c\sqrt{N/M(N - M)}, \, \overline{\Phi}(w(t, 1))\right\};$$

2) *If* $\beta \ge \gamma$ *and for some fixed natural n*

$$u(t, n, \gamma) \to -\infty, \quad u(t, n + 1, \gamma) \to u_0, \quad u(t, n + 2, \gamma) \to \infty,$$

where u_0 *is a constant, then there exists a constant c such that*

$$|\mathbf{P}\{\nu(t) = n + 1\} - \Phi(w(t, n + 1))| \le \max\left\{c/\sqrt{M}, \, \Phi(w, (t, n))\right\},$$

$$\left|\mathbf{P}\{\nu(t) = n + 2\} - \overline{\Phi}(w(t, n + 1))\right| \le \max\left\{c/\sqrt{M}, \, \overline{\Phi}(w, (t, n + 2))\right\}.$$

THEOREM 6. *Let* $N \to \infty, \; Nt/(Mm) < c, \; m/N \to 0$. *Then the following assertions are valid.*

1) *If* $Nt/(Mm)$ *is an integer, and*

a) $t^{3/2}/M \to \delta < \infty$, *then there exists a constant c such that*

$$\left|\mathbf{P}\{\nu(t) = Nt/(Mm)\} - \Phi(-t^{3/2}(1 - Mm/(Nt))/(2M))\right| \le \max\left\{cM^{-2/3},\right.$$

$$\Phi(w(t, Nt/(Mm)) - 1), \left|\Phi(w(t, Nt/(Mm)))\right.$$

$$-\Phi\big(-t^{3/2}(1-Mm/(Nt))/(2M)\big)\Big|\Big\},$$

$$\Big|\mathbf{P}\{\nu(t)=Nt/(Mm)+1\}-\overline{\Phi}\big(-t^{3/2}(1-Mm/(Nt))/(2M)\big)\Big|$$

$$\le\max\Big\{c/M^{2/3},\overline{\Phi}(w(t,Nt/(Mm)+1)),$$

$$\Big|\Phi\big(w(t,Nt/(Mm))\big)-\Phi\big(-t^{3/2}(1-Mm/(Nt))/(2M)\big)\Big|\Big\},$$

b) $t^{3/2}/M\to\infty$, *then there exists a constant* c *such that*

$$1-\mathbf{P}\{\nu(t)=Nt/(Mm)+1\}\le\max\{c/\sqrt{t},\overline{\Phi}(w(t,Nt/(Mm)+1))\}.$$

2) *If* $Nt/(Mm)$ *is not an integer*, $t/M\ge m/N$, *then there exists a constant* c *such that*

a) *for* $0\le\sqrt{t}\{Nt/(Mm)\}<c$

$$\big|\Phi\,(w\,(t,[Nt/(Mm)]))-\mathbf{P}\{\nu(t)=[Nt/(Mm)]\}\big|$$
$$\le\max\big\{c/\sqrt{t},\ \Phi\big(w(t,[Nt/(Mm)])\big)\big\},$$

$$\big|\overline{\Phi}\,(w\,(t,[Nt/(Mm)]))-\mathbf{P}\{\nu(t)=[Nt/(Mm)]+1\}\big|$$
$$\le\max\big\{c/\sqrt{t},\ \overline{\Phi}\big(w(t,[Nt/(Mm)])\big)\big\},$$

b) *for* $\min\big(\sqrt{t}\{Nt/(Mm)\},\sqrt{t}(1-\{Nt/(Mm)\}\big)\to\infty$

$$1-\mathbf{P}\{\nu(t)=[Nt/(Mm)]+1\}\le\max\big\{c/\sqrt{t},\ \overline{\Phi}\big(w(t,[Nt/(Mm)]+1)\big)\big\},$$

c) *for* $0\le\sqrt{t}(1-\{Nt/(Mm)\})<c_2<\infty$

$$\big|\Phi\,(w\,(t,\langle Nt/(Mm)\rangle))-\mathbf{P}\{\nu(t)=[Nt/(Mm)]+1\}\big|$$
$$\le\max\big\{c/\sqrt{t},\ \Phi\big(w(t,[Nt/(Mm)])\big)\big\},$$

$$\big|\overline{\Phi}\,(w\,(t,\langle Nt/(Mm)\rangle))-\mathbf{P}\{\nu(t)=[Nt/(Mm)]+2\}\big|$$
$$\le\max\big\{c/\sqrt{t},\ \overline{\Phi}\big(w(t,[Nt/(Mm)]+2)\big)\big\}.$$

THEOREM 7. *Let* N, M, $M-t$, $N-m\to\infty$ *and let one of the following conditions be fulfilled:*

$$\big\{t/M\to0,\ m/N\to0,\ Nt/(Mm)\to\infty,\ t^2/(2M)+t(N-M)/M\to\infty\big\},$$

$$\{t/M \to \gamma,\ 0 < \gamma < 1,\ m/N \to 0\},$$

$$\left\{t/M \to 1,\ M(N-m)/(N(M-t)) \to \infty\right\}.$$

Then there exists a constant c such that

$$\left|\mathbf{P}\left\{(\nu(t)-A)(M-t)\log\frac{N}{N-m}/B \le u\right\} - \Phi(w(t,n))\right| \le c/\sqrt{\operatorname{Var}\mu_0(n,M)},$$

where

$$n = \left[A + (uB/(M-t))\log\left(N/(N-m)\right)\right],$$

$$\mathbf{E}\,\mu_0(n,M) = M(1-m/N)^n,$$

$$\operatorname{Var}\mu_0(n,M) = \mathbf{E}\,\mu_0(n,M)\Bigg[1-\left(1-\frac{m}{N-1}\right)^n$$

$$+\mathbf{E}\,\mu_0(n,M)\left(\left(1-\frac{m}{(N-1)(N-m)}\right)^n-1\right)\Bigg].$$

1.3. *On the method of proofs*

We reduce the problem of establishing the limit behaviour of the r. v. $\nu(t)$ to the same problem for the r. v. $\mu_0(n,M)$ using the relation

$$\mathbf{P}\{\nu(t) \le n\} = \mathbf{P}\{\mu_0(n,M) \le M-t\}.$$

Under the conditions of Theorem 1 the distribution of the r.v $\mu_0(n,M)$ is asymptotically binomial. The asymptotics for r. v. $\mu_0(n,M)$ were analyzed by the method suggested in (Vatutin and Michailov, 1982), where the r. v. $\mu_0(n,M)$ is represented as a sum of random summands taking values 0 and 1. In (Enatskaya and Khakimullin, 1989) we demonstrated the possibility of application of this method for the cases considered in the present paper.

1.4. *Auxiliary results*

LEMMA 1. *There exists a constant c such that*

$$\sup_{-\infty<x<\infty}\left|\mathbf{P}\left\{\frac{\mu_0(n,M)-\mathbf{E}\,\mu_0(n,M)}{\sqrt{\operatorname{Var}\mu_0(n,M)}} \le x\right\} - \Phi(x)\right| \le \frac{c}{\sqrt{\operatorname{Var}\mu_0(n,M)}}.$$

The proof of this lemma is similar to the proof of the corresponding lemma from (Vatutin and Mikhailov, 1982) and we omit it.

1.5. *Proofs of theorems*

Proofs of Theorems 1–6 are given in (Enatskaya and Khakimullin, 1989, 1990).

Proof of Theorem 7. Using the relation between distributions of r. v. $\nu(t)$ and $\mu_0(n, M)$ (see Subsection 1.3) and Lemma 1 we have

$$\left|\mathbf{P}\{\nu(t) \le n\} - \Phi(w(t,n))\right|$$

$$= \left|\mathbf{P}\{\mu_0(n,M) \le M - t\} - \Phi(w(t,n))\right| \le c/\sqrt{\operatorname{Var}\mu_0(n,M)}.$$

The existence of the constant c follows from Lemma 1.
Under the conditions of Theorem 7

$$n = \left[A + (Bu/(M-t))\log\left(N/(N-m)\right)\right] \to \infty.$$

If n is chosen in such a way, then

$$\operatorname{Var}\mu_0(n,M) \to \infty, \qquad w(t,n) \to u. \tag{1}$$

As an example, we proof that conditions (1) are fulfilled if

$$t/M \to \gamma, \qquad 0 < \gamma < 1, \qquad m/N \to 0.$$

For the remaining cases the proof of the theorem is carried out similarly. It is not difficult to show that

$$B^2 = M(1-\gamma)(\gamma + (M/N)(1-\gamma)\log(1-\gamma))(1+o(1)) \to \infty,$$

$$B^2/(M-t) = O(1/\sqrt{M}) \to 0;$$

$$n = A(1 + O(B/(M-t))) = (N/m)\log(1-\gamma)(1+o(1)) \to \infty,$$

$$\mathbf{E}\,\mu_0(n,M) = M - t - uB + o(B + B^2/(M-t)),$$

$$\operatorname{Var}\mu_0(n,M) = \mathbf{E}\,\mu_0(n,M)\left(1 - (\mathbf{E}\,\mu_0(n,M)/M)\left(1 - \frac{mn}{(N-1)(N-m)}\right.\right.$$

$$\left.\left. + O\left(\frac{m^2 n}{N^2(N-m)^2}\right)\right) - \frac{nm\mathbf{E}\,\mu_0(n,M)}{(N-1)(N-m)}\left(1 + O\left(\frac{m}{N(N-m)}\right)\right)\right)$$

$$= B^2(1 + O(1/\sqrt{M})) = B^2(1+o(1)) \to \infty,$$

$$w(t,n) = (uB + o(B))/\sqrt{\operatorname{Var}\mu_0(n,M)} = u(1+o(1)).$$

For the completion of the proof, it is sufficient to see that

$$\mathbf{P}\{\nu(t) \le n\} = \mathbf{P}\left\{\frac{\nu(t) - A}{B}(M-t)\log\frac{N}{N-m} \le u\right\}.$$

2. LIMIT DISTRIBUTIONS FOR WAITING TIME $\nu_{1,k}$ IN THE CASE $k>1$

2.1. *Notation*

Let us introduce the following notation: b_i, θ_i $(i = 1, 2, \ldots)$ are positive constants,

$$t_+/N_+ = \min_{1\le i\le k} t_i/N_i, \quad t^+/N^+ = \max_{1\le i\le k} t_i/N_i,$$

$$\{(t_+, N_+), (t^+, N^+)\} \in \{(t_i, N_i),\ i = 1, \ldots, k\},$$

$$d_{i+} = N(t_+/N_+ - t_i/N_i)/\sqrt{N_+ - t_+}, \quad d_i^+ = N(t^+/N^+ - t_i/N_i)/\sqrt{N^+ - t^+},$$

$$\Delta_i = 1 - t_i/N_i, \quad \Delta_+ = 1 - t^+/N^+, \quad \Delta^+ = 1 - t_+/N_+, \quad u = \Delta_+/\Delta^+;$$

$$w^{(1)} = \left(\nu^{(1)} - \log \Delta^+/\log(1 - m/N)\right) \sqrt{N_+ - t_+} \log(N/(N-m)),$$

$$w^{(k)} = \left(\nu^{(k)} - \log \Delta_+/\log(1 - m/N)\right) \sqrt{N^+ - t^+} \log(N/(N-m)),$$

$$Q(T, n, s) = (T - \mathbf{E}\,\mu_0(n, s)) / \sqrt{\operatorname{Var} \mu_0(n, s)},$$

where T, n, s are positive integers;

$$\omega_i = \lim N_i/N, \quad i = 1, \ldots, k; \quad \omega_+ = \lim N_+/N, \quad \omega^+ = \lim N^+/N,$$

$$D_{i+} = d_{i+}\sqrt{\omega_+\omega_i} + x\sqrt{\omega_i/\omega_+},$$

$$D_i^+ = d_i^+\sqrt{\omega^+\omega_i} + y\sqrt{\omega_i/\omega^+},$$

where x and y are fixed; $s_i = N_i - t_i$, $i = 1, \ldots, k$;

$$a_l = \max_{j\in\{1,\ldots,l\}} s_{i_j}; \quad b_l = \min_{j\in\{1,2,\ldots,k\}\setminus\{i_1,\ldots,i_l\}} s_{i_j},$$

where i, $l = 1, \ldots, k$; $\bar\eta = (\eta_1, \ldots, \eta_k)$, $\eta_i = N_i - \mu_0(n, N_i)$, $i = 1, \ldots, k$;

$$a_N(n) = (1 - m/N)^n; \quad \sigma_N^2(n) = N a_N(n)(1 - a_N(n));$$

$$R(x, y) = -\frac{y + o(1)}{\sqrt{N^+ - t^+}} + \frac{x + o(1)}{\sqrt{N_+ - t_+}}.$$

We use also the notation introduced in Section 1.

2.2. *Results*

If $k > 1$, the number of parameters of the scheme increases and the number of different relations between them also increases. The variety of variants of the asymptotic behaviour (if $N \to \infty$) can be described by combinations of the following conditions:

1)
- 1a) $N_i/N \to 0$,
- 1b) $0 < c_1 < N_i/N < c_2 < 1$,
- 1c) $N_i/N \to 1$;

2)
- 2a) $m/N \to 0$,
- 2b) $0 < c_3 < m/N < c_4 < 1$,
- 2c) $m/N \to 1$;

3)
- 3a) $t_i/N \to 0$,
- 3b) $0 < c_5 < t_i/N < c_6 < 1$, for all $i = 1, \dots, k$.
- 3c) $t_i/N \to 1$

The variants 1) will be taken into account if we divide all sets of groups on three types of groups corresponding to cases 1a), 1b), 1c). Thus, it suffices to consider cases 2) and 3). The case of finite M includes variantes 1a), 2a), 2b), 2c), 3b), 3c). The case 1b) (as $M \to \infty$) where the sizes of groups are proportional to the total number of cells N looks interesting for practical applications.

THEOREM 1. *Let $N \to \infty$, $1 < M \leq c$, $m/N \to \delta$.*

1) *If x and y, $x \leq y$, are finite positive numbers and $\delta = 0$, then*

$$\mathbf{P}\left\{\nu^{(1)} \log \frac{N}{N-m} \leq x,\ \nu^{(k)} \log \frac{N}{N-m} \leq y\right\} = \prod_{i=1}^{k} B(s_i, N_i, e^{-y})$$

$$-\prod_{i=1}^{k}\left(\sum_{s_i+1 \leq \alpha_i < N_i} b(\alpha_i, N_i, e^{-x}) B(s_i, \alpha_i, e^{x-y})\right) + o(1);$$

2) *If $0 < \delta < 1$ and $n_1 < n_2$ (n_1 and n_2 are integers), then*

$$\mathbf{P}\left\{\nu^{(1)} \leq n_1,\ \nu^{(k)} \leq n_2\right\} = \prod_{i=1}^{k} B(s_i, N_i, (1-\delta)^{n_2})$$

$$-\prod_{i=1}^{k}\left(\sum_{s_i+1 \leq \alpha_i < N_i} b(\alpha_i, N_i, (1-\delta)^{n_1}) B(s_i, \alpha_i, (1-\delta)^{n_2-n_1})\right) + o(1);$$

3) *If $\delta = 1$, then*

$$\mathbf{P}\left\{\nu^{(1)} \leq 1,\ \nu^{(k)} \leq 1\right\} = \mathbf{P}\{\nu^{(k)} = 1\} = 1 + o(1).$$

THEOREM 2. *Let $N \to \infty$, $N_i/N \to \omega_i$, $0 < \omega_i < 1$, $i = 1, \dots, k$; $m/N \to \delta$, and let there exist a natural l and a set of indices $i_1, \dots, i_l$ such that $a_l < \infty$, $b_l \to \infty$. Then the following assertions are valid.*

1) *If $\delta = 0$, then for any positive numbers x and y, $x \leq y$,*

$$\mathbf{P}\left\{\nu^{(1)}\log\frac{N}{N-m}-\log N\le x,\ \nu^{(k)}\log\frac{N}{N-m}-\log N\le y\right\}$$

$$=\prod_{j=1}^{l}\Pi_{s_{ij}}(\omega_{i_j}e^{-y})-\prod_{j=1}^{l}\left(\sum_{s_{ij}+1\le\alpha_{ij}\le N_{ij}}\pi_{\alpha_{ij}}(\omega_{i_j}e^{-x})B(s_{i_j},\alpha_{i_j},e^{x-y})\right)+o(1). \tag{1}$$

2) *If* $0<\delta<1$, d_1, d_2 *are integers, and there exists*

$$\Delta=\lim_{N\to\infty}\left\{\log_{(1-m/N)}N\right\},$$

then

$$\mathbf{P}\left\{\nu^{(1)}+\log_{(1-m/N)}N\le d_1,\ \nu^{(k)}+\log_{(1-m/N)}N\le d_2\right\}$$

$$=\prod_{j=1}^{l}\Pi_{s_{ij}}\left(\omega_{i_j}(1-\delta)^{d_2+\Delta}\right)$$

$$-\prod_{j=1}^{l}\left(\sum_{N_{ij}\le\alpha_{ij}\le s_{ij}+1}\pi_{\alpha_{ij}}\left(\omega_{i_j}(1-\delta)^{d_1+\Delta}\right)\right)B(s_{i_j},\alpha_{i_j},(1-\delta)^{d_2-d_1})+o(1). \tag{2}$$

THEOREM 3. *Let* $N\to\infty$, $N_i/N\to\omega_i$, $0<\omega_i<1$, $i=1,\dots,k$, $\sum_{i=1}^{k}\omega_i=1$, *and let the levels* $t_1,\dots,t_k$ *be chosen in such a way that*

$$\min_{i=1,\dots,k}t_i/N_i\to1,\quad \min\left\{(N_+-t_+),\,(N^+-t^+)\right\}\to\infty,$$

$$m\max_{i=1,\dots,k}\left(\sqrt{N_+-t_+}/N,\ \sqrt{N^+-t^+}/N\right)\to0,\qquad 0\le u<c<1.$$

Then for any fixed x *and* y, $x<y$,

$$\mathbf{P}\left\{w^{(1)}<x,\ w^{(k)}<y\right\}=\prod_{i=1}^{k}\Phi(D_i^+)$$

$$-\prod_{i=1}^{k}\int_{D_{i+}}^{\infty}\varphi(u_i)\Phi\left(D_i^+/\sqrt{1-u}-u_i\sqrt{u/(1-u)}\right)du_i+o(1). \tag{3}$$

COROLLARY 1. *Let the conditions of Theorem 3 be fulfilled and the levels* $t_1,\dots,t_k$ *and sizes of groups* $N_1,\dots,N_k$ *be chosen in such a way that*

$$\Delta_i=(1/N)\left(\theta_i\psi(N)+c_i\sqrt{\psi(N)}+o\left(\sqrt{\psi(N)}\right)\right), \tag{4}$$

where

$$0 < \theta_1 = \cdots = \theta_s < \theta_{s+1} \leq \cdots \leq \theta_{k-p-1} < \theta_{k-p} = \cdots = \theta_k;$$

$$0 < c_1 \leq c_2 \leq \cdots \leq c_s, \quad c_{k-p} \leq c_{k-p+1} \leq \cdots \leq c_k, \qquad 1 \leq s \leq k-1,$$

$$0 \leq p = p(s) \leq k-2,$$

$$\psi(N) \to \infty, \quad \psi(N) = o(N).$$

Then

$$\mathbf{P}\left\{w^{(1)} < x,\ w^{(k)} < y\right\} = \prod_{i=1}^{s} \Phi\left(\frac{c_i - c_1}{\sqrt{\theta_1}}\sqrt{\omega_i} + y\sqrt{\omega_i/\omega_1}\right)$$

$$-\prod_{i=1}^{s} \int_{-\infty}^{\infty} \varphi(v_i)\Phi\left((c_i - c_1)\sqrt{\frac{\omega_i\theta_k}{\theta_1(\theta_k - \theta_1)}} + y\sqrt{\frac{\omega_i\theta_k}{\omega_1(\theta_k - \theta_1)}} - v_i\sqrt{\frac{\theta_1}{\theta_k - \theta_1}}\right) dv_i$$

$$\times \prod_{i=k-p}^{k} \overline{\Phi}\left(\frac{c_i - c_k}{\sqrt{\theta_k}}\sqrt{\omega_i} + x\sqrt{\omega_i/\omega_k}\right) + o(1). \tag{5}$$

COROLLARY 2. *Let the conditions of Theorem 3 be fulfilled, the levels $t_1, \ldots, t_k$ be chosen in such a way that $u \to 0$. Then*

$$\mathbf{P}\left\{w^{(1)} < x,\ w^{(k)} < y\right\} = \mathbf{P}\left\{w^{(1)} < x\right\}\mathbf{P}\left\{w^{(k)} < y\right\} + o(1). \tag{6}$$

COROLLARY 3. *Let under the conditions of Corollary 2 the levels $t_1, \ldots, t_k$ be chosen in such a way that (4) is fulfilled. Assume also that $\theta_1 = 0$ and the parameters $c_1, \ldots, c_s$ are such that $0 < c_1 = c_2 = \cdots = c_l < c_{l+1} \leq c_{l+2} \leq \cdots \leq c_s$ and Δ_i for $i = 1, \ldots, l$ may be represented in the form*

$$\Delta_i = (1/N)\left(c_i\sqrt{\psi(N)} + b_i\sqrt[4]{\psi(N)} + o\left(\sqrt[4]{\psi(N)}\right)\right),$$

where $\psi(N) \to \infty$, $\psi(N) = o(N)$. Then

$$\mathbf{P}\left\{w^{(1)} < x\right\} = 1 - \prod_{i=k-p}^{k} \Phi\left(\frac{c_i - c_k}{\sqrt{\theta_k}}\sqrt{\omega_i} + x\sqrt{\omega_i/\omega_1}\right) + o(1), \tag{7}$$

$$\mathbf{P}\left\{w^{(k)} < y\right\} = \prod_{i=1}^{l} \Phi\left(\frac{b_i - b_1}{\sqrt{c_1}}\sqrt{\omega_i} + y\sqrt{\omega_i/\omega_1}\right) + o(1), \tag{8}$$

and for $l = 1$

$$\mathbf{P}\left\{w^{(k)} < y\right\} = \Phi(y) + o(1). \tag{9}$$

THEOREM 4. *Let under the conditions of Theorem 3 the levels $t_1, \dots, t_k$ be chosen in such a way that $u \to 1$. Then (for any fixed x and y)*

$$\mathbf{P}\left\{w^{(1)} < x,\, w^{(k)} < y\right\} = \prod_{i=1}^{k} \Phi(D_i^+) - \prod_{i=1}^{k} \left(\Phi(D_i^+) - \Phi(D_{i+})\right) + o(1). \tag{10}$$

COROLLARY 4. *Let under the conditions of Theorem 4 the sizes of groups $N_1, \dots, N_k$ and the levels $t_1, \dots, t_k$ be chosen in such a way that*

$$\Delta_i = (1/N)\left(\theta_i \psi(N) + c_i\sqrt{\psi(N)} + o\left(\sqrt{\psi(N)}\right)\right), \tag{11}$$

where $\psi(N) \to \infty$, $\psi(N) = o(N)$. Let $0 < \theta_1 = \dots = \theta_s < \theta_{s+1} \le \dots \le \theta_{k-p-1} < \theta_{k-p} = \dots = \theta_k$, $0 < c_1 \le c_2 \le \dots \le c_s$, $c_{k-p} \le c_{k-p+1} \le \dots \le c_k$, $1 \le s \le k-1$, $0 \le p = p(s) \le k-2$. Then

$$\mathbf{P}\left\{w^{(1)} < x,\, w^{(k)} < y\right\} = \prod_{i=1}^{s} \Phi(D_i^+) - \prod_{i=1}^{s} \Phi(D_i^+) \prod_{i=k-p}^{k} \overline{\Phi}(D_{i+})(1 + o(1)). \tag{12}$$

COROLLARY 5. *Let the conditions of Corrolary 4 be fulfilled except of the conditions on sequences $\theta_1, \dots, \theta_k$ and $c_1, \dots, c_k$ for which $0 < \theta_1 = \theta_2 = \dots = \theta_k = \theta$; $c_1 \le c_2 \le \dots \le c_k$. Then*

$$\mathbf{P}\left\{w^{(1)} < x,\, w^{(k)} < y\right\} = \prod_{i=1}^{k} \Phi\left(\frac{c_i - c_1}{\sqrt{\theta}}\sqrt{\omega_i} + y\sqrt{\frac{\omega_i}{\omega_1}}\right)$$

$$-\prod_{i=1}^{k}\left(\Phi\left(\frac{c_i - c_1}{\sqrt{\theta}}\sqrt{\omega_i} + y\sqrt{\frac{\omega_i}{\omega_1}}\right) - \Phi\left(\frac{c_i - c_1}{\sqrt{\theta_k}}\sqrt{\omega_i} + x\sqrt{\frac{\omega_i}{\omega_k}}\right)\right) + o(1). \tag{13}$$

2.3. *On the method of proofs*

In the case $k > 1$, in order to analyze the distributions of r. v.'s $\nu^{(1)}$, $\nu^{(k)}$ and $\nu_{1,k}$ we use the following obvious relations:

$$\mathbf{P}\left\{\nu^{(1)} > n\right\} = \mathbf{P}\left\{\mu_0(n, N_i) > N_i - t_i,\, i = 1, \dots, k\right\}, \tag{14}$$

$$\mathbf{P}\left\{\nu^{(k)} \le n\right\} = \mathbf{P}\left\{\mu_0(n, N_i) \le N_i - t_i,\, i = 1, \dots, k\right\}, \tag{15}$$

$$\mathbf{P}\left\{\nu^{(1)} < n_1,\, \nu^{(k)} < n_2\right\} = \mathbf{P}\left\{\nu^{(k)} < n_2\right\} - \mathbf{P}\left\{\nu^{(1)} \ge n_1,\, \nu^{(k)} < n_2\right\}. \tag{16}$$

As to the connection of the r. v. $\nu_{1,k}$ with the random vectors $\bar{\mu}_0(n_1) = \{\mu_0(n_1, N_i), i = 1, \dots, k\}$ and $\bar{\mu}_0(n_2) = \{\mu_0(n_2, N_i),\, i = 1, \dots, k\}$ it is convenient to use the representation

$$\mathbf{P}\left\{\nu^{(1)} \ge n_1,\, \nu^{(k)} < n_2\right\} = \mathbf{P}\{\mu_0(n_1, N_i)$$

$$\ge N_i - t_i,\, \mu_0(n_2, N_i) < N_i - t_i,\, i = 1, \dots, k\}. \tag{17}$$

The joint distribution of the vectors $\bar{\mu}_0(n_1)$ and $\bar{\mu}_0(n_2)$ can be obtained from the relation

$$\mathbf{P}\{\mu_0(n_1, N_i) = \alpha_i,\, \mu_0(n_2, N_i) = \beta_i,\, i = 1, \dots, k\}$$

$$= \mathbf{P}\{\mu_0(n_1, N_i) = \alpha_i,\, i = 1, \dots, k\} \tag{18}$$

$$\times \mathbf{P}\{\mu_0(n_2 - n_1, \alpha_i) = \beta_i,\, i = 1, \dots, k\} = \Delta_{\alpha_1, \dots, \alpha_k;\, \beta_1, \dots, \beta_k}(n_1, n_2),$$

which was proved in (Enatskaya and Khakimullin, 1991).
Under the conditions of Theorems 3 and 4 a local normal limit approximation is valid for the vector $\{\mu_0(n_2, N_i),\, i = 1, \dots, k\}$ (see Lemma 8). The components of the mentioned vector are asymptotically independent. We show that the components of the vector $\{\mu_0(n_2 - n_1, \alpha_i),\, i = 1, \dots, k\}$ are also asymptotically normal and independent. To prove Theorems 3 and 4 we use the representation

$$\mathbf{P}\left\{\nu^{(1)} \ge n_1,\, \nu^{(k)} < n_2\right\} = \sum_{\substack{N_i - t_i \le \alpha_i \le N_i, \\ i=1,\dots,k}} \; \sum_{\substack{0 < \beta_i < N_i - t_i, \\ i=1,\dots,k}} \Delta_{\alpha_i, \dots, \alpha_k;\, \beta_1, \dots, \beta_k}(n_1, n_2)$$

and the well-known way of obtaining integral theorems with the help of local theorems.

2.4. *Auxiliary results*

LEMMA 1. *Under the conditions of the considered scheme the relation*

$$\mathbf{P}\left\{\mu_0(n_1, N_i) = \alpha_i,\, \mu_0(n_2, N_i) = \beta_i,\, i = 1, \dots, k\right\}$$

$$= \mathbf{P}\left\{\mu_0(n_1, M) = \alpha\right\} \mathbf{P}\left\{\mu_0(n_2 - n_1, \alpha) = \beta\right\} \left(\prod_{i=1}^{k} \binom{N_i}{\alpha_i} \binom{\alpha_i}{\beta_i} \left[\binom{M}{\alpha}\binom{\alpha}{\beta}\right]^{-1}\right)$$

is valid, where α_i, β_i *are integers and* $\alpha = \sum_{i=1}^{k} \alpha_i$, $\beta = \sum_{i=1}^{k} \beta_i$.

A proof is given in (Enatskaya and Khakimullin, 1991).

LEMMA 2. *Let* $N \to \infty$, $N_i/N \to \omega_i$, $0 < \omega_i < 1$, $i = 1, \dots, k$; $N - m \to \infty$, $m \to \infty$, $\mathbf{E}\,\mu_0(n, M) \to \lambda$, $\lambda > 0$. *Then the components of the vector* $\bar{\eta}$ *are asymptotically independent and*

$$\mathbf{P}\left\{\mu_0(n, N_i) < x\right\} \to \Pi_{[x]}(\lambda\omega_i), \qquad i = 1, \dots, k.$$

See the proof in (Ivchenko and Khonov, 1991).

LEMMA 3. *Let* $N \to \infty$, $N_i/N \to \omega_i$, $0 < \omega_i < 1$, $i = 1, \dots, k$; $\sum_{i=1}^{k} \omega_i = 1$, $a_l < \infty$, $b_l \to \infty$, $m/N \to \gamma$.

1) *If* $\gamma = 0$, *then for any fixed real* x *and* y

$$\mathbf{P}\left\{\nu^{(1)} \log \frac{N}{N-m} - \log N \le x\right\} \to 1 - \prod_{j=1}^{l} \left(1 - \Pi_{s_{i_j}}(\omega_{i_j} e^{-x})\right),$$

$$\mathbf{P}\left\{\nu^{(k)} \log \frac{N}{N-m} - \log N \le y\right\} \to \prod_{j=1}^{l} \Pi_{s_{i_j}}(\omega_{i_j} e^{-y});$$

2) *If* $0 < c_3 < \gamma < c_4 < 1$, $\lim \left\{\log N / \log \frac{N}{N-m}\right\} = \sigma$, $0 < \sigma < 1$, *then*

$$\mathbf{P}\left\{\nu^{(1)} + \log_{(1-m/N)} N \le d\right\} \to 1 - \prod_{j=1}^{l} \Pi_{s_{i_j}}\left(\omega_{i_j}(1-\gamma)^{d-\sigma}\right),$$

$$\mathbf{P}\left\{\nu^{(k)} + \log_{(1-m/N)} N \le d\right\} \to \prod_{j=1}^{l} \Pi_{s_{i_j}}\left(\omega_{i_j}(1-\gamma)^{d-\sigma}\right),$$

where d *is a fixed integer. The limit distributions of r. v.* $\nu^{(1)}$ *and* $\nu^{(k)}$ *exist only if a finite limit of* $\{-\log_{(1-m/N)} N\}$ *exists.*

Proof. In case 1 under the conditions of Lemma 3 for $n = [(x + \log N)/\log(N/(N-m))]$,

$$\mathbf{E}\,\mu_0(n, N) \to \lambda = e^{-x}.$$

In case 2, for $n = [-\log_{(1-m/N)} N]$,

$$\mathbf{E}\,\mu_0(n+d, N) \to \lambda = (1-\gamma)^{d-\sigma}.$$

Therefore in the situation of Lemma 2 the components of the vector $\bar{\mu}_0(n) = \{\mu_0(n, N_i),\ i = 1, \dots, k\}$ in case 1 and of the vector $\bar{\mu}_0(n+d) = \{\mu_0(n+d, N_i),\ i = 1, \dots, k\}$ in case 2 are asymptotically independent and have in limit the Poisson distributions with parameters $\lambda\omega_i$, $i = 1, \dots, k$. These parameters correspond to the groups of cells of sizes $N_1, \dots, N_k$. Therefore the statement of the lemma follows from (14) and (15).

LEMMA 4. *Let $N \to \infty$, $N_i/N \to \omega_i$, $0 < \omega_i < 1$, $\operatorname{Var}\mu_0(n, N_i) \to \infty$, $i = 1, \dots, k$. Then, uniformly with respect to $\bar\omega = (\omega_1, \dots, \omega_k)$ and $\bar u = (u_1, \dots, u_k)$, $|u_i| < c$, $i = 1, \dots, k$, and $\left|\sum_{i=1}^k u_i\right| \sigma_N(n)/\operatorname{Var}\bar\eta \le c$, we have*

$$\mathbf{P}\{\bar\eta = N\bar\omega a_N(n) + \bar u\sigma_N(n)\}$$

$$= \frac{\exp\left\{-\frac{1}{2}\sum_{i=1}^k(u_i^2/\omega_i) + \left(\sigma_N^2(n)/(\operatorname{Var}\bar\eta) - 1\right)\left(\sum_{i=1}^k u_i\right)^2\right\}}{\sigma_N^{k-1}(n)\sqrt{(2\pi)^k\omega_1\cdots\omega_k\operatorname{Var}\bar\eta}}$$

$$\times\left(1 + O\left(\frac{1}{\sqrt{\operatorname{Var}\bar\eta}}\right)\right).$$

The proof is given in (Ivchenko and Khonov, 1991).

LEMMA 5. *Let $N \to \infty$, $m/N \to 0$, $N_i/N \to \omega_i$, $0 < \omega_i < 1$, $i = 1, \dots, k$; $\sum_{i=1}^k \omega_i = 1$.*
1) If $t^+/N^+ \to 1$, $N^+ - t^+ \to \infty$, $N^+/N \to \omega^+$, $\sqrt{N^+ - t^+}m/N \to 0$, then for any fixed u

$$\mathbf{P}\left\{w^{(k)} < u\right\} = \prod_{i=1}^k \Phi(D_i^+)(1 + o(1)). \tag{19}$$

2) If $t_+/N_+ \to 1$, $N_+ - t_+ \to \infty$, $N_+/N \to \omega_+$, $\sqrt{N_+ - t_+}m/N \to 0$, then for any fixed u

$$\mathbf{P}\left\{w^{(1)} < u\right\} = 1 - \prod_{i=1}^k \overline{\Phi}(D_{i+})(1 + o(1)). \tag{20}$$

Proof. Under the conditions of the Lemma 5, the following relations are valid:

$$n = \left[\frac{\log(1 - t^+/N^+)}{\log(1 - m/N)} + \frac{u}{\sqrt{N^+ - t^+}\log(N/(N-m))}\right];$$

$$\mathbf{E}\,\mu_0(n, N_i) = N_i(1 - t^+/N^+) - uN_i\sqrt{N^+ - t^+}/N^+(1 + O(1)),$$

$$\operatorname{Var}\mu_0(n, N_i) = N_i(1 - t^+/N^+)(1 + o(1)), \tag{21}$$

$$\sigma_N^2 = N(1 - t^+/N^+)(1 + o(1)), \qquad i = 1, \dots, k.$$

Therefore $\sigma_N^2/\operatorname{Var}\mu_0(n, N_i) \to 1$ and under the conditions of the Lemma 4 the components of the vector

$$\{Q(\mu_0(n, N_i); n, N_i),\ i = 1, \dots, k\}$$

are asymptotically independent and normal. The assertion of Lemma 5 follows now from (21) and (14), (15).
Relation (20) can be proved similarly.
Under the conditions of Theorems 3 and 4 the following groups of lemmas give asymptotic expansions for $\mathbf{E}\,\mu_0(n,L)$ and $\operatorname{Var}\mu_0(n,L)$, where $1 < L \le N$.
Represent the integers n_1 and n_2, $n_1 < n_2$, in the form

$$n_1 = \log(1 - t_+/N_+)/\log(1 - m/N) - (x + o(1))/\left(\sqrt{N_+ - t_+}\log(1 - m/N)\right),$$

$$n_2 = \log(1 - t^+/N^+)/\log(1 - m/N) - (y + o(1))/\left(\sqrt{N^+ - t^+}\log(1 - m/N)\right),$$

where x and y are fixed, $x < y$.

LEMMA 6. *Under the conditions of Theorems 3 and 4*

$$\mathbf{E}\,\mu_0(n_1, N_i) = N_i(1 - t_+/N_+)(1 + o(1)), \tag{22}$$

$$\operatorname{Var}\mu_0(n_1, N_i) = N_i(1 - t_+/N_+)(1 + o(1)), \tag{23}$$

$$\operatorname{Var}\mu_0(n_2, N_i) = N_i(1 - t^+/N^+)(1 + o(1)), \tag{24}$$

$$\operatorname{Var}\mu_0(n_2, N_i) = N_i(1 - t^+/N^+)(1 + o(1)), \tag{25}$$

$$\operatorname{Var}\mu_0(n_1, N) = N(1 - t_+/N_+)(1 + o(1)), \tag{26}$$

$$\operatorname{Var}\mu_0(n_2, N) = N(1 - t^+/N^+)(1 + o(1)), \tag{27}$$

$$\sigma_N^2(n_1) = N(1 - t_+/N_+)(1 + o(1)), \quad \sigma_N^2(n_2) = N(1 - t^+/N^+)(1 + o(1)). \tag{28}$$

The proof of Lemma 6 follows directly from the formulas for $\mathbf{E}\,\mu_0(n,L)$ and $\operatorname{Var}\mu_0(n,L)$ from (Enatskaya and Khakimullin, 1989) and for $\sigma_N^2(n)$ from 1.1. We have

$$\mathbf{E}\,\mu_0(n_1, N_i) = N_i(1 - m/N)^{n_1} = N_i(1 - t_+/N_+)\exp\left\{-(x + o(1))/\sqrt{N_+ - t_+}\right\}.$$

To complete the proof of (22), note that under the conditions of Theorem 3 for $|x| < \infty$ and $x/\sqrt{N_+ - t_+} \to 0$ the asymptotic behaviour of $\operatorname{Var}\mu_0(n_1, N_i)$ may be obtained from the corresponding formula from (Enatskaya and Khakimullin, 1989). Under the conditions of Theorems 3 and 4

$$\operatorname{Var}\mu_0(n_1,N_i)=N_i(1-t_+/N_+)\Bigg(1-(1-t_+/N_+)+N_i(1-t_+/N_+)$$

$$\times\left(\exp\left(-\frac{n_1m}{N(N-m)}+O\left(\frac{n_1m}{N(N-m)^2}\right)\right)-1\right)\Bigg)(1+o(1)).$$

It is easy to verify that

$$\frac{n_1m}{N(N-m)}=\frac{1}{\log(1-m/N)}\frac{m}{N(N-m)}\left(\log\Delta^++\frac{x+o(1)}{\sqrt{N_+-t_+}}\right)$$

$$=\frac{1}{N-m}\left(\log\Delta^++\frac{x+o(1)}{\sqrt{N_+-t_+}}\right)+o(1)\to 0$$

(since $N^{-1}\log\Delta^+\to 0$ and $|x|<\infty$). Therefore (23) is valid.
Equalities (24)–(27) are established similarly.
Equality (28) follows from the formulas

$$\sigma_N^2(n)=Na_N(n)(1-a_N(n));$$

$$a_N(n_1)=1-(1-t_+/N_+)(1+o(1)),\quad a_N(n_2)=1-(1-t^+/N^+)(1+o(1)).$$

COROLLARY 6. *Under the conditions of Theorem 3*

$$\sigma_N^2(n_j)/\operatorname{Var}\mu_0(n_j,N)\to 1,\qquad j=1,2.$$

The assertion follows from (25)–(28).

LEMMA 7. *Under the conditions of Theorem 3 the following equalities are valid uniformly with respect to integers*

$$\alpha_i=\mathbf{E}\,\mu_0(n_1,N_i)+u_i\sqrt{\operatorname{Var}\mu_0(n_1,N_i)}(1+o(1))\tag{29}$$

such that u_i lies in any finite interval:

$$\mathbf{E}\,\mu_0(n_2-n_1,\alpha_i)=N_i(1-t^+/N^+)(1+o(1)),\tag{30}$$

$$\operatorname{Var}\mu_0(n_2-n_1,\alpha_i)=N_i(1-t^+/N^+)(1-u)(1+o(1)),\tag{31}$$

$$\operatorname{Var}\mu_0(n_2-n_1,\alpha)=N(1-t^+/N^+)(1-u)(1+o(1)),\tag{32}$$

$$\sigma_\alpha^2(n_2-n_1)=N(1-t^+/N^+)(1-u)(1+o(1)).\tag{33}$$

Proof. Since $\alpha_i = \mathbf{E}\,\mu_0(n_1, N_i)(1 + o(1))$, for $n = n_2 - n_1$

$$\mathbf{E}\,\mu_0(n, \alpha_i) = \alpha_i(1 - m/N)^{n_2 - n_1} = N_i(1 - m/N)^{n_2}(1 + o(1))$$
$$= \mathbf{E}\,\mu_0(n_2, N_i)(1 + o(1)). \qquad (34)$$

Now (30) follows from (24) and (34).
We have

$$n_2 - n_1 = \big(\log u - R(x, y)\big) / \log(1 - m/N).$$

Then

$$\operatorname{Var}\mu_0(n_2 - n_1, \alpha_i) = N_i(1 - t^+/N^+)(1 - (1 - m/(N)))^{n_2 - n_1}$$

$$+\alpha_i \left(1 - \frac{m}{N}\right)^{n_2 - n_1} \left(\exp\left(-\frac{(n_2 - n_1)m}{N(N - m)} + O\left(\frac{(n_2 - n_1)m}{N(N - m)^2}\right)\right) - 1\right). \qquad (35)$$

Now (31) follows from (34) and (35) because

$$\frac{\alpha_i(n_2 - n_1)m}{N(N - m)} = \frac{\omega_i \Delta^+}{1 - m/N}(\log u + o(1))(1 + o(1)) \to 0.$$

Similarly,

$$\operatorname{Var}\mu_0(n_2 - n_1, \alpha) = \alpha u(1 - u)(1 + o(1)) = N(1 - t^+/N^+)(1 - u)(1 + o(1)),$$

where $\alpha = \sum_{i=1}^{k} \alpha_i$. Further,

$$a_N(n_2 - n_1) = 1 - (1 - m/N)^{n_2 - n_1} = 1 + u(1 + o(1)).$$

$$\sigma_\alpha^2(n_2 - n_1) = \alpha u(1 - u)(1 + o(1)) = N\Delta_+(1 - u)(1 + o(1)).$$

Thus equalities (30)–(33) are proved.

COROLLARY 7. *Under the conditions of Theorem 3*

$$\sigma_\alpha^2(n_2 - n_1)/\operatorname{Var}\mu_0(n_2 - n_1, \alpha) \to 1.$$

The assertion follows from (32) and (33).

LEMMA 8. *Let the conditions of Lemma 7 be fulfilled. Then the following representation is valid uniformly with respect to* $\bar{\omega} = (\omega_1, \dots, \omega_k)$ *and* $\bar{z} = (z_1, \dots, z_k)$ *such that* $\mathbf{E}\,\mu_0(n_2 - n_1, \alpha_i) + z_i\sigma_\alpha(n_2 - n_1)$ *are integers and* $|z_i| \le c(n)$, $i = 1, \dots, k$, *where* $c(n) \to \infty$, *and* $c(n) = o(\sqrt[6]{N(1-t^+/N^+)(1-u)})$ *and* $|\sum_{i=1}^k z_i|\sigma_\alpha(n_2-n_1)/\sqrt{\operatorname{Var}\mu_0(n_2-n_1,\alpha)} < c(n)$:

$$\mathbf{P}\{\mu_0(n_2-n_1,\alpha_i) = \mathbf{E}\,\mu_0(n_2-n_1,N_i) + z_i\sigma_\alpha(n_2-n_1),\, i=1,\dots,k\}$$

$$= \frac{\exp\left\{-\frac{1}{2}\sum_{i=1}^k z_i/\omega_i + \left(\sigma_\alpha^2(n_2-n_1)/\operatorname{Var}\mu_0(n_2-n_1,\alpha) - 1\right)\left(\sum_{i=1}^k z_i\right)^2\right\}}{\sigma_\alpha^{k-1}(n_2-n_1)\sqrt{(2\pi)^k\omega_1\cdots\omega_k\operatorname{Var}\mu_0(n_2-n_1,\alpha)}}$$

$$\times(1+o(1)). \tag{36}$$

The proof repeats almost literally the proof of Theorem 1 from (Ivchenko and Khonov, 1991) and we omit it.

COROLLARY 8. *Under the conditions of Lemma 8, the following representation is valid uniformly with respect to* $\bar{\omega} = (\omega_1,\dots,\omega_k)$ *and* $\bar{v} = (v_1,\dots,v_k)$ *such that* $\mathbf{E}\,\mu_0(n_2-n_1,\alpha_i) + v_i\sqrt{\operatorname{Var}\mu_0(n_2-n_1,\alpha_i)}$ *are integers and* $|v_i| < c(n)$, $i = 1,2,\dots,k$:

$$\mathbf{P}\left\{\mu_0(n_2-n_1,\alpha_i) = \mathbf{E}\,\mu_0(n_2-n_1,\alpha_i) + v_i\sqrt{\operatorname{Var}\mu_0(n_2-n_1,\alpha_i)},\, i=1,\dots,k\right\}$$

$$= \prod_{i=1}^k \varphi(v_i)(1+o(1))/\sqrt{\operatorname{Var}\mu_0(n_2-n_1,\alpha_i)}. \tag{37}$$

Proof. The equalities

$$\operatorname{Var}\mu_0(n_2-n_1,\alpha_i)/\left(\sigma_\alpha^2(n_2-n_1)\right) = \omega_i(1+o(1)), \qquad i=1,\dots,k,$$

follow from (31) and (33). Therefore in (36) we set $z = v_i\sqrt{\omega_i}$. Then, taking into account Corollary 7, we obtain (37).

LEMMA 9. *Let* $N \to \infty$, $mn/N \to \delta$, $0 < \delta < \infty$, $1 \le M < \infty$. *Then the r. v.* $\mu_0(n,M)$ *has in limit the binomial distribution with the parameters* M *and* $e^{-\delta}$ *in the case* $m/N \to 0$ *and with the parameters* M *and* $(1-\sigma)^n$ *in the case* $m/N \to \sigma$, $0 < \sigma < 1$.

The proof is given in (Enatskaya and Khakimullin, 1989).

LEMMA 10. *The following equalities are valid uniformly with respect to integers* $\alpha_i = \mathbf{E}\,\mu_0(n_1,N_i) + u_i\sqrt{\operatorname{Var}\mu_0(n_1,N_i)}$ *such that* u_i *lies in any finite interval:*

$$\mathbf{E}\,\mu_0(n_2-n_1,\alpha_i) = N_i\Delta_i(1+o(1)), \tag{38}$$

$$\operatorname{Var}\mu_0(n_2-n_1,\alpha_i) = N_i\Delta_i(1-u+R(x,y))(1+o(1)), \tag{39}$$

$$\operatorname{Var}\mu_0(n_2-n_1,\alpha)=N_i\Delta_+(1-u+R(x,y))(1+o(1)), \tag{40}$$

$$\sigma_\alpha^2(n_2-n_1)=N\Delta_+(1-u+R(x,y))(1+o(1)). \tag{41}$$

Proof. Since $u_0 = o\left(\sqrt[6]{\operatorname{Var}\mu_0(n_1,N_i)}\right)$, $\operatorname{Var}\mu_0(n,N_i) \le \mathbf{E}\,\mu_0(n,N_i)$ and $\alpha_i = \mathbf{E}\,\mu_0(n_1,N_i)(1+o(1))$,

$$\mathbf{E}\,\mu_0(n_2-n_1,\alpha_i)=\alpha_i(1-m/N)^{n_2-n_1}$$

$$=N_i(1-m/N)^{n_2}(1+o(1))=\mathbf{E}\,\mu_0(n_2,N_i)(1+o(1)). \tag{42}$$

Then (38) follows from (42) and (24).
We consider now the following representations for n_2-n_1 and $(1-m/N)^{n_2-n_1}$:

$$n_2-n_1=(\log u-R(x,y))/\log(1-m/N),$$

$$(1-m/N)^{n_2-n_1}=u\exp\{R(x,y)\}=u(1-R(x,y))(1+o(1)). \tag{43}$$

Therefore (39) follows from (43) and (35). Representation (40) may be proved similarly. The asymptotic analysis of $\sigma_\alpha^2(n_2-n_1)$ is carried out directly:

$$\sigma_\alpha^2(n_2-n_1)=\alpha(1-m/N)^{n_2-n_1}(1-(1-m/N)^{n_2-n_1})$$

$$=\alpha u\exp\{R(x,y)\}(1-u\exp\{R(x,y)\})=\alpha(1-u-R(x,y))(1+o(1)).$$

Now we obtain (41) taking into account that $\alpha=\sum_{i=1}^k\alpha_i$ and

$$\sum_{i=1}^k u_i\sqrt{\operatorname{Var}\mu_0(n_1,N_i)}=o(\mathbf{E}\,\mu_0(n_1,N_i)).$$

COROLLARY 9. *Under the conditions of Theorem 4*

$$\sigma_\alpha^2(n_2-n_1)/\operatorname{Var}\mu_0(n_2-n_1,\alpha)\to 1. \tag{44}$$

Relation (44) follows from (40) and (41).

2.5. *Proof of theorems and corollaries*

Proof of Theorem 1. Proof is given in (Enatskaya and Khakimullin, 1991).

Proof of Theorem 2. Verify the validity of Lemma 2 for $\mu_0(n, N_i)$ in case 1 of Theorem 2 and show that the limit $\lim_{N\to\infty} \mathbf{E}\,\mu_0(n, N_i)$ exists.
Indeed, if

$$n = [(\log N + x + o(1))/(-\log(1 - m/N))]$$

and x is any fixed positive number, then

$$\mathbf{E}\,\mu_0(n, N_i) = N_i(1 - m/N)^n = N_i \exp\{-\log N - x + o(1)\} = \omega_i e^{-x} + o(1),$$

i.e., the conditions of Lemma 2 are fulfilled.
Let now

$$n_1 = [(\log N + x + o(1))/(-\log(1 - m/N))],$$

$$n_2 = [(\log N + y + o(1))/(-\log(1 - m/N))],$$

where $0 \le x < y < \infty$. We find the limits of each summand in (16) under conditions of the theorem using (17), representation (16), and Lemma 2:

$$\mathbf{P}\left\{\nu^{(k)} \le \frac{\log N + y + o(1)}{-\log(1 - m/N)}\right\} = \mathbf{P}\{\mu_0(n_2, N_i) \le s_i,\ i = 1, \ldots, k\}$$

$$= \sum_{0\le\alpha_i\le s_i,\ i=1,\ldots,k} \mathbf{P}\{\mu_0(n_2, N_i) = \alpha_i,\ i = 1, \ldots, k\}.$$

Using the fact (proved in (Ivchenko and Khonov, 1991)) that the components of the vector $\mu_0(n, N_i)$, $i = 1, \ldots, k$, are asymptotically independent and have the Poisson distributions with parameters $\omega_1 e^{-x}, \ldots, \omega_k e^{-x}$ it is easy to obtain the following equalities:

$$\mathbf{P}\left\{\nu^{(k)} \le n_2\right\} = \sum_{\substack{0\le\alpha_{i_j}\le s_{i_j}\\ j=1,\ldots,l}} \prod_{j=1}^{l} \pi_{\alpha_{i_j}}(\omega_{i_j} e^{-y}) + o(1)$$

$$= \prod_{j=1}^{l} \Pi_{s_{i_j}}(\omega_{i_j} e^{-y}) + o(1). \tag{45}$$

For the second summand in (16), similarly,

$$\mathbf{P}\left\{\nu^{(1)} > n_1,\ \nu^{(k)} \le n_2\right\}$$

$$= \sum_{\substack{s_{i_j}+1\le\alpha_{i_j}\le N_{i_j}\\ 0\le\beta_{i_j}\le s_{i_j}}} \mathbf{P}\left\{\mu_0(n_1, N_{i_j}) = \alpha_{i_j},\ i = 1, \ldots, k,\ j = 1, \ldots, l\right\}$$

$$\times \mathbf{P}\left\{\mu_0(n_2 - n_1, \alpha_{i_j}) = \beta_{i_j},\ i_j = 1, \ldots, k\right\} + o(1).$$

Therefore by Theorem 1 from (Enatskaya and Khakimullin, 1989) and Lemma 9,

$$\sum_{\substack{0\le\beta_{i_j}\le s_{i_j}\\ j=1,\dots,l}} \mathbf{P}\left\{\mu_0(n_2-n_1,\alpha_{i_j})=\beta_{i_j},\ i_j=1,\dots,k,\ j=1,\dots,l\right\}$$

$$=\prod_{j=1}^{l} B(s_{i_j},\alpha_{i_j},e^{x-y})+o(1).$$

From the last equality and (43) it follows that

$$\mathbf{P}\left\{\nu^{(1)}>n_1,\ \nu^{(k)}\le n_2\right\}=\sum_{\substack{s_{i_j}+1\le\alpha_{i_j}\le N_{i_j}\\ j=1,\dots,l}}\prod_{j=1}^{l}\pi_{\alpha_{i_j}}(\omega_{i_j}e^{-x})B(s_{i_j},\alpha_{i_j},e^{x-y})+o(1)$$

$$=\prod_{j=1}^{l}\sum_{s_{i_j}+1\le\alpha_{i_j}\le N_{i_j}}\pi_{\alpha_{i_j}}(\omega_{i_j}e^{-x})B(s_{i_j},\alpha_{i_j},e^{x-y})+o(1). \tag{46}$$

Thus, (1) follows from (45) and (46).
To prove case 2 of the theorem set

$$\Delta=\lim_{N\to\infty}\left\{-\log_{(1-m/N)}N\right\},\qquad n_1=[-\log_{(1-m/N)}N].$$

We have from (16) that for any d_1 and d_2, $0<d_1<d_2<\infty$,

$$\mathbf{P}\left\{\nu^{(1)}\le n_1+d_1,\ \nu^{(k)}\le n_1+d_2\right\}$$

$$=\mathbf{P}\left\{\nu^{(k)}\le n_1+d_2\right\}-\mathbf{P}\left\{\nu^{(1)}>n_1+d_1,\ \nu^{(k)}\le n_1+d_2\right\}.$$

Reasonings similar to the proof of case 1 lead to the following equality for the first summand in (16) (if the limit of $\{-\log_{(1-m/N)}N\}$ exists):

$$\mathbf{P}\left\{\nu^{(k)}\le n_1+d_2\right\}=\sum_{0\le\alpha_i\le s_i}\mathbf{P}\left\{\mu_0(n_1+d_2,N_i)=\alpha_i,\ i=1,\dots,k\right\}$$

$$=\sum_{0\le\alpha_i\le s_{i_j}}\prod_{j=1}^{l}\pi_{\alpha_{i_j}}(\omega_{i_j}(1-\delta)^{d_2+\Delta})+o(1)=\prod_{j=1}^{l}\Pi_{s_{i_j}}(\omega_{i_j}(1-\delta)^{d_2+\Delta})+o(1). \tag{47}$$

For the second summand, the similar conditions yield:

$$\mathbf{P}\left\{\nu^{(1)}>n_1+d_1,\ \nu^{(k)}\le n_1+d_2\right\}$$

$$=\sum_{\substack{s_{i_j}+1\le\alpha_{i_j}\le N_{i_j}\\ 0\le\beta_{i_j}\le s_{i_j}}}\mathbf{P}\left\{\mu_0(n_1+d_1,N_i)=\alpha_i,\ \mu_0(n_1+d_2,N_i)=\beta_i,\ i=1,\dots,k\right\}$$

$$= \prod_{j=1}^{l} \sum_{s_{i_j}+1 \le \alpha_{i_j} \le N_{i_j}} \pi_{\alpha_{i_j}}(\omega_{i_j}(1-\delta)^{d_1+\Delta}) B(s_{i_j}, \alpha_{i_j}, (1-\delta)^{d_2-d_1}) + o(1). \quad (48)$$

Thus, assertion (2) follows from (48) and (47). The proof is completed.

Proof of Theorem 3. We sequentially apply formula (16), Lemma 3 and (18) and obtain that

$$\mathbf{P}\left\{\nu^{(1)} < n_1,\, \nu^{(k)} < n_2\right\} = \mathbf{P}\left\{w^{(1)} < x,\, w^{(k)} < y\right\}$$

$$= \mathbf{P}\left\{w^{(k)} < y\right\} - \mathbf{P}\left\{w^{(1)} \ge x,\, w^{(k)} < y\right\}$$

$$= \prod_{i=1}^{k} \Phi(D_i^+) - \sum_{N_i - t_i < \alpha_i < N_i} \sum_{0 \le \beta_i \le N_i - t_i} \mathbf{P}\left\{\mu_0(n, N_i) = \alpha_i,\, i = 1, \dots, k\right\}$$

$$\times \mathbf{P}\left\{\mu_0(n_2 - n_1, \alpha_i) = \beta_i,\, i = 1, \dots, k\right\} + o(1). \quad (49)$$

Consider the expression

$$A_{\alpha_1,\dots,\alpha_k} = \sum_{0 \le \beta_i \le N_i - t_i,\, i=1,\dots,k} \mathbf{P}\left\{\mu_0(n_2 - n_1, \alpha_i) = \beta_i,\, i = 1, \dots, k\right\}.$$

Choose α_i as in Lemma 7 and represent $A_{\alpha_i,\dots,\alpha_k}$ in the form $A_{\alpha_1,\dots,\alpha_k} = A_1 + A_2$, where

$$A_1 = \sum_{\mathbf{E}\,\mu_0(n_2-n_1,\alpha_i) - z_i\sqrt{\mathrm{Var}\,\mu_0(n_2-n_1,\alpha_i)} \le \beta_i \le N_i - t_i} \mathbf{P}\{\mu_0(n_2 - n_1, \alpha_i)$$

$$= \beta_i, i = 1, \dots, k\}$$

and $z_i = o\left(\sqrt[6]{N\Delta_+(1-u)}\right)$; $z_i \to \infty$.
We estimate the asymptotic behaviour of A_1 taking into account Corollary 6:

$$A_1 = \sum_{-z_i \le Q(\beta_i; n_2 - n_1, \alpha_i) < Q(N_i - t_i; n_2 - n_1, \alpha_i)} \mathbf{P}\{Q(\mu_0(n_2 - n_1, \alpha_i); n_2 - n_1, \alpha_i)$$

$$= Q(\beta_i; n_2 - n_1, \alpha_i),\, i = 1, \dots, k\}$$

$$= \prod_{i=1}^{k} \Phi\left(Q(N_i - t_i; n_2 - n_1, \alpha_i)\right)(1 + o(1)). \quad (50)$$

Under the conditions of Theorem 3 taking into account (24) and (31) we can write $Q(N_i - t_i; n_2 - n_1, \alpha_i)$ in the following form:

$$Q(N_i - t_i; n_2 - n_1, \alpha_i)$$

$$= \frac{N_i - t_i - \mathbf{E}\,\mu_0(n_2, N_i) - u_i(1 - m/N)^{n_2 - n_1}\sqrt{\operatorname{Var}\mu_0(n_1, N_i)}}{\sqrt{\operatorname{Var}\mu_0(n_2, N_i)}}$$

$$\times\sqrt{\frac{\operatorname{Var}\mu_0(n_2, N_i)}{\operatorname{Var}\mu_0(n_2 - n_1, \alpha_i)}} = (D_i^+ - u_i(1/\sqrt{u})u(1 + o(1)))\,\sqrt{N_i\Delta_+/(N_i\Delta_+(1-u))}$$

$$= D_i^+/\sqrt{1-u} - u_i\sqrt{u/(1-u)} + o(1) = g_i(u_i) + o(1), \tag{51}$$

where

$$g_i(u_i) = D_i^+/\sqrt{1-u} - u_i\sqrt{u/(1-u)}.$$

To estimate A_2, note that there exists $i_0 \in \{1, 2, \ldots, k\}$ such that $z_{i_0} \to \infty$, $z_{i_0} = o\left(\sqrt[6]{N^+ - t^+}\right)$. Therefore we can apply the obvious inequality:

$$A_2 < \mathbf{P}\left\{\mu_0(n_2 - n_1, \alpha_{i_0}) < \mathbf{E}\,\mu_0(n_2 - n_1, \alpha_{i_0}) - z_{i_0}\sqrt{\operatorname{Var}\mu_0(n_2 - n_1, \alpha_{i_0})}\right\}.$$

The r.v. $\mu_0(n_2 - n_1, \alpha_{i_0})$ is asymptotically normal. The last fact follows from Lemma 3 from (Enatskaya and Khakimullin, 1990) because, in our case, $\operatorname{Var}\mu_0(n, L) \to \infty$ by (31). This is a sufficient condition of the asymptotic normality for the r.v. $\mu_0(n, L)$. Therefore $A_2 \to 0$ in view of the choice of z_{i_0}.
Now from (50) we obtain that

$$A_{\alpha_1, \ldots, \alpha_k} = A_1 + o(1) = \prod_{i=1}^{k} \Phi(g_i(u_i)) + o(1) \tag{52}$$

uniformly for z_i in the mentioned area of z_i, $i = 1, \ldots, k$. The asymptotic behaviour of

$$\sum_{\substack{N_i - t_i \le \alpha_i \le N_i,\\ i=1,\ldots,k}} A_{\alpha_1, \ldots, \alpha_k} \mathbf{P}\left\{\mu_0(n_1, N_i) = \alpha_i,\ i = 1, \ldots, k\right\}$$

is investigated similarly, using (37) and Lemma 3 from (Enatskaya and Khakimullin, 1990).
Thus, Lemma 8 and (51) imply that

$$\mathbf{P}\left\{\nu^{(1)} \ge n_1,\ \nu^{(k)} < n_2\right\}$$

$$= \sum_{N_i - t_i \le \alpha_i \le N_i,\, i=1,\ldots,k} \varphi(u_i)\Phi(g_i(u_i))(1 + o(1))/\sqrt{\operatorname{Var}\mu_0(n_1, \alpha_i)}$$

$$= \prod_{i=1}^{k} \int_{D_{i+}}^{\infty} \varphi(u_i)\Phi(g_i(u_i))\,du_i + o(1).$$

Thus, (3) follows from (19), (20) and (16).

Proof of Corollary 1. We use the explicit form of Δ_i given in (4) and the following relations:

$$d_{i+} = \sqrt{\frac{N}{\omega_1}}\frac{\Delta_i - \Delta_k}{\sqrt{\Delta_k}} = \frac{c_i - c_k}{\sqrt{\theta_k \omega_1}} + o(1), \qquad i = k-p, \dots, k, \tag{53}$$

$$d_i^+ = \sqrt{\frac{N}{\omega_k}}\frac{\Delta_i - \Delta_1}{\sqrt{\Delta_1}} = \frac{c_i - c_1}{\sqrt{\theta_1 \omega_k}} + o(1), \qquad i = 1, \dots, s, \tag{54}$$

$$d_{i+} \to -\infty, \quad d_i^+ \to \infty, \qquad i = s+1, \dots, k-p-1. \tag{55}$$

Hence by virtue of (53)–(55) D_{i+} has the following asymptotic behaviour:

$$D_{i+} = \frac{c_i - c_k}{\sqrt{\theta_k}}\sqrt{\omega_i} + x\sqrt{\omega_i/\omega_k} + o(1), \qquad i = k-p, \dots, k,$$

$$D_{i+} \to -\infty, \qquad i = 1, \dots, k-p-1.$$

Similarly,

$$\frac{D_i^+}{\sqrt{1-u}} = \frac{c_i - c_1}{\sqrt{\theta_1}}\sqrt{\frac{\omega_1\theta_k}{(\theta_k - \theta_1)}} + y\sqrt{\frac{\omega_i\theta_k}{\omega_1(\theta_k - \theta_1)}} + o(1), \qquad i = 1, \dots, s,$$

$$D_i^+ \to \infty \quad \text{for} \quad i = s+1, \dots, k.$$

Therefore the following equality is true:

$$\int_{D_{i+}}^{\infty} \varphi(u_i)\Phi\left(\frac{D_i^+}{\sqrt{1-u}} - u_i\sqrt{u(1-u)}\right) du_i$$

$$= \begin{cases} 1 + o(1) & \text{for } i = s+1, \dots, k-p-1, \\ \overline{\Phi}\left((c_i - c_k)\sqrt{\frac{\omega_i}{\theta_k}} + x\sqrt{\frac{\omega_i}{\omega_k}}\right) + o(1) & \text{for } i = k-p, \dots, k, \\ \int_{-\infty}^{\infty} \varphi(v_i)\Phi\Bigg((c_i - c_k)\sqrt{\frac{\omega_i\theta_k}{\theta_1(\theta_k - \theta_1)}} & \\ \quad + y\sqrt{\frac{\omega_i\theta_k}{\omega_1(\theta_k - \theta_1)}} - v_i\sqrt{\frac{\theta_1}{\theta_k - \theta_1}}\Bigg) dv_i + o(1) & \text{for } i = 1, \dots, s. \end{cases}$$

This implies assertion (5).

Proof of Corollary 2. If $u = 0$, from (3) we obtain that

$$\mathbf{P}\left\{w^{(1)} < x,\, w^{(k)} < y\right\} = \prod_{i=1}^{k} \Phi(D_i^+) - \prod_{i=1}^{k} \Phi(D_i^+)\overline{\Phi}(D_{i+}) + o(1)$$

$$= \prod_{i=1}^{k} \Phi(D_i^+)\left(1 - \prod_{i=1}^{k} \overline{\Phi}(D_{i+})\right) + o(1).$$

Finally, notice that from Lemma 5 it follows that

$$\mathbf{P}\left\{w^{(1)} < x\right\} = 1 - \prod_{i=1}^{k} \overline{\Phi}(D_{i+}) + o(1); \quad \mathbf{P}\left\{w^{(k)} < y\right\} = \prod_{i=1}^{k} \Phi(D_i^+) + o(1).$$

This implies assertion (6) of Corollary 2.

Proof of Corollary 3. The conditions of this corollary are stronger than the conditions of Corollary 2, therefore in this case the r. v. $w^{(1)}$ and $w^{(k)}$ are also asymptotically independent. By (3)

$$\mathbf{P}\left\{w^{(k)} < y\right\} = \prod_{i=1}^{k} \Phi(D_i^+) + o(1).$$

Notice that under the conditions of Corollary 3

$$\begin{aligned} d_i^+ &= (b_i - b_1)/\sqrt{\omega_1 c_1} + o(1) \quad \text{for } i = 1, \dots, l, \\ d_i^+ &\to \infty \quad \text{for } i = l+1, \dots, k, \\ d_{i+} &= (c_1 - c_k)/\sqrt{\theta_k \omega_k} + o(1) \quad \text{for } i = k-p, \dots, k, \\ d_{i+} &\to -\infty \quad \text{for } i = 1, \dots, k-p-1. \end{aligned}$$

Therefore the asymptotic behaviour of D_{i+} and D_i^+ is desribed by the following relations:

$$\begin{aligned} D_{i+} &= \frac{c_i - c_k}{\sqrt{\theta_k}}\sqrt{\omega_i} + x\sqrt{\omega_i/\omega_k} + o(1) \quad \text{for } i = k-p, \dots, k, \\ D_{i+} &\to -\infty \quad \text{for } i = 1, \dots, k-p-1, \end{aligned} \tag{56}$$

$$\begin{aligned} D_i^+ &= \frac{b_i - b_1}{\sqrt{c_1}}\sqrt{\omega_i} + y\sqrt{\omega_i/\omega_1} + o(1) \quad \text{for } i = 1, \dots, l, \\ D_i^+ &\to \infty \quad \text{for } i = l+1, \dots, k. \end{aligned} \tag{57}$$

Thus, (7) and (8) follow from (56) and (57). Corollary 3 is proved.

Proof of Theorem 4. The proof of the theorem is similar to that of Theorem 3. Therefore we indicate only the details by which they differ. Let us transform the expression $Q(N_i - t_i; n_2 - n_1, \alpha_i)$ using (29), Lemmas 6 and 10:

$$Q(N_i - t_i, n_2 - n_1, \alpha_i)$$

$$= \frac{N_i - t_i - \mathbf{E}\,\mu_0(n_2, N_i) - u_i(1 - m/N)^{n_2 - n_1}\sqrt{\operatorname{Var}\mu_0(n_1, N_i)}(1 + o(1))}{\sqrt{\operatorname{Var}\mu_0(n_2 - n_1, \alpha_i)}}$$

$$= \left(N_i - t_i - N_i(1 - t^+/N^+)\exp\left(-\frac{y + o(1)}{\sqrt{N^+ - t^+}}\right)\right.$$

$$\left. - u u_i \exp\{R(x, y)\}\sqrt{\operatorname{Var}\mu_0(n_1, N_i)}\right)$$

$$\times \left[N_i(1 - t^+/N^+)(1 - u - R(x, y)(1 + o(1)))\right]^{-1/2}$$

$$= \frac{N_i(t^+/N^+ - t_i/N_i)}{\sqrt{N_i(1 - t^+/N^+)(1 - u - R(x, y)(1 + o(1)))}} + \frac{y\sqrt{N_i/N^+} - u_i(1 + o(1))}{\sqrt{1 - u - R(x, y)(1 + o(1))}}$$

$$= \frac{(D_i^+ - u_i)(1 + o(1))}{\sqrt{1 - u - R(x,y)}}. \tag{58}$$

By Corollary 8 we obtain that

$$A_1 = \sum_{\substack{-x_i \le Q(\beta_i; n_2 - n_1, \alpha_i) \\ \le Q(N_i - t_i; n_2 - n_1, \alpha_i)}} \mathbf{P}\{Q(\mu_0(n_2 - n_1, \alpha_i); n_2 - n_1, \alpha_i)$$

$$= Q(\beta_i; n_2 - n_1, \alpha_i),\ i = 1, \ldots, k\}$$

$$= \prod_{i=1}^{k} \Phi(Q(N_i - t_i; n_2 - n_1, \alpha_i))(1 + o(1)). \tag{59}$$

Omitting the reasonings similar to those of Theorem 3 and using (59) and (37), for the second summand in (49) we have

$$\mathbf{P}\left\{\nu^{(1)} \ge n_1,\ \nu^{(k)} < n_2\right\}$$

$$= \sum_{N_i - t_i \le \alpha_i \le N_i,\ i=1,\ldots,k} A_1 \mathbf{P}\{\mu_0(n_1, N_i) = \alpha_i,\ i = 1, \ldots, k\}(1 + o(1))$$

$$= \sum_{N_i - t_i \le \alpha_i \le N_i,\ i=1,\ldots,k} \prod_{i=1}^{k} \Phi\left((D_i^+ - u_i)\frac{1 + o(1)}{\sqrt{1 - u - R(x,y)}}\right)$$

$$\times \mathbf{P}\left\{\mu_0(n_1, N_i) = \alpha_i,\ i = 1, \ldots, k\right\}(1 + o(1))$$

$$= \prod_{i=1}^{k} \left(\sum_{N_i - t_i \le \alpha_i \le N_i,\ i=1,\ldots,k} \varphi(u_i)\Phi\left((D_i^+ - u_i)/\sqrt{1 - u - R(x,y)}\right) + o(1)\right)$$

$$\times \left[\operatorname{Var} \mu_0(n_1, N_i)\right]^{-1/2}$$

$$= \prod_{i=1}^{k} \left(\sum_{D_i^+ \le u_i \le D_i^+,\ i=1,\ldots,k} \varphi(u_i)/\sqrt{\operatorname{Var} \mu_0(n_1, N_i)} + o(1)\right)$$

$$= \prod_{i=1}^{k} \left(\Phi(D_i^+) - \Phi(D_{i+})\right) + o(1). \tag{60}$$

Now (10) follows from (60) and (49).

Proof of Corollary 4. From (11) and from the relations between Δ_i and d_{i+}, d_i^+ we obtain that

$$d_{i+} = \sqrt{N/\omega_1}(\Delta_i - \Delta_k)/\sqrt{\Delta_k}, \quad d_i^+ = \sqrt{N/\omega_1}(\Delta_i - \Delta_1)/\sqrt{\Delta_1},$$

where

$$\Delta_i - \Delta_k = \left((\theta_i - \theta_k)\psi(N) + (c_i - c_k)\sqrt{\psi(N)} + o\left(\sqrt{\psi(N)}\right)\right)/N,$$

$$\Delta_i - \Delta_1 = \left((\theta_i - \theta_1)\psi(N) + (c_i - c_1)\sqrt{\psi(N)} + o\left(\sqrt{\psi(N)}\right)\right)/N.$$

Therefore

$$d_{i+} = \frac{1}{\sqrt{\omega_k}} \frac{(\theta_i - \theta_k)\psi(N) + (c_i - c_k)\sqrt{\psi(N)} + o\left(\sqrt{\psi(N)}\right)}{\sqrt{\psi(N)\theta_k(1 + o(1))}},$$

$$d_i^+ = \frac{1}{\sqrt{\omega_k}} \frac{(\theta_i - \theta_1)\psi(N) + (c_i - c_1)\sqrt{\psi(N)} + o\left(\sqrt{\psi(N)}\right)}{\sqrt{\psi(N)\theta_1(1 + o(1))}}.$$

Hence d_{i+} and D_{i+} tend to $-\infty$ for $i < k - p$, and d_i^+ and D_i^+ tend to ∞ for $i < s$. Thus, (10) can be represented in the following form:

$$\mathbf{P}\left\{w^{(1)} < x,\, w^{(k)} < y\right\} = \prod_{i=1}^{s} \Phi(D_i^+) - \prod_{i=1}^{s} \Phi(D_i^+) \prod_{i=k-p}^{k} \overline{\Phi}(D_{i+})(1 + o(1)).$$

Thus the proof is complete.

Proof of Corollary 5. The proof of (13) can be obtained by calculation of d_{i+} and d_i^+ with regard for the conditions of the corollary.
Indeed,

$$d_{i+} = \frac{1}{\sqrt{\omega_k}} \frac{c_i - c_k}{\sqrt{\theta}}(1 + o(1)), \quad d_i^+ = \frac{1}{\sqrt{\omega_1}} \frac{c_i - c_1}{\sqrt{\theta}}(1 + o(1));$$

as to D_{i+} and D_i^+ one can use the last representations and (10) from Theorem 4.

REFERENCES

Enatskaya, N. J. and Khakimullin, E. R. (1989). The rate of convergence of the distribution of waiting time to the limit in the equiprobable scheme of allocation of particles by series. In: *Probab. Processes Appl.*, Moscow Institute of Electronic Engineering, Moscow, pp. 123–130 (in Russian).

Enatskaya, N. J. and Khakimullin, E. R. (1990). Limit theorems for waiting time in the equiprobable scheme of allocation of particles by series. In: *Probab. Problems Discrete Math.*, Moscow Institute of Electronic Engineering, Moscow, pp. 63–68 (in Russian).

Enatskaya, N. J. and Khakimullin, E. R. (1991). Asymptotic behaviour of joint distribution of waiting times in the equiprobable scheme of allocation of particles by series. In: *Probab. Processes Appl.*, Moscow Institute of Electronic Engineering, Moscow, pp. 91–96 (in Russian).

Ivchenko, G. I. and Khonov, S. A. (1991). On an asymptotic estimation for stratified finite populations. *Discrete Math. Appl.* **3**, 311–320.

Ivchenko, G. I. and Medvedev, Yu. I. (1966). The asymptotic behaviour of the number of particle complexes in a classical allocation problem. *Theory Probab. Appl.* **11**, 619–626.

Vatutin, V. A., and Mikhailov, V. G. (1982). Limit theorems for number of empty cells in equiprobable scheme for group allocation of particles. *Theory Probab. Appl.* **27**, 734–742.

Probabilistic Methods in Discrete Mathematics, pp. 186 – 199
V. F. Kolchin et al. (Eds.)
1993 TVP/VSP

LIMIT THEOREMS FOR *U*-STATISTICS OF DEPENDENT RANDOM VARIABLES

A. B. GORTCHAKOV
TVP Laboratories, Krasikov Street 32, 117418 Moscow, Russia

ABSTRACT

Let $X_1^{(n)},\dots,X_n^{(n)}$ be a sequence of random variables (r. v.'s) and let $\zeta_\alpha=\zeta_\alpha^{(n)}$, $\alpha\in W_r=\{(i_1,\dots,i_r)\colon 1\leqslant i_1<\dots<i_r\leqslant n\}$, be real r. v.'s such that ζ_α, $\alpha=(i_1,\dots,i_r)\in W_r$, are measurable with respect to $\sigma(X_{i_1}^{(n)},\dots,X_{i_r}^{(n)})$. The r. v. $\xi=\sum_{\alpha\in W_r}\zeta_\alpha$ is called the U-statistic.
In this paper we obtain sufficient conditions of convergence (as $n\to\infty$) of the distribution of the r. v. ξ to the normal and Poisson distributions under the assumption that distributions of all r. v.'s $(X_1^{(n)}\dots,X_n^{(n)},\zeta_\alpha,\alpha\in W_r)$ and the value r depend on n and $X_1^{(n)},\dots,X_n^{(n)}$ is a sequence of r. v.'s satisfying the strong mixing condition.

1. INTRODUCTION

Let $X_1^{(n)},\dots,X_n^{(n)}$ be a sequence of random variables (r. v.'s) and let $\zeta_\alpha=\zeta_\alpha^{(n)},\alpha\in W_r=\{(j_1,\dots,j_r)|1\leqslant j_1<\dots<j_r\leqslant n\}$ be real r. v.'s such that for any $\alpha=(i_1,\dots,i_r)\in W_r$ the random variable ζ_α is measurable with respect to $\mathcal{F}_\alpha=\sigma(X_{i_1}^{(n)},\dots,X_{i_r}^{(n)})$ and $\mathrm{E}\zeta_\alpha^2$ exists.
The r. v. $\xi=\sum_{\alpha\in W_r}\zeta_\alpha$ is called the U-statistic. In the present paper we consider the case where all quantities (random and nonrandom) depend on n (triangular array). In such a case many results concerning the asymptotic behaviour, as $n\to\infty$, of the distribution of the r. v. ξ were obtained under the assumption that $X_1^{(n)},\dots,X_n^{(n)}$ are independent r. v.'s. Under this assumption the most general sufficient conditions of asymptotic normality of the r. v. ξ are given in (Ronzhin, 1979) and the most general sufficient conditions of convergence of the distribution of the r. v. ξ to the Poisson distribution are given in (Zubkov, 1977) and (Grusho, 1987) (see also (Brown and Silverman, 1979)). An extensive bibliography is contained in the monograph (Korolyuk and Borovskikh, 1989). In (Mikhailov, 1989) and (Khashimov, 1990) sufficient conditions of asymptotic normality of the r. v. are given under the assumption that $X_1^{(n)},\dots,X_n^{(n)}$ is a sequence of s-dependent r. v.'s and r is constant. In the present paper we obtain sufficient conditions of convergence of the distribution of the r. v. ξ to the normal and Poisson distributions under the assumption that $X_1^{(n)},\dots,X_n^{(n)}$ is a sequence of r. v.'s satisfying the strong mixing condition and $r\to\infty$. The case where $r\to\infty$ was considered

in (Frees, 1989) and (Korolyuk and Borovskikh, 1990) provided the distributions of the r. v.'s $X_1^{(n)}, \ldots, X_n^{(n)}$ do not depend on n.

2. MAIN RESULTS

Further $X_1^{(n)}, \ldots, X_n^{(n)}$ is a sequence of r. v.'s satisfying the strong mixing condition with mixing coefficient $\varphi(s) = \varphi^{(n)}(s)$.
It is well-known that if $\mathbf{E}\,\zeta_\alpha^2$ exists for any $\alpha \in W_r$, then there exist nonrandom $C_\alpha = C_\alpha^{(n)}$, $\alpha \in W_r$, such that the r. v. $\sum_{\alpha\in W_r} \zeta_\alpha \mathbf{I}(|\zeta_\alpha| > C_\alpha)$, where $\mathbf{I}(R)$ is the indicator of an event R, converges in probability to zero as $n \to \infty$. Thus without loss of generality we assume that there exist nonrandom numbers $C_\alpha = C_\alpha^{(n)}$, $\alpha \in W_r$, such that $|\zeta_\alpha| \leqslant C_\alpha$ a. s. for any $\alpha \in W_r$.
We can consider the elements of the set W_r as subsets of $\{1, \ldots, n\}$. Let

$$\rho(\alpha, \beta), \qquad \alpha = (i_1, \ldots, i_l) \in W_l, \quad \beta = (j_1, \ldots, j_t) \in W_t,$$

denote the distance between sets α and β, i.e., $\min\{|i_k - j_m|\colon 1 \leqslant k \leqslant l, 1 \leqslant m \leqslant t\}$, and let $m(\beta_1, \ldots, \beta_l)$ denote the union of sets $\beta_i \in W_r, 1 \leqslant i \leqslant l$, i.e., the element $(j_1, \ldots, j_t) \in W_t$ such that $\min\{\rho(j_k, \beta_s)\colon 1 \leqslant s \leqslant l\} = 0$ for any $1 \leqslant k \leqslant t$ and $\min\{\rho(j, \beta_s)\colon 1 \leqslant s \leqslant l\} > 0$ for any $j \in \{1, \ldots, n\}\backslash\{j_1, \ldots, j_t\}$.
Let $T(t,s) = T^{(n)}(t,s)$, $t \in \{1, \ldots, n\}$, $s \in \{0, \ldots, n-1\}$, be nonrandom values such that for any $k \leqslant t$ and $\beta_1, \ldots, \beta_k$ from $\widetilde{W}_r = \{\alpha\colon \alpha \in W_r, \operatorname{Var}\zeta_\alpha \neq 0\}$

$$\sum_{\alpha\in W_r(m(\beta_1,\ldots,\beta_k),s)} \mathbf{E}\left(|\zeta_\alpha| \,\middle|\, \mathcal{F}_{m(\beta_1,\ldots,\beta_k)}\right) \leqslant T(t,s) \quad \text{a. s.},$$

where

$$W_r(\beta, s) = \{\alpha\colon \alpha \in \widetilde{W}_r, \rho(\alpha,\beta) \leqslant s\}, \qquad \beta \in W_t, \quad 1 \leqslant t \leqslant n.$$

Since the values C_α, $\alpha \in W_r$, exist, the values $T(t,s)$, $t \in \{1, \ldots, n\}$, $s \in \{0, \ldots, n-1\}$, exist as well.
For the first time the values $T(t,s)$ were introduced in (Mikhailov, 1986) and he obtained sufficient conditions of asymptotic normality of the r. v. ξ in terms of $\operatorname{Var}\xi$, $T(t,s)$ and $\sum_{\alpha\in W_r} \mathbf{E}\,|\zeta_\alpha|$ under the assumption that $\varphi(l) = 0$ for $l > s$ and r is a constant (see also (Mikhailov, 1989)). We generalize this conditions to the case where $\varphi(l) > 0$ for all $l \neq 0$, and $r \to \infty$.

THEOREM 1. *Suppose that $\mathbf{E}\,\zeta_\alpha = 0$ for any $\alpha \in W_r$ and there exist $s = s^{(n)} \geqslant 0$ and a constant $t > 2$ such that for any fixed $k \geqslant t$*

$$\lim_{n\to\infty} \Big(\sum_{\alpha\in\widetilde{W}_r} (\operatorname{Var}\xi)^{-1/2} C_\alpha\Big)^k r\varphi(s+1) = 0, \tag{1}$$

$$\lim_{n\to\infty} (\operatorname{Var}\xi)^{-k/2} \Big(\sum_{\alpha,\beta\in\widetilde{W}_r\colon\, \rho(\alpha,\beta)\leqslant s} \mathbf{E}\,|\zeta_\alpha\zeta_\beta|\Big) \prod_{i=2}^{k-1} T(i,s) = 0. \tag{2}$$

Then moments of r. v. $(\mathrm{Var}\,\xi)^{-1/2}\xi$ *converge to the moments of the standard normal distribution, as* $n \to \infty$.

THEOREM 2. *Suppose that for any* $\alpha \in W_r$

$$\mathbf{P}\{\zeta_\alpha = 0\} = 1 - \mathbf{P}\{\zeta_\alpha = 1\},$$

$$\lim_{n\to\infty} \mathbf{E}\,\zeta_\alpha = 0, \quad \lim_{n\to\infty} \mathbf{E}\,\xi = \lambda < \infty,$$

and there exists $s = s^{(n)} \geqslant 0$ *such that for any fixed* $k > 1$ *condition* (1) *is valid and*

$$\lim_{n\to\infty} \Big(\sum_{\alpha,\beta\in\widetilde{W}_r:\ \rho(\alpha,\beta)\leqslant s,\ \alpha\neq\beta} \mathbf{E}\,\zeta_\alpha\zeta_\beta \Big) \prod_{i=2}^{k-1} (T(i,s) - i) = 0.$$

Then for any fixed $k \in \mathbf{N}_0 = \{0,1,2,\dots,\}$,

$$\lim_{n\to\infty} \mathbf{P}\{\xi = k\} = \frac{\lambda^k e^{-\lambda}}{k!}. \tag{3}$$

Notice that conditions of Theorem 2 are more general than the sufficient conditions of convergence of the distribution of the r. v. ξ to the Poisson distribution given in (Grusho, 1987).

At first we prove five general theorems about the convergence of the moments and cumulants of sums of dependent r. v.'s. Then we show that Theorems 1 and 2 are consequences of these theorems.

3. NOTATION

Let us introduce some notation. Let $\xi_j = \xi_j^{(n)}$ ($j \in V^{(n)} = \{1,\dots,L^{(n)}\}$) be a sequence of r. v.'s such that $\mathbf{E}\,|\xi_j|^k$ exists for any $k > 0$ and $j \in V$. Let us consider random vector (r. vec.)

$$\boldsymbol{\xi} = \left(\sum_{j=1}^{t_1^{(n)}} \xi_j,\ \sum_{j=t_1^{(n)}+1}^{t_2^{(n)}} \xi_j, \dots, \sum_{j=t_{l-1}^{(n)}+1}^{L^{(n)}} \xi_j \right) = (\eta_1,\dots,\eta_l),$$

where l is a constant independent of n and $(t_1^{(n)},\dots,t_{l-1}^{(n)})$ is a set of integers such that $1 \leqslant t_1^{(n)} < \dots < t_{l-1}^{(n)} < L^{(n)}$.

Further all the quantities (random and nonrandom) depend on n (if we do not specify opposite), but we omit the index n.

Let $M^k = \{(m_1,\dots,m_k):\ m_j \in M,\ 1 \leqslant j \leqslant k\}$, where M is an arbitrary set. Let $\boldsymbol{\nu} = (\nu_1,\dots,\nu_k)$ be an arbitrary element from $\mathbf{N}_0^k$, we write $|\boldsymbol{\nu}|$ for $\sum_{j=1}^k \nu_j$, $\boldsymbol{\nu}!$ for $\prod_{i=1}^k \nu_j!$ $(0! = 1)$ and $b(\boldsymbol{\nu})$ for

$$(b_1(\boldsymbol{\nu}),\dots,b_{|\boldsymbol{\nu}|}(\boldsymbol{\nu})) = (\underbrace{1,\dots,1}_{\nu_1},\dots,\underbrace{k,\dots,k}_{\nu_k}).$$

The set of $S_{\xi}(\mathbf{r})$, $\mathbf{r} = (r_1, \ldots, r_l) \in \mathbf{N}_0^l$,

$$S_{\xi}(\mathbf{r}) = \frac{\partial^{|\mathbf{r}|}}{i^{|\mathbf{r}|} \prod_{j=1}^{l} (\partial z_j)^{r_j}} \left(\operatorname{Ln} \mathbf{E} \exp \left\{ i \sum_{j=1}^{l} z_j \eta_j \right\} \right) \Big|_{z_1 = \cdots = z_l = 0}, \tag{4}$$

where simbol Ln denote the principal branch of logarithm, is called the cumulant of the random vector (r. vec.) $\boldsymbol{\xi}$. It is clear that if $\mathbf{E}|\xi_j|^k$ exists for any $k > 0$ and $j \in V$, then the values from the right-hand side of (4) exist and the definition of $S_{\xi}(\mathbf{r})$ is correct.

Let us represent the cumulant $S_{\xi}(\mathbf{r})$ in terms of sums of values $\mathbf{E}_{\boldsymbol{\nu}} = \prod_{j=1}^{L} \xi_j^{\nu_j}$, $\boldsymbol{\nu} = (\nu_1, \ldots, \nu_L) \in \mathbf{N}_0^L$.

Let $V_k = \{\boldsymbol{\nu}\colon \boldsymbol{\nu} \in \mathbf{N}_0^L,\ |\boldsymbol{\nu}| = k\}$ and let $A(\mathbf{r})$ ($\mathbf{r} = (r_1, \ldots, r_l) \in \mathbf{N}_0^l$) be the subset of $V_{|\mathbf{r}|}$ such that $\boldsymbol{\mu} = (\mu_1, \ldots, \mu_L) \in A(\mathbf{r})$ if and only if $\sum_{k=t_{j-1}+1}^{t_j} \mu_k = r_j$ ($t_0 = 1$, $t_l = L$) for any $1 \leqslant j \leqslant l$.

Let $m^{(1)}(W)$ ($W \subset V_k$) denote the sum

$$\sum_{\boldsymbol{\mu} \in W} \sum_{s=1}^{k} \frac{(-1)^{s-1}}{s} \sum_{\boldsymbol{\nu}_1 + \cdots + \boldsymbol{\nu}_s = \boldsymbol{\mu}}^{L} \prod_{r=1}^{s} (\boldsymbol{\nu}_r!)^{-1} \mathbf{E}_{\boldsymbol{\nu}_r},$$

where the simbol $\sum_{\boldsymbol{\nu}_1 + \cdots + \boldsymbol{\nu}_s = \boldsymbol{\mu}}^{r}$ denote the sum over all $\boldsymbol{\nu}_j \in \widetilde{\mathbf{N}}_0^r$, $1 \leqslant j \leqslant s$, such that $\boldsymbol{\nu}_1 + \cdots + \boldsymbol{\nu}_s = \boldsymbol{\mu} \in \widetilde{\mathbf{N}}_0^r$, $\widetilde{\mathbf{N}}_0^r = \mathbf{N}_0^r \backslash \{0, \ldots, 0\}$.

It is well known that

$$(\mathbf{r}!)^{-1} S_{\xi}(\mathbf{r}) = m^{(1)}(A(\mathbf{r})), \qquad \mathbf{r} \in \widetilde{\mathbf{N}}_0^l.$$

We can enlarge the set in the brackets of the right-hand side of the last equality. If $\boldsymbol{\nu} = (\nu_1, \ldots, \nu_L) \in \widetilde{\mathbf{N}}_0^L$, then let $V(\boldsymbol{\nu})$ denote $\{j\colon \nu_j \neq 0,\ j \in V\}$. Let $\mathbf{N}_k$ be the subset of V_k, such that $\boldsymbol{\mu} \in \mathbf{N}_k$ if and only if there exists a set I, $I \neq \varnothing$, $I \neq V(\boldsymbol{\mu})$, $I \subset V(\boldsymbol{\mu})$, such that $\operatorname{Cov}\left(\prod_{j \in I} \xi_j^{\nu_j}, \prod_{j \in V \backslash I} \xi_j^{\nu_j}\right) = 0$ for any $\boldsymbol{\nu} \leqslant \boldsymbol{\mu}$, where the relation $\boldsymbol{\nu} \leqslant \boldsymbol{\mu}$, $\boldsymbol{\nu} = (\nu_1, \ldots, \nu_L) \in \mathbf{N}_0^L$, $\boldsymbol{\mu} = (\mu_1, \ldots, \mu_L) \in \mathbf{N}_0^L$, means that $\nu_j \leqslant \mu_j$ for any $1 \leqslant j \leqslant L$.

Zhurbenko (Zhurbenko, 1970) showed that

$$m^{(1)}(\{\boldsymbol{\mu}\}) = 0 \tag{5}$$

for any $\boldsymbol{\mu} \in \mathbf{N}_k$, and hence

$$(\mathbf{r}!)^{-1} S_{\xi}(\mathbf{r}) = m^{(1)}(A(\mathbf{r}) \backslash \mathbf{N}_{|\mathbf{r}|}).$$

Using the last equality, authors of the papers (Ronzhin, 1979; Petrovskaya and Leontovich, 1982; Janson, 1988; Baldi and Rinott, 1989; Mikhailov, 1991) obtained sufficient conditions of asymptotic normality of the r. v.

$$\eta = \sum_{j \in V} \xi_j$$

in terms of the so-called "dependency graph".
Let us consider the sequence of r. v.'s ζ_α, $\alpha \in W_r$ (see Section 1). Let $\varphi(l) = 0$ for any $l > s$. Let $\mathbf{W}_k(s)$ denote the subset of V_k such that $\boldsymbol{\mu} \in \mathbf{W}_k(s)$ if and only if there exists I, $I \neq \varnothing$, $I \neq V(\boldsymbol{\mu})$, $I \subset V(\boldsymbol{\mu})$, such that

$$\rho\Big(m(\{\alpha\colon \alpha \in I\}),\ m(\{\alpha\colon \alpha \in V(\boldsymbol{\mu})\backslash I\})\Big) > s.$$

It is clear that $\mathbf{W}_k(s) \subset \mathbf{N}_k$ and hence $m^{(1)}(\mathbf{W}_k(s)) = 0$.
However, if $\varphi(l) \neq 0$ for any $l > 0$, then equality (5) for $\boldsymbol{\mu} \in \mathbf{W}_k(s)$ is not true. At the same time we see that if condition (1) is true as $n \to \infty$, then

$$\lim_{n\to\infty} m^{(1)}(\mathbf{W}_k(s)) = 0. \tag{6}$$

Let us introduce two more sets. Let $g(\boldsymbol{\nu}) = g^{(n)}(\boldsymbol{\nu})$ be a non-negative function from $\widetilde{\mathbf{N}}_0^L$ to the set of real numbers $\mathbf{R}$. Then let $\mathbf{N}_k^{(1)}(g)$ be the subset of V_k, such that $\boldsymbol{\mu} \in \mathbf{N}_k^{(1)}(g)$ if and only if there exists a set I, $I \neq \varnothing$, $I \neq V(\boldsymbol{\mu})$, $I \subset V(\boldsymbol{\mu})$, such that

$$\operatorname{Cov}\left(\prod_{j\in I} \xi_j^{\nu_j},\ \prod_{j\in V\backslash I} \xi_j^{\nu_j}\right) \leqslant g(\boldsymbol{\nu})$$

for any $\boldsymbol{\nu} = (\nu_1, \dots, \nu_L) \leqslant \boldsymbol{\mu}$.
Let $\mathbf{N}_k^{(2)}(g)$ denote the subset of V_k such that $\boldsymbol{\mu} \in \mathbf{N}_k^{(2)}(g)$ if and only if there exists a set I, $I \neq \varnothing$, $I \neq \{1, \dots, k\}$, $I \subset \{1, \dots, k\}$, such that

$$\operatorname{Cov}\left(\prod_{j\in I} \xi_{b_j(\boldsymbol{\mu})},\ \prod_{j\in\{1,\dots,k\}\backslash I} \xi_{b_j(\boldsymbol{\mu})}\right) \leqslant g(\boldsymbol{\mu}).$$

Further $N_k^{(j)}(g) = V_k\backslash\mathbf{N}_k^{(j)}(g)$, $j = 1, 2$, $k \geqslant 2$, and $N_1^{(j)}(g) = V$, $j = 1, 2$. We write $\boldsymbol{\mu}_k$ for $\sum_{\boldsymbol{\nu}\in V_k} |\mathbf{E}\,\boldsymbol{\nu}|/\boldsymbol{\nu}!$, $M_k^{(j)}$ for $\sum_{\boldsymbol{\nu}\in N_k^{(j)}(g)} |\mathbf{E}\,\boldsymbol{\nu}|/\boldsymbol{\nu}!$, $j = 1, 2$, and G_k for $\sum_{\boldsymbol{\nu}\in V_k} g(\boldsymbol{\nu})/\boldsymbol{\nu}!$. Let us prove three theorems on the convergence of cumulants and moments.

4. CONVERGENCE OF CUMULANTS AND MOMENTS OF SUMS OF DEPENDENT RANDOM VARIABLES

THEOREM 3. *Suppose that there exists a non-negative function $g(\boldsymbol{\nu}) = g^{(n)}(\boldsymbol{\nu})$ from $\widetilde{\mathbf{N}}_0^L$ to $\mathbf{R}$ such that for any $\boldsymbol{\nu}_1, \boldsymbol{\nu}_2 \in \widetilde{\mathbf{N}}_0^L$*

$$g(\boldsymbol{\nu}_1)g(\boldsymbol{\nu}_2) \leqslant g(\boldsymbol{\nu}_1 + \boldsymbol{\nu}_2), \tag{7}$$

and for any fixed $t \in \mathbf{N} = \mathbf{N}_0\backslash\{0\}$ and $(k_1, \dots, k_t) \in \mathbf{N}^t$

$$\lim_{n\to\infty} G_{k_1} \prod_{i=2}^{t} M_{k_i}^{(2)} = 0. \tag{8}$$

Then for any fixed $k \in \mathbf{N}$ and any $W \subset \mathbf{N}_k^{(1)}(g)$

$$\lim_{n\to\infty} m^{(1)}(W) = 0.$$

Notice that the function

$$g_0(\boldsymbol{\nu}) = 2^{|\boldsymbol{\nu}|}\min\{1, 4r\varphi(s+1)\} \prod_{\alpha\in W_r} C_\alpha^{\nu_\alpha}, \qquad \boldsymbol{\nu} = (\nu_\alpha, \alpha\in W_r), \tag{9}$$

satisfies condition (7) and if condition (1) is fulfilled, the function $g_0(\boldsymbol{\nu})$ satisfies condition (8). It is well known that $\mathbf{W}_k(s) \subset \mathbf{N}_k^{(1)}(g)$ and condition (6) follows from Theorem 3.
The following theorem is a consequence of Theorem 3.

THEOREM 4. *Suppose that the conditions of Theorem 3 are satisfied and for any fixed $\mathbf{r} \in \widetilde{\mathbf{N}}_0^l$ there exists a set $W(\mathbf{r}) \subset A(\mathbf{r})$ such that $N_{|\mathbf{r}|}^{(1)}(g) \cap A(\mathbf{r}) \subset W(\mathbf{r})$ and $\lim_{n\to\infty} m^{(1)}(W(\mathbf{r}))$ exists and is finite. Then for any fixed $\mathbf{r} \in \widetilde{\mathbf{N}}_0^l$*

$$\lim_{n\to\infty} (\mathbf{r}!)^{-1} S_{\boldsymbol{\xi}}(\mathbf{r}) = \lim_{n\to\infty} m^{(1)}(W(\mathbf{r})).$$

Proof of Theorem 3. If $\boldsymbol{\mu} = (\mu_1, \dots, \mu_L) \in \mathbf{N}_k^{(1)}(g)$, then there exists a set I, $I \neq \emptyset, I \neq V(\boldsymbol{\mu})$, $I \subset V(\boldsymbol{\mu})$ such that

$$\operatorname{Cov}\left(\prod_{i\in I} \xi_i^{\nu_i}, \prod_{i\in V\setminus I} \xi_i^{\nu_i}\right) \leqslant g(\boldsymbol{\nu})$$

for any $\boldsymbol{\nu} = (\nu_1, \dots, \nu_L) \leqslant \boldsymbol{\mu}$. From these sets we choose a set such that the value $\sum_{i\in I} 2^i$ attains the minimum. Denote this set by $I(\boldsymbol{\mu})$. If $\boldsymbol{\mu} \in \mathbf{N}_k^{(1)}(g)$, $\boldsymbol{\nu} = (\nu_1, \dots, \nu_L) \leqslant \boldsymbol{\mu}$, then let

$$P_{\boldsymbol{\mu}}(\boldsymbol{\nu}) = \mathbf{E} \prod_{i\in I(\boldsymbol{\mu})} \xi_i^{\nu_i}\, \mathbf{E} \prod_{i\in V\setminus I(\boldsymbol{\mu})} \xi_i^{\nu_i},$$

and let $L_{\boldsymbol{\mu}}(\boldsymbol{\nu}) = (\mathbf{E}_{\boldsymbol{\nu}} - P_{\boldsymbol{\mu}}(\boldsymbol{\nu}))$.
Zhurbenko (Zhurbenko, 1970) showed that for $\boldsymbol{\mu} \in \mathbf{N}_k^{(1)}(g)$ and $\boldsymbol{\nu}_r \in \widetilde{\mathbf{N}}_0^L$, $1 \leqslant r \leqslant s$, such that $\sum_{r=1}^s \boldsymbol{\nu}_r = \boldsymbol{\mu}$

$$\prod_{r=1}^s \mathbf{E}_{\boldsymbol{\nu}_r} = \sum_{j=1}^s \prod_{i=1}^{j-1} P_{\boldsymbol{\mu}}(\boldsymbol{\nu}_i) L_{\boldsymbol{\mu}}(\boldsymbol{\nu}_j) \prod_{i=j+1}^s \mathbf{E}_{\boldsymbol{\nu}_i} + \prod_{i=1}^s P_{\boldsymbol{\mu}}(\boldsymbol{\nu}_i),$$

and

$$\sum_{s=1}^k \frac{(-1)^{s-1}}{s} \sum_{\boldsymbol{\nu}_1+\dots+\boldsymbol{\nu}_s=\boldsymbol{\mu}}^{L} \prod_{r=1}^s (\boldsymbol{\nu}_r)^{-1} P_{\boldsymbol{\mu}}(\boldsymbol{\nu}_r!) = 0.$$

Using these two last equalities we obtain that for any $W \subset \mathbf{N}_k^{(1)}(g)$

$$|m^{(1)}(W)| \leqslant \delta(k) \sum_{s=2}^{k} \frac{1}{s} \sum_{\nu_1+\cdots+\nu_s=k}^{L} \prod_{r=1}^{s-1} (|\nu_r|+1),$$

where $\delta(m) = \max\{G_{k_1} \prod_{i=2}^{t} M_{k_i}, 1 \leqslant t \leqslant m, k_1+\cdots+k_t = m, k_i \in \mathbf{N}, 1 \leqslant i \leqslant t\}$. It follows that if

$$\lim_{n\to\infty} \delta(k) = 0, \tag{10}$$

then

$$\lim_{n\to\infty} m^{(1)}(W) = 0.$$

Let us prove equality (10). Let $F(\mathbf{k})$, $\mathbf{k} = (k_1, \ldots, k_t) \in \mathbf{N}^t, |\mathbf{k}| = r$, denote a subset of V_r such that $\boldsymbol{\mu} \in F(\mathbf{k})$ if and only if there exist $\boldsymbol{\nu}_i \in N_k^{(2)}(g)$, $1 \leqslant i \leqslant t$, such that $\boldsymbol{\nu}_1 + \cdots + \boldsymbol{\nu}_t = \boldsymbol{\mu}$ and

$$\left|\mathbf{E}_{\boldsymbol{\mu}} - \prod_{i=1}^{t} \mathbf{E}_{\boldsymbol{\nu}_i}\right| \leqslant t! \sum_{I \subset \{1,\ldots,t\}} \prod_{i \in I} |\mathbf{E}_{\boldsymbol{\nu}_i}| g\left(\sum_{i \in \{1,\ldots,t\} \setminus I} \boldsymbol{\nu}_i\right).$$

By induction we obtain the following lemma.

LEMMA. *For any fixed $r \in \mathbf{N}$ and any $\boldsymbol{\mu} \in V_r$ there exists $\mathbf{k} = (k_1, \ldots, k_t) \in \mathbf{N}^t$ such that $|\mathbf{k}| = r$, $k_1 \leqslant \cdots \leqslant k_t$ and $\boldsymbol{\mu} \in F(\mathbf{k})$.*

Using Lemma we obtain that

$$\mu_m \leqslant \sum_{\boldsymbol{\beta} \in Z_m} \sum_{\boldsymbol{\mu} \in F(b(\boldsymbol{\beta}))} \frac{|\mathbf{E}_{\boldsymbol{\mu}}|}{\boldsymbol{\mu}!},$$

where

$$Z_m = \left\{(\beta_1, \ldots, \beta_m) \colon \sum_{i=1}^{m} i\beta_i = m,\ \beta_i \in \mathbf{N}_0,\ 1 \leqslant i \leqslant m\right\}.$$

From the definition of $F(\mathbf{k})$ it follows that

$$\sum_{\boldsymbol{\nu} \in F(b(\boldsymbol{\beta}))} \frac{|\mathbf{E}_{\boldsymbol{\nu}}|}{\boldsymbol{\nu}!} \leqslant \prod_{i=1}^{m} (M_i^{(2)})^{\beta_i} + \Delta(m) 2^{|\boldsymbol{\beta}|} |\boldsymbol{\beta}|!,$$

where

$$\Delta(m) = \max\left\{G_{k_1} \prod_{i=2}^{t} M_{k_i}^{(2)} \colon 1 \leqslant t \leqslant m,\ k_1 + \cdots + k_t = m,\ k_i \in \mathbf{N},\ 1 \leqslant i \leqslant t\right\}.$$

Thus if condition (8) is true, then condition (10) is also true. Theorem 3 (and Theorem 4) is proved.

The sum $m^{(1)}(W)$ is inconvenient to calculations, therefore we consider the sum $m^{(2)}(W) = \sum_{\boldsymbol{\nu} \in W} (\boldsymbol{\nu}!)^{-1} \mathbf{E}_{\boldsymbol{\nu}}$.

Let us obtain sufficient conditions of convergence of moments of the r. vec. ξ in terms of the sums $m^{(2)}(W)$ and the sets $N_k^{(2)}(g)$, $k \geqslant 2$. It is well known that

$$(\mathbf{r}!)^{-1}\prod_{i=1}^{l}\eta_i^{r_i} = m^{(2)}(A(\mathbf{r})), \qquad \mathbf{r} = (r_1,\ldots,r_l) \in \widetilde{\mathbf{N}}_0^l, \tag{11}$$

and if $\mathbf{P}\{\xi_i = 0\} = 1 - \mathbf{P}\{\xi_i = 1\}$ for any $i \in V$, then

$$(\mathbf{r}!)^{-1}\prod_{i=1}^{l}\eta_i^{[r_i]} = m^{(2)}(A(\mathbf{r}) \cap V_{|\mathbf{r}|}^{(3)}), \qquad \mathbf{r} = (r_1,\ldots,r_l) \in \widetilde{\mathbf{N}}_0^l, \tag{12}$$

where

$$V_k^{(3)} = \{\boldsymbol{\mu} = (\mu_1,\ldots,\mu_L)\colon \boldsymbol{\mu} \in V_k, \mu_j \leqslant 1,\ 1 \leqslant j \leqslant L\}, \quad k \geqslant 1.$$

Therefore we will consider the sums $m^{(2)}(A_i(\mathbf{r}))$, $i = 2,3$, where $A_2(\mathbf{r}) = A(\mathbf{r})$ and $A_3(\mathbf{r}) = A(\mathbf{r}) \cap V_{|\mathbf{r}|}^{(3)}$.
In what follows $N_k^{(3)}(g) = N_k^{(2)}(g) \cap V_k^{(3)}$, $V_k^{(2)} = V_k$, $k \geqslant 1$. Let $H^{(i)}(\mathbf{k})$, $\mathbf{k} = (k_1,\ldots,k_t) \in \mathbf{N}^t$, $i = 2,3$, denote the set

$$\{(\boldsymbol{\nu}_1,\ldots,\boldsymbol{\nu}_t)\colon \boldsymbol{\nu}_j \in N_{k_j}^{(i)}(g),\ 1 \leqslant j \leqslant t,\ \boldsymbol{\nu}_j = (\nu_1^{(j)},\ldots,\nu_L^{(j)})\}.$$

Let $K^{(i)}(\mathbf{k})$ denote the subset of $H^{(i)}(\mathbf{k})$ such that $(\boldsymbol{\nu}_1,\ldots,\boldsymbol{\nu}_t) \in K^{(i)}(k)$ if and only if $\nu_r^{(m)}\nu_r^{(j)} = 0$ for any $m \neq j$ and $1 \leqslant r \leqslant L$ and for any $\boldsymbol{\nu} \in N_{|\boldsymbol{\nu}|}^{(2)}(g)$ such that $\boldsymbol{\nu} \leqslant \sum_{i=1}^{t}\boldsymbol{\nu}_i$ there exists $r, 1 \leqslant r \leqslant t$, such that $\boldsymbol{\nu} \leqslant \boldsymbol{\nu}_r$.
Let us prove the following theorem for fixed $i \in \{2,3\}$.

THEOREM 5. *Suppose that there exists a non-negative function $g(\boldsymbol{\nu}) = g^{(n)}(\boldsymbol{\nu})$ from $\widetilde{\mathbf{N}}_o^L$ to $\mathbf{R}$ such that condition (7) is fulfilled, for any fixed $\mathbf{r} \in \widetilde{\mathbf{N}}_0^l$*

$$\lim_{n\to\infty} m^{(2)}(A(\mathbf{r}) \cap N_{|\mathbf{r}|}^{(i)}(g)) = l_i(\mathbf{r}) < +\infty,$$

and for any fixed $t \in \mathbf{N}$ and $\mathbf{k} = (k_1,\ldots,k_t) \in \mathbf{N}^t$

$$\lim_{n\to\infty}\Bigg(\sum_{\boldsymbol{\nu}\in V_{k_1}^{(i)}}\frac{g(\boldsymbol{\nu})}{\boldsymbol{\nu}!}\Bigg)\prod_{r=2}^{t}\Bigg(\sum_{\boldsymbol{\nu}\in N_{k_r}^{(i)}(g)}\frac{|\mathbf{E}_{\boldsymbol{\nu}}|}{\boldsymbol{\nu}!}\Bigg) = 0. \tag{13}$$

$$\lim_{n\to\infty}\sum_{(\boldsymbol{\nu}_1,\ldots,\boldsymbol{\nu}_t)\in H^{(i)}(\mathbf{k})\setminus K^{(i)}(\mathbf{k})}\prod_{r=1}^{t}\frac{|\mathbf{E}_{\boldsymbol{\nu}_r}|}{\boldsymbol{\nu}_r!} = 0. \tag{14}$$

Then for any fixed $\mathbf{r} \in \widetilde{\mathbf{N}}_0^l$

$$\lim_{n\to\infty} m^{(2)}(A_i(\mathbf{r})) = \sum_{s=1}^{|\mathbf{r}|}\frac{1}{s!}\sum_{\boldsymbol{\nu}_1+\cdots+\boldsymbol{\nu}_s=\mathbf{r}}\prod_{j=1}^{s} l_i(\boldsymbol{\nu}_j). \tag{15}$$

Notice that the well-known Sevastyanov's theorem (see (Sevastyanov, 1972)) is a consequence of Theorem 5.

Proof of Theorem 5. Notice that if $t \neq m$ and $\mathbf{k}_1 \in \mathbf{N}^t, \mathbf{k}_2 \in \mathbf{N}^m$ or $\mathbf{k}_1, \mathbf{k}_2 \in \mathbf{N}^t$ but $\mathbf{k}_1 \neq \mathbf{k}_2$, then $K^{(i)}(\mathbf{k}_1) \cap K^{(i)}(\mathbf{k}_2) = \varnothing$. Here $\widetilde{K}^{(i)}(\mathbf{k})$, $\mathbf{k} = (k_1, \dots, k_t) \in \mathbf{N}^t$, is the subset of $V_{|\mathbf{k}|}$ such that $\boldsymbol{\mu} \in \widetilde{K}^{(i)}(\mathbf{k})$ if and only if there exist $\boldsymbol{\nu}_j \in N_{k_j}^{(i)}(g)$, $1 \leqslant j \leqslant t$, such that $\boldsymbol{\nu}_1 + \dots + \boldsymbol{\nu}_t = \boldsymbol{\mu}$ and $(\boldsymbol{\nu}_1, \dots, \boldsymbol{\nu}_t) \in K^{(i)}(g)$.
Using this assertion and Lemma we obtain that

$$m^{(2)}(A_i(\mathbf{r})) = \left(\sum_{\beta \in Z_{|\mathbf{r}|}} \sum_{\boldsymbol{\nu} \in \widetilde{K}^{(i)}(b(\beta)) \cap A(\mathbf{r})} \frac{\mathbf{E}_{\boldsymbol{\nu}}}{\boldsymbol{\nu}!} \right) + R, \tag{16}$$

where

$$|R| \leqslant \sum_{\beta \in Z_{|\mathbf{r}|}} \sum_{\boldsymbol{\nu} \in (F(b(\beta)) \setminus \widetilde{K}^{(i)}(b(\beta))) \cap V_{|\mathbf{r}|}^{(i)}} \frac{|\mathbf{E}_{\boldsymbol{\nu}}|}{\boldsymbol{\nu}!}.$$

Let $b^{(i)}(\mathbf{r}, \mathbf{k})$, $\mathbf{r} \in \widetilde{\mathbf{N}}_0^l$, $\mathbf{k} = (k_1, \dots, k_t) \in \mathbf{N}^t$, denote

$$\sum_{\boldsymbol{\nu}_1 + \dots + \boldsymbol{\nu}_t = \mathbf{r}}^{l} \prod_{s=1}^{t} l_i(\boldsymbol{\nu}_s) I(|\boldsymbol{\nu}_s| = k_s).$$

It is easy to see that

$$\lim_{n \to \infty} \sum_{(\boldsymbol{\nu}_1, \dots, \boldsymbol{\nu}_t) \in H^{(i)}(\mathbf{k}) \cap A(\mathbf{r})} \prod_{s=1}^{t} \frac{\mathbf{E}_{\boldsymbol{\nu}_s}}{\boldsymbol{\nu}_s} = b^{(i)}(\mathbf{r}, \mathbf{k}). \tag{17}$$

From the definition of $F(k)$, conditions (13) and (14) and equality (17) we obtain that $\lim_{n \to \infty} R = 0$,

$$\lim_{n \to \infty} \sum_{\boldsymbol{\nu} \in \widetilde{K}^{(i)}(b(\beta)) \cap A(\mathbf{r})} \frac{\mathbf{E}_{\boldsymbol{\nu}}}{\boldsymbol{\nu}!} = \frac{b^{(i)}(\mathbf{r}, b(\beta))}{\beta!}, \qquad \beta \in Z_{|\mathbf{r}|}. \tag{18}$$

Using these two last equalities and equality (16) we obtain (15). Theorem 5 is proved.

Remark. Let us consider the case, where the r. v.'s ξ_j, $j \in V$, are complex r. v.'s such that $\mathbf{E}\,|\xi_j|^k$ exists for any $k > 0$ and for any $1 \leqslant j \leqslant l$ the r. v. η_j is a real r. v. If $\varepsilon = \varepsilon_1 + i\varepsilon_2$, where $\varepsilon_1, \varepsilon_2$ are real r. v.'s, then $\mathbf{E}\,\varepsilon = \mathbf{E}\,\varepsilon_1 + i\mathbf{E}\,\varepsilon_2$. Let us define sets $N_k^{(i)}(g)$, $H^{(j)}(\mathbf{k})$, $K^{(j)}(\mathbf{k})$, $j = 2, 3$, and $F(\mathbf{k})$ as these sets are defined in the case where the r. v.'s $\xi_j, j \in V$, are a real r. v.'s. From the proofs of Lemma and Theorem 5 we easily see that they are true in this case as well.

5. PROOFS OF THE MAIN THEOREMS

Let us consider the case, where the moments of the r. v. $\eta = \sum_{i \in V} \xi_i$ converge to the moments of the normal or the Poisson distributions.

Let

$$R_k^{(i)} = \bigcup_{\mu \in N_k^{(i)}(g)} \Pi(\mu), \qquad k \in \mathbf{N}, \quad i = 1,2,3,$$

where $\Pi(\mu)$, $\mu = (\mu_1, \ldots, \mu_L) \in V_k$, is the subset of V^K such that $(a_1, \ldots, a_k) \in \Pi(\mu)$ if and only if $\sum_{i=1}^k I(a_i = j) = \mu_j$ for any $1 \leqslant j \leqslant L$.
Suppose that for any $k, m \in \mathbf{N}$, $\sigma = (a_1, \ldots, a_k) \in V^k$, $j \in V$ and $i = \{1,2,3\}$ there exist $U^{(i)}(\sigma, m) \subset V$ and a nonrandom $Q^{(i)}(k,m,j) = Q^{(i,n)}(k,m,j)$ such that $(a_1, \ldots, a_{k+m}) \notin R_{k+m}^{(i)}$ for any $(a_{k+1}, \ldots, a_{k+m}) \in (U^{(i)}(\sigma, m))^m$, and if $\sigma \in V^k(j) \backslash V^k(j+1)$, $V(j) = \{j, j+1, \ldots, L\}$, then

$$\sum_{s \in V^{(i)}(\sigma,m) \cap V(j)} \mathbf{E}\left(|\xi_s| \, |\sigma(\xi_{a_1}, \ldots, \xi_{a_k})\right) \leqslant Q^{(i)}(k,m,j) \quad \text{a.s.},$$

where $V^{(i)}(\sigma, m) = V \backslash U^{(i)}(\sigma, m)$.
Notice that for the r. v.'s ζ_α, $\alpha \in W_r$ (see Section 1), for $g(\nu) = g_0(\nu)$ (see (9)) and any $k, m \in \mathbf{N}$ and $\sigma = (\alpha_1, \ldots, \alpha_k) \in W_r^k$

$$V^{(i)}(\sigma, m) \subset W_r\big(m(\alpha_1, \ldots, \alpha_k), s\big), \qquad i = 1,2,$$

$$V^{(3)}(\sigma, m) \subset W_r\big(m(\alpha_1, \ldots, \alpha_k), s\big) \backslash \{\alpha_1, \ldots, \alpha_k\},$$

and hence for any $k, m \in \mathbf{N}$ and $j \in V$

$$Q^{(i)}(k,m,j) \leqslant T(k,s), \qquad i = 1,2, \tag{19}$$

$$Q^{(3)}(k,m,j) \leqslant T(k,s) - k. \tag{20}$$

Let $\mathbf{E}\,\xi_i = 0$ for any $i \in V$, $\operatorname{Var}\eta = 1$, and

$$B(t,j) = \min\{\mathbf{E}\,|\xi_j| Q^{(1)}(1, t-1, j),\ B_0(t,j)\},$$

where

$$B_0(t,j) = \sum_{i \in \bigcup_{s=1}^{t-1} V^{(2)}(j,s) \bigcap V(j)} \mathbf{E}\,|\xi_j \xi_i|.$$

THEOREM 6. *Suppose that the condition of Theorem 3 are satisfied and there exists a constant $k \geqslant 3$ such that for any fixed $t \geqslant k$*

$$\lim_{n \to \infty} \sum_{j=1}^{L} B(t,j) \prod_{i=2}^{t-1} Q^{(1)}(i, t-i, j) = 0. \tag{21}$$

Then moments of the r. v. η converge to moments of the standard normal distribution as $n \to \infty$.

It is easy to see that Theorem 1 follows from equality (6), inequality (19), and Theorem 6.

Let

$$\max\{Q^{(1)}(m,t-m,j) | 1 \leqslant m \leqslant t-1,\ j \in N\} \leqslant C_t Q_n,$$

where C_t depends only on t and Q_n depends only on n. Mikhailov (see (Mikhailov, 1991)) proved that the condition

$$\lim_{n\to\infty} \Big(\sum_{i\in V} \mathbf{E}|\xi_i|\Big)^{\alpha} Q_n^{2-\alpha} = 0 \tag{22}$$

for some fixed $\alpha, 0 < \alpha \leqslant 2/3$, is a sufficient condition for the convergence of the moments of the r. v. η to the moments of the standard normal distribution under the assumption that $g(\boldsymbol{\nu}) = 0$ for any $\boldsymbol{\nu} \in \widetilde{\mathbf{N}}_0^L$. Obviously, condition (22) follows from condition (21) and Theorem 2 from (Mikhailov, 1991) is a consequence of Theorem 6.

Proof of Theorem 6. Let

$$N_{t,j}^{(i)}(g) = \{\boldsymbol{\mu} = (\mu_1,\dots,\mu_L)\colon \boldsymbol{\mu} \in N_t^{(i)}(g),\ \mu_1 = \dots = \mu_{j-1} = 0,\ \mu_j \neq 0\},$$
$$j \in V, \quad i = 1,2,3.$$

Obviously,

$$m^{(1)}(N_t^{(i)}(g)) = \sum_{j=1}^{L} m^{(1)}(N_{t,j}^{(i)}(g)). \tag{23}$$

From the definitions of $V^{(1)}(\sigma,m)$ and $N_{t,j}^{(1)}(g)$ it follows that for any $\boldsymbol{\mu} \in N_{t,j}^{(1)}(g)$ and $\boldsymbol{\nu}_r \in \mathbf{N}_0^L$, $1 \leqslant r \leqslant s$, such that $\boldsymbol{\nu}_1 + \dots + \boldsymbol{\nu}_s = \boldsymbol{\mu}$ there exists a permutation $\pi(1),\dots,\pi(t)$ of numbers $1,\dots,t$ such that $b_{\pi(1)}(\boldsymbol{\mu}) = j$

$$b_{\pi(r+1)}(\boldsymbol{\mu}) \in V^{(1)}\Big(\big(b_{\pi(1)}(\boldsymbol{\mu}),\dots,b_{\pi(r)}(\boldsymbol{\mu})\big), t-r\Big) \cap V(j), \qquad r = 1,\dots,t-1,$$

and

$$\prod_{r=1}^{s} |\mathbf{E}_{\boldsymbol{\nu}_r}| \leqslant \mathbf{E}\,|\xi_j| \prod_{r=2}^{t} \mathbf{E}\Big(|\xi_{b_{\pi(r)}}(\boldsymbol{\mu})| \,\big|\, \sigma(\xi_{b_{\pi(1)}}(\boldsymbol{\mu}),\dots,\xi_{b_{\pi(r-1)}}(\boldsymbol{\mu}))\Big).$$

From the last inequality we obtain that

$$m^{(1)}(N_{t,j}^{(1)}(g)) \leqslant \mathbf{E}\,|\xi_j| \Big(\prod_{r=1}^{t} Q^{(1)}(r,t-r,j)\Big) 2^{t-1}. \tag{24}$$

Let us prove that

$$\varlimsup_{n\to\infty} m^{(1)}(N_{t,j}^{(1)}(g)) \leqslant 2^{t-1} \varlimsup_{n\to\infty} B_0(t,j) \prod_{r=2}^{t} Q^{(1)}(r,t-r,j). \tag{25}$$

Let

$$\mathfrak{N}(s,t,j) = \Big\{(\boldsymbol{\mu},\boldsymbol{\nu}_1,\dots,\boldsymbol{\nu}_s)\colon \boldsymbol{\nu}_i \in \widetilde{\mathbf{N}}_0^L, 1 \leqslant i \leqslant s,\ \sum_{i=1}^{s} \boldsymbol{\nu}_i = \boldsymbol{\mu} \in N_{t,j}^{(1)}(g)\Big\},$$

$$\mathfrak{N}_1(s,t,j)=\Big\{(\boldsymbol{\mu},\boldsymbol{\nu}_1,\ldots,\boldsymbol{\nu}_s)\colon (\boldsymbol{\mu},\boldsymbol{\nu}_1,\ldots,\boldsymbol{\nu}_s)\in\mathfrak{N}(s,t,j),\ \sum_{i=1}^{s} I(|\boldsymbol{\nu}_i|=1)=0\Big\}.$$

Obviously,

$$\sum_{s=1}^{t}\frac{(-1)^{s-1}}{s}\sum_{(\boldsymbol{\mu},\boldsymbol{\nu}_1,\ldots,\boldsymbol{\nu}_s)\in\mathfrak{N}(s,t,j)\setminus\mathfrak{N}_1(s,t,j)}\prod_{r=1}^{s}\frac{\mathbf{E}_{\boldsymbol{\nu}_r}}{\boldsymbol{\nu}_r!}=0.$$

Let $\mathfrak{N}_2(s,t,j)$ denote the subset of $\mathfrak{N}_1(s,t,j)$ such that $(\boldsymbol{\mu},\boldsymbol{\nu}_1,\ldots,\boldsymbol{\nu}_s)\in\mathfrak{N}_2(s,t,j)$ if and only if there exists $r, 1\leqslant r\leqslant s$, such that $(b_1(\boldsymbol{\nu}_r)=j$ and $b_i(\boldsymbol{\nu}_r)\notin V^{(2)}(j,|\boldsymbol{\nu}_r|-1)$ for any $i\in\{2,\ldots,|\boldsymbol{\nu}_r|\}$. Hence $|\mathbf{E}_{\boldsymbol{\nu}_r}|\leqslant g(\boldsymbol{\nu}_r)$. By virtue of the method that was used in the proof of Theorem 3 we obtain that

$$\lim_{n\to\infty}\sum_{s=1}^{t}\frac{(-1)^{s-1}}{s}\sum_{(\boldsymbol{\mu},\boldsymbol{\nu}_1,\ldots,\boldsymbol{\nu}_s)\in\mathfrak{N}_2(s,t,j)}\prod_{r=1}^{s}\frac{\mathbf{E}_{\boldsymbol{\nu}_r}}{\boldsymbol{\nu}_r!}=0.$$

It is clear that for any $(\boldsymbol{\mu},\boldsymbol{\nu}_1,\ldots,\boldsymbol{\nu}_s)\in\mathfrak{N}_1(s,t,j)\setminus\mathfrak{N}_2(s,t,j)$ there exists a permutation $\pi(1),\ldots,\pi(t)$ of the numbers $1,\ldots,t$ such that

$$b_{\pi(1)}(\boldsymbol{\mu})=1,\quad b_{\pi(2)}(\boldsymbol{\mu})\in\bigcup_{r=1}^{t-1}V^{(2)}(j,r)\ \cap V(j),$$

$$b_{\pi(r+1)}(\boldsymbol{\mu})\in V^{(1)}\Big(\big(b_{\pi(1)}(\boldsymbol{\mu}),\ldots,b_{\pi(r)}(\boldsymbol{\mu})\big),t-r\Big)\cap V(j),\qquad r=2,\ldots,t-1,$$

and

$$\prod_{r=1}^{s}|\mathbf{E}_{\boldsymbol{\nu}_r}|\leqslant\mathbf{E}\,|\xi_j\xi_{b_{\pi(2)}(\boldsymbol{\mu})}|\prod_{r=3}^{t}\mathbf{E}\left(|\xi_{b_{\pi(r)}(\boldsymbol{\mu})}|\,\big|\,\sigma(\xi_{b_{\pi(1)}(\boldsymbol{\mu})},\ldots,\xi_{b_{\pi(r-1)}(\boldsymbol{\mu})})\right).$$

This implies inequality (25). From inequalities (24) and (25) and condition (21) we obtain that $\lim_{n\to\infty}S_\eta(t)=0$ for any $t\geqslant k$. Now can be easily seen that Theorem 6 follows from Theorem 1 from (Janson, 1988). Theorem 6 (and Theorem 1) is proved.

Let us give sufficient conditions of convergence of the distribution of the r. v. η to the Poisson distribution.

THEOREM 7. *Suppose that* $\mathbf{P}\{\xi_i=0\}=1-\mathbf{P}\{\xi_i=1\}$ *for any* $i\in V$ *there exists a non-negative function* $g(\boldsymbol{\nu})=g^{(n)}(\boldsymbol{\nu})$ *from* $\widetilde{\mathbf{N}}_0^L$ *to* $\mathbf{R}$ *such that conditions (7) and (13) are fulfilled for* $i=3$, *for any* $i\in V$

$$\lim_{n\to\infty}\mathbf{E}\,\xi_i=0,\tag{26}$$

$$\lim_{n\to\infty}\mathbf{E}\,\eta=\lambda<\infty,\tag{27}$$

and for any fixed $k\geqslant 2$

$$\lim_{n\to\infty}\sum_{j=1}^{L}\Big(\sum_{i\in V^{(3)}(j,t-1)\cap V(j)}\mathbf{E}\,\xi_j\xi_i\Big)\prod_{i=2}^{t-1}Q^{(3)}(i,t-i,j)=0.\tag{28}$$

Then for any fixed $k \in \mathbf{N}_0$

$$\lim_{n\to\infty} \mathbf{P}\{\eta = k\} = \frac{\lambda^k e^{-\lambda}}{k!}. \tag{29}$$

It is easy to see that Theorem 2 follows from inequality (20) and Theorem 7.

Proof of Theorem 7. It is easy to see that for any $(a_1,\dots,a_k) \in R_k^{(3)} \cap (V^k(j)\backslash V^k(j+1))$ there exists a permutation $\pi(1),\dots,\pi(k)$ of numbers $1,\dots,k$ such that

$$a_{\pi(1)} = j, \quad a_{\pi(r+1)} \in V^{(3)}\big((a_{\pi(1)},\dots,a_{\pi(r)}),t-r\big) \cap V(j), \qquad r = 1,\dots,k-1,$$

and

$$\mathbf{E}\prod_{i=1}^{k} \xi_{a_i} \leqslant \mathbf{E}\xi_{a_{\pi(1)}}\xi_{a_{\pi(2)}} \prod_{i=3}^{k} \mathbf{E}\big(\xi_{a_{\pi(i)}} | \sigma(\xi_{a_{\pi(1)}},\dots,\xi_{a_{\pi(i-1)}}\big).$$

From the last inequality and equality (28) it follows that for any fixed $k \geqslant 2$

$$\lim_{n\to\infty} m^{(2)}(N_k^{(3)}(g)) = 0,$$

and for any fixed $t \in \mathbf{N}, \mathbf{k} = (k_1,\dots,k_t) \in \mathbf{N}^t$, such that $\sum_{i=1}^t I(k_i > 1) > 0$

$$\lim_{n\to\infty} \sum_{(\nu_1,\dots,\nu_t)\in H^{(3)}(\mathbf{k})} \prod_{i=1}^{t} \mathbf{E}_{\nu_i} = 0.$$

It follows from conditions (26) and (27) that for any fixed $t \in \mathbf{N}$

$$\lim_{n\to\infty} \sum_{(i_1,\dots,i_t)\in H^{(3)}(b(t))\backslash K^{(3)}(b(t))} \prod_{j=1}^{t} \mathbf{E}\xi_{i_j} = 0.$$

Using the three last equalities and Theorem 5 for $i = 3$, we obtain (29). Theorem 7 (and Theorem 2) is proved.

Acknowledgement. I am grateful to V. G. Mikhailov whose valuable suggestions improved this paper.

REFERENCES

Baldi, P. and Rinott, Y. (1989). On normal approximations of distributions in terms of dependency graphs. *Ann. Probab.* **17**, 1646–1650.

Brown, T. S. and Silverman, B. W. (1979). Rates of Poisson convergence for *U*-statistics. *J. Appl. Probab.* **16**, 428–432.

Frees, E. W. (1989). Infinite order U-statistics. *Scandinavian J. Statist. Theory.* **16**, 29–46.

Grusho, A. A. (1987). On the problem of distributions of U-statistics. *Theory Probab. Appl.* **32**, 369–371.

Janson, S. (1988). Normal convergence by higher semiinvariants with applications to sums of dependent random variables and random graphs. *Ann. Probab.* **16**, 305–312.

Khashimov, Sh. A. (1990). Asymptotic normality of generalized Mises functional for m-dependent random variables. In: *Theory of Random Processes and Applications*, Naukova Dumka, Kiev, pp. 141–148 (in Russian).

Korolyuk, V. S. and Borovskikh, Yu. V. (1989). *Theory of U-statistics.* Naukova Dumka, Kiev (in Russian).

Korolyuk, V. S. and Borovskikh, Yu. V. (1990). U-statistics for increasing degrees. *Doklady AN UkrSSR* **3**, 3–7 (in Russian).

Mikhailov, V. G. (1986). On asymptotic normality of U-statistics with non-negative kernels. *Proc. Steklov Math. Inst.* **177**, 105–113 (in Russian).

Mikhailov, V. G. (1989). On asymptotic normality of symmetric decomposable statistics in heterogeneous scheme. *Discrete Math. Appl.* **1**, 15–27 (in Russian).

Mikhailov, V. G. (1991). On a Janson's theorem. *Theory Probab. Appl.* **36**, 173–175.

Petrovskaya, M. V. and Leontovich, A. M. (1982). The central limit theorem for a sequence of random variables with a slowly growing number of dependences. *Theory Probab. Appl.* **27**, 815–826.

Ronzhin, A. F. (1979). On limit theorems for sums of dependent random variables. *Theory Probab. Appl.* **24**, 549–559.

Sevast'yanov, B. A. (1972). Poisson limit law for a scheme of sums of dependent random variables. *Theory Probab. Appl.* **17**, 695–700.

Zhurbenko, I. G. (1970). On the estimation of mixed semi-invariants for a certain class of random processes. *Theor. Probab. Appl.* **15**, 525–528.

Zubkov, A. M. (1977). Inequalities for the distribution of a sum of functions of independent random variables. *Math. Notes* **22**, 745–758 (in Russian).

Probabilistic Methods in Discrete Mathematics, pp. 200 – 211
V. F. Kolchin et al. (Eds.)
1993 TVP/VSP

LIMIT THEOREMS FOR DECOMPOSABLE STATISTICS OF DEPENDENT RANDOM VARIABLES

A. B. GORTCHAKOV
TVP Laboratories, Krasikov Street 32, 117418 Moscow, Russia

ABSTRACT

Let $X_1^{(n)}, \ldots, X_n^{(n)}$ be a sequence of positive integer-valued random variables (r. v.'s). Let ν_v, $v \in \mathbf{N} = \{1, 2, \ldots\}$, be the number of occurences of the outcome v in the sequence $X_1^{(n)}, \ldots, X_n^{(n)}$ and let $f_v(k)$, $v \in \mathbf{N}$, be some functions from $\mathbf{N} \cup \{0\}$ to the set of real numbers.
The r. v. $L = \sum_{v \in \mathbf{N}} f_v(\nu_v)$ is called a decomposable statistic. Mikhailov obtained an expression of L in terms of sums of U-statistics. Using this expression, in this paper we obtain sufficient conditions of convergence as $n \to \infty$ of the distributions of L to the normal and Poisson distributions under the assumption that the distributions of r. v.'s $X_1^{(n)}, \ldots, X_n^{(n)}$ and the functions $f_v(k)$, $v \in \mathbf{N}$, depend on n, $X_1^{(n)}, \ldots, X_n^{(n)}$ being a sequence of r. v.'s with strong mixing property and $a = \max\{\mathbf{E}\,\nu_v | v \in \mathbf{N}\} = o(\log n)$.

1. INTRODUCTION

Let $X_1^{(n)}, \ldots, X_n^{(n)}$ be a sequence of positive integer-valued random variables (r. v.'s). Further all quantities (random and nonrandom) depend on n unless otherwise specified, but we omit the index n.
Let ν_v, $v \in \mathbf{N} = \{1, 2, \ldots\}$, be the number of occurences of the outcome v in the sequence $X_1, \ldots, X_n$ and let $f_v(k)$, $v \in V$, be some functions from $\mathbf{N}_0 = \mathbf{N} \cup \{0\}$ to the set $\mathbf{R}$ of real numbers. Let $V = \{v \,|\, v \in \mathbf{N},\ \nu_v > 0\}$.
The r. v. $L = \sum_{v \in V} f_v(\nu_v)$ is called a decomposable statistic. Many results about asymptotic (as $n \to \infty$) distributions of decomposable statistics were obtained under the assumption that $X_1, \ldots, X_n$ are independent identically distributed r. v.'s. They were summarized in (Kolchin et al., 1978) and (Ivanov et al., 1984).
Set

$$R_n = \sum_{v \in V} f_v(0). \tag{1}$$

Let

$$\xi_r(\nu, v) = \sum_{1 \le i_1 < \cdots < i_r \le \nu} I(X_{i_1} = X_{i_2} = \cdots = X_{i_r} = v), \qquad r \le \nu \le n, \quad v \in V,$$

$$\xi_r = \sum_{v \in V} \xi_r(n, v),$$

where $I(R)$ is the indicator of a random event R.

Zubkov and Mikhailov (see (Zubkov and Mikhailov, 1974, 1979)) obtained sufficient conditions of the convergence of distributions of the r.v. L to the Poisson distribution under the assumption that $X_1, \ldots, X_n$ is a stationary sequence of s-chains, i.e., if $X_t = (Y_t, \ldots, Y_{t+s-1})$, $t = 1, 2, \ldots, n$, where $Y_1, \ldots, Y_{n+s-1}$ are independent positive integer-valued identically distributed r.v.'s. Mikhailov (Mikhailov, 1974) extended this result to a nonstationary case.

Belyaev (Belyaev, 1964, 1965) studied the probability of nonoccurrence of a chain from a given set of s-chains in the case where $Y_1, \ldots, Y_{n+s-1}$ is a stationary Markov chain.

It is easy to see that if condition (1) is valid, then we obtain

$$L = R_n + \sum_{t=1}^{n} \sum_{v \in V} C_v(t)\, \xi_t(n, v), \tag{2}$$

where

$$C_v(t) = \sum_{i=0}^{t} (-1)^{t-i} \binom{t}{i} f_v(i), \qquad \binom{t}{i} = \frac{t!}{i!(t-i)!}, \quad 0! = 1.$$

Using equality (2), Mikhailov obtained in (Mikhailov, 1989) sufficient conditions of asymptotic normality of the r.v. L under the assumption that $X_1, \ldots, X_n$ is a sequence of s-chains, s is a constant, $Y_1, \ldots, Y_{n+s-1}$ are independent positive integer-valued r.v.'s.

Also using equality (2), Dyakonova and Mikhailov (Dyakonova and Mikhailov, 1991) obtained more general sufficient conditions of asymptotic normality of the r.v. L under the assumption that $X_1, \ldots, X_n$ are independent r.v.'s.

Using equality (2) and results from (Gortchakov, 1992), we obtain sufficient conditions of convergence of the distributions of the r.v. L to the normal and Poisson distributions under the assumption that $X_1, \ldots, X_n$ is a sequence of r.v.'s. with strong mixing property.

2. MAIN THEOREMS

In what follows $X_1, \ldots, X_n$ is a sequence of positive integer-valued r.v.'s with strong mixing property with coefficient $\alpha(s) = \alpha^{(n)}(s)$. Let us introduce the following notation. Let $P_X = \max\{\mathbf{P}\{X_j = v\} | 1 \le j \le n,\ v \in V\}$,

$$\beta(\tau) = \max\left\{|\mathbf{P}\{X_j = v|A\} - \mathbf{P}\{X_j = v\}| \colon A \in V_0(j, \tau),\ v \in V,\ 1 \le j \le n\right\},$$

$$\gamma(\tau) = \max\left\{|\mathbf{P}\{X_j = v|A\} - \mathbf{P}\{X_j = v\}| \colon A \in V_1(j, \tau),\ v \in V,\ 1 \le j \le n\right\},$$

where $V_0(j, \tau)$, $1 \le j \le n$, $1 \le \tau \le n-1$, is the set of events $\{A = (X_{j_1} = v_1, \ldots, X_{j_t} = v_t) \colon 1 \le t \le n-1,\ 1 \le j_1 < \ldots < j_t \le n,\ v_i \in V,\ \min\{|j - j_k| \colon 1 \le k \le t\} \ge \tau,\ \mathbf{P}\{A\} > 0\}$, $V_1(j, \tau)$, $1 \le j \le n$, $1 \le \tau \le j-1$, is the set of events

$$\{A = (X_{j_1} = v_1, \ldots, X_{j_t} = v_t)\colon 1 \le t \le j - \tau,$$

$$1 \le j_1 < \cdots < j_t \le j - \tau,\ v_i \in V,\ \mathbf{P}\{A\} > 0\}.$$

Let
1. $a = \max\{\nu_v\colon v \in V\}$;
2. $\gamma = \sum_{\tau=1}^{n-1} \gamma(\tau)$;
3. $p_0 = \min\{1, P_X + \gamma(1)\}$;
4. $p_1 = \min\{1, P_X + \beta(1)\}$;
5. $G_r = a^{-1} \log \mathbf{E}\,\xi_r$;
6. $g_r = \begin{cases} \min\{G_r, \mathbf{E}\,\xi_r, \gamma^{-1}\} & \text{if } \gamma \ne 0, \\ \min\{G_r, \mathbf{E}\,\xi_r\} & \text{if } \gamma = 0, \end{cases}$
7. $m_r = r - 1 +](\log g_r)^{-1} \log \mathbf{E}\,\xi_r[$, where $]z[$ is the smallest integer which is greater than z.

Suppose that

$$|f_v(k)| \le Ce^{dk}, \qquad v \in V, \quad k \in \mathbf{N}_0, \tag{3}$$

where C and d are positive constants.
The following three theorems are valid.

THEOREM 1. *If $n \to \infty$ and the distributions of $X_1, \ldots, X_n$ and functions $f_v(k)$, $v \in V$, vary in such a way that conditions (1) and (3) are fulfilled uniformly in n and there exists $s = s^{(n)} \ge 1$ such that for any fixed $l > 0$ and some constants $\omega > 0$ and $r \ge 1$*

$$f_v(k) = 0, \qquad v \in V, \quad k = 0, 1, 2, \ldots, r-1, \tag{4}$$

$$\mathbf{E}\,\xi_r \to \infty, \quad (\mathbf{E}\,\xi_r)^{\omega} = O(\operatorname{Var} L), \tag{5}$$

$$g_r \to \infty, \tag{6}$$

$$\limsup_{n\to\infty} (\log s)(\log \mathbf{E}\,\xi_r)^{-1} < \omega/2, \tag{7}$$

$$\limsup_{n\to\infty} m_r (\log(1 + \sqrt{p_1} s))(\log \mathbf{E}\,\xi_r)^{-1} < \omega/2, \tag{8}$$

$$n\beta(s)(\log \mathbf{E}\,\xi_r)^{-1} \to 0, \tag{9}$$

$$m_r \left(\sum_{k=r}^{m_r} (\mathbf{E}\,\xi_r)^{-\omega/2} \binom{n}{k} \right)^{l} \alpha(s) \to 0, \tag{10}$$

then for any fixed $x \in \mathbf{R}$

$$\lim_{n\to\infty} \mathbf{P}\left\{\frac{L-\mathbf{E}\,L}{\sqrt{\operatorname{Var} L}} < x\right\} = \frac{1}{\sqrt{2\pi}}\int_{-\infty}^{x} e^{-t^2/2}\,dt. \tag{11}$$

THEOREM 2. *If $n \to \infty$ and for $r = 1$ all the conditions of Theorem 1 are fulfilled with the exception of condition (4), then (11) is true.*

THEOREM 3. *If $n \to \infty$ and for $r = 2$ all the conditions of Theorem 1 are fulfilled with the exception of condition (4) and there exist values $f_0 = f_0^{(n)}$, $f_1 = f_1^{(n)}$ such that $f_v(0) = f_0$ and $f_v(1) = f_1$ for any $v \in V$, then (11) is true.*

Remark 1. In (Dyakonova and Mikhailov, 1991) Theorems 1–3 have been proved under the assumption that $X_1, \dots, X_n$ are independent r. v.'s, $\omega = 1$ and $a = O(1)$ as $n \to \infty$. In the case where $X_1, \dots, X_n$ is a sequence of s-chains Mikhailov proved Theorem 2 (Mikhailov, 1974) under the assumption that s is a constant and $n\,P_X = O(1)$. It is easy to see that $a \le n\,P_X$ and the results of (Mikhailov, 1989) are consequences of Theorem 2.

Consider the case where there exists $f_0 = f_0^{(n)}$ such that $f_v(0) = f_0$ for any $v \in V$. Notice that in this case if condition (1) is true, then either $f_0 = 0$ or for any n there exists an integer T_n (T_n depends on N) such that $|V| = T_n$, where $|M|$ is the cardinality of M.

The following two theorems are valid.

THEOREM 4. *If $n \to \infty$ and the distributions of $X_1, \dots, X_n$ and functions $f_v(k)$, $v \in V$, vary in such a way that conditions (1) and (3) are fulfilled uniformly in n and there exist $s = s^{(n)} \ge 1$ and $f = f^{(n)} \ne 0$ such that for some constant $r \ge 2$ condition (4) is fulfilled, $f_v(r) = f$ for any $v \in V$, and for any fixed $t \ge 0$ values γ, sa^{r-1}, $s(n\beta(s))^{r-1}$ and $p_0 s(\sqrt{p_1})^{-1}$ are $o((\sqrt{p_1}s)^t)$ and*

$$\lim_{n\to\infty} \mathbf{E}\,\xi_r = \lambda < \infty, \tag{12}$$

$$\lim_{n\to\infty} n^t \alpha(s) = 0, \tag{13}$$

then for any fixed $k \in \mathbf{N}_0$

$$\lim_{n\to\infty} \mathbf{P}\left\{\frac{L}{f} = k\right\} = \frac{\lambda^k e^{-\lambda}}{k!}. \tag{14}$$

THEOREM 5. *If $n \to \infty$ and for $r = 2$ all the conditions of Theorem 4 are fulfilled with the exception of condition (4) and there exist $f_0 = f_0^{(n)}$, $f_1 = f_1^{(n)}$ such that $f_v(0) = f_0$, $f_v(1) = f_1$ for any $v \in V$ and $f \ne 2f_1 - f_0$, then for any fixed $k \in \mathbf{N}_0$*

$$\lim_{n\to\infty} \mathbf{P}\left\{\frac{L - R_n - (f_1 - f_0)n}{f - 2f_1 + f_0} = k\right\} = \frac{\lambda^k e^{-\lambda}}{k!}. \tag{15}$$

Remark 2. In the case where $X_1, \ldots, X_n$ is a sequence of s-chains Mikhailov proved Theorem 4 under the assumption that condition (12) is fulfilled and $nP_X = o(s^t)$ for any $t \geq 0$. It is easy to see that $p_1 = 1$, $p_0 \leq 2(P_X)^{1/s}$ and the result of (Mikhailov, 1974) is a consequence of Theorem 2.

3. PROOFS OF THE MAIN THEOREMS

Obviously, $\sum_{v \in V} f_v(0)$ is a nonrandom value. Moreover, from the conditions of Theorems 3 and 5 it follows that $\sum_{v \in V} C_v(1)\xi_1(n,v) = (f_1 - f_0)n$, and it follows from condition (4) that $C_v(t) = 0$ for any $v \in V$ and $t \leq r-1$. Therefore, we will consider the r. v.

$$L_r = \sum_{t=r}^{n} \sum_{v \in V} C_v(t)\xi_t(n,v),$$

where r is a constant, $r \leq n$.

Let us introduce the following notation.

Let $W_t = \{(i_1, \ldots, i_t) | 1 \leq i_1 < \cdots < i_t \leq n\}$, $1 \leq t \leq n$, and for $\vec{i} = (i_1, \ldots, i_t) \in W_t$, $\vec{j} = (j_1, \ldots, j_m) \in W_m$

$$\rho(\vec{i}, \vec{j}) = \min\{|i_k - j_l| \,|\, 1 \leq k \leq t,\ 1 \leq l \leq m\},$$

and let $m(\vec{i}_1, \ldots, \vec{i}_l)$, $\vec{i}_k \in W_{r_k}$, $1 \leq k \leq l$, denote the element $\vec{j} = (j_1, \ldots, j_t) \in W_t$ such that $\min\{\rho(j_k, \vec{i}_m) | 1 \leq m \leq l\} = 0$ for any $1 \leq k \leq t$ and $\min\{\rho(j, \vec{i}_m) | 1 \leq m \leq l\} > 0$ for any $j \in \{1, \ldots, n\} \setminus \{j_1, \ldots, j_t\}$.

Let

$$Q(t,l,s-1) = Q^{(n)}(t,l,s-1), \qquad t,l,s \in I_n = \{1, \ldots, n\},$$

be nonrandom values such that for any $m \leq t$ and $\vec{i} \in W_m$

$$\sum_{\vec{j} \in W_l(\vec{i}, s-1)} \sum_{v \in V} \mathbf{E}(\eta_{\vec{j}}(v) | \mathcal{F}_{\vec{i}}) \leq Q(t,l,s-1) \quad \text{a.s.},$$

where $\eta_{\vec{j}}(v) = I(X_{j_1} = \cdots = X_{j_l} = v)$, $\vec{j} = (j_1, \ldots, j_l) \in W_l$,

$$W_l(\vec{i}, s) = \{\vec{j} | \rho(\vec{i}, \vec{j}) \leq s,\ \vec{j} \in W_l\},$$

$$\mathcal{F}_{\vec{i}} := \sigma(X_{i_1}, \ldots, X_{i_m}), \qquad \vec{i} = (i_1, \ldots, i_m) \in W_m.$$

It is clear that such values $Q(t,l,s-1)$, $t,l,s \in I_n$, exist.

First we prove five auxiliary lemmas. Let

$$\eta_{\bar{\imath}} = \sum_{v \in V} \eta_{\bar{\imath}}(v).$$

The two following lemmas are true.

LEMMA 1. *If $n \to \infty$ and the distributions of the r. v.'s $X_1, \ldots, X_n$ and functions $f_v(k)$, $v \in V$, vary in such a way that conditions (1) and (3) are fulfilled uniformly in n and there exist $m = m^{(n)} \geq r$ and $s = s^{(n)} \geq 1$ and a constant $k \geq 3$ such that the r. v.*

$$\bar{\varepsilon}_m = \left(\sqrt{\operatorname{Var} L_r}\right)^{-1} \left(L_{m+1} - L_r - \mathbf{E}\left(L_{m+1} - L_r\right)\right)$$

converge (as $n \to \infty$) to zero in probability and for any fixed $l \geq k$

$$m \left(\left(\sqrt{\operatorname{Var} L_r}\right)^{-1} \sum_{t=r}^{m} \binom{n}{t} \right)^{l} \alpha(s) \to 0, \tag{16}$$

$$\left(\sqrt{\operatorname{Var} L_r}\right)^{-l} \left(\sum_{t=r}^{m} (1+e^d)^t \mathbf{E}\, \xi_t \right) \left(\sum_{t=r}^{m} Q((l-1)m,\ t,\ s-1)(1+e^d)^t \right)^{l-1} \to 0, \tag{17}$$

then for any fixed $x \in \mathbf{R}$

$$\lim_{n \to \infty} \mathbf{P}\left\{ \frac{L_r - \mathbf{E}\, L_r}{\sqrt{\operatorname{Var} L_r}} < x \right\} = \frac{1}{\sqrt{2\pi}} \int_{-\infty}^{x} e^{-t^2/2}\, dt. \tag{18}$$

LEMMA 2. *If $n \to \infty$ and the distributions of the r. v.'s $X_1, \ldots, X_n$ and functions $f_v(k)$, $v \in V$, vary in such a way that conditions (1) and (3) are fulfilled uniformly in n and there exist $s = s^{(n)} \geq 1$ and $f = f^{(n)} \neq 0$ such that the r. v. $\varepsilon_{r+1} = L_{r+1} - L_r$ converges (as $n \to \infty$) to zero in probability, conditions (12) and (13) are fulfilled, $C_v(r) = f$ for any $v \in V$ and for any fixed $k \geq 2$*

$$\lim_{n \to \infty} \left(\sum_{\bar{\imath} \in W_r} \sum_{\bar{\jmath} \in W_r(\bar{\imath}, s-1) \setminus \{\bar{\imath}\}} \mathbf{E}\, \eta_{\bar{\imath}} \eta_{\bar{\jmath}} \right) \left(Q((k-1)r,\ r,\ s-1) \right)^{k-2} = 0, \tag{19}$$

then for any fixed $k \in \mathbf{N}_0$

$$\lim_{n \to \infty} \mathbf{P}\left\{ \frac{L_r}{f} = k \right\} = \frac{\lambda^k e^{-\lambda}}{k!}. \tag{20}$$

From condition (3) we obtain that

$$C_v(t) \leq C(1+e^d)^t. \tag{21}$$

From the last inequality we obtain that Lemma 1 is a consequence of Theorem 1 and Lemma 2 is a consequence of Theorem 2 from (Gortchakov, 1993).
Let us obtain upper bounds for $\mathbf{E}\,\xi_t$, $r+1 \le t \le n$, $Q(t,\, l,\, s-1)$, $t,\, l,\, s \in I_n$, and

$$M(r) = \sum_{\bar{\imath}\in W_r,\, \bar{\jmath}\in W_r(\bar{\imath},s-1)\setminus\{\bar{\imath}\}} \mathbf{E}\,\eta_{\bar{\imath}}\eta_{\bar{\jmath}}.$$

Let

$$N(r) = \sum_{\nu=1}^{r-1} \binom{2r-\nu}{\nu}\binom{2r-2\nu}{r-\nu}\mathbf{E}\,\xi_{2r-\nu}, \quad M_s = a + n\beta(s),$$

$$G(s,r) = 2es(2r-1)\sqrt{p_1}.$$

LEMMA 3. *For any $i \in I_{n-r}$ and $t,\, l,\, s \in I_n$*

$$\mathbf{E}\,\xi_{r+i} \le \mathbf{E}\,\xi_r \frac{(a+i\gamma)^i}{i!}, \tag{22}$$

$$Q(t,l,s-1) \le \frac{2e(t+l-1)M_s^{l-1}(s-1)}{(l-1)!} + \frac{2(t+2e(t+l-1)(s-1)\sqrt{p_1}+M_s)^l}{l!}, \tag{23}$$

$$M(r) \le N(r) + \frac{\mathbf{E}\,\xi_r}{r!}\left[\frac{2p_0}{p_1}(G(s-1,r))^{r-2} + \sum_{l=1}^{r-1}\binom{r}{l}M_s^{r-l}(G(s-1,r))^l\right]. \tag{24}$$

Proof. First we obtain inequality (22). It is easy to see that

$$\mathbf{E}\,\xi_l(n,v) \le \sum_{\nu=l-1}^{n-1} \mathbf{E} \sum_{\bar{\imath}\in W_{l-1}(\nu)} \eta_{\bar{\imath}}(v) \sum_{\mu=1}^{n-\nu} \mathbf{P}\{X_{\nu+\mu} = v|\mathcal{F}_{\bar{\imath}}\}, \qquad l = r+i,$$

where $W_t(\nu) = \{(i_1,\ldots,i_t)|1 \le i_1 < \cdots < i_t = \nu\}$.
From the definition of $\gamma(\tau)$ and the last equality we obtain that

$$\mathbf{E}\,\xi_l(n,v) \le \sum_{\nu=l-1}^{n-1} \mathbf{E} \sum_{\bar{\imath}\in W_{l-1}(\nu)} \eta_{\bar{\imath}}(v) \sum_{\mu=1}^{n-\nu} (\mathbf{P}\{X_{\nu+\mu} = v\} + \gamma(\mu)).$$

The right-hand side of the last inequality is equal to

$$\sum_{\nu=l-1}^{n-1} \mathbf{E}\,\xi_{l-1}(\nu,v)\big(\mathbf{P}\{X_{\nu+1} = v\} + \gamma(n-\nu)\big).$$

By induction on i and ν we obtain that the last value is not greater than

$$\mathbf{E}\,\xi_r(n,v)\sum_{\mu=0}^{i}\binom{i}{\mu}\frac{\gamma^{\mu}(\mathbf{E}\,\nu_v)^{i-\mu}}{(i-\mu)!}.$$

Summing the last values over all v from V, we obtain inequality (22).

Let $K_{\vec{j}}(k,u,l)$, $0 < k+u \leq l$, be the subset of $W_l(\vec{j},s)$, $\vec{j}=(j_1,\dots,j_t)\in W_t$, $s\geq 0$, such that $\vec{i}=(i_1,\dots,i_l)\in K_{\vec{j}}(k,u,l)$ if and only if there exists a partition $\{J_1,J_2,J_3\}$ of I_l such that $|J_1|=k$, $J_2=\{\nu_1,\dots,\nu_u\}$, $\{i_\omega|\omega\in J_1\}\subset\{j_1,\dots,j_t\}$, for any $\mu\in\{1,\dots,u\}$

$$i_{\nu_\mu}\in \big(W_1(\vec{j},s)\cup\bigcup_{q=1}^{\mu-1}W_1(i_{\nu_q},s)\big)\backslash\{j_1,\dots,j_t\},$$

and for any $\mu\in J_3$

$$i_\mu\notin W_1(m(\vec{j},(i_1,\dots,i_{\mu-1},i_{\mu+1},\dots,i_l)),s).$$

Choosing k and u, we obtain an upper bound for the value

$$T_{\vec{j}}(k,u,l)=\sum_{\vec{i}\in K_{\vec{j}}(k,u,l)}\sum_{v\in V}\mathbf{E}\,(\eta_{\vec{i}}(v)|\mathcal{F}_{\vec{j}}).$$

First we consider the case $k=0$. From the definitions of p_1, $\beta(\tau)$ and M_s and from the total probability formula we obtain that

$$T_{\vec{j}}(0,u,l)\leq p_1^{u-1}\left|K_{\vec{j}}(0,u,u)\right|M_{s+1}^{l-u}((l-u)!)^{-1}. \tag{25}$$

Let us obtain an upper bound for $|K_{\vec{j}}(0,u,u)|$. It is easy to see that

$$|K_{\vec{j}}(0,u,u)|\leq\sum_{u_1+\dots+u_t=u}\prod_{i=1}^{t}|K_{j_i}(0,u_i,u_i)|,\quad |K_j(0,0,0)|=1,\qquad j\in I_n,$$

where $\sum_{u_1+\dots+u_t=u}$ denote the sum over all $u_i\in\mathbf{N}_0$, $i\in I_t$, such that $u_1+\dots+u_t=u$.

It is clear that for any j and $\vec{k}=(k_1,\dots,k_r)\in K_j(0,r,r)$ there exists a unique partition $F_1,\dots,F_{l(\vec{k})}$ of I_r such that for any $\mu\in F_1$

$$k_\mu\in W_1(j,s)\backslash\{j\} \tag{26}$$

and for any $i\in\{2,\dots,l(\vec{k})\}$ and $\mu\in F_i$

$$k_\mu\in\Big(W_1(j,s)\cup\Big(\bigcup_{q=1}^{i-1}\bigcup_{\nu\in F_q}W_1(k_\nu,s)\Big)\Big)\Big\backslash\Big(W_1(j,s)\cup\Big(\bigcup_{q=1}^{i-2}\bigcup_{\nu\in F_q}W_1(k_\nu,s)\Big)\Big). \tag{27}$$

It is clear that the number of elements from $K_j(0,r,r)$, which satisfy conditions (26) and (27), is equal to $\prod_{i=1}^{l(\vec{k})}\binom{2s}{|F_i|}$. Using the Stirling formula, we obtain that

$$|K_{\bar{j}}(0,r,r)| \le (2es)^r,$$

and

$$|K_{\bar{j}}(0,u,u)| \le \binom{t+u-1}{u}(2es)^u. \tag{28}$$

From the last inequality we obtain that

$$\sum_{u=1}^{l} T_{\bar{j}}(0,u,l) \le \frac{2es(t+l-1)M_{s+1}^{l-1}}{(l-1)!} + \frac{(2es(t+l-1)\sqrt{p_1}+M_{s+1})^l}{l!}. \tag{29}$$

It is easy to see that for any $1 \le k \le \min\{t,l\}$, $0 \le u \le l-k$

$$T_{\bar{j}}(k,u,l) \le \binom{t}{k} p_{\bar{i}}^u |K_{\bar{j}}(0,u,u)| M_{s+1}^{l-k-u}((l-k-u)!)^{-1}.$$

From the last inequality and inequality (28) we see that

$$\sum_{k=1}^{l}\sum_{u=0}^{l-k} T_{\bar{j}}(k,u,l) \le \frac{1}{l!}(t+2es(t+l-1)p_1+M_{s+1})^l.$$

It is easy to see that inequality (23) follows from the last inequality and inequalities (25) and (29).
Obviously,

$$\sum_{\bar{i}\in W_r}\ \sum_{\bar{j}\in W_r(\bar{i},0)\setminus\{\bar{i}\}} \mathbf{E}\,\eta_{\bar{j}}\eta_{\bar{j}} = N(r), \tag{30}$$

for $s \ge 1$ and any $1 \le u \le r-1$, $\bar{i} \in W_r$

$$\sum_{\bar{j}\in K_{\bar{i}}(0,u,r)} \mathbf{E}\,\eta_{\bar{i}}\eta_{\bar{j}} \le \mathbf{E}\,\eta_{\bar{i}}T_{\bar{i}}(0,u,r),$$

and

$$\sum_{\bar{i}\in W_r}\ \sum_{\bar{j}\in K_{\bar{i}}(0,r,r)} \mathbf{E}\,\eta_{\bar{i}}\eta_{\bar{j}} \le 2\sum_{\bar{i}\in W_r}\ \sum_{\bar{j}\in \widetilde{K}_{\bar{i}}(0,r,r)} \mathbf{E}\,\eta_{\bar{i}}\eta_{\bar{j}},$$

where $\widetilde{K}_{\bar{i}}(0,r,r) = \{\bar{j}=(j_1,\dots,j_r): \bar{j}\in K_{\bar{i}}(0,r,r),\ i_r < j_r\}$, $\bar{i}=(i_1,\dots,i_r)\in W_r$.
It can be easily seen that

$$\sum_{\bar{j}\in \widetilde{K}_{\bar{i}}(0,r,r)} \mathbf{E}\,\eta_{\bar{i}}\eta_{\bar{j}} \le \mathbf{E}\,\eta_{\bar{i}}|\widetilde{K}_{\bar{i}}(0,r,r)|\,p_0p_1^{r-2}.$$

Using the last three inequalities, inequalities (25) and (28) and equality (30) we obtain inequality (24). Lemma 3 is proved.
Obviously, if $m = m_r$, then condition (16) coincides with condition (10). Let us prove that the other conditions of Lemma 1 follow from conditions (5)–(9). In what follows $\operatorname{Var} L = \operatorname{Var} L_r$, $\varepsilon_m = L_{m+1} - L_r$.

LEMMA 4. *If conditions (5) and (6) are valid, then the r. v. $\bar{\varepsilon}_{m_r}$ converges (as $n \to \infty$) to zero in probability.*

LEMMA 5. *If conditions (5)–(9) are valid, then for $m = m_r$ condition (17) is fulfilled.*

Proof of Lemma 4. From inequality (21) we obtain that

$$|\mathbf{E}\,\varepsilon_{m_r}| \leq \sum_{t=m_r+1}^{n} C(1+e^d)^t \mathbf{E}\,\xi_t.$$

From the definition of m_r, in view of condition (6), inequality (22) and the Stirling formula we obtain that the right-hand side of the last inequality is not greater than

$$C((1+e^d)(at_r^{-1}+\gamma))^{t_r}(1-(1+e^d)(at_r^{-1}+\gamma))^{-1}\mathbf{E}\,\xi_r,$$

where $t_r = (\log \mathbf{E}\,\xi_r)(\log g_r)^{-1}$.
It is clear that the logarithm of the last value is not greater than

$$\log \mathbf{E}\,\xi_r + t_r\big(\log \max\{G_r^{-1},\gamma\} + \log\,[(1+e^d)(\log g_r + 1)]\big) + o(\log \mathbf{E}\,\xi_r).$$

Consequently, if condition (5) is true, then the last value is $o(\log \mathbf{E}\,\xi_r)$. Lemma 4 is proved.

Proof of Lemma 5. From inequalities (21) and (22) we obtain that

$$\sum_{t=r}^{m_r}(1+e^d)^t\mathbf{E}\,\xi_t \leq (1+e^d)^r\mathbf{E}\,\xi_r \exp\{(1+e^d)(a+m_r\gamma)\}.$$

From the last inequality and condition (6) we obtain that the logarithm of the right-hand side of the last inequality is not greater than $\log \mathbf{E}\,\xi_r(1+o(1))$.
The sum

$$2elm_r(s-1)\sum_{t=r}^{m_r}\frac{M_s^{t-1}}{(t-1)!}(1+e^d)^t \qquad (M_s = a+n\beta(s)),$$

is not greater than $T_1 = O(m_r(s-1)e^{M_s(1+e^d)})$.

And the sum

$$2\sum_{t=r}^{m_r} \frac{(1+e^d)^t((l-1)m_r + 2e(s-1)l\sqrt{p_1}m_r + M_s)^t}{t!}$$

does not exceed

$$T_2 = O\left(m_r(m_r!)^{-1}((1+e^d)(lm_r(1+2es\sqrt{p_1}) + M_s))^{m_r}\right).$$

Now it is easy to see that the condition

$$\liminf_{n\to\infty} l\left((\log \max\{T_1, T_2\})(\log \mathbf{E}\,\xi_r)^{-1} - \frac{\omega}{2}\right) < -1 \tag{31}$$

is a sufficient condition for (17). It follows from conditions (6) and (9) that $\log T_1$ is not greater than $\log s + o(\log \mathbf{E}\,\xi_r)$, and $\log T_2 \le m_r \log(1+2es\sqrt{p_1}) + o(\log \mathbf{E}\,\xi_r)$. It implies that if conditions (7) and (8) are fulfilled then there exists a constant $k \ge 3$ such that for any fixed $l \ge k$ condition (31) is fulfilled. Lemma 5 is proved. Theorems 1–3 follow from Lemmas 1, 4 and 5.

Let us prove that the conditions of Lemma 2 follow from the conditions of Theorems 4 and 5. It follows from (22) that

$$|\mathbf{E}\,\varepsilon_{r+1}| \le C(ar^{-1}+\gamma)(1+e^d)(1-(ar^{-1}+\gamma)(1+e^d))^{-1}\mathbf{E}\,\xi_r.$$

Now it is clear that if $\lim_{n\to\infty} a = \lim_{n\to\infty} \gamma = 0$ and condition (12) is fulfilled, then the r. v. ε_{r+1} converges (as $n \to \infty$) to zero in probability.

It follows from inequalities (23) and (24) that if condition (12) is fulfilled and for any fixed $t \ge 0$

$$sa^{r-1}, \quad s(n\beta(s))^{r-1}, \quad p_0(\sqrt{p_1})^{-1}s$$

and γ are $o((\sqrt{p_1}s)^t)$, then condition (19) is true. Notice that if condition (4) is fulfilled and $f_v(r) = f$ for any $v \in V$, then $C_v(r) = f$ for any $v \in V$ and if $f_v(0) = f_0$, $f_v(1) = f_1$, $f_v(2) = f$ for any $v \in V$, then $C_v(2) = f - 2f_1 + f_0$ for any $v \in V$. Therefore the conditions of Lemma 2 follow from the conditions of Theorems 4 and 5. Theorems 4 and 5 are proved.

REFERENCES

Belyaev, P. F. (1964). On the probability of non-occurrence of a given number of outcomes. *Theory Probab. Appl.* 9, 489–495.

Belyaev, P. F. (1965). On the probability of non-appearence of a given number of s-tuples in compound Markov chains. *Theory Probab. Appl.* 10, 496–500.

Dyakonova, E. E. and Mikhailov, V. G. (1991). On sufficient conditions for asymptotic normality of decomposable statistics in a nonhomogeneous allocation scheme. *Discrete Math. Appl.* 3, 325–336.

Gortchakov, A. B. (1993). Limit theorems for U-statistics of dependent random variables. In: *Probabilistic Methods in Discrete Mathematics*, TVP/VSP, Moscow/Utrecht, pp. 186–199.

Ivanov, V. A., Ivchenko, G. I., and Medvedev, Yu. I. (1984). Discrete problems in probability theory. *J. Soviet Math.* **31** (in Russian).

Kolchin, V. F., Sevastyanov, B. A., and Chistyakov, V. P. (1978). *Random Allocations*. Halstead Press (Wiley), New York.

Mikhailov, V. G. (1974). Limit distributions of random variables associated with multiple long duplications in a sequence of independent trials. *Theory Probab. Appl.* **19**, 180–185.

Mikhailov, V. G. (1989). On asymptotic normality of decomposable statistics on frequencies of m-series. *Discrete Math. Appl.* **1**, 92–103 (in Russian).

Zubkov, A. M. and Mikhailov, V. G. (1974). Limit distributions of random variables, associated with long duplications in the sequence of independent trials. *Theory Probab. Appl.* **19**, 172–180.

Zubkov, A. M. and Mikhailov, V. G. (1979). Repetitions of s-tuples in a sequence of independent trials. *Theory Probab. Appl.* **24**, 269–277.

Probabilistic Methods in Discrete Mathematics, pp. 212 – 225
V. F. Kolchin et al. (Eds.)
1993 TVP/VSP

ESTIMATING FINITE STRATIFIED POPULATIONS

G. I. IVCHENKO and S. A. KHONOV

Moscow Institute of Electronic Engineering, B. Vuzovskii per. 3/12, 109028 Moscow, Russia

ABSTRACT

During the last years a number of important results in the problem of estimation of the size or the structure of a finite closed population was published. The aim of the paper is to review the literature and suggest further lines of investigations.

1. INTRODUCTION

The estimation of an unknown size N of a finite population U and also unknown sizes $N_1, \ldots, N_k$ of strata (classes) $U_1, \ldots, U_k$: $U = U_1 \cup \cdots \cup U_k$; $U_i \cap U_j = \varnothing$, $i \neq j$, $i, j = 1, \ldots, k$, is a traditional and intensively studying field of applied statistical analysis with a wide range of applications.
There is a lot of papers both theoretical and applied directions in this field; these papers differ both on the structure of studying population and on the corresponding sampling design, which selects some random subset of U as an object of study. A comprehensive description of the results related with this problem was done in (Belyaev, 1975; Voinov and Nikulin, 1989; Seber, 1986; Ivanov et al., 1985).
Our purpose is to give an analytical survey of the corresponding theoretical results which are close to problems of allocating particles into cells. This survey contains also unpublished results of the authors of this paper. It should be remarked that the selection of results and their representation reflect author's individual scientific interests.

2. THE MODEL

The generality of setting of the problem and, consequently, of the results are determined by the following items.

2.1. The structure of the examined population U is determined by the set of parameters $\mathfrak{N} = \{\overline{N} = (N_1, \ldots, N_k)\}$, that allows to cover a wide range of applied problems.

2.2. The structure of samples is determined in terms of the so-called general sampling scheme connected with sampling of s simple samples without replacement from U with given sizes $m_1, \ldots, m_s$ so that the total number of outcomes of the random experiment equals

$$\Psi(N) = \prod_{i=1}^{s} \binom{N}{m_i}$$

and each of the outcomes has the same probability.

2.3. Not only the structure $\overline{N}$ of U, but any parametric function $\tau(\overline{N})$ are estimated, point and interval estimates and tests for corresponding statistical hypotheses about parameter $\overline{N}$ and $\tau(\overline{N})$ are considered. Note that the following examples of parametric sets are typical in applications:

$$\mathfrak{N} = \{\overline{N}: N = N_1 + \cdots + N_k > 0\}$$

and

$$\mathfrak{N} = \{\overline{N}: N = N^0\},$$

where the total size $|U| = N^0$ of U is a known constant, but the structure of U is unknown and must be examined.

In the case of $\{\overline{N}: N > 0\}$, for example, it may be the problem of estimation of the specific structure of biological populations, where, as a rule, the total size of a population is unknown. The second case $\{\overline{N}: N = N^0\}$ is typical for sampling quality control problems, where the size of the examined set of production is known and equal to N^0, but its quality structure is unknown and must be examined.

In the case $k = 1$ we have an unstratified population with the (unknown) size N. The general sampling scheme (GSS) has a fixed number of observation $n = m_1 + \cdots + m_s$; in the case $m_1 = \cdots = m_s = 1$ the scheme GSS reduces to the classical sampling scheme with replacement and in the case $m_1 = n$, $s = 1$ it corresponds to the classical sampling scheme without replacement and does not include experiments with a random number of observations and nonequiprobable outcomes.

In GSS we have input data in the frequency form: let

$$\bar{\mu} = (\mu_{jr}, \quad j = 1, \ldots, k, \ r = 1, \ldots, s)$$

be the initial statistic, where μ_{jr} is the number of elements of U_j observed in s samples exactly r times, $r > 0$. We accent attention to the construction of unbiased estimates and to the finding an optimal one among them using the principles of sufficiency and completeness. In the case of incomplete statistic we use the principle of maximum likelihood. In the asymptotic approach we find consistent asymptotically normal (CAN) estimates and asymptotic confidence regions based on CAN.

In terms of the equiprobable allocation scheme of series of particles let $N_1 + \cdots + N_k$ boxes form the set of N boxes and s series of particles are allocated into the cells so that $\binom{N}{m_i}$ variants of replacement of ith series are equiprobable, $i = 1, \ldots, s$, and these s allocations are independent. Then μ_{jr} is the number of boxes contained exactly r particles of jth group, and the aim is to estimate the vector of parameters $\overline{N} = (N_1, \ldots, N_k)$ using the data vector $\bar{\mu}$.

Note that $r > 0$, i.e., the empty boxes are unobserved. Thus, in the study of this problem we may use the well-developed probability theory for allocation schemes (see (Kolchin, 1986; Kolchin et al., 1978)).

3. SUFFICIENT STATISTIC

Introduce the additional notation:

$$\eta_j = \sum_{r>0} \mu_{jr}; \quad \bar{\eta} = (\eta_1, \cdots, \eta_k); \quad \eta = \eta_1 + \cdots + \eta_k;$$

l_{jr} is a possible value of μ_{jr};

$$l_j = \sum_{r>0} l_{jr}; \quad \bar{l} = (l_1, \ldots, l_k); \quad l = l_1 + \cdots + l_k,$$

$$0 \le l_j \le N_j, \qquad j = 1, \ldots, k, \quad r = 1, \ldots, s;$$
$$\mathfrak{L}_{(i)} = \{\bar{l} : l = i;\ 0 \le l_j \le N_j,\ j = 1, \ldots, k\},$$
$$\mathfrak{L}_{m,n|\overline{N}} = \bigcup_{i=m}^{n \wedge N} \mathfrak{L}_{(i)}; \quad \mathfrak{L}_{m,n} = \bigcup_{i=m}^{n} \mathfrak{L}_{(i)};$$

$\mathcal{L}(\bar{\eta})$ is the distribution of $\bar{\eta}$,

$$n \wedge N = \min(n, N), \quad \mathfrak{N}_{(i)} = \{\overline{N} : N = i\},$$

$$\mathbf{I} f(\bar{x}) = f(\bar{x}); \quad \mathbf{A}_j f(\bar{x}) = f(x_1, \ldots, x_{j-1}, x_j + 1, x_{j+1}, \ldots, x_k);$$
$$\Delta_j = \mathbf{A}_j - \mathbf{I}; \quad \Delta f(x) = f(x+1) - f(x), \quad \Delta^k f(0) = \Delta^k f(x)|_{x=0};$$
$$x_{i,s}(N) = \Delta^i \Psi(N-1)/\Psi(N); \quad (x)_k = x(x-1)\cdots(x-k+1);$$

$\bar{x} \prec \bar{y}$, if $x_i \le y_i$ for all i and there exists $j \in \{1, \ldots, k\}$ such that $x_j < y_j$.
The answer to the key question about sufficiency of a statistic and its distribution in the scheme under consideration is contained in the following theorem.

THEOREM 1 (Khonov, 1987 a). *The vector statistic* $\bar{\eta} = (\eta_1, \ldots, \eta_k)$ *with the distribution*

$$\mathbf{P}_{\overline{N}}\{\bar{\eta} = \bar{l}\} = \prod_{j=1}^{k} \binom{N_j}{l_j} \Delta^l \Psi(0)/\Psi(N), \qquad \bar{l} \in \mathfrak{L}_{m,n|\overline{N}},$$

and the moments

$$\mathbf{E}_{\overline{N}}((\eta_1)_{i_1} \cdots (\eta_k)_{i_k}) = (N_1)_{i_1} \cdots (N_k)_{i_k} \Delta^{i_1 + \cdots + i_k} \Psi(N - i_1 - \cdots - i_k)/\Psi(N)$$

is sufficient for the vector parameter $\overline{N} = (N_1, \ldots, N_k)$.

4. COMPLETENESS OF SUFFICIENT STATISTIC

The next principal question is the question of completeness of the sufficient statistic. The answer depends on the form of the parameter set $\mathfrak{N}$.

THEOREM 2. *The sufficient statistic $\bar{\eta}$ is complete if and only if the parameter set with linear structure has the form $\mathfrak{N} = \bigcup_{i=m}^{M} \mathfrak{N}_{(i)}$ for some M, $m \le M \le n$, and for any $\overline{N} \in \mathfrak{N}$ there exists $\overline{N}^* \in \mathfrak{N}_{(m)}$ such that $\overline{N}^* \prec \overline{N}$.*
In particular, if $\mathfrak{N} = \mathfrak{N}_{(N^0)}$, $N^0 > m$, then $\bar{\eta}$ is complete only if $s = 1$; if $s > 1$, then there exists a free (subordinated) statistics $\eta = \eta_1 + \cdots + \eta_k$ with the distribution and the moments of the following form:

$$\mathbf{P}_{\overline{N}}\{\eta = l\} = N^0 \Delta^l \Psi(0)/\Psi(N), \qquad m \le l \le n \wedge N^0,$$

$$\mathbf{E}_{\overline{N}}(\eta)_i = (N^0)_i \Delta^i \Psi(N^0 - i)/\Psi(N^0) = (N^0)_i x_{i,s}(N^0).$$

5. ESTIMATION IN THE CASE OF COMPLETE STATISTICS AND FINDING OF THE UNIFORM MINIMUM VARIANCE UNBIASED ESTIMATE (UMVUE)

Let $\tau(\overline{N})$ be the estimated parametric function. Then the following theorems are valid.

THEOREM 3 (Khonov, 1987 a). *If the parameter set has the form*

$$\mathfrak{N} = \bigcup_{i=m}^{M} \mathfrak{N}_{(i)} \text{ for some } M, \qquad m \le M \le n,$$

then UMVUE exists for any parametric function $\tau(\overline{N})$ and has the form

$$\widehat{\tau}(\bar{\eta}) = \frac{1}{\Delta^{\eta}\Psi(0)} \left(\sum_{i_1+\cdots+i_k \ge m} \prod_{j=1}^{k} (-1)^{\eta_j - i_j} \binom{\eta_j}{i_j} \mathbf{A}_j^{i_j}) f(\bar{0}) \right),$$

where $f(\overline{N}) = \tau(\overline{N})\Psi(N)$, or the equivalent form

$$\widehat{\tau}(\bar{\eta}) = \frac{1}{\Delta^{\eta}\Psi(0)} \left(\left(\Delta_1^{\eta_1} \cdots \Delta_k^{\eta_k} - \sum_{i_1+\cdots+i_k < m} \prod_{j=1}^{k} (-1)^{\eta_j - i_j} \binom{\eta_j}{i_j} \mathbf{A}_j^{i_j} \right) f(\bar{0}) \right).$$

If the parametric function satisfies the condition

$$\tau(\bar{x}) = 0 \qquad \forall \bar{x} \notin \mathfrak{N}$$

(for example, parametric function is bounded), then UMVUE has the following simple form:

$$\widehat{\tau}(\bar{\eta}) = \Delta_1^{\eta_1} \cdots \Delta_k^{\eta_k} f(\bar{0})/\Delta^{\eta}\Psi(0). \tag{1}$$

THEOREM 4 (Khonov, 1987 a). *Let the parameter set has the form*

$$\mathfrak{N} = \bigcup_{i=m}^{M} \mathfrak{N}_{(i)} \text{ for some } M > n.$$

Then UMVUE exists only for the parametric functions of the form

$$\tau(\overline{N}) = f(\overline{N})/\Psi(N),$$

where $f(\overline{N})$ *is a polynomial of degree* n, *and*

$$\tau(\bar{x}) = 0 \qquad \forall \bar{x} \in \bigcup_{i=0}^{m-1} \mathfrak{N}_{(i)}, \quad \bar{x} = (x_1, \ldots, x_s), \quad x_i = 0, 1, 2, \ldots \quad \forall i.$$

This condition can be reduced to the condition

$$f(\bar{x}) = 0 \qquad \forall \bar{x} \in \bigcup_{i=n+1}^{M} \mathfrak{N}_{(i)}$$

for $M > 2n$. *The estimates for such function have form* (1).

In particular, if the number of strata k equals 1 then the corresponding theorems for one-dimensional parametric functions $\tau(N)$ dependent on the size of the population see in (Ivchenko, 1984).
If the parametric set has the structure

$$\mathfrak{N} = \mathfrak{N}_{(N^0)}, \qquad N^0 > m,$$

then the sufficient statistic is complete and the estimate $\hat{\tau}(\bar{\eta}) = \eta_j N^0/n$ for N_j, $j = 1, \ldots, k$, is UMVUE if $s = 1$.
Berg (Berg, 1976) noted a great calculation complexity of the above estimates even in the simple cases $k = 1$ and $m_1 = \cdots = m_s$.

6. ESTIMATION WITH INCOMPLETE STATISTICS

If the sufficient statistic $\bar{\eta}$ is not complete, then the unbiased estimate is not unique in the class of all functions which depend on it. But by a restriction of the considered class of estimated functions it is possible to construct the unique unbiased estimate for any function from this restricted class. Another possible way is to construct the maximum likelihood estimate (MLE).

6.1. Let

$$\mathcal{A}_r = \left\{ \bar{T}_r(\bar{\eta}) = \left(T_{rj}(\bar{\eta}) = \sum_{i_1+\cdots+i_k=r} a^{(j)}_{i_1,\ldots,i_k} \prod_{\nu=1}^{k} (\eta_\nu)_{i_\nu},\ j = 1, \ldots, k \right) \right\}$$

be the class of polynomial functions dependent on the statistic $\bar{\eta}$.

THEOREM 5 (Khonov, 1987b). *In the class $\mathcal{A}_r$ of the parametric functions for any $r < n \wedge N$ a unique unbiased estimate for the function of the form*

$$\tau_{i_1,\dots,i_k}(\overline{N}) = \sum_{i_1+\dots+i_k=i\le r} b_{i_1,\dots,i_k}(N_1)_{i_1}\cdots(N_k)_{i_k}$$

exists. This estimate is equal to

$$\tau^*_{i_1,\dots,i_k}(\bar{\eta}) = \sum_{i_1+\dots+i_k=i\le r} b_{i_1,\dots,i_k}(\eta_1)_{i_1}\cdots(\eta_k)_{i_k}\frac{(\eta-1)_{r-i}\Psi(N)}{(N-1)_{r-i}\Delta^r\Psi(N-r)}.$$

In particular, for the parameter $\overline{N} \in \mathfrak{N}$ the unique unbiased estimate

$$\overline{T}^*_r(\bar{\eta}) = \frac{\bar{\eta}(\eta-1)_{r-1}\Psi(N)}{(N-1)_{r-1}\Delta^r\Psi(N-r)} = \frac{\bar{\eta}(\eta-1)}{(N-1)_{r-1}x_{r,s}(N)}$$

exists in each class $\mathcal{A}_r$ and its variance has the following representation:

$$\mathrm{Var}_{\overline{N}}(T^*_{rj}(\bar{\eta})) = N(N_j)_2K^{(2)}_{r,s}(N) + NN_jK^{(1)}_{r,s}(N) - N_j^2, \tag{2}$$

where $j = 1,\dots,k$,

$$K^{(1)}_{r,s}(N) = (r-1)!\sum_{i=0}^{r-1}\binom{r-1}{i}\frac{x_{r+i,s}(N)}{i!\,x^2_{r,s}(N)},$$

$$K^{(2)}_{r,s}(N)(N-1) = r!\sum_{i=0}^{r}\binom{r}{i}\frac{x_{r+i,s}(N)}{i!\,x^2_{r,s}(N)} - K^{(1)}_{r,s}(N),$$

$$x_{r,s}(N) = \Delta^r\Psi(N-r)/\Psi(N).$$

As was found in (Khonov, 1987b) variance (2) increases with the parameter r of the class, thus the linear estimate

$$\overline{T}^*_1(\bar{\eta}) = (T^*_{1j}(\bar{\eta}) = \eta_jN/n,\ j = 1,\dots,k)$$

has the minimum variance.

In the same article it was established that if $n \le N$, s is a constant, $1 \le s \le n$, then the unbiased estimate based on the unique sample is more precise in the sense of variance than any estimate based on more then one sample (if the total number of observations is fixed), i.e.,

$$\mathrm{Var}_1T^*_{1j}(\bar{\eta}) < \mathrm{Var}_sT^*_{1j}(\bar{\eta}), \qquad s > 1.$$

6.2. Maximum likelihood estimation. The problem of obtaining MLE in the most general case for the sizes of stratified population, where the total size of population may be both known or unknown, was solved in (Khonov, 1989). The solution is given in the following theorems.

THEOREM 6. *The MLE $\widetilde{N}$ for the total size N of population is calculated from the relations*

$$\widetilde{N}x_{1,s}(\widetilde{N}) < l \leq (\widetilde{N}+1)x_{1,s}(\widetilde{N}+1) \quad \text{if} \quad m < l < n,$$

$$\widetilde{N} = m \quad \text{if} \quad l = m, \tag{3}$$

and this estimate is unique.
If $l = n$, then the MLE is undetermined ($\widetilde{N} = \infty$).
If $(\widetilde{N}+1)x_{1,s}(\widetilde{N}+1)$ is an integer, then $\widetilde{N}+1$ is the MLE too.

THEOREM 7. *If the total size M of the population is known, then the MLE $\widetilde{N}_1, \ldots, \widetilde{N}_k$ for the structure components $N_1, \ldots, N_k$, $\sum_{j=1}^{k} N_j = M$, are calculated from the relations*

$$(M+1)\eta_j/\eta - 1 \leq \widetilde{N}_j < (M+1)\eta_j/\eta, \qquad j = 1, \ldots, k-1,$$

$$\widetilde{N}_k = M - \widetilde{N}_1 - \cdots - \widetilde{N}_{k-1} \tag{4}$$

and these estimates are unique.
If $l(\widetilde{N}_j+1)/(M+1)$ is an integer, then $\widetilde{N}_j+1$ is the MLE for N_j too, $j = 1, \ldots, k$, $k > 1$.

THEOREM 8. *If the total size of the population is unknown, then the MLE $\widetilde{N}_1, \ldots, \widetilde{N}_k$ for the structure components $N_1, \ldots, N_k$ are calculated from* (3) *and* (4) *with $M = \widetilde{N}$.*

For special cases these results were published in different forms in (Driml and Ullrich, 1967) for the case of the simple sample with replacement, in (Ivchenko and Timonina, 1981) for the case $k = 1$ and $m_1 = \cdots = m_s = m$ in GSS, and in (Leite et al., 1988) for the same case $k = 1$.

7. CONFIDENCE ESTIMATION (Ivchenko and Medvedev, 1990)

Instead of central confidence intervals in the discrete case it is usual to consider the construction of the confidence set D_γ by including into it the values of statistic η, whose probabilities are the greatest (for each value of the estimated parameter). Then the boundaries t_1, t_2, N_L, N_H are found from the condition

$$\mathbf{P}_N\{N_L < N < N_H\} = \mathbf{P}_N\{t_1 < \eta < t_2\} = \sum \mathbf{P}_N\{\eta = l\} \geq \gamma, \qquad N \in \mathfrak{N}.$$

At the same time we require that the sum should include those l, whose probabilities are the greatest, and the interval (t_1, t_2) should be the smallest one meeting this condition. In some cases where the statistic is asymmetrically distributed this method gives the shortest confidence intervals.

The general technique for construction of a confidence region is to assign to each value of $\overline{N} \in \mathfrak{N}$ a subset $\mathfrak{H}(\overline{N}) \subset \mathfrak{L}_{m,n|\overline{N}}$ such that the condition

$$\mathbf{P}_{\overline{N}}\{\bar{\eta} \in \mathfrak{H}(\overline{N})\} \geq \gamma \tag{5}$$

is satisfied. Then we obtain a family of subsets $\{\mathfrak{H}(\overline{N}), \overline{N} \in \mathfrak{N}\}$ in the sample space. Now for every $\bar{l} \in \mathfrak{L}_{m,n|\overline{N}}$ we define a subset $\mathfrak{G}(\bar{l}) \in \mathfrak{N}$ according to the rule

$$\mathfrak{G}(\bar{l}) = \{\overline{N}\colon \bar{l} \in \mathfrak{H}(\overline{N})\},$$

and thus in the parametric set we find a family of subsets

$$\{\mathfrak{G}(\bar{l}),\ \bar{l} \in \mathfrak{L}_{m,n|\overline{N}}\} \subset \mathfrak{N}.$$

The events $\{\bar{l} \in \mathfrak{H}(\overline{N})\}$ and $\{\overline{N} \in \mathfrak{G}(\bar{l})\}$ are equivalent, since by construction one implies another, therefore, their probabilities coincide for every value $\overline{N}$. Taking into account (5), we see that for $\mathfrak{G}(\bar{l})$ the corresponding inequalities hold. Hence, $\mathfrak{G}(\bar{l})$ is the required γ-confidence region for the estimated parameter.

8. ASYMPTOTIC ANALYSIS

We suppose that the total population size N and the strata sizes $N_1, \dots, N_k$ are large and as $N \to \infty$ the relations

$$N_j = N\beta_j, \qquad \beta_j \in [\delta_1, 1-\delta_1], \quad \delta_1 > 0, \quad j = 1, \dots, k \tag{6}$$

hold (the parameter k is supposed to be fixed). As to the parameters of the GSS we suppose that

$$\max_{1\leq i\leq s} m_i = m < N, \qquad s \geq 2, \quad N - m \to \infty, \quad n - m \to \infty.$$

For a more detailed analysis we choose the following two cases, which are the most interesting for applications:

$$1 \leq m \leq C < \infty, \quad s \to \infty, \quad n = \alpha N, \quad 0 < \alpha_1 \leq \alpha \leq \alpha_2 < \infty; \tag{7}$$

$$s \text{ is fixed}, \quad m_i \to \infty, \quad m_i = \gamma_i N, \quad \gamma_i \in [\delta_2, 1-\delta_2], \qquad \delta_2 > 0, \quad i = 1, \dots, s. \tag{8}$$

The formulas for the first moments of the statistic $\bar{\eta}$

$$\mathbf{E}\,\eta_j = N_j x_{1,s}(N) = N_j\left(1 - \prod_{i=1}^{s}(1 - m_i/N)\right);$$

$$\operatorname{cov}(\eta_i, \eta_j) = N_i(1 - x_{1,s}(N))\big(\delta_{ij}x_{1,s}(N-1) \\ -N_j(x_{1,s}(N-1) - x_{1,s}(N))\big); \qquad i,j = 1, \dots, k, \tag{9}$$

follow from (Khonov, 1987 a); here δ_{ij} is the Kronecker delta.

Let

$$\mathbf{E}\bar{\eta} = N\bar{\beta}a(1+O(1/N)), \quad \operatorname{Var}\bar{\eta} = NB(1+O(1/N)), \tag{10}$$

where

$$\bar{\beta} = (\beta_1, \dots, \beta_k), \quad B = \|b_{ij}\|_1^k,$$

$$b_{ij} = \beta_i(\beta_j b + (\delta_{ij} - \beta_j)a(1-a)); \qquad a = \lim_{N\to\infty} x_{1,s}(N),$$

$$b = \lim_{N\to\infty}(1 - x_{1,s}(N))(x_{1,s}(N) - (N-1)(x_{1,s}(N-1) - x_{1,s}(N))).$$

In particular, for case (7)

$$a = 1 - \exp(-\alpha), \quad b = (1-a)(1-(1-a)(1+\alpha)),$$

and for case (8)

$$a = 1 - \prod_{i=1}^{s}(1-\gamma_i), \quad b = (1-a)\left(1-(1-a)\left(1+\sum \gamma_i/(1-\gamma_i)\right)\right).$$

THEOREM 9. *Let conditions* (6) *hold and* $\operatorname{Var}\eta \to \infty$ *as* $N \to \infty$. *Then as* $N \to \infty$

$$\mathbf{P}\{\bar{\eta} = N\bar{\beta}x_{1,s}(N) + \bar{u}\sigma_N)\} = \frac{1 + O\left(1/\sqrt{\operatorname{Var}\eta}\right)}{\sigma_N^{k-1}\sqrt{(2\pi)^k\beta_1\cdots\beta_k\operatorname{Var}\eta}}$$

$$\times \exp\left\{-\frac{1}{2}\sum_{j=1}^{k}\frac{u_j^2}{\beta_j} + (\sigma_N^2/\operatorname{Var}\eta - 1)\left(\sum_{j=1}^{k}u_j\right)^2\right\}$$

uniformly with respect to all vectors $\bar{l}$ *with integer-valued components such that* $\bar{l} = N\bar{\beta}x_{1,s}(N) + \bar{u}\sigma_N$, $\bar{u} = (u_1, \dots, u_k)$, $\bar{\beta} = (\beta_1, \dots, \beta_k)$, $|u_j| \le C < \infty$ *for all* j, $|u_1 + \cdots + u_k|\sigma_N/\sqrt{\operatorname{Var}\eta} \le C < \infty$, $\sigma_N^2 = Nx_{1,s}(N)(1 - x_{1,s}(N))$.

The situation, where the relations similar to (10) hold, i. e., where $\operatorname{Var}\eta \asymp \sigma_N^2 \asymp N$, is of special interest for statistical applications. Corresponding results under such a kind of conditions can be represented in the following form: as $N \to \infty$

$$\mathbf{P}\{\bar{\eta} = N\bar{\beta}a + \bar{u}\sqrt{N}\} = (2\pi N)^{-k/2}|B|^{-1/2}\exp\left\{-\frac{1}{2}\bar{u}'B^{-1}\bar{u}\right\}\left(1 + O\left(\frac{1}{\sqrt{N}}\right)\right).$$

COROLLARY. *Under the conditions of Theorem 9*

$$\mathcal{L}(\bar{\eta}) \sim \mathcal{N}(N\bar{\beta}a, NB).$$

The complete spectrum of local limit theorems for $\bar{\eta}$ is obtained by Ivchenko and Khonov (Ivchenko and Khonov, 1991) with the help of results of (Vatutin and Mikhailov, 1982) for one-dimensional statistic η.

The results mentioned above can be applied to the problem of estimating the unknown parameters, which determine the structure of the examined population. Let condition (6), (10) be fulfilled a priori and the parameters of GSS satisfy conditions (7) or (8). Under these assumptions Corollary to Theorem 9 gives a basis for corresponding statistical inferences and makes it possible to reduce the problem to specific variations of the problem of estimating parameters of a multidimensional normal distribution from one observation of the corresponding random vector.

In the case where the total population size N is known, the population structure is completely determined by the parameter vector $\bar{\beta}$ and, as follows from Theorem 9, the statistic $\widehat{\beta}_N = \bar{\eta}/(Na)$ is an CAN $\mathcal{N}(\bar{\beta}, B/(Na^2))$ estimate of $\bar{\beta}$. Moreover, since the matrix $B = B(\bar{\beta})$ is a smooth function of $\bar{\beta}$ we obtain that

$$\mathcal{L}(\widehat{\beta}_N) \sim \mathcal{N}(\bar{\beta}, B_N/(Na^2)), \qquad N \to \infty, \quad B_N = B(\widehat{\beta}_N).$$

This result from (Ivchenko and Khonov, 1991) completely solves the problem, because it allows to construct asymptotic confidence regions for all parameters $\beta_1, \ldots, \beta_k$ or for any part of them in usual manner.

In the case where the total population size is unknown, the population structure is determined by the parameters $\bar{\beta}$ and α. For this case the assertion of Theorem 9 can be written in the form

$$\mathcal{L}(\bar{\eta}) \sim \mathcal{N}(nG(\alpha)\bar{\beta}, nB/\alpha),$$

where

$$G(\alpha) = \big(1 - \exp(-\alpha)\big)/\alpha$$

for case (7) and

$$G(\alpha) = \frac{1}{\alpha}\Big(1 - \prod_{i=1}^{s}(1 - \alpha\rho_i)\Big), \qquad \rho_i = \frac{m_i}{n}, \quad i = 1, \ldots, s,$$

for case (8).

As the function $G(\alpha)$ is monotone, then the inverse function $G^{-1}(\alpha)$ exists and the following assertion is true.

THEOREM 10 (Ivchenko and Khonov, 1991). *Under the conditions of Theorem 9 the statistic*

$$\widehat{\alpha}_n = G^{-1}(\eta/n)$$

is a CAN $\mathcal{N}(\alpha, \sigma^2(\alpha)/n)$ estimate of the parameter α, where $\sigma^2(\alpha) = b/(\alpha(G'(\alpha))^2)$; moreover, the random variable $\sqrt{n}(\widehat{\alpha}_n - \alpha)/\sigma(\widehat{\alpha}_n)$ is uniformly asymptotical normal $\mathcal{N}(0,1)$, and an asymptotic $(1-\delta)$-confidence interval for α has the form

$$\big(\widehat{\alpha}_n - t_{\delta/2}\sigma(\widehat{\alpha}_n)/\sqrt{n}, \widehat{\alpha}_n + t_{\delta/2}\sigma(\widehat{\alpha}_n)/\sqrt{n}\big),$$

where $\Phi(-t_\delta) = \delta$ and $\Phi(x)$ is the distribution function of the standard normal law $\mathcal{N}(0,1)$.

THEOREM 11 (Ivchenko and Khonov, 1991). *Under the conditions of Theorem 9 the statistic $\widehat{\beta}_n = \bar{\eta}/\eta$ is a CAN $\mathcal{N}(\bar{\beta}, \Sigma_n/n)$ estimate of $\bar{\beta}$, where*

$$\Sigma = \psi(\alpha)\|\beta_i(\delta_{ij} - \beta_j)\|_1^k, \quad \psi(\alpha) = \big(1 - \alpha G(\alpha)\big)/G(\alpha),$$

and the subscript n of the matrix Σ means that the estimates $\widehat{\alpha_n}, \widehat{\beta}_n$ are substituted into the matrix Σ for the unknown parameters α and $\bar{\beta}$.

Khonov (Khonov, 1990) proved that these estimates are also asymptotically MLE. Now consider the problem of simultaneous estimation of all coordinate of $\bar{\beta}$. Let $\bar{\beta}^0 = (\beta_1, \ldots, \beta_{k-1})$ and

$$H = H(\bar{\beta}^0) = \|\beta_i(\delta_{ij} - \beta_j)\|_1^{k-1}.$$

As a result of direct verifications we obtain that

$$|H| = \beta_1 \cdots \beta_k > 0, \quad H^{-1} = \|\delta_{ij}/\beta_i + 1/\beta_k\|_1^{k-1},$$

$$(\widehat{\beta^0}_n - \bar{\beta}^0)' H^{-1} (\widehat{\beta^0}_n - \bar{\beta}^0) = \sum_{i=1}^{k} (\widehat{\beta}_{in} - \beta_i)^2/\beta_i, \tag{11}$$

$$\mathcal{L}\left(\frac{n}{\psi(\widehat{\alpha}_n)} \sum_{i=1}^{k} \frac{(\widehat{\beta}_{in} - \beta_i)^2}{\widehat{\beta}_{in}}\right) \to \chi^2(k-1).$$

Now let $\chi^2_{\gamma,k-1}$ be the γ-quantile of the distribution $\chi^2(k-1)$ and

$$\mathfrak{G}_{\gamma_n} = \left\{\bar{\beta}: \sum_{i=1}^{k} \frac{(\widehat{\beta}_{in} - \beta_i)^2}{\widehat{\beta}_{in}} < \frac{\chi^2_{\gamma,k-1}\psi(\widehat{\alpha}_n)}{n}, \quad \sum_{j=1}^{k} \beta_j = 1\right\}.$$

Then from (11) it follows that

$$\mathbf{P}\{\bar{\beta} \in \mathfrak{G}_{\gamma_n}\} \to \gamma,$$

i.e., $\mathfrak{G}_{\gamma_n}$ is an asymptotic γ-confidence set for $\bar{\beta}$, which is the intersection of the random ellipsoid centered at the point $\widehat{\beta}_n$ and the hyperplane $\sum_{i=1}^{k} \beta_i = 1$.
Suppose that it is necessary to construct a confidence set for $\bar{d} = Z\bar{\beta}^0$, where Z is a given $r \times (k-1)$ matrix, the rank Z is equal to $r \le k-1$. Then (see (Ivchenko and Khonov, 1991)) an asymptotic γ-confidence ellipsoid for $\bar{d}$ has the form

$$\mathfrak{G}_{Z\gamma_n} = \{\bar{d}: (\widehat{d}_n - \bar{d})' H_{Z_n}^{-1} (\widehat{d}_n - \bar{d}) < \chi^2_{\gamma,r}\psi(\widehat{\alpha}_n)/n\},$$

where $H_{Z_n} = ZH_nZ'$, $H_n = H(\widehat{\beta^0}_n)$.
In particular, for $r = 1$ the asymptotic $(1-\delta)$-confidence interval for a given linear combination

$$d = \bar{\lambda}'\bar{\beta}^0 = \sum_{i=1}^{k-1} \lambda_i \beta_i$$

has the form

$$\bar{\lambda}'\widehat{\beta^0}_n \pm t_{\delta/2}\sqrt{\psi(\widehat{\lambda}_n)(\bar{\lambda}' H_n \bar{\lambda})/n}\,.$$

Another equivalent representation for this set was given by the same authors in the form

$$\mathfrak{G}_{\gamma_n} = \left\{\bar{\beta}^0: |\bar{\lambda}'(\widehat{\beta^0}_n - \bar{\beta}^0)| < \sqrt{\chi^2_{\gamma,k-1}\psi(\widehat{\alpha}_n)(\bar{\lambda}' H_n \bar{\lambda})/n} \quad \forall \bar{\lambda}\right\},$$

i.e., the system of confidence intervals for all linear combinations $\{\bar{\lambda}'\bar{\beta}^0\}$ of the form

$$\left\{(\bar{\lambda}'\widehat{\beta^0}_n \pm \sqrt{\chi^2_{\gamma,k-1}\psi(\widehat{\alpha}_n)(\bar{\lambda}'H_n\bar{\lambda})/n}\,)\right\}$$

has the asymptotic confidence level γ. If one wants to estimate simultaneously a finite number of such linear combinations that usually happens in practice then the system of corresponding confidence intervals will have the asymptotic confidence level not less than γ.

In particular, the corresponding system of confidence intervals for the differences $\{\beta_i - \beta_j,\ j < i\}$ has the form

$$\left\{\left(\widehat{\beta}_{in} - \widehat{\beta}_{jn} \pm \sqrt{\chi^2_{\gamma,k-1}\psi(\widehat{\alpha}_n)(\widehat{\beta}_{in} + \widehat{\beta}_{jn} - (\widehat{\beta}_{in} - \widehat{\beta}_{jn})^2)/n}\right),\ j < i\right\}.$$

Note that Theorems 10 and 11 make it possible to construct CAN estimates and asymptotic confidence intervals for any smooth functions $\tau(\alpha, \bar{\beta})$. For this aim the statistic $\tau(\widehat{\alpha}_n, \widehat{\beta}_n)$ should be taken as the corresponding estimate.

Besides, the confidence regions can be used in standard manner for constructing a goodness-of-fit test in the problem of testing a simple hypothesis about the values of parameters of the scheme.

In this direction Ivchenko and Timonina (Ivchenko and Timonina, 1980, 1982) obtained some results for unstratified population based on the asymptotical normality of linear statistics in the case of equal sizes of samples in GSS.

As an illustration we note the following facts.

Let

$$m_1 = \cdots = m_s = m, \qquad p = m/N$$

and a priori

$$0 < p_1 \le p \le p_2 < 1 \quad \text{as } m, N \to \infty.$$

Let also denote

$$b_s(p) = (1 - q^s)/p, \quad q = 1 - p,$$

$$\xi_m = b^{-1}{}_s(\eta/m), \qquad \varphi_s(p) = p^3q^s/(1 - q^s - spq^{s-1}).$$

Then the critical region

$$\mathfrak{S}_\gamma(p_0) = \left\{|\xi_m - p_0| > t_{\gamma/2}\sqrt{\varphi_s(p_0)/m}\right\}$$

generates the test for the simple hypothesis

$$H_0\colon p = p_0, \qquad 0 < p_0 < 1,$$

and the critical region

$$\mathfrak{S}_\gamma(p_1, p_2) = \{\xi_m < p_1 - t_\gamma\sqrt{\varphi_s(p_1)/m}\} \cup \{\xi_m > p_2 + t_\gamma\sqrt{\varphi_s(p_2)/m}\}$$

may be used for testing of the composite hypothesis

$$H_0\colon p \in [p_1, p_2].$$

These tests are asymptotically unbiased, they have preassigned size γ and can differ from H_0 close alternatives of the order $O(1/\sqrt{m})$.
Analogously it follows from (11) that the critical region

$$\mathfrak{S}_\alpha = \left\{ \sum_{i=1}^{k} \frac{(\widehat{\beta}_{in} - \beta_{i0})^2}{\widehat{\beta}_{in}} > \chi^2_{1-\alpha,k-1} \frac{\Psi(\widehat{\alpha}_n)}{n} \right\}$$

allows to test the hypothesis H_0: $\bar{\beta} = \bar{\beta}_0$ for the stratified population with the preassigned test size that is equal to α.
Those who are interested in the Bayesian and minimax approaches to multiple capture-mark-recapture survey we refer to (Tribula and Wilczynski, 1985; Tribula, 1985; Smith, 1988), which contain both interesting results and references.

REFERENCES

Belyaev, Yu. K. (1975). *Probabilistic Methods of Sampling Inspection*. Nauka, Moscow (in Russian).

Berg, S. (1974). Factorial series distribution with applications to capture-recapture problems. *Scandinavian J. Statist.* 1, 145–152.

Berg, S. (1976). A note on unbiased minimum variance estimate in a capture-recapture problems. *Scandinavian J. Statist.* 3, 86–88.

Driml, M. and Ullrich, M. (1967). Maximum likelihood estimate of a number of types. *Acta Techn. CSAV* 3, 300–303.

Ivanov, V. A., Ivchenko, G. I., and Medvedev, Yu. I. (1985). Discrete problems of probability theory. *J. Soviet Math.* 35, 31–60.

Ivchenko, G. I. (1984). On some problems of statistical inference for stratified populations. *Theory Probab. Appl.* 29, 113–118.

Ivchenko, G. I. and Khonov, S. A. (1991). On asymptotic estimation for stratified finite populations. *Discrete Math. Appl.* 1, 322–320.

Ivchenko, G. I. and Medvedev, Yu. I. (1990). *Mathematical Statistics*. Mir, Moscow (in Russian).

Ivchenko, G. I. and Timonina, E. E. (1980). On estimation for sampling from a finite population. *Math. Notes* 28, 623–633 (in Russian).

Ivchenko, G. I. and Timonina, E. E. (1981). Om some problems in estimation for samples from a finite population. *Izv. Acad. Sci. Uzbek SSR, ser. phys.-math.* 4, 27–33 (in Russian).

Ivchenko, G. I. and Timonina, E. E. (1982). On estimation of size of finite population. *Theory Probab. Appl.* 31, 380–383.

Khonov, S. A. (1987 a). On estimation for stratified finite population in general scheme of sampling. In: *Probab. Proc. Appl.*, MIEE, pp. 41-49 (in Russian).

Khonov, S. A. (1987 b). On unbiased estimation for consistency of stratified population in general sampling scheme. In: *Probab. Problems Discrete Math.*, Moscow, MIEE, pp. 111–120 (in Russian).

Khonov, S. A. (1989). MLE for strata sizes of a finite population. In: *Probab. Proc. Appl.*, MIEE, Moscow, pp. 144–147 (in Russian).

Khonov, S. A. (1990). Asymptotical maximum likelihood estimates for stratified population parameters. In: *Probab. Problems Discrete Math.*, Moscow, MIEE, pp. 127–129 (in Russian).

Kolchin, V. F. (1986). *Random Mappings*. Optimization Software, New York.

Kolchin, V. F., Sevastyanov, B. A. and Chistyakov, V. P. (1978). *Random Allocations*. Wiley, New York.

Leite, J. G. et al. (1988). A note on the exact maximum likelihood estimate. *Biometrika* **75**, 178–180.

Seber, G. A. F. (1986). A review of estimating animal abundance. *Biometrics* **42**, 267–292.

Smith, P. J. (1988). Bayesian methods for multiple capture-recapture surveys. *Biometrics* **44**, 1177–1189.

Tribula, S. (1985). Minimax estimation of the parameters of the multivariate hypergeometric distribution. *Zastosowania Math. Appl.* **18**, 559–570.

Tribula, S. and Wilczynski, M. (1985). Estimation under sampling from a finite population. *Zastosowania Math. Appl.* **18**, 571–576.

Vatutin, V. A. and Mikhailov, V. G. (1982). Limit theorems for the number of empty cells in the equiprobable allocation scheme of series of particles. *Theory Probab. Appl.* **27**, 684–692.

Voinov, V. G. and Nikulin, M. S. (1989). *Unbiased Estimators and Their Applications*. Nauka, Moscow (in Russian).

Probabilistic Methods in Discrete Mathematics, pp. 226 – 239
V. F. Kolchin et al. (Eds.)
1993 TVP/VSP

ASYMPTOTIC EXPANSIONS FOR PERMUTATION TESTS WITH SEVERAL SAMPLES

Sh. ISMATULLAEV and Sh. MIRAKHMEDOV
Tashkent University, Vuzgorodok, 700095 Tashkent, Uzbekistan

ABSTRACT

Let $\{a_{nk},\ k = 1 \div n;\ n = 2,3,\dots\}$ be a triangular array of real numbers and $(R_{n1},\dots,R_{nn})$ be a r. v. having a uniform distribution on the set of all permutations of degree n. Denote by X_{nj}, $j = 0 \div r$, a sample sum of size n_j in the scheme of sampling without replacement from the finite population $\{a_{n1},\dots,a_{nn}\}$.
Under certain conditions an asymptotic expansion for the distribution function of

$$V_n = \sum_{k=0}^{r} p_k^{-1} X_{nk}^2,$$

where $p_k = n_k/n$, is established.

1. INTRODUCTION

Let $\{a_{nk},\ k = 1 \div n,\ n = 2,3,\dots\}$ be a triangular array of real numbers such that

$$\sum_{k=1}^{n} a_{nk} = 0, \quad \sum_{k=1}^{n} a_{nk}^2 = 1. \tag{1}$$

Suppose that the r. v. $(R_{n1},\dots,R_{nn})$ has a uniform distribution on the set of all permutations of degree n. Let

$$X_{nj} = \sqrt{\frac{n-1}{n}} \sum_{k=b_{j-1}}^{b_j} a_{nR_{nk}},$$

where $b_{-1} = 0$, $b_j = n_0 + \dots + n_j$, $j = 0 \div r$, and $n_0,\dots,n_r$ are natural numbers such that $n_0 + \dots + n_r = n$.
We note that X_{nj} is a sample sum of size n_j in the scheme of sampling without replacement from the finite population $\{a_{n1},\dots,a_{nn}\}$.
Denote $p_k = n_k/n$. We consider the r. v.

$$V_n = \sum_{k=0}^{r} p_k^{-1} X_{nk}^2.$$

The random variable V_n is used as a statistics in the homogeneity test for $r+1$ samples. The well-known rank statistics of chi-square type, in particular, the Kruskall–Wallis statistics (see (Rosen, 1965; Hajek and Šidak, 1967)) are special cases of V_n. It can be shown that V_n converges in distribution to a chi-squared variable χ_r^2 with r degrees of freedom, if $b_n = \max_k |a_{nk}| \to 0$ and $n \min_i p_i \to \infty$ as $n \to \infty$. This follows from the results of (Rosen, 1965) and (Hajek and Šidak, 1967, Chapter V), where a functional limit theorem for partial sums of the $\{a_{n1}, \ldots, a_{nn}\}$ is given. An estimation of the remainder term of order $(\min_{0\leq j\leq r} n_j)^{-1}$ in this limits theorem was obtained in (Huskova, 1980). An asymptotical expansion with two terms for the distribution function of V_n is obtained by Robinson (see (Robinson, 1980)) under the conditions $\min_{0\leq j\leq r} p_j \geq \delta > 0$ and $\mathbf{I}^{(s)}$ (see below). The remainder term given in (Robinson, 1980) has the form $C(\varepsilon, \beta)A_{(n5)}^{1-\varepsilon}$, where $\varepsilon > 0$ is an arbitrary constant, $A_{nl} = \sum_{k=1}^{n} |a_{nk}|^l$. In what follows $C(\cdot)$ denotes a positive value which depends on its arguments.

In the present paper we give an asymptotic expansion with $\varkappa - 3$ terms and a uniform remainder of the form $C(\varkappa, \beta)A_{n\varkappa}$ under condition $\mathbf{I}^{\varkappa}$, $\varkappa \geq 5$.

The method of the proof is a development of Robinson's method; namely, we use an integral representation of the characteristic function (ch. f.) of the r. v. $(X_{n0}, \ldots, X_{nr})$ (see (2)), which is an analog of well-known Erdös–Rényi formula. In constant to (Robinson, 1980) we note that the integrand in that representation can be considered as the ch. f. a sum of independent nondegenerated r. v.'s in $\mathbf{R}^{2r}$. This made in possible to use the method of ch. f. (Bhattacharya and Ranga Rao, 1976). A part of the results was announced in (Ismatullaev and Mirakhmedov, 1986).

2. MAIN RESULTS

It is clear that by analogy with the case of derivation of asymptotic expansions for a distribution of a sum of independent r. v. we need some conditions on the arithmetic structure of the numbers $a_{n1}, \ldots, a_{nn}$. A condition of such a kind is the following one.

Condition $\mathbf{I}^{(s)}$. For given $C_0 > 0$, there exist positive constants C_1, β and ε depending on C_0 only such that for any fixed x the number of indices k for which

$$|a_{nk}t - x - 2\pi l| > \varepsilon A_{ns}^{1/(s-2)} \log A_{ns}^{-1}$$

is greater than δn for all $t \in [C_0 A_{n3}^{-1}, C_1 A_{ns}^{-1}]$ and $l = 0, \pm 1, \ldots$.

Conditions like $\mathbf{I}^{(s)}$ are commonly encountered in problems of obtaining asymptotic expansions in the scheme of sampling without replacement (see (Robinson, 1980; Mirakhmedov, 1983)).

The following two theorems are refinements of the results from (Huskova, 1980) and (Robinson, 1980) respectively.

Denote $F_n(x) = \mathbf{P}\{V_n < x\}$.

THEOREM 1. *If* $\min_{0\le j\le r} p_j \ge \delta > 0$, *then the inequality*

$$\sup_x \left|F_n(x) - \mathbf{P}\{\chi_r^2 < x\}\right| \le C(\delta)A_{n_3}$$

holds.

THEOREM 2. *If* $\min_{0\le j\le r} p_j \ge \delta > 0$ *and for some* $\varkappa \ge 4$ *condition* $\mathbf{I}^{(\varkappa)}$ *is fulfilled, then uniformly with respect to all* $x > 0$

$$\left|F_n(x) - \widehat{G}_{n\varkappa}(x)\right| \leqslant C(x,\delta)A_{n\varkappa},$$

where $\widehat{G}_{n\varkappa}(x) = \mathbf{P}\{\chi_r^2 < x\} + G_{n\varkappa}(x)$; *the algorithm of obtaining* $\widehat{G}_{n\varkappa}(x)$ *is given in the proof below. In particular*

$$G_{n4}(x) = 0, \quad G_{n5}(x) = \left[\frac{1}{24}\Big(A_{n4} - \frac{3}{n}\Big)\Big(\sum_{j=0}^{r}\frac{1}{p_j} - r^2 - 4r - 1\Big)\right.$$

$$\times\Big(\frac{3x}{r+2} - 3\Big) - \frac{r(r+2)}{4n}\Big(\frac{x}{r+2} - 1\Big) + \frac{1}{24}A_{n3}^2\Big(5\sum_{j=0}^{r}\frac{1}{p_j} - 3r^2 - 12r - 5\Big)$$

$$\left.\times\Big(\frac{x^2}{(r+4)(r+2)} - \frac{2x}{r+2} + 1\Big)\right]\frac{x^{r/2}e^{-x/2}}{2^{r/2}\Gamma((r+2)/2)}.$$

Remark. In case $\varkappa = 5$ Theorem 2 refines the result of (Robinson, 1980).

COROLLARY. *If the assumptions of Theorem 2 are fulfilled for* $\varkappa = 4$, *then*

$$\left|F_n(x) - \mathbf{P}\{\chi_r^2 < x\}\right| \le C(\delta)A_{n4}.$$

3. PROOF OF THEOREM 2

We use the following notation:

$$\psi = (\psi_1,\dots,\psi_r), \quad \tau = (\tau_1,\dots,\tau_r), \quad \widetilde{\psi} = (\psi_0,\psi_1,\dots,\psi_r),$$

$$\widetilde{\tau} = (\tau_0,\tau_1,\dots,\tau_r), \quad (\widetilde{\psi},\widetilde{\tau}) = (\psi_0,\dots,\psi_r,\tau_0,\dots,\tau_r);$$

$\langle\cdot\,,\,\cdot\rangle$ is an inner product; for vector $\alpha \in \mathbf{R}^r$ with integer non-negative components $|\alpha| = \alpha_1 + \dots + \alpha_r$, $\tau^\alpha = \tau_1^{\alpha_1}\cdots\tau_r^{\alpha_r}$,

$$\varphi_k(\widetilde{\psi},\widetilde{\tau}) = \sum_{j=0}^{r} p_j \exp\left\{i(\psi_j n^{-1/2} + \tau_j a_{nk})\right\}.$$

In (Robinson, 1980) for the ch. f. $f_n(t)$ of r. v. $(X_{n0}, \dots, X_{nr})$ the following Erdös–Rényi type integral representation was obtained:

$$f_n(t) = B_n \int_{K(\psi)} \Phi_n(\widehat{\psi}, \widehat{\tau})\, d\psi, \tag{2}$$

where

$$\Phi_n(\widehat{\psi}, \widehat{\tau}) = \prod_{k=1}^{n} \varphi_k(\widehat{\psi}, \widehat{\tau}), \quad \tau_j = t_j\sqrt{1-n^{-1}}, \qquad j = 0 \div r;$$

$\widetilde{t} = (t_0, \dots, t_r)$ is such that $\sum_{j=0}^{r} p_j t_j = 0$,

$$B_n = \left[p_0 (2\pi\sqrt{n})^r n! \prod_{j=0}^{r} p_j^{n_j}/n_j! \right]^{-1},$$

$$K(\psi) = \left\{ \psi : \; |\psi_j - \psi_0| \le \pi\sqrt{n}, \; j = 1 \div r, \; \sum_{j=0}^{r} p_j \psi_j = 0 \right\}.$$

It is evident that $\varphi_k(\widetilde{\psi}, \widetilde{\tau})$ is the ch. f. of r. v. ξ_{kn} with the distribution

$$\mathbf{P}\{\xi_{kn} = b_{kn}(j)\} = p_j, \qquad j = 0 \div r,$$

where $b_{kn}(j)$ is the $2(r+1)$-dimensional vector whose $(j+1)$th and $(r+j+2)$th coordinates are equal to $1/\sqrt{n}$ and a_{nk}, respectively, and all the others are equal to zero. Therefore, $\Phi_n(\widetilde{\psi}, \widetilde{\tau})$ is the ch. f. of the sum $S_n = \xi_{1n} + \cdots + \xi_{nn}$ of independent r. v.'s. It is easy to show that the covariance matrix W of S_n degenerates. Thus, for the asymptotic analysis of $\Phi_n(\widetilde{\psi}, \widetilde{\tau})$ we cannot directly use the well-known asymptotic expansions for ch. f.'s. of sums of independent r. v.'s. By this reason we obtain another interpretation of $\Phi_n(\widetilde{\psi}, \widetilde{\tau})$. But the first interpretation will be used as well.

It should be noted that the coordinates of vectors $\widetilde{\psi}$ and $\widetilde{\tau}$ in (2) satisfy the conditions

$$\sum_{j=0}^{r} p_j \tau_j = \sum_{j=0}^{r} p_j \psi_j = 0. \tag{3}$$

Without loss of generality we can assume that $p_0 \ge p_1 \ge \cdots \ge p_r$.

If we find from (3) expressions for t_0 and ψ_0 and substitute them to the formula for $\varphi_k(\widetilde{\psi}, \widetilde{\tau})$, we obtain that

$$\varphi_k(\widetilde{\psi}, \widetilde{\tau}) = \mathbf{E} \exp\{i\langle Y_k, v\rangle\}, \tag{4}$$

where $v = (\psi, \tau)$, $\quad Y_k = (\overline{\nu}_{k1}, \dots, \overline{\nu}_{kr}, \overline{\mu}_{k1}, \dots, \overline{\mu}_{kr})$,

$$\overline{\nu}_{kj} = \nu_{kj} - p_j p_0^{-1} \nu_{k0}, \quad \overline{\mu}_{kj} = \mu_{kj} - p_j p_0^{-1} \mu_{k0}, \qquad j = 0 \div r,$$

and ν_{kj}, μ_{kj} are the $(j+1)$th and $(r+j+2)$th coordinates of r. v. ξ_{kn} respectively.

It is easily checked that $\mathbf{E}\,\overline{\nu}_{kj} = \mathbf{E}\,\overline{\mu}_{kj} = 0$ and the covariance matrix Λ of $Z_n = Y_1 + \cdots + Y_n$ is equal to

$$\begin{bmatrix} \Lambda_r & 0 \\ 0 & \Lambda_r \end{bmatrix}; \quad \Lambda_r = (\lambda_{ij}), \qquad i,j = 1 \div r, \quad \lambda_{ij} = \begin{cases} p_i(1 + p_i/p_0), & \text{if } i = j \\ p_i p_j / p_0, & \text{if } i \neq j. \end{cases}$$

Denoting $\theta_k(v) = \mathbf{E}\exp\{i\langle Y_k, v\rangle\}$, according to what has been said we conclude that in region (3) the function $\widetilde{\Phi}_n(\widetilde{\psi}, \widetilde{\tau})$ coincides with the ch. f. of Z_n which is equal to $\Theta_n(v) = \prod_{k=1}^{n} \theta_k(v)$. Note that det $\Lambda > 0$. This fact is significant for all our consequent reasoning.

Now we turn to finding an asymptotic expansion for the ch. f. $f_n(t)$ in (2). It is known (see (Bhattacharya and Ranga Rao, 1976, Section 9)) that the ch. f. $\Theta_n(v)$ can be approximated by the expression ($\varkappa \geq 3$ is an integer)

$$\Psi_{n\varkappa}^{**}(v) = e^{-\frac{\langle v, \Lambda v\rangle}{2}} \Big(1 + \sum_{\nu=1}^{\varkappa-3} \frac{\overline{Q}_{\nu n}(v)}{(\sqrt{n})^{\nu}}\Big), \tag{5}$$

where $\overline{Q}_{\nu n}(v)$ is a polynomial in $\psi_j, \tau_j, j = 1 \div r$, of degree 3ν, whose coefficients depend on cumulants of $\sqrt{n}Z_n$, and the smallest degree of its monomials is equal to $\nu + 2$.

Define the function $\Psi_{n\varkappa}^{*}(\tau)$ by

$$\Psi_{n\varkappa}^{*}(\tau) = \int_{\mathbf{R}^r} \Psi_{n\varkappa}^{**}(\psi, \tau)\, d\psi,$$

then, taking into account (5) and the relation

$$\langle(\widetilde{\psi}, \widetilde{\tau}), W(\widetilde{\psi}, \widetilde{\tau})\rangle = \langle v, \Lambda v\rangle = T_{\psi}^2 + T_{\tau}^2, \tag{6}$$

with $T_x^2 = \sum p_j x_j^2$, which is true in region (3), we obtain that

$$\Psi_{n\varkappa}^{*}(\tau) = \frac{p_0 (2\pi)^{r/2}}{\sqrt{p_0 \cdots p_r}} e^{T_\tau^2/2} \Big(1 + \sum_{\nu=1}^{\varkappa-3} n^{-\nu/2} Q_{\nu n}(\tau)\Big); \tag{7}$$

where $Q_{\nu n}(\tau) = \sum_{0 \leq |\alpha| \leq 3\nu} q_{\alpha\nu} \tau^{\alpha}$, and $q_{\alpha\nu}$ depends on p_j, $j = 1 \div r$, and cumulants of $\sqrt{n}Z_n$. According to inequality (9.12) in (Bhattacharya and Ranga Rao, 1976) it follows that $|q_{\alpha\nu}| \leq C n^{\nu/2} A_{n\varkappa}^{\nu/(\varkappa-2)}$. Since $\tau_j = t_j(1 - n^{-1})^{1/2}$, the Taylor expansion argument yields

$$n^{-\nu/2} Q_{\nu n}(\tau) = n^{-\nu/2} \sum_{k=0}^{(\varkappa-\nu-3)/2} n^{-k} Q_{k\nu n}(t) + \theta A_{n\varkappa} Q_{\nu}(t), \tag{8}$$

where $Q_{k\nu n}(t)$ is a polynomial with coefficients depending on $p_1, \ldots, p_r$ and the cumulants of $\sqrt{n}Z_n$, which can be bounded from above by $A_{n\varkappa}^{\nu/(\varkappa-2)} n^{\nu/2}$; $Q_{\nu}(t)$ is a polynomial whose coefficients do not depend on n, and (as usual) θ is some variable with $|\theta| \leq 1$.

Now we transform the exponential factor in (7). It is clear that for $T_t^2 \le C\sqrt{n}$

$$\exp\Big\{-\frac{1}{2}T_r^2\Big\} = \exp\Big\{-\frac{1}{2}T_t^2\Big\}\Big(\sum_{s=0}^{m}\frac{T_t^{2s}}{s!(2n)^s} + \frac{C\theta}{(2n)^{m+1}}T_t^{2(m+1)}\Big). \tag{9}$$

Finally using the refinement of the Stirling formula

$$k! = \sqrt{2\pi k}k^k \exp\Big\{-k + \sum_{\nu=1}^{l}\frac{B_{2\nu}}{2\nu(2\nu-1)}k^{2\nu-1} + \frac{C(l)\theta}{k^{2l}}\Big\}$$

(here $B_{2\nu}$ is the Bernoulli number), and not complicated computations, we represent B_n from (2) as

$$B_n = \frac{\sqrt{p_0\cdots p_r}}{p_0(2\pi)^{r/2}}\Big(1 + \sum_{\nu=1}^{l}\frac{A_\nu(p)}{n^\nu} + \frac{C(l)\theta}{n^{l+1}}\Big(\sum_{j=0}^{r}p_j^{1-2l}\Big)^{l+1}\Big). \tag{10}$$

It is easily checked that

$$A_1(p) = 1 - \sum_{j=0}^{r}\frac{1}{p_j}, \quad A_2(p) = A_1^2(p).$$

It follows from (2), (5), (7)–(10) that $f_n(t)$ can be approximated by

$$\Psi_{n\varkappa}(t) = e^{-T_t^2/2}\Big(1 + \sum_{m=1}^{\varkappa-3}n^{-m/2}\sum\nolimits^{*} A_j(p)\frac{T_t^{2s}}{s!2^s}Q_{k\nu n}(t)\Big), \tag{11}$$

where the summation in $\sum^*$ is taken over all integers $\nu \ge 1,\ k \ge 1,\ j \ge 0,\ s \ge 0$ such that $\nu + 2(k+j+s) = m$.
Let G_{nl}^r be the generalized measure with Fourier–Stieltjes transform $\Psi_{nl}(t)$ and let F_n^r be the distribution with ch. f. $f_n(t)$,

$$S_u = \Big\{(x_0,\ldots,x_r):\ \sum_{j=0}^{r}x_j^2/p_j \le u,\ \sum_{j=0}^{r}x_j = 0\Big\},$$

$$\widehat{G}_{nl}(u) = G_{nl}^r(S_u), \quad \Delta_n(u) = F_n(u) - \widehat{G}_{n\varkappa}(u).$$

Evidently $F_n(u) = F_n^r(S_u)$ and therefore $\Delta_n(u) = (F_n^r - G_{n\varkappa}^r)(S_u)$. In what follows we use the notation

$$D_t^\alpha = D_{t_1}^{\alpha_1}\cdots D_{t_r}^{\alpha_r}, \qquad D_{t_j}^{\alpha_j} = \partial^{\alpha_j}/\partial t_j^{\alpha_j}.$$

LEMMA 1. *For any $L > 0$ the inequality*

$$|\Delta_n(u)| \le C(r) \max_{0\le|\alpha|\le r+\varkappa+1} \int_{\|t\|\le L} |D_t^\alpha(f_n(t) - \Psi_{n\varkappa}(t))|\, dt + \frac{C(r)}{L} \tag{12}$$

holds.

This assertion follows directly from Corollary 11.5 and Lemma 11.6 of (Bhattacharya and Ranga Rao, 1976).

Putting $L = C_1 A_{n\varkappa}^{-1}\sqrt{p_r}/2$ in Lemma 1 (C_1 from assumption $\mathbf{I}^{(\varkappa)}$), we rewrite the integral from (12) as $J_1 + J_2$, where J_1 and J_2 are the integrals over regions $\|t\| \le L_0$ and $L_0 < \|t\| \le L$, respectively; and $L_0 = [4(r+1)^2 A_{n3}]^{-1}$ For the sake of definiteness suppose that $L_0 < L$ (otherwise the proof is simpler). Denoting

$$\widehat{B}_n = \left(p_0(2\pi)^r\right)^{-1/2}\sqrt{p_1\cdots p_r}\left(1 + \sum_{\nu=1}^{[(\varkappa-3)/2]} n^{-\nu}A_\nu(p)\right),$$

and taking into consideration the definition of vectors v, t, ψ, τ we find that

$$\begin{aligned} J_1 \le & B_n \int_{\|t\|\le L_0}\int_{\psi\in K(\psi)} |D_t^\alpha[\Theta_n(v) - |\Psi_{n\varkappa}^{**}]|\, dv \\ & + |B_n - \widehat{B}_n| \int_{\|t\|\le L_0}\int_{\psi\in K(\psi)} |D_t^\alpha \Psi_{n\varkappa}^{**}(v)|\, dv \\ & + \widehat{B}_n \int_{\|t\|\le L_0}\int_{\psi\in \overline{K}(\psi)} |D_t^\alpha \Psi_{n\varkappa}^{**}(v)|\, dv \\ & + \int_{\|t\|\le L_0} |D_t^\alpha(\widehat{B}_n \Psi_{n\varkappa}^{*}(\tau) - \Psi_{n\varkappa}(t))|\, dt = J_{11} + J_{12} + J_{13} + J_{14}, \end{aligned} \tag{13}$$

where $\overline{K}(\psi) = \mathbf{R}^r \setminus K(\psi)$.

In order to estimate the integrals from (13) we need the following inequality from (Bhattacharya and Ranga Rao, 1976, p. 85, Lemma 9.5): for $0 \le |\alpha| \le 3\nu$, $\nu = 0 \div \varkappa - 2$

$$|D_t^\alpha Q_{\nu n}(v)| = C(\alpha,\nu)(1 + \|v\|^{3\nu+|\alpha|})\rho_\varkappa^{\nu/(\varkappa-2)}, \tag{14}$$

where

$$\rho_\varkappa = \frac{1}{n}\sum_{k=1}^{n} \mathbf{E}\,\|\sqrt{n}Y_k\|^\varkappa \le Cn^{(\varkappa-2)/2}A_{n\varkappa}. \tag{15}$$

A simple computation based on (10), (14) and the estimates

$$A_{nk} \ge n^{-(k-2)/2}, \qquad k \ge 2,$$

gives

$$J_{12} \le C(\varkappa,\delta)A_{n\varkappa}. \tag{16}$$

Using inequalities (14), (15) and relation (6), after not complicated computations, by virtue of presence of the exponential factor in the definition of function $\Psi_{n\varkappa}^{**}(v)$, we obtain that

$$J_{13} \le C_3(r,\varkappa)A_{n\varkappa}. \tag{17}$$

In order to estimate the integral J_{14} we write the following equalities:

$$\widehat{B}_n\Psi^*_{n\varkappa}(\tau) - \Psi_{n\varkappa}(t) = \widehat{B}_n[\Psi^*_{n\varkappa}(\tau) - \widehat{\Psi}^*_{n\varkappa}(\tau)] + \widehat{B}_n[\widehat{\Psi}^*_{n\varkappa}(t) - \Psi_{n\varkappa}(t)],$$

where

$$\widehat{\Psi}^*_{n\varkappa}(t) = \sum_{m=1}^{\varkappa-3} n^{-m/2} \sum{}' \frac{T_t^{2s}}{s!2^s} Q_{k\nu n}(t),$$

and the summation in $\sum'$ is taken over all integers $k \geq 1,\ \nu \geq 1,\ s \geq 0$ such that $\nu + 2(s+k) = m$; and

$$\Psi^*_{n\varkappa}(\tau) - \widehat{\Psi}^*_{n\varkappa}(t) = \left[e^{-T_\tau^2/2} - e^{-T_t^2/2} \sum_{s=0}^{[(\varkappa-3)/2]} \frac{T_t^{2s}}{s!(2n)^s}\right] \sum_{\nu=0}^{\varkappa-3} \frac{Q_{\nu n}(\tau)}{n^{\nu/2}}$$

$$+ e^{-T_t^2/2} \sum_{s=0}^{[(\varkappa-3)/2]} \frac{T_t^{2s}}{s!(2n)^s} \left[\sum_{\nu=0}^{\varkappa-3} \frac{Q_{\nu n}(\tau)}{n^{\nu/2}} - \sum_{\nu=0}^{\varkappa-3} n^{-\nu/2} \sum_{k=0}^{[(\varkappa-\nu-3)/2]} n^{-k} Q_{k\nu n}(t)\right]$$

$$+ \left[e^{-T_t^2/2} \sum_{s=0}^{[(\varkappa-3)/2]} \frac{T_t^{2s}}{s!(2n)^s} \sum_{\nu=0}^{\varkappa-3} \sum_{k=2}^{[(\varkappa-\nu-3)/2]} n^{-\nu/2-k} Q_{k\nu n}(t) - \widehat{\Psi}^*_{n\varkappa}(t)\right].$$

Using these equalities, definition (11) of $\Psi_{n\varkappa}(t)$, formulas (7)–(10), inequalities (14), (15) and noting that $D_t^\alpha(v) = 0$ for $|\alpha| > 3\nu$, after tedious but not complicated computations we get the estimate

$$J_{14} \leq C_4(r,\varkappa) A_{n\varkappa}. \tag{18}$$

Now we consider the integral J_{11}.

LEMMA 2. *Let $l \geq 3$ be an integer. There exist $C_5(r,l)$, $C_6(r,l)$ such that for $\|(t,\psi)\| \leq C_5(r,l)\sqrt{p_r/p_0}A_{nl}^{-1/l}$ and any α, $0 \leq |\alpha| \leq l$, we have*

$$\left|D_t^\alpha\left[\prod_{k=1}^n \theta_k(t,\psi) - \Psi^{**}_{nl}(t,\psi)\right]\right| \leq C_6(r,l) p_r^{-l/2} A_{nl}$$

$$\times \left[\|(t,\psi)\|^{(l-|\alpha|)/2} + \|(t,\psi)\|^{(3(l-2)+|\alpha|)/2}\right] \exp\left\{-\frac{1}{4}(T_t^2 + T_\psi^2)\right\}.$$

Proof. Remind that $\prod_{k=1}^n \theta_k(t,\psi)$ is the ch. f. of

$$Z_n = (\overline{Y}_1 + \cdots + \overline{Y}_n)/\sqrt{n},$$

where $\overline{Y}_1 = \sqrt{n}Y_i$, $i = 1 \div n$, have a nonsingular covariance matrix Λ. It is easy to see that

$$\langle(t,\psi), \Lambda(t,\psi)\rangle = \sum_{j=1}^r p_j(t_j^2 + \psi_j^2) + p_0^{-1}\left[\Big(\sum_{j=1}^r p_j t_j\Big)^2 + \Big(\sum_{j=1}^r p_j\psi_j\Big)^2\right]. \tag{19}$$

Hence

$$p_r\|(t,\psi)\|^2 \le \langle (t,\psi), \Lambda(t,\psi)\rangle \le (r+1)p_0\|(t,\psi)\|^2.$$

Therefore, denoting $\lambda_{\min}$ and $\lambda_{\max}$ the minimal and maximal eigenvalues of Λ, we have

$$p_r \le \lambda_{\min} \le \lambda_{\max} \le (r+1)p_0. \tag{20}$$

It is clear that

$$\mathbf{E}\,\|\overline{Y}_m\|^l \le (\sqrt{n})^l p_0 \sum_{\nu=1}^{r}\Big(\frac{p_\nu}{p_0}+\frac{p_\nu^l}{p_0^l}\Big) r^{(l-2)/2}\Big[\Big(\frac{1}{\sqrt{n}}\Big)^l + |a_{nm}|^l\Big]$$

$$\le 2p_0 r^{l/2}[1+n^{l/2}|a_{nm}|^l].$$

Therefore, taking into account (20) and the inequality

$$\Big(\frac{1}{n}A_{ns}\Big)^{1/s} \le \Big(\frac{1}{n}A_{nl}\Big)^{1/l}, \qquad 2\le s\le l,$$

and denoting B the matrix such that $B^2=\Lambda^{-1}$, we obtain the estimates

$$\frac{1}{n}\sum_{m=1}^{n}\mathbf{E}\,\|B\overline{Y}_m\|^l \le \lambda_{\min}^{-l/2}\frac{1}{n}\sum_{m=1}^{n}\mathbf{E}\,\|\overline{Y}_m\|^l$$

$$\le a\lambda_{\min}^{-l/2}p_0 r^{l/2}(1+n^{(l-2)/2}A_{nl}) \le 4p_r^{-l/2}p_0 r^{l/2}n^{(l-2)/2}A_{nl}. \tag{21}$$

The assertion of Lemma 2 follows now from inequality (9.55) of (Bhattacharya and Ranga Rao, 1976, p.96) and relations (6), (20), (21).

LEMMA 3. *If for vectors $\tilde{t}$ and $\tilde{\psi}$ the condition* (3) *is fulfilled, and $|t_j|\le[2(r+1)^2A_{n3}]^{-1}$, $|\psi_j|\le\sqrt{n}/(2(r+1)^2)$, $j=1\div r$, then for any α*

$$\Big|D_t^\alpha\prod_{k=1}^{n}\theta_k(t,\psi)\Big| \le C(\alpha,r)\omega^{|\alpha|}\exp\Big\{-\frac{1}{4}(T_t^2+T_\psi^2)\Big\},$$

where $\omega=\max\{1,\|(t,\psi)\|\}$.

Proof. Since the r. v. $\overline{\nu}_{kj}$ takes values $1/\sqrt{n}$, 0, $-p_j/(p_0\sqrt{n})$, and $\overline{\mu}_{kj}$ takes values a_{nk}, 0, $-p_ja_{nk}/p_0$ with probabilities $p_j, 1-p_0-p_j, p_0$, respectively,

$$\|Y_k\|^2 \le 2r(1+1/n). \tag{22}$$

Since $\mathbf{E}\,Y_k=0$,

$$|D_{t_j}\theta_k(t,\psi)| \le 4(\|t\|^2+\|\psi\|^2)\mathbf{E}\,\|Y_k\|^2, \qquad j=1\div r. \tag{23}$$

Taking into account (2), for any vector β with integer non-negative elements such that $|\beta|\ge 2$, we have

$$|D_t^\beta\theta_k(t,\psi)| \le [2r(1+1/n)]^{(|\beta|-2)/2}\mathbf{E}\,\|Y_k\|^2. \tag{24}$$

Put $l_k = \max(2r^{(|\beta|-2)/2}, 4\omega)\mathbf{E}\,\|Y_k\|^2$. Let N_ν be a subset of $N = \{1,2,\ldots,n\}$ with ν elements, and (N_ν) is the population of all N_ν.
Noting that $\sum_{k=1}^{n} \mathbf{E}\,\|Y_k\|^2 \le 4$ we obtain the estimates

$$\sum_{(n_\nu)} \prod_{k\in N_\nu} l_k \le \Big(\sum_{k=1}^{n} l_k\Big)^{\nu} \le C(k,\nu)\omega^{\nu}. \tag{25}$$

By standard arguments one can obtain the inequality

$$|\varphi_k(\widetilde{t},\widetilde{\psi})|^2 \le \exp\Big\{-\frac{1}{n}\mathbf{E}\,\langle(\widetilde{t},\widetilde{\psi}),\xi_{kn}\rangle^2 + \frac{2}{3}\mathbf{E}\,|\langle(\widetilde{t},\widetilde{\psi}),\xi_{kn}\rangle|^3\Big\}. \tag{26}$$

Reminding that $\theta_k(t,\psi) = \varphi_k(\widetilde{t},\widetilde{\psi})$ and taking into account (6), (26) and the inequality $\sup_{a\ge 0}\exp\{a^2 - 2a^3/3\} \le e^{1/3}$, we obtain that

$$\prod_{k\in N/N_\nu} |\theta_k(t,\psi)|^2 \le e^{r/3}\exp\Big\{-\langle(\widetilde{t},\widetilde{\psi}),\, W(\widetilde{t},\widetilde{\psi})\rangle + \frac{2}{3}\sum_{k=1}^{n}\mathbf{E}\,|\langle(\widetilde{t},\widetilde{\psi}),\xi_{kn}\rangle|^3\Big\}$$

$$\le e^{r/3}\exp\Big\{-T_t^2 - T_\psi^2 + \frac{2(r+1)^2}{3}\Big[A_{n3}\sum_{j=0}^{r} p_j|t_j|^3 + \frac{1}{\sqrt{n}}\sum_{j=0}^{r} p_j|\psi_j|^3\Big]\Big\}.$$

By the assumptions of Lemma 3

$$|t_0| \le r/(2(r+1)^2A_{n3}), \quad |\psi_0| \le r\sqrt{n}/(2(r+1)^2).$$

Therefore

$$\prod_{k\in N/N_\nu} |\theta_k(t,\psi)|^2 \le e^{r/3}\exp\Big\{-\frac{1}{2}(T_t^2 + T_\psi^2)\Big\}. \tag{27}$$

Using the Leibnitz rule of differentiation of a product, it is easy to see that

$$\Big|D_t^{\alpha}\prod_{k=1}^{n}\theta_k(t,\psi)\Big| \le \sum_{1\le\nu\le|\alpha|} C(\nu,\alpha)\sum_{(N_\nu)}\prod_{k\in N/N_\nu}\Big|\theta_k(t,\psi)\Big|\prod_{m=1}^{\nu}\Big|D_t^{\beta_m}\theta_{j_m}(t,\psi)\Big|,$$

where $\beta_1,\ldots,\beta_\nu$ are vectors with integer non-negative elements such that

$$|\beta_j| \ge 1, \qquad \beta_1 + \cdots + \beta_\nu = \alpha, \qquad j_m \in N_\nu = \{j_1,\ldots,j_\nu\}.$$

Applying relations (22)–(26), we obtain the assertion of Lemma 3.

LEMMA 4. *If condition (3) is fulfilled for $\tilde{t}$ and $\tilde{\psi}$, $\psi \in K(\psi)$, $\|t\| \le [4(r+1)^2 A_{n3}]^{-1}$, and $|\psi_j| \ge \sqrt{n}/(2(r+1)^2)$ for some $0 \le j \le r$, then for any α*

$$\Big|D_t^\alpha \prod_{k=1}^n \theta_k(t,\psi)\Big| \le C(r,\alpha)\omega^{|\alpha|} \exp\Big\{-C_1 p_r^2 n - C_2 p_r^2 \|t\|^2\Big\},$$

where C_1 and C_2 are constants.

Proof. Let $K_\alpha = \{k: \ |a_{nk}| < dA_{n3}\}$ and $|K_\alpha|$ denote the number of elements of K_d. It is clear that

$$1 = \sum_{k=1}^n a_{nk}^2 \ge \sum_{k \in K_d} a_{nk}^2 \ge (n - |K_d|)d^2 A_{n3}^2 \ge (n - |K_d|)d^2 n^{-1}.$$

Hence

$$|K_d| \ge \frac{d^2-1}{d^2} n. \tag{28}$$

Under the conditions of Lemma $|t_j| \le [4(r+1)^2 A_{n3}]^{-1}$ for all $1 \le j \le r$, therefore, if $k \in K_{3/2}$, then

$$\max_{0 \le j \le r} |\psi_j n^{-1/2} + \tau_j a_{nk}| > 1/(8(r+1)^2).$$

Taking into account this inequality and the pattern of the proof of inequality (14) in (Robinson, 1980) we obtain that

$$|\theta_k(t,\psi)|^2 = |\varphi_k(\tilde{t},\tilde{\psi})|^2 \le \exp\Big\{-2p_r^2\Big(1 - \cos\frac{1}{8(r+1)^2}\Big)\Big\} \le \exp\Big\{-\frac{p_r^2}{96(r+1)^4}\Big\}.$$

It follows directly from this and (28) that

$$\prod_{k \in N \setminus N_\nu} |\theta_k(t,\psi)| \le \prod_{k \in K_{3/2} \cap (N - N_\nu)} |\theta_k(t,\psi)| \le \exp\Big\{-\frac{p_r^2(5n/9 - \nu)}{192(r+1)^4}\Big\}$$

$$\le C(r,\alpha)\exp\Big\{-\frac{5p_r^2 n}{18 \cdot 192(r+1)^4} - \frac{9p_r^2}{192(r+1)^2}\|t\|^2\Big\}. \tag{29}$$

The proof of Lemma 4 is completed by applying the Leibnitz rule of differentiation of a product and successive using of inequalities (26), (23), (24), (29) and $\sum_{k=1}^n \mathbf{E}\,\|Y_k\|^2 \le 4$.
Let $C_5(r,l)$ be the constant from Lemma 2, and

$$L_{r,l} = C_5(r,l)\sqrt{p_r/p_0}\,A_{nl}^{-1/l}.$$

Denote

$$D_0 = \{(t,\psi): \ \|t\| \le L_0, \ \psi \in K(\psi)\},$$

$$D_1 = \{(t,\psi): \ \|(t,\psi)\| \le L_{r,\varkappa+r+1}\},$$

$$D_2 = \{(t,\psi):\ |t_j| \le L_0,\ |\psi_j| \le \sqrt{n}/2(r+1)^2,\ j = 1 \div r\} \setminus D_1,$$

$$D_3 = \{(t,\psi):\ \|t\| \le L_0,\ |\psi_j| > C(r)\sqrt{n}\ \text{ for some } j = 1 \div r,\ \psi \in K(\psi)\}.$$

Evidently $D_0 \subseteq D_1 \cup D_2 \cup D_3$. Therefore, denoting Λ_i the part of integral J_{11} taken over the region D_i we can write $J_{11} \le \Delta_1 + \Delta_2 + \Delta_3$.
Since $\Phi_n(\widetilde{t},\widetilde{\psi}) = \prod_{k=1}^{n} \theta_k(t,\psi)$, we have

$$\Delta_1 = \int_{D_1} \left|D_t^\alpha[\Phi_n(\widetilde{t},\widetilde{\psi}) - \Psi_{n\varkappa}^{**}(t,\psi)]\right| dt\, d\psi$$

$$\le \int_{D_1} \left|D_t^\alpha\Big[\prod_{k=1}^{n} \theta_k(t,\psi) - \Psi_{n,\varkappa+r+1}^{**}(t,\psi)\Big]\right| dt\, d\psi$$

$$+ \int_{D_1} \left|D_t^\alpha[\Psi_{n,\varkappa+r+1}^{**}(t,\psi) - \Psi_{n\varkappa}^{**}(t,\psi)]\right| dt\, d\psi.$$

Applying Lemma 2 with $l = \varkappa + r + 1$ to the first integral in the last inequality, we obtain the upper estimate $C(r,\varkappa,\delta)A_{n\varkappa}$ for this integral. After some computations we obtain from (6), (19), (20) that the similar estimation is held for the second integral as well. Thus

$$\Delta_1 \le C(r,\varkappa,\delta)A_{n\varkappa}.$$

To estimate Δ_2 and Δ_3, one can use Lemma 3 and 4, respectively. Using (19), it is easy to verify that

$$\Delta_2 \le C(r,\varkappa,\delta)A_{n\varkappa}, \qquad \Delta_3 \le C(r,\varkappa,\delta)A_{n\varkappa}.$$

It follows from estimates of Δ_1, Δ_2 and Δ_3 that

$$J_{11} \le C(r,\varkappa,\delta)A_{n\varkappa}. \tag{30}$$

In view of (13), (16), (17), (18) and (30), we obtain the estimate

$$J_1 \le C(r,\varkappa,\delta)A_{n\varkappa}. \tag{31}$$

In order to estimate the integral J_2, we need the following assertion.

LEMMA 5. *Let assumptions* $\mathrm{I}^{(\varkappa)}$ *and* (3) *be satisfied, then for any* α

$$\int_{\frac{1}{2}L_0 \le \|t\| \le L} |D_t^\alpha f_n(t)|\, dt \le C(r,\alpha)L^{r(1+|\alpha|)} \exp\left\{-\frac{\varepsilon\delta p_r^2 n}{2}\right\}.$$

Proof. It is clear that $p_r\|t\|^2 \le T_t^2 \le \|t\|^2$. Therefore, from the inequality $\|t\| \ge [4(r+1)^2 A_{n3}]^{-1}$ it follows that $T_t \ge \sqrt{p_r}[4(r+1)^2 A_{n3}]^{-1}$. Hence $|t_j| \ge \sqrt{p_r}[4(r+1)^2 A_{n3}]^{-1}$ for some $0 \le j \le r$. From this and condition (3) it follows that there exist j_1 and j_2, $0 \le j_1, j_2 \le r$, such that $|t_{j_1} - t_{j_2}| \ge \sqrt{p_r}[4(r+1)^2 A_{n3}]^{-1}$.

Note that $T_t \leq L$ as $\|t\| \leq L$. Hence $|t_{j_1} - t_{j_2}| \leq 2\sqrt{p_r}L$. Since $L = C_1\sqrt{p_r}A_{n\varkappa}^{-1}/2$, for t in the domain of integration of J_2, there exist t_{j_1} and t_{j_2} such that

$$\frac{\sqrt{p_r}}{4(r+1)^2 A_{n3}} \leq |t_{j_1} - t_{j_2}| \leq C_1 A_{n\varkappa}^{-1}. \tag{32}$$

Evidently

$$|\theta_k(t,\psi)|^2 = \sum_{i,j=0}^{r} p_i p_j \cos(\gamma_{ik} - \gamma_{jk}) = 1 - \sum_{i,j=0}^{r} p_i p_j (1 - \cos(\gamma_{ik} - \gamma_{jk}))$$

$$\leq 1 - 2p_r^2 \sum_{i,j=0}^{r} \sin^2 \frac{\gamma_{ik} - \gamma_{jk}}{2} \leq 1 - 2p_r^2 \sin^2 \frac{\gamma_{j_1 k} - \gamma_{j_2 k}}{2},$$

where $\gamma_{lk} = \psi_l n^{-1/2} + \tau_l a_{nk}$.
According to (32) we can apply condition $\mathbf{I}^{(\varkappa)}$, therefore for at least δn values of k we have

$$|\theta_k(t,\psi)|^2 \leq 1 - p_r^2 \sin^2\left(\frac{\varepsilon}{2} A_{n\varkappa}^{1/(\varkappa-2)} \log A_{n\varkappa}^{-1}\right). \tag{33}$$

Denoting by K the set of indices k of such kind, we have

$$\prod_{k\in N/N_\nu} |\theta_k(t,\psi)| \log \prod_{k\in K\cap(N\setminus N_\nu)} \exp\Big\{-\frac{1}{2}p_r^2\varepsilon^2 A_{n\varkappa}^{2/(\varkappa-2)} \log^2 A_{n\varkappa}^{-1}\Big\}$$

$$\leq \exp\Big\{-\frac{p_r^2\varepsilon^2}{2} A_{n\varkappa}^{2/(\varkappa-2)} \log^2 A_{n\varkappa}^{-1}(\delta n - \nu)\Big\} = C(\nu)\exp\{-\frac{p_r^2\varepsilon^2\delta}{2}\log^2 A_{n\varkappa}^{-1}\}. \tag{34}$$

The proof of Lemma 5 is completed by applying the Leibnitz formula of differentiation of a product and inequalities (24) and (34).
Evidently

$$J_2 \leq \int_{\frac{1}{2}L_0 \leq \|t\| \leq L} |D_t^\alpha f_n(t)|\, dt + \int_{\frac{1}{2}L_0 \leq \|t\| \leq L} |D_t^\alpha \Psi_{n\varkappa}(t)|\, dt. \tag{35}$$

In view of the inequality $A_{n\varkappa} \geq n^{-(\varkappa-2)/2}$ and the assertion of Lemma 5 we obtain that the first integral in (35) has the bound $C(\varkappa,\alpha)A_{n\varkappa}$. From the definition of $\Psi_{n\varkappa}(t)$ one can easily show that the same bound is true for the second integral of (35).
Thus

$$J_2 \leq C(\varkappa,\delta)A_{n\varkappa}. \tag{36}$$

Substituting estimates (31) and (36) to the inequality of Lemma 1 we complete the proof of Theorem 2.
The proof of Theorem 1 is similar to the proof of Theorem 2.
Since $G_{n4}^{(x)} \equiv 0$, assertion of Corollary follows from the assertion of Theorem 2.

REFERENCES

Bhattacharya, R. N. and Ranga Rao, R. (1976). *Normal Approximation and Asymptotic Expansions.* Wiley, New York.

Bickel, P. and van Zwet, W. (1978). Asymptotic expansions for the power of distribution-free tests in the two sample problem. *Ann. Statist.* **6**, 937–1004.

Hajek, J. and Šidák (1967). *Theory of Rank Tests.* Academic Press, New York.

Huskova, M. (1980). Some asymptotic results on the multivariate rank statistics. In: *Multivariate Analysis*, V. North-Holland, Amsterdam, pp. 223–237.

Ismatullaev, Sh. A. and Mirakhmedov, Sh. A. (1986). On asymptotic expansion for the distribution of permutation test. *Dokl. AN UzbSSR* **5**, 12–13 (in Russian).

Mirakhmedov, Sh. A. (1983). An asymptotic expansion for a sample sum from a finite population. *Theory Probab. Appl.* **28**, 492–501.

Robinson, J. (1978). An asymptotic expansion for the samples from a finite population. *Ann. Statist.* **6**, 1005–1011.

Robinson, J. (1980). An asymptotic expansion for permutation tests with several samples. *Ann. Statist.* **8**, 851–864.

Rosen, B. (1965). Limit theorems for sampling from a finite population. *Ark. Math.* **4**, 383–424.

Probabilistic Methods in Discrete Mathematics, pp. 240 – 250
V. F. Kolchin et al. (Eds.)
1993 TVP/VSP

ON APPROXIMATION OF WEIGHTED SUMS OF RANDOM VARIABLES WITH TWO-DIMENSIONAL INDICES

K. A. JAFAROV
Sankt-Petersburg University, Department of Mathematics and Mechanics, Bibliotechnaya Pl. 2, St. Peterhof, 198904 Sankt-Petersburg, Russia

ABSTRACT

We prove some probability inequalities useful for approximations of weighted sums of independent and identically distributed random variables with two-dimensional indices. As consequences of these approximations we obtain estimates for the Prohorov–Levy distance between weighted partial sums and two-parameter Wiener processes. We also use these results to prove the functional law of the iterated logarithm for Cesàro sums in two-dimensional case.

Let $X = \{X_{ij},\ i,j \geqslant 1\}$ be a sequence of independent and identically distributed random variables such that

I) $\mathbf{E}X = 0$, $\mathbf{E}X^2 = 1$.

Suppose that $h(t,x,s,y)$ is a real-valued function defined on $\{(t,x,s,y):\ t,x,s,y \geqslant 0,\ t > x,\ s > y\}$ and satisfying the following conditions:

II) $h(t,x,s,y) = f(t,x)g(s,y)$;

III) there exist constants α, β such that

$$f(\lambda t, \lambda x) = \lambda^{\alpha} f(t,x), \qquad \lambda > 0, \quad 0 < x < t < \infty,$$
$$g(\mu s, \mu y) = \mu^{\beta} g(s,y), \qquad \mu > 0, \quad 0 < y < s < \infty;$$

IV) $\int_0^1 f^2(1,x)dx < \infty$, $\int_0^1 g^2(1,y)dy < \infty$.

Put

$$Z_{nm}(t,s) = n^{-1/2-\alpha} m^{-1/2-\beta} \sum_{1 \leqslant k \leqslant nt-1} \sum_{1 \leqslant l \leqslant ms-1} h(n,k,m,l) X_{kl}, \quad 0 \leqslant t,s \leqslant 1,$$

$$Y(t,s) = \begin{cases} 0, & t,s \leqslant 1\,, \\ \sum\limits_{1 \leqslant k \leqslant t-1} \sum\limits_{1 \leqslant l \leqslant s-1} h(t,k,s,l) X_{kl}, & t,s > 1\,. \end{cases}$$

First of all we obtain some probability inequalities useful for our approximations of Z_{nm} and Y. Then by means of these inequalities we estimate the Prohorov–Levy

distance between Z_{nm} and two-parameter Wiener process (with transformed time parameter), we get also a functional law of the iterated logarithm for Cesàro sums. It is necessary to point out that Horváth (Horváth, 1986) considered similar problems in the one-dimensional case.
Further on, we shall use the notations

$$\xi(t,s) = o(\eta(t,s)) \quad \text{a.s.}$$

and

$$\xi(t,s) = O(\eta(t,s)) \quad \text{a.s.}$$

for

$$\mathbf{P}\{\lim_{t,s\to\infty} \xi(t,s)/\eta(t,s) = 0\} = 1$$

and

$$\mathbf{P}\{\limsup_{t,s\to\infty} |\xi(t,s)/\eta(t,s)| < \infty\} = 1$$

respectively.
Let $S(n,m) = \sum_{1\leqslant k\leqslant n}\sum_{1\leqslant l\leqslant m} X_{kl}$. The next well-known assertions are the basis of our investigation (see, for example, (Wichura, 1973; Major, 1976; Zinchenko, 1984)).

THEOREM 1. *If I) is satisfied, then it is possible to define $\{X_{ij}\}$ together with a two-parameter Wiener process $\{W(t,s),\ t,s \geqslant 0\}$ without changing of their distributions on the wider probability space so that*

$$|S(n,m) - W(n,m)| = o\big((nm\log\,\log nm)^{1/2}\big) \quad \text{a.s.,} \tag{1}$$

and if additionally $\mathbf{E}\,|X|^{\nu} < \infty$ for some $\nu > 2$, then

$$|S(n,m) - W(n,m)| = o\big((nm)^{1/\nu}\big) \quad \text{a.s.} \tag{2}$$

and

$$\mathbf{P}\Big\{\max_{\substack{1\leqslant k\leqslant n\\ 1\leqslant l\leqslant m}} |S(k,l) - W(k,l)| > (nm)^{(1+\varepsilon)/\nu}\Big\} < D_{nm}(nm)^{-\varepsilon} \tag{3}$$

for any $\varepsilon \in (0,(\nu-2)/2)$, where D_{nm} is a sequence of positive numbers tending to zero as $n,m \to \infty$.

To formulate our main results we need to introduce six numerical sequences:

$$r_1(n) = \sup_{0\leqslant t\leqslant 1}\Big|\int_0^t f^2(1,x)dx - \frac{1}{n}\sum_{1\leqslant k\leqslant nt-1} f^2\Big(1,\frac{k}{n}\Big)\Big|,$$

$$r_1^*(m) = \sup_{0\leqslant s\leqslant 1}\Big|\int_0^s g^2(1,y)dy - \frac{1}{m}\sum_{1\leqslant l\leqslant ms-1} g^2\Big(1,\frac{l}{m}\Big)\Big|,$$

$$r_2(n,m) = \max_{\substack{1\leqslant k\leqslant n-1\\ 1\leqslant l\leqslant m-1}}\Big|h\Big(1,\frac{k}{n},1,\frac{l}{m}\Big)\Big|,$$

$$r_3(n,m) = \sum_{1\leqslant k\leqslant n-2}\sum_{1\leqslant l\leqslant m-2}\Big|h\Big(1,\frac{k}{n},1,\frac{l}{m}\Big) - h\Big(1,\frac{k+1}{n},1,\frac{l}{m}\Big)$$

$$-h\Big(1,\frac{k}{n},1,\frac{l+1}{m}\Big) + h\Big(1,\frac{k+1}{n},1,\frac{l+1}{m}\Big)\Big|,$$

$$r_4(n,m) = \sum_{1\leqslant k\leqslant n-2}\sum_{1\leqslant l\leqslant m-1}\Big|h\Big(1,\frac{k}{n},1,\frac{l}{m}\Big) - h\Big(1,\frac{k+1}{n},1,\frac{l}{m}\Big)\Big|,$$

$$r_5(n,m) = \sum_{1\leqslant k\leqslant n-1}\sum_{1\leqslant l\leqslant m-2}\Big|h\Big(1,\frac{k}{n},1,\frac{l}{m}\Big) - h\Big(1,\frac{k}{n},1,\frac{l+1}{m}\Big)\Big|.$$

THEOREM 2. *If* I) - IV) *are satisfied and* $\mathbf{E}|X|^\nu < \infty$ *for some* $\nu > 2$, *then it is possible to define a sequence of Wiener processes* $\{W_{nm}(t,s),\ t,s \geqslant 0\}$ *such that for all* $\varepsilon \in (0,(\nu-2)/2)$ *and* $x > 0$

$$\mathbf{P}\Big\{\sup_{0\leqslant t,s\leqslant 1}\Big|Z_{nm}(t,s) - W_{nm}\big(\rho(t),\tau(s)\big)\Big| > (nm)^{(1+\varepsilon)/\nu-1/2}\sum_{i=2}^{5} r_i(n,m)$$

$$+2\Big(r_1(n)\big(x+\log\big(1/r_1(n)\big)\big)\Big)^{1/2} + 2\Big(r_1^*(m)\big(x+\log\big(1/r_1^*(m)\big)\big)\Big)^{1/2}$$

$$+2\Big(r_1(n)r_1^*(m)\big(x+\log\big(\log^2\big(1/\big(r_1(n)r_1^*(m)\big)\big)/\big(r_1(n)r_1^*(m)\big)\big)\big)\Big)^{1/2}\Big\}$$

$$\leqslant D_{nm}(nm)^{-\varepsilon} + De^{-x},$$

where D *is a positive constant and* D_{nm} *is the same sequence as in Theorem* 1, $\rho(t) = \int_0^t f^2(1,x)dx,\ \tau(s) = \int_0^s g^2(1,y)dy$.

Proof. Let us use the Hardy transformation for the double series

$$n^{\alpha+1/2}m^{\beta+1/2}Z_{nm}(t,s) = \sum_{1\leqslant k\leqslant[nt]-2}\sum_{1\leqslant l\leqslant[ms]-2}\big(h(n,k,m,l) - h(n,k+1,m,l)$$

$$-h(n,k,m,l+1) + h(n,k+1,m,l+1)\big)S(k,l)$$

$$+\sum_{1\leqslant k\leqslant[nt]-2}\Big(h(n,k,m,[ms]-1) - h(n,k+1,m,[ms]-1)\Big)S(k,[ms]-1)$$

$$+\sum_{1\leqslant l\leqslant[ms]-2}\Big(h(n,[nt]-1,m,l) - h(n,[nt]-1,m,l+1)\Big)S([nt]-1,l)$$

$$+h(n,[nt]-1,m,[ms]-1)S([nt],[ms]),$$

where $[x]$ denotes the integer part of x. In view of (3) we have

$$\mathbf{P}\Big\{\max_{\substack{1\leqslant k\leqslant n-1\\ 1\leqslant l\leqslant m-1}}\Big|\sum_{1\leqslant i\leqslant k-1}\sum_{1\leqslant j\leqslant l-1}\big(h(n,i,m,j) - h(n,i+1,m,j)$$

$$-h(n,i,m,j+1)+h(n,i+1,m,j+1))\big(S(i,j)-W(i,j)\big)\Big|$$

$$+\max_{\substack{1\leqslant k\leqslant n-1\\1\leqslant l\leqslant m-1}}\Big|\sum_{1\leqslant i\leqslant k-1}\big(h(n,i,m,l)-h(n,i+1,m,l)\big)\big(S(i,l)-W(i,l)\big)\Big|$$

$$+\max_{\substack{1\leqslant k\leqslant n-1\\1\leqslant l\leqslant m-1}}\Big|\sum_{1\leqslant j\leqslant l-1}\big(h(n,k,m,j)-h(n,k,m,j+1)\big)\big(S(k,j)-W(k,j)\big)\Big|$$

$$+\max_{\substack{1\leqslant k\leqslant n-1\\1\leqslant l\leqslant m-1}}\Big|h(n,k,m,l)\big(S(k,l)-W(k,l)\big)\Big|>(nm)^{(1+\varepsilon)/\nu}n^{\alpha}m^{\beta}\sum_{i=2}^{5}r_i(n,m)\Big\}$$

$$\leqslant D_{nm}(nm)^{-\varepsilon}.$$

Let $W^*_{kl}=W(k,l)-W(k-1,l)-W(k,l-1)+W(k-1,l-1)$ be the increments of the Wiener process defined in Theorem 1. Increments $W^*_{kl}, 1\leqslant k\leqslant n-1, 1\leqslant l\leqslant m-1$, are independent standard normal variables, so

$$\mathbf{E}\sum_{1\leqslant i\leqslant k}\sum_{1\leqslant j\leqslant l}h(n,i,m,j)W^*_{ij}=0$$

and

$$n^{-1-2\alpha}m^{-1-2\beta}\mathbf{E}\Big(\sum_{1\leqslant i\leqslant k}\sum_{1\leqslant j\leqslant l}h(n,i,m,j)W^*_{ij}\Big)\Big(\sum_{1\leqslant z\leqslant p}\sum_{1\leqslant s\leqslant q}h(n,z,m,s)W^*_{zs}\Big)$$

$$=\rho_n\Big(\frac{k}{n}\wedge\frac{p}{n}\Big)\tau_m\Big(\frac{l}{m}\wedge\frac{q}{m}\Big),$$

where

$$t\wedge s=\min(t,s),\quad \rho_n\Big(\frac{k}{n}\Big)=\frac{1}{n}\sum_{1\leqslant i\leqslant k}f^2\Big(1,\frac{i}{n}\Big),$$

$$\tau_m\Big(\frac{l}{m}\Big)=\frac{1}{m}\sum_{1\leqslant j\leqslant l}g^2\Big(1,\frac{j}{m}\Big).$$

It is easy to see that $\{W_{nm}(\rho_n(k/n),\tau_m(l/m)),\ 1\leqslant k\leqslant n-1,\ 1\leqslant l\leqslant m-1\}$ and $\{n^{-1/2-\alpha}m^{-1/2-\beta}\sum_{1\leqslant i\leqslant k}\sum_{1\leqslant j\leqslant l}h(n,i,m,j)W^*_{ij},\ 1\leqslant k\leqslant n-1,\ 1\leqslant l\leqslant m-1\}$ are identically distributed, hence

$$\mathbf{P}\Big\{\max_{\substack{1\leqslant k\leqslant n-1\\1\leqslant l\leqslant m-1}}\Big|W_{nm}\Big(\rho_n\Big(\frac{k}{n}\Big),\tau_m\Big(\frac{l}{m}\Big)\Big)-Z_{nm}\Big(\frac{k}{n},\frac{l}{m}\Big)\Big|$$

$$>(nm)^{(1+\varepsilon)/\nu-1/2}\sum_{i=2}^{5}r_i(n,m)\Big\}<D_{nm}(nm)^{-\varepsilon}. \tag{4}$$

It follows from IV) that $\rho_n([nt]/n)\to\rho(t)$, and $\tau_m([ms]/m)\to\tau(s)$ uniformly and the rates of convergence are $r_1(n)$ and $r^*_1(m)$ respectively.

Using Lemmas 1.11.1 and 1.13.1.from (Csörgö and Révész, 1981) we obtain for every $x > 0$

$$\mathbf{P}\Big\{\sup_{0\leqslant t,s\leqslant 1}\Big|W_{nm}(\rho_n(t),\tau_m(s)) - W_{nm}(\rho(t),\tau(s))\Big|$$

$$> 2\Big(r_1(n)(x+\log(1/r_1(n)))\Big)^{1/2} + 2\Big(r_1^*(m)(x+\log(1/r_1^*(m)))\Big)^{1/2}$$

$$+2\Big(r_1(n)r_1^*(m)(x+\log(\log^2(1/(r_1(n)r_1^*(m)))/(r_1(n)r_1^*(m))))\Big)^{1/2}\Big\}$$

$$\leqslant \mathbf{P}\Big\{\sup_{0\leqslant z_1\leqslant r_1(n)}\sup_{0\leqslant t,s\leqslant 1}|W_{nm}(t+z_1,s)-W_{nm}(t,s)|$$

$$+\sup_{0\leqslant z_2\leqslant r_1^*(m)}\sup_{0\leqslant t,s\leqslant 1}\Big|W_{nm}(t,s+z_2)-W_{nm}(t,s)\Big| + \sup_{0\leqslant t,s\leqslant 1}\Big|W_{nm}(R_{\rho\tau})\Big|$$

$$> 2\Big(r_1(n)(x+\log(1/r_1(n)))\Big)^{1/2} + 2\Big(r_1^*(m)(x+\log(1/r_1^*(m)))\Big)^{1/2}$$

$$+2\Big(r_1(n)r_1^*(m)(x+\log(\log^2(1/r_1(n)r_1^*(m))/(r_1(n)r_1^*(m))))\Big)^{1/2}\Big\} \leqslant De^{-x},$$

where D is a positive absolute constant, domain $R_{\rho\tau} = [x_1,x_2]\times[y_1,y_2] \subset [0,1]^2$ is such that $(y_2-y_1)(x_2-x_1) \leqslant r_1(n)r_1^*(m)$ and $W(R_{\rho\tau}) = W(x_2,y_2) - W(x_2,y_1) - W(x_1,y_2) + W(x_1,y_1)$. Hence, using (4) we immediately obtain the assertion of Theorem 2.

Let us now consider an approximation of $Y(t,s)$. It is possible to represent the process $Y(t,s)$ as a stochastic integral

$$Y(t,s) = \int_0^{t-1}\int_0^{s-1} h(t,x,s,y)dS(x,y), \qquad t,s>1.$$

We shall approximate Y by the Gaussian process

$$\Gamma(t,s) = \int_0^t\int_0^s h(t,x,s,y)dW(x,y),$$

where W is the two-parameter Wiener process defined in Theorem 1.

Suppose that

V) for all $0 < t,x,s,y < \infty$ there exist $\frac{\partial f(t,x)}{\partial x}$, $\frac{\partial g(s,y)}{\partial y}$ and

$$\int_1^{t-1}\Big|\frac{\partial f(t,x)}{\partial x}\Big|dx < \infty, \quad \int_1^{s-1}\Big|\frac{\partial g(s,y)}{\partial y}\Big|dy < \infty.$$

Denote also

$$r_6(t,s) = |h(t,t-1,s,s-1)| + \int_1^{t-1}\Big|\frac{\partial h(t,x,s,s-1)}{\partial x}\Big|dx$$

$$+\int_1^{s-1}\left|\frac{\partial h(t,t-1,s,y)}{\partial y}\right|dy+\int_1^{t-1}\int_1^{s-1}\left|\frac{\partial^2 h(t,x,s,y)}{\partial x\partial y}\right|dxdy,$$

$$r_7(t,s)=\Big(\log ts\iint\limits_C h^2(t,x,s,y)dxdy\Big)^{1/2},\quad C=(0,t]\times(0,s]\backslash(1,t-1]\times(1,s-1].$$

THEOREM 3. *If conditions* I) - V) *are satisfied, then*

$$|Y(t,s)-\Gamma(t,s)|=o\big(r_6(t,s)(ts\log\log ts)^{1/2}\big)+O\big(r_7(t,s)\big)\quad \text{a.s.,} \tag{5}$$

and if, in addition, $\mathbf{E}|X|^\nu<\infty$ *for some* $\nu>2$, *then*

$$|Y(t,s)-\Gamma(t,s)|=o\big((ts)^{1/\nu}r_6(t,s)\big)+O\big(r_7(t,s)\big)\quad \text{a.s.} \tag{6}$$

Proof. We prove only (5). The proof of (6) is analogous. Note that according to (1) and Theorem 1.12.1 from (Csörgö and Révész, 1981)

$$\sup_{\substack{k\le x\le k+1\\ l\le y\le l+1}}|W(x,y)-S(k,l)|=o\big((kl\log\log kl)^{1/2}\big)\quad \text{a.s.}$$

Let us estimate the difference $Y(t,s)-\Gamma(t,s)$:

$$|Y(t,s)-\Gamma(t,s)|\leqslant\left|\int_1^{t-1}\int^{s-1}\frac{\partial^2 h(t,x,s,y)}{\partial x\partial y}(W(x,y)-S(x,y))dxdy\right|$$

$$+\left|\int_1^{t-1}\frac{\partial h(t,x,s,s-1)}{\partial x}(W(x,s-1)-S(x,s-1))dx\right|$$

$$+\left|\int_1^{s-1}\frac{\partial h(t,t-1,s,y)}{\partial y}(W(t-1,y)-S(t-1,y))dy\right|$$

$$+|h(t,t-1,s,s-1)|\,|W(t,s)-S(t,s)|+\left|\iint\limits_C h(t,x,s,y)dW(x,y)\right|.$$

Using the Borel–Cantelli lemma and the properties of the normal distribution (see, for example, (Zimmermann, 1972, Theorem 4)) we have

$$\left|\iint\limits_C h(t,x,s,y)dW(x,y)\right|=O\Big((\log ts\iint\limits_C h^2(t,x,s,y)dxdy)^{1/2}\Big)\quad \text{a.s.}$$

This relation and (1) prove (5).
Let $R(t,s,u,v)=\int_0^{t\wedge u}\int_0^{s\wedge v}h(t,x,s,y)h(u,x,v,y)dxdy$ be the covariance function of the process $\Gamma(t,s)$. According to II) $R(t,s,u,v)=\sigma_1(t,u)\sigma_2(s,v)$, where $\sigma_1(t,u)$ and $\sigma_2(s,v)$ are the covariance functions of Gaussian processes $\Gamma_1(t)=\int_0^t h(t,x)dW(x)$ and $\Gamma_2(s)=\int_0^s g(s,y)dW(y)$ correspondingly, W is a one-parameter Wiener process.

H. Oodaira (Oodaira, 1972) proved the law of the iterated logarithm in the Strassen form for Gaussian processes. We take advantage of this assertion and its generalization for the multiparameter case (see (Lai, 1974)). Suppose that σ_1 and σ_2 satisfy the following conditions used by H. Oodaira:

VI) for any $T > 0$, there exists a positive nondecreasing function $\varphi(\varepsilon, T)$ $(\varepsilon > 0)$ such that

$$\left|\sigma_i(z+\varepsilon, z+\varepsilon) - 2\sigma_i(z+\varepsilon, z) + \sigma_i(z,z)\right| \leqslant \varphi(\varepsilon, T), \qquad i = 1,2,$$

for all $z, z+\varepsilon \in [0,T]$,

$$\lim_{\varepsilon\to 0} \varphi(\varepsilon, T) = 0, \quad \limsup_{T\to\infty}(\varphi(1,T))^{-1/2}\int_1^\infty (\varphi(e^{-u^2}, T))^{1/2} du < \infty,$$

$$\sigma_i^2(T,T)/\varphi(1,T) \nearrow \infty, \qquad T\to\infty, \quad i = 1,2.$$

Suppose also that

VII) $\lim_{\delta\to 0}\sup_{0\leqslant t\leqslant 1}\int_0^\delta f^2(t,x)dx = \lim_{\delta\to 0}\sup_{0\leqslant s\leqslant 1}\int_0^\delta g^2(s,y)dy = 0.$

It is possible to represent the reproducing kernel Hilbert space $H(R)$ with the kernel $R(t,s,u,v)$ as the direct product $H_1(\sigma_1)\otimes H_2(\sigma_2)$, where $H_1(\sigma_1)$ and $H_2(\sigma_2)$ are reproducing kernel Hilbert spaces with the kernels $\sigma_1(t,u)$ and $\sigma_2(s,v)$ respectively.

Using these conclusions and Theorem 2 from (Lai, 1974) we obtain that the set

$$K = \left\{\psi \in H(R) : \ \|\psi\|_{H(R)} \leqslant (R(1,1,1,1))^{1/2}\right\}$$

is the set of limit points of the sequence

$$\left\{\Gamma(nt, ms)(2R(n,m,n,m)\log\log nm)^{-1/2}, 0 \leqslant t, s \leqslant 1\right\}.$$

Hence by the Theorem 3 we get

COROLLARY. *Let* I)–VII) *be satisfied and*

$$r_7(n,m) = o\big((R(n,m,n,m)\log\log nm)^{1/2}\big).$$

If, furthermore, $nmr_6^2(n,m) = O(R(n,m,n,m))$ *or* $\mathbf{E}\,|X|^\nu < \infty$ *for some* $\nu > 2$ *such that*

$$(nm)^{1/\nu} r_6(n,m) = O\big((R(n,m,n,m)\log\log nm)^{1/2}\big),$$

then the set of limit points of $\{Y(nt,ms)(2R(n,m,n,m)\log\log nm)^{-1/2},\ 0 \leqslant t,s \leqslant 1\}$ *coincides with* K *almost surely.*

Now we go on to the application of these results for the estimation of Prohorov–Levy distances and to the ascertainment of the functional law of the iterated logarithm for Cesáro sums.

Let

$$h(t,x,s,y) = (t-x)^{\alpha^*}(s-y)^{\beta^*}. \tag{7}$$

Let us denote by $S_{nm}(t,s)$ the process $Z_{nm}(t,s)$ with weight function (7). It is easy to see that II)–IV) hold with $\alpha = \alpha^*, \beta = \beta^*$ (let us keep notations α and β for simplicity) and

$$\rho(t) = \int_0^t (1-x)^{2\alpha} dx = \frac{1}{2\alpha+1}\left(1-(1-t)^{2\alpha+1}\right) \quad \text{if} \quad \alpha > -\frac{1}{2},$$

$$r(s) = \int_0^s (1-y)^{2\beta} dy = \frac{1}{2\beta+1}\left(1-(1-s)^{2\beta+1}\right) \quad \text{if} \quad \beta > -\frac{1}{2}.$$

If $\alpha, \beta \geqslant 0$, then $r_1(n) \leqslant 1/n$, $r_1^*(m) \leqslant 1/m$, $r_2(n,m) \leqslant 1$, $r_i(n,m) \leqslant 2$, $i = 3,4,5$, and if $-1/2 < \alpha, \beta < 0$, then $r_1(n) = O(n^{-(1+2\alpha)})$, $r_1^*(m) = O(m^{-(1+2\beta)})$, $r_2(n,m) \leqslant n^{-\alpha}m^{-\beta}, r_i(n,m) \leqslant 2n^{-\alpha}m^{-\beta}$, $i = 3,4,5$.
Let the probability measure $\mathbf{P}_{nm}$ be induced by $S_{nm}(t,s)$ and let $\mathbf{W}_{\rho r}$ be induced by $W_{nm}(\rho(t), r(s))$ on the Skorokhod space $D([0,1]^2)$. It is shown in (Komlós, Major, and Tusnády, 1976) that the Prohorov–Levy distance $d(\mathbf{P}_{nm}, \mathbf{W}_{\rho r})$ satisfies the inequality

$$d(\mathbf{P}_{nm}, \mathbf{W}_{\rho r}) \leqslant \inf_{\varepsilon>0} \left\{\varepsilon + \mathbf{P}\{ \sup_{0\leqslant t,s\leqslant 1} |S_{nm}(t,s) - W_{nm}(\rho(t), r(s))| > \varepsilon\}\right\},$$

and this together with Theorem 2 yields

THEOREM 4. *Let I) be satisfied and* $\mathbf{E}\,|X|^\nu < \infty$ *for some* $\nu > 2$. *If* $h(t,x,s,y)$ *is defined by* (7), *then*

$$d(\mathbf{P}_{nm}, \mathbf{W}_{\rho r}) = \begin{cases} o\left((nm)^{-(\nu-2)/2(\nu+1)}\right), & \alpha, \beta \geqslant 0, \\ o\left(n^{-(\nu(1+2\alpha)-2)/2(\nu+1)} m^{-(\nu(1+2\beta)-2)/2(\nu+1)}\right), & -\frac{1}{2} < \alpha, \beta < 0. \end{cases}$$

Furthrmore, if $h(t,x,s,y)$ is defined by (7), then it is easy to calculate that

$$r_6(t,s) = O\left((1+t^\alpha)(1+s^\beta)\right), \quad r_7(t,s) = O\left(t^\alpha s^\beta (\log\log ts)^{1/2}\right),$$

$$R(t,s,u,v) = \int_0^{t\wedge u} \int_0^{s\wedge v} (t-x)^\alpha (u-x)^\alpha (s-y)^\beta (v-y)^\beta dx dy = \sigma_1(t,u)\sigma_2(s,v). \tag{8}$$

H. Oodaira (Oodaira, 1972) proved that σ_1 and, σ_2 from (8) satisfy VI) and, therefore, we can use the corollary.

THEOREM 5. *Let I) be satisfied. If* $\alpha, \beta > -1/2$, *then the sets of limit points of*

$$\left\{ \left(\frac{2nm \log\log nm}{(2\alpha+1)(2\beta+1)}\right)^{-1/2} n^{-\alpha} m^{-\beta} \times \sum_{1\leqslant k\leqslant nt-1} \sum_{1\leqslant l\leqslant ms-1} (nt-k)^\alpha (ms-l)^\beta X_{kl},\ 0 \leqslant t,s \leqslant 1 \right\} \tag{9}$$

and

$$\left\{\left(\frac{2nm\log\log nm}{(2\alpha+1)(2\beta+1)}\right)^{-1/2} t^\alpha s^\beta \times \sum_{1\leqslant k\leqslant nt-1}\sum_{1\leqslant l\leqslant ms-1}\left(1-\frac{k}{nt}\right)^\alpha\left(1-\frac{l}{ms}\right)^\beta X_{kl},\ 0\leqslant t,s\leqslant 1\right\} \tag{10}$$

coincide with K almost surely.

It is necessary to point out that Cesáro sums

$$S^*_{nm}(t,s) = (nm)^{-\frac{1}{2}}\frac{1}{\binom{n+\alpha}{n}\binom{m+\beta}{m}} \times \sum_{1\leqslant k\leqslant nt-1}\sum_{1\leqslant l\leqslant ms-1}\binom{n-k+\alpha}{n-k}\binom{m-l+\beta}{m-l}X_{kl},\qquad 0\leqslant t,s\leqslant 1$$

cannot be written by means of a weight function satisfying III) and IV). That is why we shall approximate $S^*_{nm}(t,s)$ by $S_{nm}(t,s)$ for which III) and IV) are true. We know (see (Horváth, 1986)) that

$$\binom{k-i+\alpha}{k-i}\Big/\binom{n+\alpha}{n} = \left(\frac{k-i}{n}\right)^\alpha + O\left(\frac{1}{k}\left(\frac{k}{n}\right)^\alpha\right),\quad 1\leqslant i\leqslant k,\ 1\leqslant k\leqslant n,$$

$$\binom{l-j+\beta}{l-j}\Big/\binom{m+\beta}{m} = \left(\frac{l-j}{m}\right)^\beta + O\left(\frac{1}{l}\left(\frac{l}{m}\right)^\beta\right),\quad 1\leqslant j\leqslant l,\ 1\leqslant l\leqslant m.$$

Using this fact for $k=n, l=m$ we obtain

$$\sup_{0\leqslant t,s\leqslant 1}\left|S_{nm}(t,s)-S^*_{nm}(t,s)\right| = O\Big(A(n,m)\sum_{1\leqslant k\leqslant n-1}\sum_{1\leqslant l\leqslant m-1}|X_{kl}|\Big),$$

where $A(n,m)\to 0$, $n,m\to\infty$. Chebyshev inequality yields

$$\mathbf{P}\Big\{A(n,m)\sum_{1\leqslant k\leqslant n-1}\sum_{1\leqslant l\leqslant m-1}|X_{kl}| > (nm)^{-1/2}(1+\mathbf{E}\,|X|)\Big\}$$

$$\leqslant \mathbf{P}\Big\{\sum_{1\leqslant k\leqslant n-1}\sum_{1\leqslant l\leqslant m-1}(|X_{kl}|-\mathbf{E}\,|X|) > b(n,m)\Big\} \leqslant \frac{1}{b^2(n,m)}\mathbf{E}\,(|X|-\mathbf{E}\,|X|)^2,$$

where $b(n,m)\to\infty$, $n,m\to\infty$. Hence, Theorem 4 is valid for $S^*_{nm}(t,s)$.
If $\alpha,\beta\geqslant 0$, then the sets of limit points of the sequences

$$\left\{\left(\frac{2nm\log\log nm}{(2\alpha+1)(2\beta+1)}\right)^{-1/2}\frac{1}{\binom{n+\alpha}{n}\binom{m+\beta}{m}} \times \sum_{1\leqslant k\leqslant nt-1}\sum_{1\leqslant l\leqslant ms-1}\binom{[nt]-k+\alpha}{[nt]-k}\binom{[ms]-l+\beta}{[ms]-l}X_{kl},\ 0\leqslant t,s\leqslant 1\right\} \tag{11}$$

and

$$\left\{\left(\frac{2nm\log\log nm}{(2\alpha+1)(2\beta+1)}\right)^{-1/2}\sum_{1\leqslant k\leqslant nt-1}\sum_{1\leqslant l\leqslant ms-1}\left[\binom{[nt]-k+\alpha}{[nt]-k}\right.\right.$$

$$\left.\left.\times\binom{[ms]-l+\beta}{[ms]-l}\Big/\binom{[nt]+\alpha}{[nt]}\binom{[ms]+\beta}{[ms]}\right]X_{kl},\ 0\leqslant t,s\leqslant 1\right\} \tag{12}$$

coincide with the set of limit points of (9).
Indeed, we have

$$\max_{\substack{1\leqslant k\leqslant n\\1\leqslant l\leqslant m}}\left|\sum_{1\leqslant i\leqslant k-1}\sum_{1\leqslant j\leqslant l-1}\left[\binom{k-i+\alpha}{k-i}\binom{l-j+\beta}{l-j}\Big/\binom{n+\alpha}{n}\binom{m+\beta}{m}\right.\right.$$

$$\left.\left.-\left(\frac{k-i}{n}\right)^{\alpha}\left(\frac{l-j}{m}\right)^{\beta}\right]X_{ij}\right|=O\left(\max_{\substack{1\leqslant k\leqslant n-1\\1\leqslant l\leqslant m-1}}a_1(k,l)\sum_{1\leqslant i\leqslant k}\sum_{1\leqslant j\leqslant l}|X_{ij}|\right),$$

$$\max_{\substack{1\leqslant k\leqslant n\\1\leqslant l\leqslant m}}\left|\sum_{1\leqslant i\leqslant k-1}\sum_{1\leqslant j\leqslant l-1}\left[\binom{k-i+\alpha}{k-i}\binom{l-j+\beta}{l-j}\Big/\binom{k+\alpha}{k}\binom{l+\beta}{l}\right.\right.$$

$$\left.\left.-\left(1-\frac{i}{k}\right)^{\alpha}\left(1-\frac{j}{l}\right)^{\beta}\right]X_{ij}\right|=O\left(\max_{\substack{1\leqslant k\leqslant n-1\\1\leqslant l\leqslant m-1}}a_2(k,l)\sum_{1\leqslant i\leqslant k}\sum_{1\leqslant j\leqslant l}|X_{ij}|\right),$$

where $a_1(k,l)\to 0,\ a_2(k,l)\to 0,\ k,l\to\infty$.
By the strong law of large numbers

$$a_2(k,l)\sum_{1\leqslant i\leqslant k}\sum_{1\leqslant j\leqslant l}|X_{ij}|=O(1)\quad \text{a.s.}$$

Observe also that

$$\sup_{0\leqslant t,s\leqslant 1}\left|\sum_{1\leqslant k\leqslant nt-1}\sum_{1\leqslant l\leqslant ms-1}\left[([nt]-k)^{\alpha}([ms]-l)^{\beta}-(nt-k)^{\alpha}(ms-l)^{\beta}\right]X_{kl}\right|$$

$$=o(n^{\alpha+1/2}m^{\beta+1/2}(\log\log nm)^{1/2})\quad \text{a.s.},$$

$$\sup_{0\leqslant t,s\leqslant 1}t^{\alpha}s^{\beta}\left|\sum_{1\leqslant k\leqslant nt-1}\sum_{1\leqslant l\leqslant ms-1}\left[\left(1-\frac{k}{[nt]}\right)^{\alpha}\left(1-\frac{l}{[ms]}\right)^{\beta}\right.\right.$$

$$\left.\left.-\left(1-\frac{k}{nt}\right)^{\alpha}\left(1-\frac{l}{ms}\right)^{\beta}\right]X_{kl}\right|=o((nm\log\log nm)^{1/2})\quad \text{a.s.}$$

These estimates prove that the sets of limit points of (11) and (12) are the same as that of (9) and (10).

REFERENCES

Csörgö, M. and Révész, P. (1981). *Strong Approximations in Probability and Statistics*. Academic Press, New York.

Horváth, L. (1986). Strong approximations of weighted sums of random variables. *J. London Math. Soc.* **33**, 567–576.

Komlós, J., Major, P., and Tusnády, G. (1976). An approximation of partial sums of independent random variables and the sample distribution function II. *Z. Wahrscheinlichkeitstheor. verw. Geb.* **34**, 33–58.

Lai, T. L. (1974). Reproducing kernel Hilbert spaces and the law of the iterated logarithm for Gaussian processes. *Z. Wahrscheinlichkeitstheor. verw. Geb.* **29**, 7–19.

Major, P. (1976). Approximation of partial sums of independent identically distributed random variables when the summands have only two moments. *Z. Wahrscheinlichkeitstheor. verw. Geb.* **35**, 221–230.

Oodaira, H. (1972). On Strassen's version of the law of the iterated logarithm for Gaussian processes. *Z. Wahrscheinlichkeitstheor. verw. Geb.* **21**, 289–299.

Wichura, M. (1973). Some Strassen-type laws of the iterated logarithm for multiparameter stochastic processes with independent increments. *Ann. Probab.* **1**, 272–295.

Zimmermann, G. (1972). Some sample function properties of the two-parameter Gaussian processes. *Ann. Math. Statist.* **43**, 1235–1246.

Zinchenko, N. M. (1984). On approximation of sums of random variables with multiindices. *Ukrainian Math. J.* **36**, 149–154.

Probabilistic Methods in Discrete Mathematics, pp. 251 – 265
V. F. Kolchin et al. (Eds.)
1993 TVP/VSP

ON A LOWER BOUND FOR THE ISOPERIMETRIC NUMBER OF CUBIC GRAPHS

A. V. KOSTOCHKA and L. S. MELNIKOV
Institute of Mathematics, Siberian Branch of Russia Academy of Sciences, Universitetskii Prospect 4, 630090 Novosibirsk-90, Russia

ABSTRACT

It is proved that almost every cubic simple graph G on n vertices has the isoperimetric number not less than 1/4.95.

1. INTRODUCTION

Graphs under consideration are finite and may have loops and multiple edges. Graphs without loops and multiple edges will be called *simple*.

Let G be a graph. If $V_1 \subseteq V(G)$ and $V_2 = V(G)\backslash V_1$, then $E_G(V_1, V_2)$ denotes the set of edges of G having one end in V_1 and the other end in V_2. A *cut* (V_1, V_2) of G is the partition $V(G) = V_1 \cup V_2$ with $0 < |V_1| \le |V_2|$. The quantity

$$i(G) = \min\Big\{|E_G(V_1, V_2)|/|V_1|\colon (V_1, V_2) \text{ is a cut of } G\Big\}, \tag{1.1}$$

is called the *isoperimetric number* of G. The isoperimetric number was introduced by Buser (Buser, 1978), who later showed (Buser, 1984) that for every even $n \geq 4$ there exists a simple cubic graph G on n vertices with isoperimetric number $i(G) \geq 1/128$.

A bisection of a graph G is a partition (V_1, V_2) of $V(G)$ into two sets of almost equal cardinality, i.e., a partition with $\big||V_1| - |V_2|\big| \le 1$. The *bisection width* $\mathrm{bw}(G)$ of a graph G is defined as the minimum number of edges between the vertex sets of bisections (V_1, V_2), shortly

$$\mathrm{bw}(G) = \min\Big\{|E_G(V_1, V_2)|\colon (V_1, V_2) \text{ is a bisection of } G\Big\}. \tag{1.2}$$

From definitions (1.1) and (1.2) the following inequality for these characteristics follows

$$i(G) \le \frac{2}{n}\mathrm{bw}(G). \tag{1.3}$$

The attention to the isoperimetric number $i(G)$ and the bisection width $\mathrm{bw}(G)$ may be explained by various interesting interpretations of these numbers (a measure of connectivity of graphs) and some practical applications, for example, in

VLSI designs (see (Goldberg and Gardner, 1984; Mohar, 1989) and the references therein). The problems of finding the isoperimetric number $i(G)$ and the bisection width $\mathrm{bw}(G)$ of a graph are known to be NP-hard (Mohar, 1989; Garey at al., 1976). Moreover, it was shown in (MacGregor, 1978; Bui et al., 1987) that the problem of finding the bisection width is NP-hard even within the class of graphs with maximum degree 3. Generalizations have been considered in (Barnes and Hoffman, 1984; Buser, 1978).

In (Bui et al., 1987) an algorithm was given for transforming a regular graph G with n vertices into a cubic graph G^* with $O(n^6)$ vertices such that any minimum bisection of G^* uses only edges of the graph G. Therefore we are interested in the examination of cubic graphs.

Clark and Entringer (Clark and Entringer, 1988) obtained that $\sup \mathrm{bw}(G) \leq (n+138)/3$ for cubic graphs and also showed that for almost every r-regular simple graph G on n vertices $\mathrm{bw}(G) \geq c_r n$, where $c_r \to r/4$ as $r \to \infty$ (in particular, $c_3 \geq 1/11$).

The present paper is a continuation of (Kostochka and Melnikov, 1991), where it is proved that, for any integer q ($q \geq 2$) and for any simple q-regular graph G on n vertices,

$$\mathrm{bw}(G) \leq (q-2)n/4 + O(q\sqrt{n}\log n). \tag{1.4}$$

As a corollary for any simple cubic graph G on n vertices,

$$\mathrm{bw}(G) \leq n/4 + O(\sqrt{n}\log n).$$

In (Kostochka and Melnikov, 1991) it was announced that for any sufficiently large n there exists a cubic graph G on n vertices such that $\mathrm{bw}(G) \geq n/9.9$ or a bit stronger that $i(G) \geq 1/4.95$. Bollobás (Bollobás, 1988) also estimated the isoperimetric number of regular graphs and in particular he proved that for almost every cubic simple graph $i(G) > 2/11$. The main goal of this paper is to present a complete proof of the following result: *For almost every cubic simple graph G on n vertices $i(G) \geq 1/4.95$.*

All graph theoretic notions not defined here, may be found in (Harary, 1969).

2. MINIMAL CUTS IN MARKED CUBIC GRAPHS

For a cut (V_1, V_2) of a graph G we denote

$$V_{i,j} = \left\{v \in V_i \,\middle|\, |E_G(v, V_{3-i})| = j\right\}, \quad V_{i,j,k} = \left\{v \in V_{i,j} \,\middle|\, \deg_{G(V_{i,j})} v = k\right\}.$$

LEMMA 2.1. *Let G be a cubic graph with $i(G) \leq 2/9$ and (V_1, V_2) be a cut such that $i(G) = |E_G(V_1, V_2)|/|V_1|$ (recall that by definition $|V_1| \leq |V_2|$). Then*

$$V_{i,2} = V_{i,3} = \varnothing, \qquad i = 1, 2.$$

Proof. Suppose that $v \in V_{i,2} \cup V_{i,3}$. Since $i(G) \leq 2/9$, we have $|V_1| \geq 9$. Denoting $V_i' = V_i \setminus \{v\}$, $V_{3-i}' = V_{3-i} \cup \{v\}$ we obtain

$$\frac{|E_G(V_1', V_2')|}{\min\{|V_1'|, |V_2'|\}} \leq \frac{|E_G(V_1, V_2)|}{|V_1| - 1} \leq \frac{i(G)|V_1| - 1}{|V_1| - 1} < i(G).$$

LEMMA 2.2. *Under conditions of Lemma 2.1*

1) $V_{i,1,3} = V_{i,1,2} = \emptyset, \quad i = 1,2;$

2) *if* $|V_2| \geq |V_1| + 3$, *then* $V_{2,1,1} = \emptyset$;

3) *if* $G(V_{3-i,1,1})$ *has more than five edges, then any vertex from* V_i *is adjacent to at most one vertex from* $V_{i,1,1}$.

Proof. Suppose that $v \in V_{i,1,2} \cup V_{i,1,3}$. Then $|E_G(V_1, V_2)| \geq 3$ and hence $|V_1| \geq \lceil 3 \cdot 9/2 \rceil = 14$. Let w_1 and w_2 be the vertices of $V_{i,1}$ adjacent to v. Putting $V_i' = V_i \backslash \{v, w_1, w_2\}$, $V_{3-i}' = V_{3-i} \cup \{v, w_1, w_2\}$ we see that $|E_G(V_1', V_2')| \leq |E_G(V_1, V_2)| - 1$ and

$$\frac{|E_G(V_1', V_2')|}{\min\{|V_i| - 3, |V_{3-i}| + 3\}} \leq \frac{|E_G(V_1, V_2)| - 1}{|V_1| - 3} \leq \frac{i(G)|V_1| - 1}{|V_1| - 3} < i(G).$$

Now, let $|V_2| \geq |V_1| + 3$, $v \in V_{2,1,1}$ and let $w \in V_{2,1}$ be adjacent to v. Denote $V_1' = V_1 \cup \{v, w\}$, $V_2' = V_2 \backslash \{v, w\}$. Then

$$|E_G(V_1', V_2')| = |E_G(V_1, V_2)|$$

and

$$\frac{|E_G(V_1', V_2')|}{\min\{|V_1'|, |V_2'|\}} = \frac{|E_G(V_1, V_2)|}{\min\{|V_1| + 2, |V_2| - 2\}} \leq \frac{i(G)|V_1|}{|V_1| + 1} < i(G).$$

At last, suppose that

$$\left|E(G(V_{3-i,1,1}))\right| > 5, \tag{2.1}$$

and vertices v_1, $v_2 \in V_{i,1,1}$ are adjacent to $v \in V_i$. It follows from above that $v \in V_{i,0}$. By (2.1), $|E_G(V_1, V_2)| > 10$, and $|V_1| > 10 \cdot 9/2 = 45$. If $(v_1, v_2) \in E(G)$, then transferring v_1, v_2, and v from V_i into V_{3-i}, we decrease the number of edges in the cut. Let

$$(v_1, v_2) \notin E(G), \quad (v_1, x_2) \in E(G), \quad (x_1, v_2) \in E(G), \quad \{x_1, x_2\} \subset V_{1,1}.$$

By (2.1) there is an edge $(z_1, z_2) \in E(G(V_{3-i,1,1}))$ whose ends are not adjacent to v_1, v_2, x_1, and x_2. Setting

$$V_i' = V_i \backslash \{v, v_1, v_2, x_1, x_2\} \cup \{z_1, z_2\}, \quad V_{3-i}' = V(G) \backslash V_i',$$

we see that

$$|E_G(V_1', V_2')| \leq |E_G(V_1, V_2)| - 1$$

and

$$\frac{|E_G(V_1', V_2')|}{\min\{|V_i| - 3, |V_{3-i}| + 3\}} \leq \frac{|E_G(V_1, V_2)| - 1}{|V_1| - 3} = \frac{i(G)|V_1| - 1}{|V_1| - 3} < i(G).$$

Now we need some definitions.

Let us call a cubic graph *marked* if the ends of any edge $e = (v, w)$ have marks $l(e, v)$ and $l(e, w)$ from $\{-1, 0, 1\}$ such that for any vertex u and the edges e_1, e_2, e_3 incident to u we have $\{l(e_1, u), l(e_2, u), l(e_3, u)\} = \{-1, 0, 1\}$.
We say that a cut (V_1, V_2) of a marked cubic graph G is *optimal* if

(i) $i(G) = |E_G(V_1, V_2)|/|V_1|$;

(ii) for this cut the expression

$$\delta(V_1, V_2) = \sum_{\{e=(v,w)|v\in V_1, w\in V_2\}} (l(e, v) - l(e, w))$$

takes the minimum value over all the cuts for which (i) holds.
Let $e = (v, w)$ be an edge of a marked cubic graph G, $N_G(v) = \{w, w_1, w_2\}$, $N_G(w) = \{v, v_1, v_2\}$. The value

$$l(e, v) + l(e, w) + l((v, w_1), w_1) + l((v, w_2), w_2) + l((w, v_1), v_1) + l((w, v_2), v_2)$$

is called the *weight* $W(e)$ *of* e.
Suppose that (V_1, V_2) is a cut of a marked cubic graph G,

$$\{v, w, v_2, w_2\} \subset V_2, \quad \{v_1, w_1\} \subset V_1,$$
$$\{(v, w), (v, w_1), (v, w_2), (v_1, w), (v_2, w)\} \subset E(G),$$
$$V_1' = V_1 \cup \{v, w\}, \quad V_2' = V_2 \backslash \{v, w\}.$$

Then

$$|E_G(V_1', V_2')| = |E_G(V_1, V_2)|$$

and

$$\delta(V_1', V_2') = \delta(V_1, V_2) - \Big(l((v, w_1), w_1) - l((v, w_1), v)\Big) - \Big(l((w, v_1), v_1)$$
$$-l((w, v_1), w)\Big) + \Big(l((v, w_2), v) - l((v, w_2), w_2)\Big) + \Big(l((w, v_2), w) - l((w, v_2), v_2)\Big).$$

Since

$$l((v, w_1), v) + l((v, w_2), v) + l((v, w), v)$$
$$= l((w, v_1), w) + l((w, v_2), w) + l((w, v), w) = 0,$$

we have $\delta(V_1', V_2') = \delta(V_1, V_2) - W((v, w))$. This yields the following assertion.

LEMMA 2.3. *Let* (V_1, V_2) *be the optimal cut of a marked cubic graph* G. *If* $G(V_{1,1})$ *has an edge* e_1 *with* $W(e_1) = t$, *then any edge* $e_2 \in E(G(V_{2,1}))$ *with* $W(e_2) > t$ *should be adjacent to* e_1. *In particular, at most two edges of* $G(V_{2.1})$ *can have a weight greater than* t.

3. A LOWER BOUND FOR THE ISOPERIMETRIC NUMBER OF CUBIC GRAPHS

We will show that for almost all cubic graphs G the inequality $i(G) \geq 1/4.95$ holds. To do this, Bollobás' configurations (Bollobás, 1985) will be used.

For an even n a $(3, n)$-configuration τ is a set X of $3n$ labelled vertices partitioned into n 3-element subsets with some perfect matching M on these $3n$ vertices. We may assume that $X = \{x_{ij} | \ 1 \le i \le n, \ j \in \{-1, 0, 1\}\}$.
There is a simple one-to-one correspondence between the set of $(3, n)$-configurations and the set of cubic marked graphs (c.m.g.) on n labelled vertices. We can simply merge every set $X_i = \{x_{i,-1}, x_{i,0}, x_{i,1}\}$ to the vertex v_i of degree 3. Bollobás showed that simply cubic marked graphs correspond to constant (nonzero) part of the set of $(3, n)$-configurations as $n \to \infty$.
We will identify cubic marked graphs on n labelled vertices with corresponding $(3, n)$-configurations.
A cut (V_1, V_2) $(|V_1| \le |V_2|)$ of c.m.g. G is called *convenient* if (in the notions of the previous section)

(i) $V_{i,2} = V_{i,3} = \emptyset, \quad i = 1, 2;$

(ii) $V_{i,1,3} = V_{i,1,2} = \emptyset, \quad i = 1, 2;$

(iii) if $|V_2| \ge |V_1| + 3$, then $V_{2,1,1} = \emptyset$;

(iv) if $G(V_{3-i,1,1})$ has more than five edges, then any vertex from V_i is adjacent to at most one vertex from $V_{i,1,1}$;

(v) if $G(V_{1,1})$ has an edge e_1 with $W(e_1) = t$, then any edge $e_2 \in E(G(V_{2,1}))$ with $W(e_2) > t$ is adjacent to e_1;

(vi) if $G(V_{2,1})$ has an edge e_2 with $W(e_2) = t$, then any edge $e_1 \in E(G(V_{1,1}))$ with $W(e_1) < t$ is adjacent to e_2.

Lemmas 2.1–2.3 imply that any c.m.g. G with $i(G) \le 1/4.95$ has some convenient cut such that

$$\big|E_G(V_1, V_2)\big|/|V_1| \le 1/4.95. \tag{3.1}$$

We can introduce a probability distribution on the set of all $(3n-1)!!$ of $(3, n)$-configurations letting $\mathbf{P}\{\tau\} = 1/(3n-1)!!$ for every τ. Let us estimate the probability of existence of a convenient cut satisfying (3.1) in $(3, n)$-configurations.
Theorem VII.32 from (Bollobás, 1985) implies that almost every (a.e.) random $(3, n)$-configuration is connected.

LEMMA 3.1. *The probability that there exists a cut (V_1, V_2) with $|E_G(V_1, V_2)| = 1$; $4 \le |V_1| \le |V_2|$ in a random $(3, n)$-configuration is $O(1/n)$.*

Proof. Since in this case $|V_1|$ should be odd, we estimate the required probability by the value

$$d_1 = \frac{1}{(3n-1)!!} \sum_{i=2}^{[n/4]-1} \binom{n}{2i+1} (6i+3)(3n-6i-3)(6i+1)!!(3n-6i-5)!!$$
$$= \frac{n!}{(3n-1)!!} \sum_{i=2}^{[n/4]-1} \frac{(6i+3)!!(3n-6i-3)!!}{(2i+1)!(n-2i-1)!}.$$

For $3 \le i \le \lceil n/4 \rceil - 1$ the ratio of ith summand to $(i-1)$th summand is equal to

$$\frac{(6i+3)(6i+1)(6i-1)(n-2i+1)(n-2i)}{(2i+1)2i(3n-6i+3)(3n-6i+1)(3n-6i-1)}$$
$$= \frac{(6i+1)(6i-1)(n-2i)}{2i(3n-6i+1)(3n-6i-1)} \le \frac{6i+1}{3n-6i+1} \le 1.$$

Hence

$$d_1 = \frac{1}{(3n-1)!!}\frac{n}{4}\binom{n}{5}(15)!!(3n-15)!!$$
$$\le((1+o(1))\frac{n}{4}\frac{n^5}{5!}(15)!!/(3n)^{-7} = O(1/n).$$

LEMMA 3.2. *The probability d_2 of existing a convenient cut (V_1, V_2) with $|E_G(V_1, V_2)| > 1$ and $V_{2,1,1} = \varnothing$ is $o(1)$.*

Proof. The following procedure gives us all the convenient cuts (V_1, V_2) with $|E_G(V_1, V_2)| = j$ and $|V_1| = i$:

1) select i vertices for V_1 ($\binom{n}{i}$ variants);
2) select end vertices for the j-matching between V_1 and V_2 ($\binom{i}{j}\binom{n-i}{j}j!$ variants);
3) mark by $-1, 0$, and 1 the ends of this matching (3^{2j} variants);
4) add marked edges in $G(V_1)$ to obtain the degrees of the vertices of V_1 to be equal to 3 ($(3i-j-1)!!$ variants);
5) add $2j$ marked edges connecting $V_{2,1}$ with $V_{2,0}$ ($\frac{(3n-3i-3j)!}{(3n-3i-5j)!}$ variants);
6) add marked edges to obtain a cubic marked graph ($(3n-3i-5j-1)!!$ variants).

So,

$$d_2 \le \frac{1}{(3n-1)!!}\sum_{j=2}^{\lceil n/9.9\rceil}\sum_{i=\lceil 4.95j\rceil}^{\lceil n/2\rceil}\binom{n}{i}\binom{i}{j}\binom{n-i}{j}3^{2j}(j!)(3i-j-1)!!$$
$$\times\frac{(3n-3i-3j)!(3n-3i-5j-1)!!}{(3n-3i-5j-1)!}$$
$$=\frac{n!}{(3n-1)!!}\sum_{j=2}^{\lceil n/9.9\rceil}\sum_{i=\lceil 4.95j\rceil}^{\lceil n/2\rceil}\frac{3^{2j}(3i-j-1)!!(3n-3i-3j)!}{j!(i-j)!(n-i-j)!(3n-3i-5j)!!} \tag{3.2}$$
$$\le\frac{\bar{c}n^{0.5}}{3^{1.5n}n^{0.5n}}\sum_{j=2}^{\lfloor n/9.9\rfloor}\sum_{i=\lfloor 4.95j\rfloor}^{\lfloor n/2\rfloor}3^{2j}\frac{(3i-j)^{1.5i-0.5j}}{j^j(i-j)^{i-j}}$$
$$\times\frac{3^{3n-3i-3j}(n-i-j)^{2(n-i-j)}}{(3n-3i-5j)^{0.5(3n-3i-5j)}}.$$

Putting

$$F(n,i,j) = \frac{(3i-j)^{1.5i-0.5j}(n-i-j)^{2(n-i-j)}}{3^{3i+j}j^j(i-j)^{i-j}(3n-3i-5j)^{0.5(3n-3i-5j)}},$$

we have

$$d_2 \le \frac{\bar{c}n^{0.5}3^{1.5n}}{n^{0.5n}}\sum_{j=2}^{\lfloor n/9.9\rfloor}\sum_{i=\lceil 4.95j\rceil}^{\lfloor n/2\rfloor}F(n,i,j). \tag{3.3}$$

Clearly,

$$\begin{aligned}\big(\log F(n,i,j)\big)'_j &= -0.5\big(1+\log(3i-j)\big) - 2\big(1+\log(n-i-j)\big) - \log 3\\ &\quad - (1+\log j) + \big(1+\log(i-j)\big) + 2.5\big(1+\log(3n-3i-5j)\big)\\ &= \log\frac{(i-j)\big(3(n-i-5j/3)\big)^{2.5}}{(3i-j)^{0.5}(n-i-j)^2 3j}.\end{aligned}$$

Since for any fixed i and j such that $i \geq 4.9j$, $j \geq 2$ the derivative of the function $\psi(n) = (n-i-5j/3)^{2.5}(n-i-j)^{-2}$ is positive for $n \geq 2i$, we have

$$\frac{(i-j)\big(3(n-i-5j/3)\big)^{2.5}}{(3i-j)^{0.5}(n-i-j)^2 3j} \geq \frac{3(i-5j/3)^{2.5}}{(i-j/3)^{0.5}(i-j)j} = \frac{3(i/j-5/3)^{2.5}}{(i/j-1/3)^{0.5}(i/j-1)}.$$

The minimum of the last expression under the condition $i/j \geq 4.9$ is attained at $i/j = 4.9$ and is equal to

$$\frac{3(4.9-5/3)^{2.5}}{(4.9-1/3)^{0.5}(4.9-1)} > 3.$$

Consequently, $(\log F(n,i,j))'_j > \log 3 > 0$ and

$$\sum_{j=2}^{\lfloor n/9.9\rfloor}\ \sum_{i=\lfloor 4.95j\rfloor}^{\lfloor n/2\rfloor} F(n,i,j) \leq \sum_{i=10}^{\lfloor n/2\rfloor} \frac{i}{4.9}F(n,i,i/4.9). \tag{3.4}$$

Put

$$\alpha = i/4.9n, \quad \Phi(\alpha,n) = \frac{3^{1.5n}}{n^{0.5(n-1)}}\alpha n F(n, 4.9\alpha n, \alpha n).$$

Then (3.3) and (3.4) imply

$$d_2 \leq \sum_{i=10}^{\lfloor n/2\rfloor} \Phi\big(i/(4.9n), n\big).$$

Let us count

$$\begin{aligned}\Phi(\alpha,n) &= \frac{3^{1.5n}\alpha(13.7\alpha n)^{0.5\cdot 13.7\alpha n}(n-5.9\alpha n)^{2(n-5.9\alpha n)}}{n^{0.5(n-3)}3^{15.7\alpha n}(3.9\alpha n)^{3.9\alpha n}(3n-19.7\alpha n)^{0.5(3n-19.7\alpha n)}(\alpha n)^{\alpha n}}\\ &= \alpha n^{1.5}\left[\frac{3^{1.5-15.7\alpha-0.5(3-19.7\alpha)}(13.7\alpha n)^{6.85\alpha}\big(n(1-5.9\alpha)\big)^{2(1-5.9\alpha)}}{n^{0.5}(3.9\alpha n)^{3.9\alpha}\big(n(1-19.7\alpha/3)\big)^{1.5-9.85\alpha}(\alpha n)^{\alpha}}\right]^n\\ &= \alpha n^{1.5}\left[\frac{3^{-5.85\alpha}13.7^{6.85\alpha}\alpha^{1.95\alpha}(1-5.9\alpha)^{2(1-5.9\alpha)}}{3.9^{3.9\alpha}(1-19.7\alpha/3)^{1.5-9.85\alpha}}\right]^n.\end{aligned}$$

Denote by $\Phi_0(\alpha)$ the expression in square brackets and put $c = 3^{-5.85}13^{6.85}3.9^{-3.9}$. Then

$$\begin{aligned}\log\Phi_0(\alpha) &= \alpha\log c + 1.95\alpha\log\alpha + 2(1-5.9\alpha)\log(1-5.9\alpha)\\ &\quad - 1.5(1-19.7\alpha/3)\log(1-19.7\alpha/3),\\ \big(\log\Phi_0(\alpha)\big)' &= \log c + 1.95 + 1.95\log\alpha - 11.8\\ &\quad - 11.8\log(1-5.9\alpha) + 9.85 + 9.85\log(1-19.7\alpha/3)\\ &= \log c + 1.95\log\alpha - 11.8\log(1-5.9\alpha) + 9.85\log(1-19.7\alpha/3),\\ \big(\log\Phi_0(\alpha)\big)'' &= \frac{1.95}{\alpha} + \frac{11.8\cdot 5.9}{1-5.9\alpha} - \frac{9.85\cdot 19.7/3}{1-19.7\alpha/3}.\end{aligned}$$

Hence, for $1/12 \le \alpha \le 1/9.8$,

$$\left(\log \Phi_0(\alpha)\right)' = \log c + 1.95 \log \frac{\alpha}{1-5.9\alpha} + 9.85 \log \frac{1-19.7\alpha/3}{1-5.9\alpha}$$

$$\ge \log c + 1.95 \log \frac{1}{12-5.9} + 9.85 \log \frac{9.8-19.7/3}{9.8-5.9} \ge 6.1 - 3.6 - 1.9 = 0.6.$$

For $0 < \alpha < 1/12$ we have $16.3\alpha/3 \le 1 - 19.7\alpha/3$ and

$$\left(\log \Phi_0(\alpha)\right)'' \ge \frac{1.95 \cdot 16.3/3 - 9.85 \cdot 19.7/3}{1-19.7\alpha/3} + \frac{11.8 \cdot 5.9}{1-5.9\alpha}$$
$$= \frac{-162.26}{3-19.7\alpha} + \frac{67.28}{1-5.8\alpha} = \frac{39.58 - 384.308\alpha}{(3-19.7\alpha)(1-5.8\alpha)} > 0.$$

Therefore, the function $\log \Phi_0(\alpha)$ is convex in the interval $0 < \alpha < 1/12$ and increases in $1/12 \le \alpha \le 1/9.8$. Thus

$$\max\left\{\Phi_0(\alpha) | 2/n \le \alpha \le 1/9.8\right\} = \max\left\{\Phi_0(2/n), \Phi_0(1/9.8)\right\}.$$

Let us estimate these values of $\Phi(\alpha)$:

$$\Phi_0(2/n) = \left[\frac{c^2(2/n)^{3.9}(1-5.9\cdot 2/n)^{2n-23.6}}{(1-19.7\cdot 2/3n)^{1.5n-19.7}}\right]^{1/n} = \left[O(n^{-3.9})\right]^{1/n},$$

$$9.8 \log \Phi_0(1/9.8) = \log c - 1.95 \log 9.8 + 2(9.8-5.9)\log(1-5.9/9.8)$$

$$-1.5(9.8-19.7/3)\log\left(1-19.7/(3\cdot 9.8)\right) \le 6.2 - 4.45 - 7.186 + 5.379 \le -0.05.$$

Hence

$$\Phi(\alpha, n) \le \alpha n^{1.5} \max\left\{O(n^{-3.9}), \exp\{-0.05n\}\right\} = o(n^{-2}).$$

Lemma 3.3. *The probability d_3 that there exists a convenient cut (V_1, V_2) with $0.5n - 1 \le |V_1| \le |V_2|$ such that $|E_G(V_1, V_2)|/|V_1| \le 1/4.95$ and either $G(V_{1,1})$ or $G(V_{2,1})$ contains less than 6 edges is $o(1)$.*

Proof. Obviously,

$$d_3 = \sum_{i=0.5n-1}^{n/2} \sum_{j=2t}^{i/4.95} \sum_{l=1}^{2} \sum_{t=0}^{5} p(i,j,l,t),$$

where $p(i,j,l,t)$ is the probability that there exists a convenient cut (V_1, V_2) with $|V_1| = i$, $|E_G(V_1, V_2)| = j$ and $|E_G(V_{l,1})| = t$.
First we estimate $p(n/2, j, 2, t)$. To do this, we repeat the first four steps of the procedure described in the proof of Lemma 3.2 and then make the following steps:

5′) put t marked independent edges on $V_{2,1}$ and $2j - 2t$ marked edges connecting $V_{2,1}$ with $V_{2,0}$ $\left(\binom{2j}{2t}(2t-1)!!\frac{(3(n-i-j))!}{(3n-3i-5j+2t)!}\right.$ ways);

6′) add marked edges to obtain a cubic marked graph ($(3n-3i-5j+2t-1)!!$ variants).

The ratio $p(n/2,j,2,t)$ to the summand with parameters $(n/2,j)$ in sum (3.2) is equal to

$$\binom{2j}{2t}(2t-1)!!\frac{(3(n-i-j))!}{(3n-3i-5j+2t)!}\frac{(3n-3i-5j+2t-1)!!}{(3n-3i-5j-1)!!}$$
$$\leq \frac{(2j)^{2t}}{2^t t!(3n-3i-5j)^t} < j^t.$$

But in the course of the proof of Lemma 3.2 it was shown that this summand does not exceed $n^2\exp\{-0.05\}$. Hence $p(n/2,j,2,t)=o(n^{-3})$ for any j and t under consideration.

Evidently, the cases $l=1$ and $i=0.5n-1$ are quite similar.

LEMMA 3.4. *The probability d_4 that $i(G)\leq 1/5.2$ is $o(1)$.*

Proof. If $i(G)\leq 1/5.2$ for some c.m.g. G, then there exists a convenient cut (V_1,V_2) with

$$|E_G(V_1,V_2)|/|V_1|\leq 1/5.2\,.$$

Denote this event by $\mathcal{B}$. Obviously,

$$\mathbf{P}\{\mathcal{B}\}\ \leq \mathbf{P}\{\mathcal{B}_1\}+\mathbf{P}\{\mathcal{B}_2\}+\mathbf{P}\{\mathcal{B}_3\},$$

where $\mathcal{B}_i$ means that $\mathcal{B}$ takes place and, in addition, $\mathcal{B}_1$ means that $|E_G(V_1,V_2)|\leq 1$, $\mathcal{B}_2$ means that $E_G(V_{2,1})=\varnothing$, and $\mathcal{B}_3=\mathcal{B}\backslash(\mathcal{B}_1\cup\mathcal{B}_2)$.

By Lemmas 3.1 and 3.2, $\mathbf{P}\{\mathcal{B}_1\}+\mathbf{P}\{\mathcal{B}_2\}=o(1)$. Suppose that $\mathcal{B}_3$ takes place. By the definition of a convenient cut we have $|V_1|\in\{0.5n,0.5n-1\}$. We consider the case $|V_1|=0.5n$. The other case is quite similar.

Let us estimate the probability p_j of the event that $\mathcal{B}_3$ takes place, $|V_1|=0.5n$, and $|E_G(V_1,V_2)|=j$. To do this, we repeat first four steps of the procedure described in the proof of Lemma 3.2 and then make the following step:

5″) add marked edges to obtain a cubic marked graph ($(1.5n-j-1)!!$ variants).

It is clear now that

$$p_j\leq\frac{1}{(3n-1)!!}\binom{n}{n/2}\left(\binom{n/2}{j}\right)^2 3^{2j}j!\left((1.5n-j-1)!!\right)^2$$
$$=\frac{n!3^{2j}\left((1.5n-j-1)!!\right)^2}{(3n-1)!!j!\left((0.5n-j)!\right)^2}$$
$$\leq\frac{(1+o(1))(2\pi n)^{0.5}(n/e)^n3^{2j}2\left((1.5n-j)/e\right)^{1.5n-j}}{(2\pi j)^{0.5}(j/e)^j2^{0.5}(3n/e)^{1.5n}2\pi(0.5n-j)\left((0.5n-j)/e\right)^{n-2j}}$$
$$\leq\frac{(1+o(1))3^{2j-1.5n}(1.5-j/n)^{1.5n-j}}{n^{0.5}(j/n)^j(0.5-j/n)^{n-2j}}.$$

Putting $\alpha = j/n$, we have

$$p_j \le \frac{(1+o(1))}{n^{0.5}}\left(\frac{3^{2\alpha-1.5}(1.5-\alpha)^{1.5-\alpha}}{\alpha^\alpha(0.5-\alpha)^{1-2\alpha}}\right)^n.$$

Put

$$\varphi(\alpha) = \frac{3^{2\alpha-1.5}(1.5-\alpha)^{1.5-\alpha}}{\alpha^\alpha(0.5-\alpha)^{1-2\alpha}}.$$

It is clear that, for $0 < \alpha \le 1/10.4$,

$$\begin{aligned}(\log\varphi(\alpha))' &= -1-\log(1.5-\alpha)+2\log 3-1-\log\alpha+2+2\log(0.5-\alpha)\\ &= \log\frac{9(0.5-\alpha)}{(1.5-\alpha)\alpha} > \log\frac{9\cdot 0.4}{1.5\cdot 0.1} > 0.\end{aligned}$$

Hence $\max\{\varphi(\alpha) | 0 < \alpha \le 1/10.4\} = \varphi(1/10.4)$ and

$$\mathbf{P}\{\mathcal{B}_3\} \le n\max\{p_j | 2 \le j \le n/10.4\} \le n\frac{(1+o(1))}{n^{0.5}}(\varphi(1/10.4))^n.$$

But $\varphi(1/10.4) < 0.999$.

LEMMA 3.5. *Let j be sufficiently large and $9.9j \le n \le 10.4j$. Then the probability d_5 of the event $\mathcal{D}_5$ that random c.m.g. G has a convenient cut (V_1, V_2) with $|V_1| = |V_2|$ and $|E_G(V_1, V_2)| = j$ does not exceed $jn^2\exp\{-0.005j\}$.*

Proof. *Case* 1. Let

$$\min\left\{W(e) | e \in E(G(V_{1,1}))\right\} \ge 0.$$

Let $p(1,n)$ be the probability of the event $\mathcal{D}_5$ provided Case 1 takes place, and $p(1,n,k,l)$ be the probability that, provided Case 1 takes place, the event $\mathcal{D}_5$ holds and, in addition, there are exactly k edges in $G(V_{1,1})$ and exactly l edges in $G(V_{2,1})$. Then

$$\begin{aligned}p(1,n) &\le \sum_{k=1}^{\lfloor j/2\rfloor}\sum_{l=1}^{\lfloor j/2\rfloor} p(1,n,k,l)\\ &\le j^2\max\{p(1,n,k,l) | 1 \le k \le q\lfloor j/2\rfloor,\ 1 \le l \le \lfloor j/2\rfloor,\ 9.9j \le n \le 10.4j\}.\end{aligned}$$

The following eleven steps give us all the required c.m.g.

1) select $n/2$ vertices for V_1 ($\binom{n}{n/2}$ variants);

2) select end vertices for the j-matching between V_1 and V_2 ($\binom{n/2}{j}j!$ variants);

3) select end vertices for the k-matching on $V_{1,1}$ ($\binom{j}{2k}(2k-1)!!$ variants);

4) select end vertices in $V_{1,0}$ for the edges whose other ends belong to $V_{1,1,1}$ ($\frac{(0.5n-j)!}{(0.5n-j-2k)!}$ variants);

5) select points for all the ends of all edges incident to $V_{1,1,1}$ in such a way that $W(e) \geq 0$ for every edge of the chosen k-matching ($(3^6 2^2 435/729)^2$ variants);

6) select points for the remaining edges of the j-matching (3^{2j-4k} variants);

7) select points in $V_{1,0}$ for the ends of edges whose other ends belong to $V_{1,1,0}$ ($\frac{(1.5n-3j-2k)!}{(1.5n-5j+2k)!}$ variants);

8) select points for the ends of edges whose both ends belong to $V_{1,0}$ ($(1.5n - 5j + 2k - 1)!!$ variants);

9) select points for the l-matching in $V_{2,1}$ ($\binom{j}{2l}(2l-1)!!2^{2l}$ variants);

10) select points in $V_{2,0}$ for the ends of edges whose other ends belong to $V_{2,1}$ ($\frac{(1.5n-3j)!}{(1.5n-5j+2l)!}$ variants);

11) select points for the ends of edges whose both ends belong to $V_{2,0}$ ($(1.5n - 5j + 2l - 1)!!$ variants).

Hence

$$p(1,n,k,l) \leq \frac{1}{(3n-1)!!}\binom{n}{n/2}\left(\binom{n/2}{j}\right)^2 j!\binom{j}{2k}(2k-1)!!\frac{(0.5n-j)!}{(0.5n-j-2k)!}$$
$$\times \left(3^6 2^2 \frac{435}{729}\right)^k 3^{2j-4k}\frac{(1.5n-3j-2k)!}{(1.5n-5j+2k)!}(1.5n-5j+2k-1)!!\binom{j}{2l}(2l-1)!!$$
$$\times 2^{2l}\frac{(1.5n-3j)!}{(1.5n-5j+2l)!}(1.5n-5j+2l-1)!!.$$

Since

$$\frac{3^{2k}(0.5n-j)!}{(0.5n-j-2k)!} \leq \frac{(1.5n-3j)!}{(1.5n-3j-2k)!},$$

we have

$$p(1,n,k,l) \leq h(n,k,l) = \frac{n!3^{2j}j!2^{2k}(435/729)^k}{(3n-1)!!j!((0.5n-j)!)^2(j-2k)!2^k k!}$$
$$\times \frac{(1.5n-3j)!(1.5n-5j+2k-1)!!j!2^{2l}(1.5n-3j)!}{(1.5n-5j+2k)!(j-2l)!2^l l!2^{0.75n-2.5j+l}(0.75n-2.5j+l)!}.$$

Note that for even j

$$\frac{h(n,j/2,l)}{h(n,j/2-1,l)} = \frac{4(435/729)}{k(1.5n-4j)} < 1$$

and

$$\frac{h(n,k,j/2)}{h(n,k,j/2-1)} = \frac{4}{l(1.5n-4j)} < 1.$$

For $1 \le l \le 2j-1$, $1 \le k \le 2j-1$,

$$h(n,k,l) \le \frac{3(2\pi n)^{0.5}(n/e)^n 3^{2j} 2^{5j-1.5n}(435/729)^k}{2^{0.5}(3n/e)^{1.5n}\big(2\pi(0.5n-j)\big)\big((0.5n-j)/e\big)^{n-2j}(2\pi k)^{0.5}}$$
$$\times \frac{2\pi(1.5n-3j)\big((1.5n-3j)/e\big)^{3n-6j}(2\pi j)^{0.5}(j/e)^j}{(k/e)^k\big(2\pi(j-2k)\big)^{0.5}\big((j-2k)/e\big)^{j-2k}\big((0.75n-5j+2k/e)\big)^{0.75n-5j+2k}}$$
$$\times \frac{1}{\big(2\pi(0.75n-5j+2k)\big)^{0.5}\big(2\pi(j-2l)\big)^{0.5}\big((j-2l)/e\big)^{j-2l}(2\pi l)^{0.5}(l/e)^l}$$
$$\times \frac{1}{\big(2\pi(0.75n-5j+2l)\big)^{0.5}\big((0.75n-5j+2l)/e\big)^{0.75n-5j+2l}}$$
$$\le \frac{n^{-0.5n}3^{1.5n-4j}j^j 2^{5j-1.5n}(0.5n-j)^{2n-4j}(435/729)^k}{(j-2k)^{j-2k}k^k(0.75n-5j+2k)^{0.75n-5j+2k}l^l(j-2l)^{j-2l}}$$
$$\times \frac{1}{(0.75n-5j+2l)^{0.75n-5j+2l}}.$$

Putting $\alpha = k/j$, $\beta = n/j$, $\gamma = l/j$, we obtain that

$$h(n,k,l)$$
$$\le \frac{1}{j}\left[\frac{\beta^{-0.5\beta}3^{1.5\beta-4}1^1 2^{5-1.5\beta}(0.5\beta-1)^{2\beta-4}(435/729)^\alpha \alpha^{-\alpha}(1-2\gamma)^{2\gamma-1}}{(1-2\alpha)^{1-2\alpha}(0.75\beta-2.5+\alpha)^{0.75\beta-2.5+\alpha}(0.75\beta-2.5+\gamma)^{0.75\beta-2.5+\gamma}\gamma^\gamma}\right]^j.$$

Thus,

$$\log\Big(\max\{p(1,n,k,l)\,|\,1 \le k < 0.5j,\ 1 \le l < 0.5j,\ 9.9j \le n \le 10.4j\}\Big)$$
$$\le -\log j + j\Big(\max\{\varphi(\alpha,\beta,\gamma)\,|\,0 < \alpha < 0.5,\ 0 < \gamma < 0.5,\ 9.9 \le \beta \le 10.4\}\Big),$$

where

$$\varphi(\alpha,\beta,\gamma) = -0.5\beta(\log\beta) + (1.5\beta-4)(\log 1.5) + \log 2 + (2\beta-4)\log(0.5\beta-1)$$
$$+\alpha\log(435/729) - (1-2\alpha)\log(1-2\alpha) - \alpha\log\alpha - (0.75\beta-2.5+\alpha)\log(0.75\beta-2.5+\alpha)$$
$$-(1-2\gamma)\log(1-2\gamma) - \gamma\log\gamma + (0.75\beta-2.5+\gamma)\log(0.75\beta-2.5+\gamma).$$

For

$$0 < \alpha < 0.5, \quad 0 < \gamma < 0.5, \quad 9.9 \le \beta \le 10.4,$$

we have

$$\big(\varphi(\alpha,\beta,\gamma)\big)'_\beta = -0.5 - 0.5\log\beta + 1.5\log 1.5 + 2 + 2\log(0.5\beta-1)$$
$$-0.75 - 0.75\log(0.75\beta-2.5+\alpha) - 0.75 - 0.75\log(0.75\beta-2.5+\gamma)$$
$$\le -0.5\log\beta + 1.5\log 1.5 + 2\log(0.5\beta-1) - 1.5\log(0.75\beta-2.5)$$
$$\le 0.5\log\big((0.5\beta-1)/\beta\big) + 1.5\log\big(1.5(0.5\beta-1)/(0.75\beta-2.5)\big)$$
$$\le 0.5\log(0.5-1/10.4) + 1.5\log\big((0.5\cdot 9.9-1)/(0.5\cdot 9.9-5/3)\big)$$
$$< -0.45 + 1.5\cdot 0.2 < 0.$$

Hence it is sufficient to verify the assertion of the lemma for $\beta = 9.9$.
Since

$$(\varphi(\alpha,\beta,\gamma))'_\alpha = 2+2\log(1-2\alpha)-1-\log\alpha-1-\log(0.75\beta-2.5+\alpha)+\log(435/729),$$

$$(\varphi(\alpha,\beta,\gamma))''_\alpha = -\frac{4}{1-2\alpha}-\frac{1}{\alpha}-\frac{1}{0.75\beta-2.5+\alpha} < 0,$$

we can conclude that for fixed β and γ the function $\varphi(\alpha,\beta,\gamma)$ is concave in α and attains the maximum at $\alpha = x_1$, where x_1 is the root of the equation $(\varphi(\alpha,\beta,\gamma))'_\alpha = 0$. Here this equation has the form

$$\log\left(435(1-2x)^2/729\right) = \log x(0.75\beta - 2.5 + x),$$

and may be reduced to

$$(4-\delta)x^2 - (0.75\beta\delta - 2.5\delta + 4)x + 1 = 0, \tag{3.5}$$

where $\delta = 435/729$. The largest root of (3.5) is greater than 1. The smallest root of (3.5) is equal to

$$2\Big(0.75\beta\delta - 2.5\delta + 4 + \left((0.75\beta\delta - 2.5\delta + 4)^2 - 4(4-\delta)\right)^{1/2}\Big)^{-1}. \tag{3.6}$$

For $\beta = 9.9$, $\delta = 435/729$, it follows from (3.6) that

$$0.0829 \le x_1 \le 0.083.$$

Analogously, under conditions of the lemma

$$(\varphi(\alpha,\beta,\gamma))'_\gamma = 2\log(1-2\gamma) - \log\gamma - \log(0.75\beta - 2.5 + \gamma),$$
$$(\varphi(\alpha,\beta,\gamma))''_\gamma = -\frac{4}{1-2\gamma}-\frac{1}{\gamma}-\frac{1}{0.75\beta-2.5+\gamma} < 0,$$

and for fixed β and α the function $\varphi(\alpha,\beta,\gamma)$ is concave in γ and attains the maximum at $\gamma = x_2$, where x_2 is the root of the equation (3.5), with $\delta = 1$. According to (3.6),

$$0.1166 \le x_2 \le 0.1167.$$

Since $(\varphi(\alpha,\beta,\gamma))'_\alpha$ does not depend on γ and $(\varphi(\alpha,\beta,\gamma))'_\gamma$ does not depend on α, we have

$$\begin{aligned}\varphi(x_1; 9.9; x_2) &\le \varphi(0.083; 9.9; 0.1166)\\ &\quad - 10^{-4}(\varphi(0.083; 9.9; 0.1166))'_\alpha + 10^{-4}(\varphi(0.083; 9.9; 0.1166))'_\gamma\\ &\le -0.0051 - 10^{-4}(-0.01) + 10^{-4}\cdot 0.01 \le -0.005.\end{aligned}$$

Thus, $p(1,n) \le j\exp\{-0.005j\}$.

Case 2. Let

$$\max\left\{W(e)\Big|\, e \in E(G(V_{2,1}))\right\} \leq 0.$$

This case is symmetrical to Case 1.

Case 3. Let

$$\min\left\{W(e)\Big|\, e \in E(G(V_{1,1}))\right\} \leq -1, \quad \max\left\{W(e)\Big|\, e \in E(G(V_{2,1}))\right\} \geq 1.$$

Then, by the definition of a convenient cut, at least $k-2$ edges $e \in E(G(V_{1,1}))$ have $W(e) \geq 1$ and at least $l-2$ edges $e \in E(G(V_{2,1}))$ have $W(e) \leq -1$. Obviously, the probability of this event less than $p(1,n)$.

THEOREM. *For almost every cubic marked graph G on n vertices the inequality* $i(G) \geq 1/4.95$ *holds.*

Proof. Let G be a c.m.g. with $i(G) < 1/4.95$. Then there is a convenient cut (V_1, V_2) ($|V_1| \leq |V_2|$) such that

$$|E_G(V_1, V_2)|/|V_1| \leq 1/4.95.$$

With probability $1 - o(1)$ such a cut contains more than one edge. Lemma 3.2 and the definition of a convenient cut imply that with probability $1 - o(1)$ either $|V_1| = |V_2|$ or $|V_1| = |V_2| - 2$. Lemmas 2.4 and 2.5 give the required estimate in the former case. The latter case is quite analogous.
Bollobás (Bollobás, 1985) proved that simple cubic marked graphs correspond to constant (nonzero) part of the set of $(3,n)$-configurations as $n \to \infty$. Besides he showed that any simple cubic graph on n labelled vertices corresponds to exactly 6^n different $(3,n)$-configurations, and any nonsimple cubic graph on n labelled vertices corresponds to less that 6^n $(3,n)$-configurations. Hence the theorem implies the following assertion.

COROLLARY. *For almost every cubic simple graph G on n vertices* $i(G) \geq 1/4.95$.

The constant 1/4.95 could be slightly improved using the same ideas, but then the computations would be too complicate.

REFERENCES

Barnes, E. R. and Hoffman, A. J. (1984). *Partitioning, Spectra and Linear Programming. Progress in Combinatorial Optimization.* Academic Press, New York.

Bollobás, B. (1985). *Random Graphs.* Academic Press, London.

Bollobás, B. (1988). The isoperimetric number of random regular graphs. *European J. Combinatorics* **9**, 241–244.

Bui, T. H., Chaudhuri, S., Leigton, F. T., and Sipser, M. (1987). Graph bisection algorithms with good average case behaviour. FOCS (1984). *Combinatorica* **7**, 171–191.

Buser, P. (1978). Cubic graphs and the first eigenvalue of a Riemann surface. *Math. Z.* **162**, 87–99.

Buser, P. (1984). On the bipartition of graphs. *Discrete Math. Appl.* **9**, 105–109.

Clark, L. H. and Entringer, R. C. (1988). The bisection width of cubic graphs. *Bull. Austral. Math. Soc.* **39**, 389–396.

Garey, M. R., Johnson, D. S., and Stockmeyer, L. (1976). Some simplified NP-complete graph problems. *Theor. Comput. Sci.* **1**, 237–267.

Goldberg, M. K. and Gardner, R. (1984). On the minimal cut problem. In: *Progress in Graph Theory*. Academic Press, San Diego, pp. 295–305.

Harary, F. (1969). *Graph Theory*. Addison-Wesley, Reading, Mass.

Kostochka, A. V. and Melnikov, L. S. (1991). On bounds of the bisection width of cubic graphs. In: *Combinatorics, Graphs, Complexity. Proceedings of the Fourth Czechoslovak Symposium on Combinatorics*, Soc. of Czechosl. Math. and Phys., Praha, pp. 151–154.

MacGregor, R. M. (1978). *On Partitioning a Graph: a Theoretical and Empiric Study*. Ph. D. Thesis, Univ. of California, Berkeley.

Mohar, B. (1989). Isoperimetric numbers of graphs. *J. Comb. Theory* **B 47**, 274–291.

Probabilistic Methods in Discrete Mathematics, pp. 266 – 270
V. F. Kolchin et al. (Eds.)
1993 TVP/VSP

A RISK FUNCTION IN GLOBAL OPTIMIZATION PROBLEM

S. V. KOVALEV and V. V. MAZALOV
Chita Institute of Natural Resources of Russia Academy of Sciences,
Butina Street 26, 672090 Chita, Russia

ABSTRACT

We suggest a random search algorithm in the global optimization problem. The Brownian motion is used as a statistical model of multipeak functions. We introduce risk and expected risk functions and define an algorithm using these functions. The aim of the article is to get the analytical form of the risk function.

1. INTRODUCTION

We consider a global optimization problem in the one-dimensional case, i. e., the analyzed dependence has a form $y = y(x)$, where $x \in \mathbf{R}^1$, but the explicit form of the function $y(x)$ is unknown. The problem is to find the point of the maximum of the function $y(x)$ at the interval $[a, b]$. Suppose the function $y(x)$ has several extremum points. We will use a random search method. Every step of the method is based on the information

$$x_1, x_2, \dots, x_n, \quad y_1, y_2, \dots, y_n, \tag{1}$$

where $a = x_1 < x_2 < \cdots < x_n = b$ are the points chosen in the experiment and $y_1, y_2, \dots, y_n$ are the values of the function in the corresponding experiment point. At each step the algorithm must define whether we need stop observations or continue and pick up the next point x for a new experiment. For convenience we denote $X_n = (x_1, \dots, x_n)$, $Y_n = (y_1, \dots, y_n)$.
There are many random search methods concerning the global optimization problem, see (Zhigliavskii and Zhilinskas, 1991).
Further we suggest a new algorithm of the random global search based on the idea of the risk functions.

2. MODEL OF THE FUNCTION $y = y(x)$

The explicit form of the function $y(x)$ is unknown. We will consider the function $y(x)$ being a trajectory of the Brownian motion with drift σ. It is well known (see, e. g., (Hida, 1980)) that the conditional density of the random variable $y(x)$ for

$x_i < x < x_{i+1}$ under the conditions $y(x_i) = y_i,\ y(x_{i+1}) = y_{i+1}$ has the Gaussian form with parameters

$$m_i = \frac{y_i(x_{i+1} - x) + y_{i+1}(x - x_i)}{x_{i+1} - x_i}, \quad s_i^2 = \frac{\sigma_i^2(x_{i+1} - x)(x - x_i)}{x_{i+1} - x_i}$$

(here σ_i is the parameter of the Brownian motion). We denote it by $\varphi(y, m_i, s_i)$.

3. RISK FUNCTIONS

Suppose that the algorithm at the nth step is in the state (1). Let y^* be the maximal value among $y_1, \ldots, y_n$, and x^* is the corresponding value of the argument. If the point of the maximum is x, then the error is equal to $|x - x^*|$. Multiplying by the probability of this event and integrating over all x we obtain the risk function

$$R(X_n, Y_n) = \int_{x_1}^{x_n} |x - x^*|\,\mathbf{P}\{y(x) > y^*\}\,dx, \tag{2}$$

where according to the choice of the function $y(x)$

$$\mathbf{P}\{y(x) > y^*\} = \int_{y^*}^{\infty} \varphi(y, m_i, s_i)\,dy, \qquad x_i < x < x_{i+1}, \quad i = 1, \ldots, n-1.$$

Note that the notion of risk function (2) can be introduced for any probabilistic model of the function $y(x)$.

Now let us introduce the expected risk if we continue the experiment at a point x for the $(n+1)$th step. We define

$$\mathbf{E}(R(X_n, Y_n), x) = \int_{-\infty}^{\infty} R\begin{pmatrix} x_1, \ldots, x_j, x, x_{j+1}, \ldots, x_n \\ y_1, \ldots, y_j, y, y_{j+1}, \ldots, y_n \end{pmatrix} \varphi(y, m_j, s_j)\,dy$$

as the expected risk at a point x within the interval (x_j, x_{j+1}).

4. DESCRIPTION OF THE GLOBAL RANDOM SEARCH ALGORITHM

Determine the algorithm of the global maximum search on the base of the introduced notion for risk functions. At the nth step we choose a point x within the interval (a, b) as the point of the next observation if the expected risk $\mathbf{E}\,(R(X_n, Y_n), x)$ attains minimum at the point x. It can be achieved by solving the equation

$$\frac{d\mathbf{E}\,(R(X_n, Y_n, x))}{dx} = 0.$$

The stopping rule can be determined using some sufficiently small number ε. We stop the observations at the states (X_n, Y_n) if $R(X_n, Y_n) < \varepsilon$.

5. SIMPLIFICATION OF RISK FUNCTION EXPRESSION

For numerical calculations it is important to simplify the expression of risk function. Present expression (2) as the sum

$$R = \sum_{i=1}^{n-1} J_i$$

of the integrals

$$J_i = \int_{x_i}^{x_{i+1}} |x - x^*| \, dx \int_{y^*}^{\infty} \varphi(y, m_i, s_i) \, dy. \tag{3}$$

We find first the integrals J_i for $x_{i+1} \leq x^*$. Then $|x - x^*| = x^* - x$. Substituting $z = (x - x_i)/(x_{i+1} - x_i)$ in the external integral and $v = (y - y_i)/(\sigma_i\sqrt{x_{i+1} - x_i})$ in the internal integral we obtain

$$J_i = (x_{i+1} - x_i)^2 \int_0^1 \left(\frac{x^* - x_i}{x_{i+1} - x_i} - z \right) dz \int_{\hat{y}}^{\infty} \varphi(v, \hat{m}_i, \hat{s}_i) \, dv,$$

where

$$\hat{m}_i = \frac{(y_{i+1} - y_i)z}{\sigma_i\sqrt{x_{i+1} - x_i}}, \quad \hat{s}_i = \sqrt{z(1 - z)}, \quad \hat{y} = \frac{y^* - y_i}{\sigma_i\sqrt{x_{i+1} - x_i}}.$$

Setting

$$\frac{x^* - x_i}{x_{i+1} - x_i} = \xi, \qquad \frac{y^* - y_i}{\sigma_i\sqrt{x_{i+1} - x_i}} = \eta, \qquad \frac{y_{i+1} - y_i}{\sigma_i\sqrt{x_{i+1} - x_i}} = k$$

and making the substitution $t = (v - kz)/\sqrt{2z(1 - z)}$ in the internal integral, we obtain

$$J_i = \frac{(x_{i+1} - x_i)^2}{\sqrt{\pi}} \int_0^1 (\xi - x) \, dx \int_w^{\infty} \exp\{-t^2\} \, dt, \tag{4}$$

where $w = (\eta - kx)/\sqrt{2x(1 - x)}$. The function $w(x)$ attains its minimum equal to $\eta^* = \sqrt{2\eta(\eta - k)}$ at the point $x = \eta/(2\eta - k)$ and the inverse function consists of the two parts

$$x_- = \frac{w^2 + \eta k - w\sqrt{w^2 - 2\eta(\eta - k)}}{k^2 + 2w^2}, \qquad 0 \leq x \leq \frac{\eta}{2\eta - k},$$

$$x_+ = \frac{w^2 + \eta k + w\sqrt{w^2 - 2\eta(\eta - k)}}{k^2 + 2w^2}, \qquad \frac{\eta}{2\eta - k} \leq x \leq 1.$$

Changing the order of integration in the iterated integral in (4) we find that

$$L = \int_0^1 (\xi - x) \, dx \int_w^{\infty} \exp\{-t^2\} \, dt = \int_{\eta^*}^{\infty} \exp\{-t^2\} \, dt \big(\xi(x_+ - x_-) - (x_+^2 - x_-^2)/2\big),$$

and substituting $w^2 - 2\eta(\eta - k) = z^2$ we obtain

$$\begin{aligned} L = {} & \xi \exp\{-2\eta(\eta - k)\} \int_0^{\infty} \frac{2z^2 \exp\{-z^2\}}{c^2 + 2z^2} dz \\ & - \exp\{-2\eta(\eta - k)\} \int_0^{\infty} \frac{2z^2(z^2 + \eta(2\eta - k)) \exp\{-z^2\}}{(c^2 + 2z^2)^2} dz, \end{aligned} \tag{5}$$

where $c = 2\eta - k$. Using the identities

$$\frac{2z^2}{c^2+2z^2} = 1 - \frac{c^2}{c^2+2z^2}, \qquad \frac{2z^2(z^2+\eta c)}{(c^2+2z^2)^2} = \frac{1}{2}\left(1 - \frac{2c(\eta-k)}{c^2+2z^2} - \frac{c^3 k}{(c^2+2z^2)^2}\right)$$

we can present (5) in the form

$$L = \xi \exp\{-2\eta(\eta-k)\}(\sqrt{\pi}/2 - c^2 K_1)$$
$$- \exp\{-2\eta(\eta-k)\}(\sqrt{\pi}/2 - 2c(\eta-k)K_1 - c^3 k K_2)/2,$$

where

$$K_1 = \int_0^\infty \frac{\exp\{-z^2\}}{c^2+2z^2}dz, \quad K_2 = \int_0^\infty \frac{\exp\{-z^2\}}{(c^2+2z^2)^2}dz. \tag{6}$$

Integrals (6) can be expressed with the help of the Gaussian type function $\Phi(x) = \int_x^\infty \exp\{-t^2/2\}dt$:

$$K_1 = \frac{\sqrt{\pi}}{2c}\exp\left\{\frac{c^2}{2}\right\}\Phi(c),$$
$$K_2 = \frac{\sqrt{\pi}}{4c}\left(\frac{1}{c^2}\exp\left\{\frac{c^2}{2}\right\}\Phi(c) - \exp\left\{\frac{c^2}{2}\right\}\Phi(c) + \frac{1}{c}\right). \tag{7}$$

Inserting (5) and (7) into (4) we finally obtain the expression of the term J_i in the expression of the risk function in the case $x_{i+1} < x^*$:

$$J_i = \frac{(x_{i+1}-x_i)^2}{2}\exp\{-2\eta(\eta-k)\}\left(\xi - \frac{1}{2} + \left(\eta - \frac{3}{4}k - \xi c\right)\exp\left\{\frac{c^2}{2}\right\}\Phi(c)\right.$$
$$\left. + \frac{ck}{4}\left(1 - c\exp\left\{\frac{c^2}{2}\right\}\Phi(c)\right)\right). \tag{8}$$

Now assume that the maximal point lies on the left of the interval (x_i, x_{i+1}). Denote it x_*. In this case $|x - x_*| = x - x_*$. Note that the integral

$$J_i = \int_{x_i}^{x_{i+1}} (x - x_*)\,dx \int_{y^*}^\infty \varphi(y, m_i, s_i)\,dy$$

with parameters $m_i(x)$, $s_i(x)$ after the substitutions $z = x_i + x_{i+1} - x$, $x^* + x_* = x_i + x_{i+1}$ transfers into the integral

$$J_i = \int_{x_i}^{x_{i+1}} (x^* - z)\,dz \int_{y^*}^\infty \varphi(t, \hat{m}_i, \hat{s}_i)\,dt \tag{9}$$

with parameters

$$m_i(z) = \frac{y_{i+1}(x_{i+1}-z) + y_i(z-x_i)}{x_{i+1}-x_i}, \quad s_i^2(z) = \frac{\sigma_i^2(z-x_i)(x_{i+1}-z)}{x_{i+1}-x_i}.$$

Integral (9) differs from (3) and consequently from (8) only in that the random variable $y(x)$ takes the values y_{i+1} at x_i and y_i at x_{i+1}. Replacing y_i by y_{i+1}, y_{i+1}

by y_i, $x^*$by $x_i + x_{i+1} - x_*$ in (8), we get the expression of integral (8) in the case $x_* \leq x_i$.
Summing up J_i, we obtain a simple expression of the risk function at the nth step of the random search. A similar expression can be obtained for the expected risk function.
These expressions are the basis of the global optimization algorithm included in the software package "Stopping".

REFERENCES

Zhigliavskii, A. and Zhilinskas, A. (1991). *Methods of Global Optimization*. Nauka, Moscow (in Russian).

Hida, T. (1980). *Brownian Motion*. Springer, New York.

Software Package "Stopping" (1989). In: *Software Information Bulletin* **2**, Comput. Center Acad. Sci. USSR, Moscow (in Russian).

Probabilistic Methods in Discrete Mathematics, pp. 271 – 286
V. F. Kolchin et al. (Eds.)
1993 TVP/VSP

ON AN EXISTENCE THEOREM FOR DECOMPOSABLE STATISTICS

E. M. KUDLAEV
Laboratory of Mathematical Statistics, Department of Mechanics and Mathematics, Moscow State University, Vorobjevy Gory, 119899 Moscow, Russia

ABSTRACT

This work is devoted to an existence theorem for decomposable statistics and some its corollaries.

1. INTRODUCTION

In this paper sufficient conditions for existence of decomposable statistics are proposed. In contrast to author's publications (Kudlaev, 1988 a, 1989 a), where a sketch of proof was given; here a complete proof of the existence theorem is considered in more wide conditions.

Decomposable (or more strictly, l-decomposable) statistics are sums of random variables

$$\zeta = \sum_{k=1}^{N} f_k(w_k, \dots, w_{k+l}), \quad 0 \le l < N, \tag{1}$$

where the indices of the random variables $w_1, \dots, w_N$ in the sum are taken modulo N, and for

$$w = (w_1, \dots, w_N) \tag{2}$$

there exist independent random variables $\theta_1, \dots, \theta_N$ such that the distribution of (2) coincides with the conditional distribution of

$$\theta = (\theta_1, \dots, \theta_N) \tag{3}$$

under the condition

$$\sum_{k=1}^{N} \theta_k = a \tag{4}$$

for some a. In (1), functions f_k, $k = 1, \dots, N$, are supposed to be Borel ones. The variables w_k involved in (2) are dependent so that for any values of k and $d > 0$,

$$0 < k < N + 1, \quad 1 < k + d < N + 1,$$

the dependence between w_k and w_{k+d} does not decrease, generally speaking, with increasing of d (it is, particularly, constant for symmetrically dependent variables

$w_1, \ldots, w_N$). This sequence of the random variables differs essentially from the sequence in which the dependence between kth and $(k+d)$th members either vanishes beginning with some value of $d = d_0$ (e.g., for the sequences of independent random variables we have $d_0 = 1$ and for the sequences of m-dependent random variables we have $d_0 = m + 1$) or for rather high values of d it can be as low as possible (e.g., for weakly dependent random variables with an appropriate choice of a mixing coefficient value).

The sums of type (1) are extremely useful and can be applied both in mathematics (in combinatorial theory, theory of graphs and random mappings of finite sets, theory of automata) and in other spheres of human knowledge like nuclear physics, statistical mechanics, biology, zoology, medicine, sociology, computer technology and even in the history of antiquity (see (Holst, 1981a, 1981b, 1981c; Ivanov et al., 1984; Kolchin, 1986; Kolchin et al., 1978; Kudlaev, 1988 a, 1989 b)). Major applications decomposable statistics (for brevity, DS) find in the mathematical statistics (in particular, in a sampling survey of finite sets as well as in a construction of goodness-of-fit tests and tests of homogeneity).

Some individual results dealing with DS have been obtained since long ago. Among them are A. De Moivre (1711), W. A. Whitworth (1897), K. Pearson (1900), A. A. Markov (1912), R. A. Fisher (1922), S. P. Wicksell (1933), A. Wald and J. Wolfowitz (1940), A. M. Mood (1940), F. Wilcoxon (1945), F. N. David (1950), B. S. Sherman (1950), G. Arfwedson (1951), D. A. Darling (1953), L. Castoldi (1954), S. Tumanyan (1956) L. Weiss (1957), G. P. Steck (1957), L. Le Cam (1958). The last three decades show a great interest to DS which is realized in a growing number of publications on the subject. The lack of place here limits the list of authors who were working on the problem in this period. They are: V. P. Chistyakov, P. Erdős, B. Harris, L. Holst, V. A. Ivanov, G. I. Ivchenko, V. F. Kolchin, Yu. I. Medvedev, C. Morris, C. J. Park, A. Rényi, B. Rosen, B. A. Sevastyanov (problems dealing with allocation of particles on cells); B. Harris, V. F. Kolchin, S. P. Lloyd, Yu. L. Pavlov, L. A. Shepp, V. E. Stepanov (problems dealing with random mappings of finite sets); S. A. Alam, S. Blumenthal, D. M. Chibisov, N. Cressie, P. Deheuvels, G. E. Del Pino, P. Hall, S. N. Kirmani, E. M. Kudlaev, R. Pyke, J. Sethuraman, L. Weiss (problems dealing with spacings); E. Edens, L. Holst, F.W. Huffer, Y. Husler, N. Kaplan, J. S. Rao, L. A. Shepp, A. F. Siegel (problems dealing with random coverage of the circle); L. Holst, E. M. Kudlaev, K. V. Mardia, J. S. Rao, R. L. Schweitzer (problems dealing with spacing-frequencies); E. M. Kudlaev (Kudlaev, 1978, 1988 a, 1989 b) (problems dealing with the additional functions dependent on spacing and spacing-frequencies).

Frequently, specialists searching for a result within some DS-scheme do not make any attempt to find some analogs for such a result within other schemes. Nevertheless, various DS-schemes have very much in common, consequently progress in one scheme can result in a success in another one (e.g., DS have been studied rather well in schemes: polynomial allocation and spacing). The primary author's idea is to detect some properties being common for all DS-schemes; the study of particular DS-schemes employs this general part, moreover such a study can be considered as a complicated autonomous problem. To find a desirable solution the author (Kudlaev, 1988 a, 1989 a, 1989 b) has introduced DS class with the values in continuous groups. On the one hand, the class involves all the known schemes and, on the other hand, a study of such general statistics is of individual interest. The appearance of continuous groups in these problems can be explained as follows. Formulas (1) and (4) are based on the commutative operations of addition in the spaces G_1 and G_2 containing sets of values of the random variables

$$f_k(w_k, \dots, w_{k+l}) \text{ and } w_k; \qquad k = 1, \dots, N,$$

correspondingly. The natural requirement for such operations (particularly, in the case of the real functions f_k of real arguments) is their continuity in some topologies; e.g., if $G_1 = \mathbf{R}$ and $G_2 = \mathbf{Z}$, the topologies are taken to be the class of all open intervals and the discrete topology correspondingly. Since a choice of a version for conditional distribution dealing with formulas (1) and (4) is not unique, the availability of such a topology in the space G_2 enables one to choose the version such that the conditional distribution is continuous (in the sense of the weak convergence) in some neighbourhood of the point a. We suggest G_1 and G_2 to be locally compact Hausdorff Abelian groups (for brevity, LCA-groups) which gives us an opportunity to use the Pontryagin duality theorem and its numerous corollaries. Besides, having declared G_1 and G_2 be complete separable metric spaces, we can make use of multiple results related to probability measures on these spaces. If the random variables X and Y take values in LCA-groups being simultaneously complete separable metric spaces, then, as Zabell (Zabell, 1979) showed in the case of absolute integrability with respect to the second argument of the joint characteristic function $\varphi_{X,Y}$ of the variables X and Y, there exists a continuous (in the sense of the weak convergence) variant of regular conditional distribution of the random variable X under the condition Y. The version of the characteristic function corresponding to this variant is proportional to the integral of $\varphi_{X,Y}$ taken over the second argument. This result is a generalization of the earlier results obtained by S. P. Wicksell, M. S. Bartlett, G. P. Steck, J. Yeh (see the references in (Zabell, 1979)) related to the "elementary" groups composed of the finite weak products of the groups $\mathbf{R}$ and $\mathbf{Z}$. In our case the roles of the variables X and Y are played by

$$\xi = \sum_{k=1}^{N} f_k(\theta_k, \dots, \theta_{k+l}), \quad \eta = \sum_{k=1}^{N} \theta_k \tag{5}$$

correspondingly.

The application of the existence theorem to DS is based on the classifying collection (this definition is introduced in (Kudlaev, 1989 b))

$$(N, a, G_2, \{\mathbf{P}_2\theta_k^{-1}; k = 1, \dots, N\}, G_1) \tag{6}$$

for this DS, where $\mathbf{P}_2\theta_k^{-1}$ is the distribution of the random variable θ_k. According to (Kudlaev, 1989 b) the classification of DS themselves is a two-stage process: first, DS-scheme is defined; secondly, we define the set of functions f_k, $k = 1, \dots, N$. For some conditions the using of the existence theorem together with (6) gives us the asymptotic distribution of concrete DS, if as $n \to \infty$ and $a = a(n) \to a_0$

$$N = N(n) \to \infty, \quad \tau(a) \to \tau(a_0)$$

for some a_0 and for some automorphism τ of LCA-group G_2 (see (Kudlaev, 1988 a, 1989 b)).
The organization of this paper is the following. In Section 2, basic definitions and notations are introduced, and in Section 3 a slight generalization of above mentioned Zabell's result on the existence of a continuous version of a regular conditional distribution is given. Section 4 is devoted to formulating and proving the existence theorem. Section 5 contains some discussion concerning the assumptions which had been made in the statement of the main result. In Section 6, some corollaries are presented.

2. BASIC DEFINITIONS AND NOTATIONS

For $j = 1$ or $j = 2$ let G_j be LCA-group with topology $\mathfrak{A}_j$, and let $\mathcal{B}_j$ be the smallest σ-algebra containing all the subsets of $\mathfrak{A}_j$. Then $\mathcal{B}_j$ is called a *Borel σ-algebra* and elements of $\mathcal{B}_j$ are called *Borel subsets.* The random variables (Borel functions) $w_1, \dots, w_N$ and the independent random variables (Borel functions) $\theta_1, \dots, \theta_N$ take their values in G_2 and are defined on some probability spaces $(\Omega_1, S_1, \mathbf{P}_1)$ and $(\Omega_2, S_2, \mathbf{P}_2)$ correspondingly, and the variables w and θ are defined by formulas (2) and (3). Let $\mathcal{B}_2^N$ be the smallest σ-algebra containing all the Borel rectangles $B = B_1 \times \cdots \times B_N$ with $B_j \in \mathcal{B}_2$, $j = 1, \dots, N$. Then $\mathcal{B}_2^N$ is called a *product of N σ-algebras* $\mathcal{B}_2$, and according to (Parthasarathy, 1980, Exercise 22.9) the variables w and θ are $(\mathcal{B}_2^N, S_j)$-measurable functions for $j = 1$ and $j = 2$ correspondingly. For any subset $B \in \mathcal{B}_2^N$ we denote

$$\mathbf{P}_1 w^{-1}(B) = \mathbf{P}_1\{\omega_1\colon\ w(\omega_1) \in B\}$$

and

$$\mathbf{Q}(y, B) = \mathbf{P}_2(\{\omega_2\colon\ \theta(\omega_2) \in B\} \mid \eta = y),$$

where η is defined by formula (5). Then $\mathbf{P}_1 w^{-1}(\cdot)$ is called a *distribution of the variable* w and $\mathbf{Q}(\cdot,\cdot)$ is called a *conditional distribution of the variable θ for given* η. We assume there exists a version $\hat{\mathbf{Q}}(\cdot,\cdot)$ of regular conditional distribution $\mathbf{Q}(\cdot,\cdot)$ (see (Parthasarathy, 1967, pp. 145–146)), and this version varies continuously in y (in the sense of the weak convergence) in some neighbourhood V of the point a. The latter means that for any $y \in V$ and for any $y_0 \in V$

$$\mathbf{Q}(y,\cdot) \Rightarrow \mathbf{Q}(y_0,\cdot) \quad \text{as} \quad y \to y_0.$$

Definition 1 (see (Kudlaev, 1988a)). Variable (1) is called an *l-decomposable statistic,* if the following conditions hold:

(D1) the function

$$f\colon\ (x_2^{(1)}, \dots, x_2^{(N)}) \mapsto \sum_{k=1}^{N} f_k(x_2^{(k)}, \dots, x_2^{(k+l)}),$$

where $x_2^{(k)} \in G_2$, $k = 1, \dots, N$ and $x_2^{(k+N)} = x_2^{(k)}$, $k = 1, \dots, l$, is a $(\mathcal{B}_1, \mathcal{B}_2^N)$-measurable function;

(D2) there exist independent random variables $\theta_1, \ldots, \theta_N$ such that for any $B \in \mathcal{B}_2^N$

$$\mathbf{P}_1 w^{-1}(B) = \hat{\mathbf{Q}}(a, B); \tag{7}$$

(D3) the version $\hat{\mathbf{P}}(\cdot, \cdot)$ of a conditional distribution of ξ for given η, which is defined by the formula

$$\hat{\mathbf{P}}(y, A) = \hat{\mathbf{Q}}(y, f^{-1}[A]) \tag{8}$$

for any $A \in \mathcal{B}_1$, varies continuously in y (in the sense of the weak convergence) in some neighbourhood V^* of the point a.

According to (D1) and to (Parthasarathy, 1980, Exercise 22.11), the variable ξ is a random one, and by formulae (7) and (8) for any $A \in \mathcal{B}_1$

$$\mathbf{P}_1 \zeta^{-1}(A) = \hat{\mathbf{Q}}(a, f^{-1}[A]) = \hat{\mathbf{P}}_2(\{\omega_2 : f(\theta(\omega_2)) \in A\} \mid \eta = a), \tag{9}$$

where $\mathbf{P}_1 \zeta^{-1}(\cdot)$ is the distribution of ζ and $\hat{\mathbf{P}}_2(\cdot \mid \cdot)$ is an appropriate version of a regular conditional distribution of ξ for given η.
In accordance with (Parthasarathy, 1967, pp. 28–29) the closed set C_μ is called *the support* or *spectrum of probability measure* μ, if C_μ is a closed set satisfying conditions: $\mu(C_\mu) = 1$ and if C is any closed set such that $\mu(C) = 1$, then $C_\mu \subseteq C$.

Definition 2 (see (Kudlaev, 1989 b)). Collection (6) is called a DS-*scheme* if G_2 is the LCA-group which is generated by the set of the supports of $\mathbf{P}_2 \theta_k^{-1}$, $k = 1, \ldots, N$, and if G_1 is the LCA-group which is generated by the support of $\mathbf{P}_1 \zeta^{-1}$.

Definition 3. A DS-scheme is called *symmetric*, if the distributions $\mathbf{P}_2 \theta_k^{-1}$ coincide for all $k = 1, \ldots, N$; in other cases the DS-scheme is called *nonsymmetric*. Every l-DS is defined by DS-scheme and by the collection of the functions f_k, $k = 1, \ldots, N$.

Definition 4. An l-DS is called *symmetric* if the appropriate DS-scheme is symmetric and if the appropriate functions f_k coincide for all $k = 1, \ldots, N$; otherwise, the l-DS is called *nonsymmetric*.

Let T be the circle group in the complex plane. For $j = 1$ or $j = 2$ let Γ_j be the set of all continuous characters γ_j, where every γ_j is a continuous map $G_j \to T$ such that

$$|\gamma_j(g_j)| = 1 \quad \text{and} \quad \gamma_j(g_j' + g_j'') = \gamma_j(g_j')\gamma_j(g_j'')$$

for all g_j, g_j' and g_j'' in G_2; Γ_j is called *the dual group to* G_j. The commutative operation in group Γ_j is defined by the formula

$$\gamma_j'\gamma_j''(g_j) = \gamma_j'(g_j)\gamma_j''(g_j)$$

for all g_j in G_j and for all γ_j', γ_j'' in Γ_j. Sets of the form

$$M(K, V) = \{\gamma_j \colon G_j \to T,\ \gamma_j[K] \subseteq V\},$$

where K is any compact set in G_j and V is any open set in T form a prebase of some topology $\mathfrak{A}_j^*$. It can be shown (Morris, 1977, Assertion 29), that the space Γ_j with $\mathfrak{A}_j^*$ becomes a LCA-group. It is well known (see, e. g., (Heyer, 1977)), that the Haar measure exists on every LCA-group G, and it is the unique (up to multiplicative constant) translation-invariant measure on appropriate σ-algebra $\mathcal{B}$. Let $\varkappa_2$ be the Haar measure on G_2 and let $\varkappa_2^*$ be the Haar measure on the dual group Γ_2. Now introduce the space Γ containing the set of all $\gamma \equiv \gamma(\gamma_1, \gamma_2)$ with the commutative operation

$$\gamma(g) = \gamma_1(g_1)\gamma_2(g_2)$$

for all γ_j in Γ_j and for all g_j in G_j, where $g = (g_1, g_2)$ and $j = 1$ or $j = 2$. If $G = G_1 \times G_2$ is the product of the LCA-group G_1 and G_2, under the product topology, then Γ becomes (Morris, 1977, Theorem 13) the LCA-group which is dual to G and $\Gamma = \Gamma_1 \times \Gamma_2$, under the product topology.

3. ZABELL'S RESULT

Suppose that X and Y are some random variables taking their values in G_1 and G_2 correspondingly. If $\mathbf{P}$ is the joint distribution of X and Y, then the corresponding joint characteristic function (c. f.) φ, the Fourier–Stieltijes transform, is given by the formula

$$\varphi(\gamma) = \int_G \prod_{j=1}^{2} \gamma_j(g_j)\, d\mathbf{P}(g) \tag{10}$$

for all γ in Γ. Now suppose that the following conditions are satisfied:

(C1) the LCA-groups G_1 and G_2 are complete separable metric spaces;

(C2) the group G_2 is generated by the support of the distribution of Y;

(C3) there exists a non-negative Borel function Δ on Γ such that it does not depend on γ_1 and the inequality

$$|\varphi(\gamma(\gamma_1, \gamma_2))| \le \Delta(\gamma_2)$$

holds for all $\gamma \in \Gamma$ and

$$\int_{\Gamma_2} \Delta(\gamma_2) d\varkappa_2^*(\gamma_2) < \infty.$$

Then in accordance with (Zabell, 1979, Theorem 3.1) the formula

$$\widehat{\varphi}_{X|Y}(\gamma_1 \mid Y = y) = \frac{\int_{\Gamma_2} \varphi(\gamma(\gamma_1, \gamma_2))\gamma_2(-y)\, d\varkappa_2^*(\gamma_2)}{\int_{\Gamma_2} \varphi(\gamma(e_1^*, \gamma_2)\gamma_2(-y)\, d\varkappa_2^*} \tag{11}$$

defines the appropriate version of the conditional c. f. $\varphi_{X|Y}$ of X for given Y, if the denominator of the right-hand side of (11) does not equal zero. In (11) e_1^* is the neutral element of the group Γ_1. Let $\hat{\mathbf{P}}_{X|Y}(\cdot,\cdot)$ be the version of a regular conditional distribution such that

$$\hat{\varphi}_{X|Y}(\gamma_1 \mid Y = y) = \int_{G_1} \gamma_1(g_1)\hat{\mathbf{P}}_{X|Y}(dg_1 \mid Y = y)$$

for all $\gamma_1 \in \Gamma_1$ and for all $y \in G_2$. If (C3) holds, then by the Fourier inversion theorem the distribution $\mathbf{P}Y^{-1}$ of the random variable Y has a bounded continuous density with respect to the Haar measure $\varkappa_2$, namely

$$(d\mathbf{P}Y^{-1}/d\varkappa_2)(y) = C \int_{\Gamma_2} \varphi(\gamma(e_1^*, \gamma_2))\gamma_2(-y)\, d\varkappa_2^*(\gamma_2). \tag{12}$$

In (12) C is a constant which can be included in $\varkappa_2^*$.

Remark 1. The right-hand side of formula (11) does not depend on the choice of a version of the Haar measure $\varkappa_2^*$.
Also according to (Zabell, 1979, Theorem 3.2) we have the following assertion.

LEMMA. *If conditions* (C1)–(C3) *are satisfied, then the version* $\hat{\mathbf{P}}_{X|Y}(\cdot,\cdot)$ *varies continuously (in the sense of the weak convergence) for all* $y \in V$, *where*

$$V = \{g_2 \in G_2 : (d\mathbf{P}Y^{-1}/d\varkappa_2)(g_2) > 0\}$$

is an open set and (12) *is used.*

Remark 2 (Zabell, 1979). In (Zabell, 1979) the case $G_1 = G_2$ is considered; the case $G_1 \neq G_2$ is analyzed similarly.

4. THE MAIN RESULT

The next statement is an improved variant of author's result (Kudlaev, 1988 a, 1989 a).

THEOREM. *Suppose that*

(A1) *the LCA-groups* G_1 *and* G_2 *are complete separable metric spaces;*

(A2) *there are independent random variables* $\theta_1, \ldots, \theta_N$ *such that the version* $\hat{\mathbf{Q}}(\cdot,\cdot)$ *of a regular conditional distribution of* θ *for given* η *varies continuously (in the sense of the weak convergence) in the first argument for all* $a \in V$, *where* V *is an open set in* $\mathfrak{A}_2$;

(A3) *for all* $k = 1, \ldots, N$ *the functions* f_k *are* $(\mathcal{B}_1, \mathcal{B}_2^{l+1})$*-measurable;*

(A4) *for all* $a \in V$

$$\hat{\mathbf{Q}}(a, D_f) = 0,$$

where D_f *is the set of discontinuity of the function* f.

Then

(S1) *there exist random variables*

$$w_1 = w_1(a), \ldots, w_N = w_N(a) \tag{13}$$

such that equation (7) *holds and the distribution of the variable*

$$w = w(a) = (w_1(a), \dots, w_N(a))$$

varies continuously (*in the sense of the weak convergence*) *in* $a \in V$;

(S2) *variable* (1) *where the random variables* $w_1, \dots, w_N$ *are defined by* (13) *is* *l-DS, and the distribution of variable* (1) *varies continuously* (*in the sense of the weak convergence*) *in* $a \in V$.

Proof. The proof contains several steps.
(1) If an operation of coordinate-wise addition is introduced, then the product group $G = G_1 \times G_2$, under the product topology, becomes a LCA-group (e. g., see (Morris, 1977, Assertions 15 and 16)). A similar statement is true for any finite product group consisting of the factors G_1 and G_2.
(2) Prove that G is a complete separable metric space. By (A1) there are metrics ρ_1 and ρ_2 which induce the topologies $\mathfrak{A}_1$ and $\mathfrak{A}_2$ correspondingly. If G_1 and G_2 are metric spaces, then their Cartesian product, under the product topology, is a metric space (e. g., see (Billingsley, 1968, Appendix 2, formula (11))) and, moreover, the corresponding metric ρ_{12} can be defined by the formula:

$$\rho_{12}((g_1', g_2'), (g_1'', g_2'')) = \max\{\rho_1(g_1', g_1''), \rho_2(g_2', g_2'')\} \tag{14}$$

for all $g_j', g_j'' \in G_j$ ($j = 1$ and $j = 2$). If the spaces G_1 and G_2 are complete, then in accordance with (14) the space G is complete as well. Besides, it is well known that the Cartesian product, under the product topology, is separable if and only if each factor is separable.
A similar statement holds for any finite Cartesian product, under the product topology, consisting of the factors G_1 and G_2.
In particular, item (2) means that G is a topological space with the second countability axiom (see (Kelley, 1957, Ch. 4, Theorem 11)).
(3) Prove that maps

$$\eta\colon \Omega_2 \to G_2 \quad \text{and} \quad \xi\colon \Omega_2 \to G_1$$

are the Borel ones. We have the following composition

$$\eta = g \odot \theta,$$

where the map $\theta\colon \Omega_2 \to G_2^N$ is defined by (3) and g is defined by the formula

$$g\colon (g_2^{(1)}, \dots, g_2^{(N)}) \mapsto \sum_{k=1}^{N} g_2^{(k)}$$

for all $g_2^{(k)} \in G_2$, $k = 1, \dots, N$. Above (see Section 2) we have seen that θ is (B_2^N, S_2)-measurable. By definition, in a topological group the operation of addition is continuous under the product topology. Consequently, for any set $B \in \mathfrak{A}_2$ its preimage $g^{-1}[B]$ belongs to the product topology of the Cartesian product G_2^N. In turn, the latter topology generates the Borel σ-algebra $\mathcal{B}(G_2^N)$, and a class of sets B generates the Borel σ-algebra $\mathcal{B}_2$. Using Assertion 22.4 (Parthasarathy, 1980), we prove the $(\mathcal{B}_2, \mathcal{B}(G_2^N))$-measurability of g. Always we have:

$$\mathcal{B}(G_2^N) \supseteq \mathcal{B}_2^N,$$

but in our case (Parthasarathy, 1980, Exercise 22.9) these σ-algebras coincide, because G_2^N is a topological space with the second countability axiom (see step (2)). In accordance with (Parthasarathy, 1980, Exercise 22.10) we obtain the $(\mathcal{B}_2, S_2)$-measurability of η.
Let

$$\xi = \bar{g} \odot h \odot \theta,$$

where

$$h\colon (g_2^{(1)}, \dots, g_2^{(N)}) \mapsto (f_1 \odot \pi_1^{(l)}(g_2^{(1)}, \dots, g_2^{(N)}), \dots, f_N \odot \pi_N^{(l)}(g_2^{(1)}, \dots, g_2^{(N)})),$$

$$\pi_k^{(l)}\colon (g_2^{(1)}, \dots, g_2^{(N)}) \mapsto (g_2^{(k)}, \dots, g_2^{(k+l)}); \qquad g_2^{(k)} \in G_2 (k = 1, \dots, N)$$

with

$$g_2^{(k+N)} = g_2^{(k)}; \qquad k = 1, \dots, N,$$

and

$$\bar{g}\colon (g_1^{(1)}, \dots, g_1^{(N)}) \mapsto \sum_{k=1}^{N} g_1^{(k)}; \qquad g_1^{(k)} \in G_1, \quad k = 1, \dots, N.$$

By (Kelley, 1957, Ch. 3) and by definition of a topological group correspondingly, the projective map $\pi_k^{(0)}$, $k = 1, \dots, N$, and the map $\bar{g}$ are continuous. Therefore, for any open sets

$$B_k \in \mathcal{B}_2, \qquad k = 1, \dots, N,$$

and for any open set $B \in \mathfrak{A}_1$ the preimages

$$(\pi_k^{(l)})^{-1}[B_k \times \cdots \times B_{k+l}], \qquad k = 1, \dots, N,$$

and

$$\bar{g}^{-1}[B]$$

are open sets in $\mathcal{B}_2^N$ and in $\mathcal{B}_1^N$ correspondingly. Then by (Parthasarathy, 1980, Assertion 22.4) the functions $\pi_k^{(l)}$, $k = 1, \dots, N$, and $\bar{g}$ are $(\mathcal{B}_2^{l+1}, \mathcal{B}_2^N)$-measurable and $(\mathcal{B}_1, \mathcal{B}_1^N)$-measurable correspondingly. In accordance with (Parthasarathy, 1980, Assertions 22.3 and 22.4, Exercise 22.9) and by (A3) the following statements are true: the maps

$$\pi_k^{(l)} \odot \theta \text{ and } f_k \odot \pi_k^{(l)} \odot \theta, \qquad k = 1, \dots, N,$$

are $(\mathcal{B}_2^{l+1}, S_2)$-measurable and $(\mathcal{B}_1, S_2)$-measurable correspondingly; the map $h \odot \theta$ is $(\mathcal{B}_1^N, S_2)$-measurable; the map $\bar{g} \odot h\theta$ is $(\mathcal{B}_1, S_2)$-measurable.
(4) Now prove (S1). Let

$$\Omega_1(a) = \left\{ (g_2^{(1)}, \dots, g_2^{(N)})\colon g_2^{(k)} \in G_2,\ k = 1, \dots, N;\ \sum_{k=1}^{N} g_2^{(k)} = a \in V \right\}.$$

By definition, a Hausdorff space possesses the separability axiom T_2, and, therefore, in this space the separability axiom T_1 holds (e. g., see (Morris, 1977, p. 9)). The latter means that in the above mentioned space every set consisting of one point only is closed one. Then by the continuity of g the preimage $g^{-1}[\{a\}]$ is a closed set under the product topology and thus

$$\Omega_1(a) \subseteq \mathcal{B}_2^N \text{ for all } a \in V.$$

For the probability space mentioned in Section 2 set

$$\Omega_1 = \Omega_1^N(a), \quad S_1 = \mathcal{B}_2^N \cap \Omega_1^N(a), \quad \mathbf{P}_1(\cdot) = \hat{\mathbf{Q}}(a, \cdot).$$

For every $a \in V$ the support of the measure $\hat{\mathbf{Q}}(a, \cdot)$ exists (see (Parthasarathy, 1967, Theorem 2.1, p. 27)) and is contained in $\Omega_1^N(a)$ (see (Parthasarathy, 1980, Assertion 46.3(i)). Let the maps

$$w_k = w_k(a)\colon\ \Omega_1^N(a) \to G_2; \qquad k = 1, \ldots, N, \tag{15}$$

be defined by the formula

$$w_k\colon\ (g_2^{(1)}, \ldots, g_2^{(N)}) \mapsto g_2^{(k)}; \qquad k = 1, \ldots, N.$$

We have

$$w_k^{-1}[B] = (\pi_k^{(0)})^{-1}[B] \cap \Omega_1^N(a) \subseteq S_1$$

for all $B \in \mathcal{B}_2$ and for all $k = 1, \ldots, N$. Thus, variables (15) are random and assumptions (D2) and (D3) hold for all $a \in V$.

(5) In order to complete the proof, the validity of (S2) must be verified. If the variable $\zeta = \zeta_a$ is defined by (1), where

$$w(a) = (w_1(a), \ldots, w_N(a))$$

is defined by (15), then

$$\zeta_a = f \odot w(a). \tag{16}$$

In accordance with steps (3) and (4), the function

$$f = \bar{g} \odot h$$

is $(\mathcal{B}_1, \mathcal{B}_2^N)$-measurable, and assumption $(D1)$ holds. By step (4) and Assertion 22.3 from (Parthasarathy, 1980), the function ζ_a is $(\mathcal{B}_1, S_1)$-measurable for all $a \in V$. Further, by (16)

$$\mathbf{P}_1\zeta_a^{-1}(A) = \hat{\mathbf{Q}}(a, f^{-1}[A]) \tag{17}$$

for all $A \in \mathcal{B}_1$ and for all $a \in V$. Let

$$a_0 \in V \quad \text{and} \quad \hat{\mathbf{Q}}(a, \cdot) \Rightarrow \hat{\mathbf{Q}}(a_0, \cdot)$$

for $a \to a_0$ and $a \in V$. Then

$$\mathbf{P}_1\zeta_a^{-1} \Rightarrow \mathbf{P}_1\zeta_{a_0}^{-1}$$

by (17), assumption (A4) and Theorem 5.1 from (Billingsley, 1968, Ch. 1). The validity of (S2) is established.

Remark 3. Put $l = 0$ in (1) and let

$$\xi_k = f_k(\theta_k), \qquad k = 1, \dots, N. \tag{18}$$

By (Sazonov and Tutubalin, 1966, pp. 7–8), in our case

$$\mathbf{P}_2(\xi, \eta)^{-1} = \mathbf{P}_2(\xi_1, \theta_1)^{-1} * \cdots * \mathbf{P}_2(\xi_N, \theta_N)^{-1},$$

where the right-hand side is the convolution of distributions of the random variables

$$(\xi_k, \theta_k), \qquad k = 1, \dots, N.$$

Then by a property of c. f. (see (Parthasarathy, 1967, p. 75))

$$\varphi_{\xi,\eta} = \prod_{k=1}^{N} \varphi_{\xi_k,\theta_k}. \tag{19}$$

In order to use the theory of summation of independent random variables, representation (19) (along with formula (11)) is of great importance.

5. DISCUSSION

By the Birkhoff–Kakutani theorem (e.g., see (Engelking, 1977, Exercise 8.1.G)), a topological group is metrizable if and only if it possesses the T_1-axiom. Consequently, a LCA-group is a metric space because this group possesses T_2-axiom (see step (4) above).

Now let us discuss assumption (A2). Let μ_k, $k = 1, \dots, N$, be arbitrary probability measures on σ-algebra $\mathcal{B}_2$. According to the Andersen–Jessen theorem (see (Loeve, 1960, Ch. 1, Section 4.2)) on the Cartesian product $(G_2^N, \mathcal{B}_2^N)$ of the Borel spaces $(G_2, \mathcal{B}_2)$, the product measure $\prod_{k=1}^{N} \mu_k$ exists. Let

$$\Omega_2 = G_2^N, \quad S_2 = \mathcal{B}_2^N,$$

and let θ_k, $k = 1, \dots, N$, be defined by the relations

$$\theta_k(\omega_2) = g_2^{(k)}, \quad \text{where} \quad \omega_2 = (g_2^{(1)}, \dots, g_2^{(N)}) \in \Omega_2, \quad k = 1, \dots, N.$$

Then the random variables $\theta_1, \dots, \theta_N$ are independent and, moreover, the variable θ_k has the distribution μ_k, $k = 1, \dots, N$. If G_2 is a separable metric space, then any measure has a support (see (Parthasarathy, 1967, p. 27, Theorem 2.1)) on $\mathcal{B}_2$. By steps (2) and (3), the space G_2^N is a complete separable metric space and the map η is a Borel one. Consequently, on the σ-algebra $\mathcal{B}_2^N$ (or on the σ-algebra $\mathcal{B}(G_2^N)$ what is the same) the regular conditional distribution of θ for given η exists (e. g., see (Parthasarathy, 1980, Assertion 46.3)). If the assumptions of the lemma mentioned above hold with $G_1 = G_2^N$, then there exists a variant (defined by formulas (10) and (11) with $\varphi = \varphi_{\theta,\eta}$) of this conditional distribution which varies continuously in the second argument (in the sense mentioned above). In this case, Γ_1 and Γ_2 are the dual groups to G_2^N and G_2 correspondingly.

For instance, (C3) holds if the point a is an atom of the distribution of η and the set $\{a\}$ is open. In particular, the latter holds if the space G_2 is a discrete topological one. It is very useful when the c. f. $\varphi = \varphi_{\theta,\eta}$ defined by (10) is a c. f. of a Gaussian distribution (in the sense of Parthasarathy, see (Parthasarathy, 1967, Ch. 4, Definition 6.1)). In particular, condition (C3) holds if $G_2 = \mathbf{R}^d$ or $G_2 = T^d$, $d > 0$, and the function Δ is the c. f. of a d-dimensional Gaussian distribution. In the general case, the c. f. of a nondegenerate Gaussian distribution may be not absolute integrable with respect to the Haar measure $\varkappa_2^*$ and it may be not absolutely continuous with respect to $\varkappa_2^*$ (see (Heyer, 1977, Ch. 5, Section 5; Parthasarathy, 1967, p. 101, Remark 2)).

6. COROLLARIES

Let us consider several special cases.

COROLLARY 1. *Let $\mathfrak{A}_2$ be a discrete topology and* (A1) *hold. Then* (S1) *and* (S2) *are true.*

Proof. In our case G_2 is an at most countable topological space; moreover, the space G_2 is the finite if and only if it is compact. Let $\{\omega_2(i);\ i = 1,2,\ldots\}$ be the space of all elementary events for G_2 and let non-negative integers $p_k(i)$, $i = 1,2,\ldots,\ k = 1,\ldots,N$, satisfy the conditions:

$$\sum_{i=1}^{\infty} p_k(i) = 1, \qquad k = 1,\ldots,N. \tag{20}$$

Introduce the probability measures μ_k which were mentioned in Section 5 by the formulas

$$\mu_k(\{\omega_2(i)\}) = p_k(i), \qquad i = 1,2,\ldots, \quad k = 1,\ldots,N.$$

By (20) for every $k = 1,\ldots,N$ the atom $\omega_2(i_k)$ of the distribution μ_k exists. Set

$$a = \sum_{k=1}^{N} \omega_2(i_k). \tag{21}$$

In accordance with Section 5 the independent random variables $\theta_1,\ldots,\theta_N$ exist and the distribution of η has the atom a. Further, the pre-image $f^{-1}[B]$ is an open set for every $B \in G_1$ because G_2^{l+1} is a discrete topological space. Thus, f is a continuous map and statements ($S1$) and ($S2$) hold.

COROLLARY 2. *Let for $j = 1,2$ the LCA-group G_j and the group $\mathbf{Z}^{b(j)}$ with some positive integer $b(j)$ be topologically isomorphic. Then the assertion of the theorem holds.*

The proof follows from the previous corollary since the discrete topological space $\mathbf{Z}^{b(j)}$ satisfies the second axiom of countability and, therefore, the space G_j is the same.

Remark 4. In order to prove the existence theorem for generalized scheme of allocation introduced in (Kolchin, 1968), it is sufficient to choose in (20) $p_1(i) = \cdots = p_N(i)$ for all $i = 1, 2, \ldots$ and to consider $\omega_2(i_k)$ in (21) as non-negative integers and then to use Corollary 2 with $b(1) = b(2) = 1$.
Now suppose that

$$G_2 = \mathbf{R}^{b(1)} \times T^{b(2)} \times \mathbf{Z}^{b(3)} \times F^{b(4)}, \tag{22}$$

where F is a finite (discrete) Abelian group, $b(j)$ are non-negative integers for $j = 1, 2, 3$, the number $b(4)$ equals 0 or 1 and $\sum_{i=1}^{4} b(i) \geq 1$. If $b(j) = 0$ in (22), then the corresponding group is omitted. Denote

$$x = (x_1, \ldots, x_{b(1)+b(2)})$$

with $x_j \in \mathbf{R}$, $j = 1, \ldots, b(1)$, and with $x_r \in \mathbf{Z}$, $r = b(1) + 1, \ldots, b(1) + b(2)$,

$$s = (s_1, \ldots, s_{b(1)}) \quad \text{with} \quad s_k \in \mathbf{R}, \qquad k = 1, \ldots, b(1),$$

$$t = (t_1, \ldots, t_{b(2)}) \quad \text{with} \quad t_m \in T, \qquad m = 1, \ldots, b(2).$$

It is well known (e. g., see (Morris, 1977, Ch. 3, Examples 2 and 3, and Theorem 13) and also (Zabell, 1979, Section 2)) that any continuous character $\gamma^{(x)}$ of the LCA-group $\mathbf{R}^{b(1)} \times T^{b(2)}$ is defined by

$$\gamma^{(x)}((s,t)) = \left(\prod_{k=1}^{b(2)} (t_k)^{x_{b(1)+k}} \right) \exp \left\{ i \sum_{j=1}^{b(1)} s_j x_j \right\}$$

for all $(s,t) \in \mathbf{R}^{b(1)} \times T^{b(2)}$, and, besides, a Haar measure on LCA-group $\mathbf{R}^{b(1)} \times \mathbf{Z}^{b(2)}$ dual to the preceding group is the product of $b(1)$-dimensional Lebesgue measure and the counting measure μ_c on $\mathbf{Z}^{b(2)}$ (i.e., $\mu_c(\{z\}) = 1$ for all $z \in \mathbf{Z}^{b(2)}$). If $\gamma_0 \in \Gamma_0$ is a continuous character on the LCA-group $\mathbf{Z}^{b(3)} \times F^{b(4)}$, then by (Morris, 1977, Theorem 13) any continuous character of LCA-group (22) is defined by $\gamma_2 = \gamma^{(x)}\gamma_0$ for some x and some γ_0. Let $\theta_1, \ldots, \theta_N$ be the random variables mentioned in Section 5 for case (22), and let $\varphi_{\theta,\eta}$ be the corresponding c. f. For some positive numbers δ_1 and δ_2 we denote

$$A_1 = \{\gamma_2\colon \ \|(x_1, \ldots, x_{b(1)})\| \geq \delta_1, \quad \|(x_{b(1)+1}, \ldots, x_{b(1)+b(2)})\| < \delta_2\},$$

$$A_2 = \{\gamma_2\colon \ \|(x_1, \ldots, x_{b(1)})\| < \delta_1, \quad \|(x_{b(1)+1}, \ldots, x_{b(1)+b(2)})\| \geq \delta_2\},$$

$$A_3 = \{\gamma_2\colon \ \|(x_1, \ldots, x_{b(1)})\| \geq \delta_1, \quad \|(x_{b(1)+1}, \ldots, x_{b(1)+b(2)})\| \geq \delta_2\},$$

$$A_4 = \{\gamma_2\colon \ \|(x_1, \ldots, x_{b(1)})\| < \delta_1, \quad \|(x_{b(1)+1}, \ldots, x_{b(1)+b(2)})\| < \delta_2\},$$

where $\|\cdot\|$ is the corresponding Euclidean norm. It is obvious that the sets A_1–A_4 are disjoint and

$$\Gamma_2 = \sum_{m=1}^{4} A_m. \tag{23}$$

COROLLARY 3. *Suppose that*

(a1) *the LCA-group G_1 satisfies* (A1) *and for the group G_2 representation* (22) *holds;*

(a2) *there exist independent random variables $\theta_1, \ldots, \theta_N$, independent of $\gamma_1 \in \Gamma_1$, numbers*

$$\alpha_m > b(m) > 0, \quad \delta_m > 0, \qquad m = 1, 2,$$

and non-negative Borel functions Δ_j defined on Γ_2 with

$$\sup_{\gamma \in \Gamma_2} \Delta_j(\gamma) \leq C_j < \infty, \qquad j = 1, 2, 3,$$

such that the inequalities

$$\|(x_1, \ldots, x_{b(1)})\|^{\alpha_1} |\varphi_{\theta,\eta}(\gamma_2)| \leq \Delta_1(\gamma_2) \quad \textit{for all } \gamma_2 \in A_1,$$

$$\|(x_{b(1)+1}, \ldots, x_{b(1)+b(2)})\|^{\alpha_2} |\varphi_{\theta,\eta}(\gamma_2) \leq \Delta_2(\gamma_2) \quad \textit{for all } \gamma_2 \in A_2,$$

$$\|(x_1, \ldots, x_{b(1)})\|^{\alpha_1} \|(x_{b(1)+1}, \ldots, x_{b(1)+b(2)})\|^{\alpha_2} \leq \Delta_3(\gamma_2)$$

hold for all $\gamma_2 \in A_3$; if $b_m = 0$ for some m, then the existence of the corresponding numbers α_m, δ_m and the function Δ_m is not required;

(a3) *conditions* (C2) (*with $Y = \eta$*), (A3), *and* (A4) *hold.*

Then (S1) *and* (S2) *are true.*

Proof. It is obvious that the LCA-group G_2 defined by (22) is a complete separable metric space. The LCA-group Γ_0 is a compact one since the topological space $\mathbf{Z}^{b(3)} \times F^{b(4)}$ is discrete (e. g., see (Morris, 1977, Theorem 12)); besides, the space Γ_2 possesses the second axiom of countability. Therefore, by (Parthasarathy, 1980, Assertion 54.2(i)) the Haar measure $\varkappa_0^*$ on Γ_0 satisfies the condition

$$\varkappa_0^*(A_4) \leq C < \infty. \tag{24}$$

Introduce the function Δ by the relations

$$\Delta(\gamma_2) = C_1 \|(x_1, \ldots, x_{b(1)})\|^{-\alpha_1} \quad \text{for all } \gamma_2 \in A_1,$$

$$\Delta(\gamma_2) = C_2 \|(x_{b(1)+1}, \ldots, x_{b(1)+b(2)})\|^{-\alpha_2} \quad \text{for all } \gamma_2 \in A_2,$$

$$\Delta(\gamma_2) = C_3 \|(x_1, \ldots, x_{b(1)})\|^{-\alpha_1} \|(x_{b(1)+1}, \ldots, x_{b(1)+b(2)})\|^{-\alpha_2} \quad \text{for all } \gamma_2 \in A_3,$$

$$\Delta(\gamma_2) = \varkappa_0^*(A_4) \quad \text{for all } \gamma_2 \in A_4.$$

The function Δ: $\Gamma_2 \to \mathbf{R}_+$ is a Borel function, because the sets A_1–A_4 belong to the σ-algebra $\mathcal{B}_2^*$ generated by the topology $\mathfrak{A}_2^*$. Assumption (a2) implies

$$|\varphi_{\theta,\eta}(\gamma(\gamma_1,\gamma_2))| \le \Delta(\gamma_2)$$

for all $\gamma_1 \in \Gamma_1$ and for all $\gamma_2 \in \Gamma_2$. Taking into account (23) and (24), we obtain that

$$\int_{\Gamma_2} \Delta(\gamma_2)\, d\varkappa_2^*(\gamma_2) = \sum_{j=1}^{4} \int_{A_j} \Delta((\gamma^{(x)}\gamma_0)) \left(\prod_{m=1}^{b(1)+b(2)} dx_m \right) d\varkappa_0^*(\gamma_0) < \infty,$$

Then by the lemma the version $\hat{\mathbf{Q}}$ exists and varies continuously (in the sense mentioned above) in the first argument on V. It is defined in the lemma with $X = \theta$ and with $Y = \eta$. The use of the theorem completes the proof.

Remark 5. The assumption on c. f. of such kind as in Corollary 3 has been considered earlier (e. g., see (Sadikova, 1966; Bikyalis, 1970; Chibisov, 1972)).

REFERENCES

Bikyalis, A. (1970). Inequalities for multivariate characteristic functions. *Lithuanian Math. J.* **10**, 5–12 (in Russian).

Billingsley, P. (1968). *Convergence of Probability Measures.* Wiley, New York.

Chibisov, D. M. (1972). On the normal approximation for a certain class of statistics. In: *Proceedings of the Sixth Berkeley Symposium on Mathematical Statisties and Probability,* **1**, Univ. California Press, Los Angeles, pp. 153–174.

Engelking, R. (1977). *General Topology.* Panstwowe Wydawnictwo Naukowe, Warszawa.

Heyer, H. (1977). *Probability Measures on Locally Compact Groups.* Springer, Berlin.

Holst, L. (1981 a). On numbers related to partition of unlike objects and occupancy problems. *European J. Combinatorics* **2**, 231–237.

Holst, L. (1981 b). Some asymptotic results for incomplete multinomial and Poisson samples. *Scandinavian J. Statist. Theory Appl.* **8**, 243–246.

Holst, L. (1981 c). Some conditional limit theorems in exponential families. *Ann. Probab.* **9**, 818–830.

Ivanov, V. A., Ivchenko, G. I., and Medvedev, Yu. I. (1984). Discrete problems in probability theory. *J. Soviet Math.* **31**,

Kelley, J. L. (1957). *General Topology.* Van Nostrand, Toronto.

Kolchin, V. F. (1968). A class of limit theorems for conditional distributions. *Lithuanian Math. J.* **8**, 53–63 (in Russian).

Kolchin, V. F. (1986). *Random Mappings.* Springer, New York.

Kolchin, V. F., Sevastyanov, B. A., and Chistyakov, V. P. (1978). *Random Allocations.* Wiley, New York.

Kudlaev, E. M. (1978). Asymptotic normality of some classes of goodness-of-fit statistics and homogeneity tests. *Soviet Math. Dokl.* **240**, 534–537 (in Russian).

Kudlaev, E. M. (1988 a). Decomposable statistics. *J. Soviet Math.* **34**.

Kudlaev, E. M. (1988 b). On statistical tests constructed by means of spacings and spacing-frequencies. In: *Probabilistic Problems in Discrete Mathematics*, MIEE, Moscow, pp. 88–91 (in Russian).

Kudlaev, E. M. (1989 a). Summation of some dependent random variables with values in locally compact Abelian groups. *Soviet Math.* **2**, 41–46 (in Russian).

Kudlaev, E. M. (1989 b). A classification of decomposable statistics. In: *Probab. Processes Appl.*, MIEE, Moscow, pp. 101–114 (in Russian).

Loeve, M. (1960). *Probability Theory*. Van Nostrand, Toronto.

Morris, S. A. (1977). *Pontryagin Duality and the Structure of Locally Compact Abelian Groups*. Cambridge Univ. Press, Cambidge.

Parthasarathy, K. R. (1967). *Probability Measures on Metric Spaces*. Academic Press, New York.

Parthasarathy, K. R. (1980). *Introduction to Probability and Measure*. Indian Statist. Inst. Press, New Delhi.

Sadikova, S. M. (1966). Certain inequalities for characteristic functions. *Theory Probab. Appl.* **11**, 500–506.

Sazonov, V. V., and Tutubalin, V. N. (1966). Probability distributions on topological groups. *Theory Probab. Appl.* **11**, 3–56.

Zabell, S. (1979). Continuous versions of regular conditional distributions. *Ann. Probab.* **7**, 159–165.

Probabilistic Methods in Discrete Mathematics, pp. 287 - 297
V. F. Kolchin et al. (Eds.)
1993 TVP/VSP

THE PROPERTY OF PHASE TRANSITIONS IN RANDOM GRAPHS

A. V. LAPSHIN
TVP Laboratories, Krasikov Street 32, 117418 Moscow, Russia

ABSTRACT

This paper continues investigations of properties of phase transitions in random graphs. For integer values of the parameter that characterizes the statistical sum of the small canonical ensemble, the thermodynamic potential is calculated, and the relation, which can be considered as an analogue of the Van Hove theorem well-known in statistical physics, is also ascertained.

1. INTRODUCTION

A relationship of combinatorial analysis and statistical physics is based on an analogue between the generating functions of some characteristics of combinatorial objects and the statistical sums of corresponding physical systems. This analogue was used by many researchers including (Minlos, 1968; Baxter, 1982; Viennot, 1985; Stepanov, 1970, 1972). In investigations of some properties of random graphs by Stepanov the analogue of a class of random graphs and the large canonical ensemble of particles in statistical physics was used. Particularly, in his papers analogues of phase transition properties for characteristics of random graph such as its density and disconnectedness were established. In the present paper the other well-known in combinatorial analysis class of random graphs is studied as an analogue of the small canonical ensemble of particles. An expression of the thermodynamic potential is obtained for integer values of the parameter, which defines the statistical sum of the small canonical ensemble, and a correspondence between the thermodynamic potentials of the studied random graphs is found. One can consider this correspondence as an analogue of the well-known in statistical physics theorem due to Van Hove.

2. MAIN DEFINITIONS AND RESULTS

Let us remind some notions and results of (Stepanov, 1970), which will be used in the present investigation.
The graph $\mathcal{G}_m(t)$ under consideration is a subgraph of the complete graph G_m with m vertices; it is obtained by a random exclusion of edges in the graph G_m such that the probability of exclusion of each edge is equal to $q = e^{-t}$. The probability that a specific edge is contained in the graph $\mathcal{G}_m(t)$ is $p = 1 - e^{-t}$, the exclusion of some

edge is realized independently of other edges. Let $p_m(t)$ be the probability that the random graph $\mathcal{G}_m(t)$ is connected and let $p_m(k \mid t)$ be the probability that the random graph $\mathcal{G}_m(t)$ contains exactly k connected components. Let $P_m(x \mid t)$ be the generating function of the distribution of the number of connected components in the random graph $\mathcal{G}_m(t)$:

$$P_m(x \mid t) = \sum_{k=1}^{\infty} x^k p_m(k \mid t).$$

Let $\kappa(G)$ and $\nu(G)$ be accordingly the number of connected components and the number of edges of a subgraph G of the complete graph G_m. Let x, h be arbitrary non-negative numbers. We assume that each subgraph G has the statistical weight $w(G) = x^{\kappa(G)} h^{\nu(G)}$, and the random graph $\mathcal{G}_m(x \mid h)$ is defined by the following condition: the probability that the random graph $\mathcal{G}_m(x \mid h)$ coincides with an arbitrary subgraph G is proportional to the weight of this subgraph:

$$\mathbf{P}\{\mathcal{G}_m(x \mid h) = G\} = \frac{w(G)}{Z_m(x \mid h)} = \frac{x^{\kappa(G)} h^{\nu(G)}}{Z_m(x \mid h)}.$$

The factor of proportionality

$$Z_m(x \mid h) = \sum_{\{G\}} w(G) = \sum_{\{G\}} x^{\kappa(G)} h^{\nu(G)}$$

is called the large statistical sum. Using the large statistical sum one can express the characteristics of the random graph $\mathcal{G}_m(x \mid h)$ such as its density $\rho_m(x \mid h)$ and its disconnectedness $\varkappa_m(x \mid h)$:

$$\rho_m(x \mid h) = \frac{1}{m} \mathbf{E}\nu(\mathcal{G}_m(x \mid h)) = h \frac{\partial}{\partial h} \frac{\log Z_m(x \mid h)}{m},$$
$$\varkappa_m(x \mid h) = \frac{1}{m} \mathbf{E}\kappa(\mathcal{G}_m(x \mid h)) = x \frac{\partial}{\partial x} \frac{\log Z_m(x \mid h)}{m}.$$

Put $q = 1 - p = e^{-t}$, $h = e^t - 1$, then

$$Z_m(x \mid h) = \sum_{\{G\}} x^{\kappa(G)} h^{\nu(G)} = e^{\binom{m}{2}t} \sum_{k=1}^{m} x^k \sum_{\{G\}_k} p^{\nu(G)} q^{\binom{m}{2} - \nu(G)}.$$

The inner sum is taken over those subgraphs of the complete graph G_m that consist of exactly k connected components. Using the equality

$$\mathbf{P}\{\mathcal{G}_m(t) = G\} = p^{\nu(G)} q^{\binom{m}{2} - \nu(G)}$$

we obtain that the inner sum is equal to the probability $p_m(k \mid t)$ that the graph $\mathcal{G}_m(t)$ consists of exactly k components. Therefore

$$Z_m(x \mid e^t - 1) = e^{\binom{m}{2}t} P_m(x \mid t).$$

Stepanov (Stepanov, 1970) proved that under the conditions $t = y/m$, $m \to \infty$ and $mh \to y$ the limit

$$\lim_{\substack{m\to\infty \\ mh\to y}} \frac{\log Z_m(x \mid h)}{m} = \chi(x,y)$$

exists, this limit is called the thermodynamic potential of the random graph $\mathcal{G}_m(x \mid h)$. In (Stepanov, 1970) it was obtained also that

$$\chi(x,y) = y/2 + G(x,y),$$

where

$$G(x,y) = \lim_{m\to\infty} P_m(x \mid y/m)/m,$$

$$G(x,y) = \log x - \frac{y}{2} + \frac{y}{2x}(1-\alpha)^2 - \frac{\alpha^2 y}{2} - \log(1-\alpha),$$

and $\alpha = \alpha(x,y)$ is the largest root of the equation

$$(1-\alpha)e^{\alpha y} - \frac{1}{\alpha} = x. \tag{1}$$

It was also found that

$$\rho(x,y) = y\frac{\partial\chi(x,y)}{\partial y} = \frac{y}{2x} + \frac{\alpha^2 y}{2}\left(1 - \frac{1}{x}\right). \tag{2}$$

In (Stepanov, 1970, 1972) the phenomena of phase transitions of the density and the disconnectedness and also the structure of the random graph $\mathcal{G}_m(x \mid h)$ were investigated asymptotically.

According to the definition of the random graph $\mathcal{G}_m(x \mid h)$ one can define the random graph $\mathcal{G}_{m,n}(x)$ that corresponds to the small canonical ensemble of particles in statistical physics, using the equality

$$\mathbf{P}\{\mathcal{G}_{m,n}(x) = G\} = \frac{x^{\kappa(G)}}{Z_{m,n}(x)},$$

where

$$Z_{m,n}(x) = \sum_{\{G\}} x^{\kappa(G)} = \binom{\binom{m}{2}}{n} \sum_{k=1}^{m} p_{m,n}(k)x^k$$

is the small statistical sum,

$$P_{m,n}(x) = \sum_k x^k p_{m,n}(k)$$

is the generating function of the distribution of the number of connected components in the random graph $\mathcal{G}_{m,n}(1) = \mathcal{G}_{m,n}$, and $p_{m,n}(k)$ is the probability that the random graph $\mathcal{G}_{m,n}$ consists of exactly k connected components.

Let us remind that the random graph $\mathcal{G}_{m,n}$ can be defined in the following way (see, for example, (Kolchin, 1986)): n arbitrary edges are selected from the set of $\binom{m}{2}$ edges of the complete graph G_m with equal probabilities.

THEOREM 2.1. *For $x = M$, where M is an arbitrary natural number,*

$$P_{m,n}(M) = \sum_{s_1+\cdots+s_M=m} \frac{m!}{s_1!\cdots s_M!} \frac{\sum_{i_n} \binom{s_i}{2}}{\binom{\binom{m}{2}}{n}}.$$

COROLLARY 2.1. *The generating function of the distribution of the number of connected components in the random graph $\mathcal{G}_{m,n}$ satisfies the relation*

$$P_{m,n}(x) = \frac{x^{m-n}}{n!} \prod_{j=1}^{n+1} (j-x) \sum_{i=0}^{n} \frac{(-1)^i \binom{n}{i} (i+1)^{n-m}}{(i+1-x)} P_{m,n}(i+1),$$

where $(j-x)/(j-x)$ is equal to 1 for $x = j$.

By analogy with the thermodynamic potential of the random graph $\mathcal{G}_m(x \mid h)$, the limit value

$$\lim_{\substack{m\to\infty \\ n/m\to\rho}} \frac{1}{m} \log \frac{Z_{n,m}(x)}{m^n},$$

where ρ is a constant, is called the thermodynamic potential of the random graph $\mathcal{G}_{m,n}(x)$ and is denoted by $g(x,\rho)$. While analyzing the states of the physical systems represented by descriptions of the small and large canonical ensembles, it is conventional to use the identity $\rho(x,y) = \rho$ (see, for example, (Minlos, 1968)), where the function $\rho(x,y)$ is defined by (2). We will proceed in the same manner while comparing values of the thermodynamic potentials of the random graphs $\mathcal{G}_m(x \mid h)$ and $\mathcal{G}_{m,n}(x)$.

THEOREM 2.2. *For $x = M$, where M is an arbitrary natural number, the equality*

$$\chi(x,y) = g(x,y) + \rho \log y$$

holds, where the variables ρ and y satisfy the relation $\rho(x,y) = \rho$.

This equality relating the thermodynamic potentials $\chi(x,y)$ and $g(x,y)$ of the random graphs $\mathcal{G}_m(x \mid h)$ and $\mathcal{G}_{m,n}(x)$ one can consider as the analogue of the Van Hove theorem well-known in statistical physics (see (Minlos, 1968)). The author of the present paper succeeded in proving this equality only for natural values of x.

3. PROOFS

Proof of Theorem 2.1. As well as in other problems of statistical physics (see (Viennot, 1985)) we assume that the variable σ_l corresponds to the vertex l of the graph $\mathcal{G}_{m,n}$ and it can take each of M possible values with equal probabilities, $\sigma_l = 1,\ldots,M$, $l = 1,\ldots,m$. The random variables σ_l, $l = 1,\ldots,m$, are independent.

Let the variables of all m vertices of the graph $\mathcal{G}_{m,n}$ take some values, and let the edges of the graph $\mathcal{G}_{m,n}$ be chosen as results of a random trial. A configuration of the edges is called a matching if each edge connects the vertices with the same value of the variables. In this case the value

$$P_{m,n}(M) = \sum_{k=1}^{m} p_{m,n}(k)M^k$$

is equal to the expectation of the number of matchings because the values of all variables from a connected component of the graph $\mathcal{G}_{m,n}$ must be the same and take one of the M possible values. Let, as a result of the random trial, exactly s_i variables among $\sigma_1, \dots, \sigma_m$ take value i, $i = 1, \dots, M$. In this case the probability that the configuration of the edges of the graph $\mathcal{G}_{m,n}$ will be a matching is equal to

$$\binom{\sum_i \binom{s_i}{2}}{n}\binom{\binom{m}{2}}{n}^{-1}.$$

Now we take into account that the number of sets of variables σ_l, $l = 1, \dots, m$, which consist of s_i variables with the same value i, $i = 1, \dots, M$, is $m!/(s_1! \cdots s_M!)$. Thus we obtain the assertion of Theorem 2.1.

Since the number of independent cycles c and the number of connected components κ in an arbitrary graph satisfy the identity $\kappa - c = m - n$ (the Euler theorem), the generating function $C_{m,n}(x)$ of the distribution of independent cycles in the random graph $\mathcal{G}_{m,n}$ can be expressed in terms of the generating function $P_{m,n}(x)$ of the distribution of connected components in such a way:

$$C_{m,n}(x) = x^{n-m} P_{m,n}(x).$$

The function $C_{m,n}(x)$ is a polynomial of degree not greater than n and the values $C_{m,n}(x)$ for natural values of x are known, therefore one can uniquely restore the function $C_{m,n}(x)$. For this purpose one can use, for example, the interpolation polynomial in the Lagrange's form for the equidistant nodes. From the statement of interpolation problem it follows that the function $C_{m,n}(x)$, and hence the function $P_{m,n}(x)$, is uniquely restored and the solution does not depend on the choice of the interpolation nodes.

Further, we will calculate the asymptotic value of $P_{m,n}(x)$ for $x = M$, as $m, n \to \infty$, $n/m \to \rho$, ρ is a constant, and this will give us a possibility to find the value of the thermodynamic potential $g(x, \rho)$ for natural values x.

THEOREM 3.1. *If $m, n \to \infty$, $n/m \to \rho$, where ρ is a constant, then for any fixed natural M the value of $P_{m,n}(M)$ is equivalent to*

$$M^{m-n}(1-2\rho)^{-(M-1)/2} \exp\{-\rho(1+\rho)(M-1)\} \quad \textit{for} \quad \rho < \rho_c, \tag{3}$$

$$M^{m-n+1}\left(\frac{2\rho}{\sigma}\right)^{1/2} \exp\left\{-\rho(1+\rho)\left(\frac{1}{\sigma}-1\right)\right\}$$
$$\times \frac{\exp\left\{m\left((\rho-1)\log \sigma M + \log 2\rho - \frac{\sigma}{2\rho}u\log u - \frac{\sigma}{2\rho}\tau(M-1)\log\tau\right)\right\}}{\sqrt{u\tau(1-\tau)^{M-2}(((M-1)\tau+u)/u\tau - M + \sigma(M-1)(u-\tau)^2/2\rho^2)}} \quad \textit{for } \rho > \rho_c, \tag{4}$$

where the parameter σ is defined from the equation $\rho(\sigma) = \rho$,

$$\rho(\sigma) = -\frac{\sigma}{2}\sqrt{\frac{M-1}{\sigma M-1}}\log\left(1 - \frac{M\sqrt{M-1}}{\sqrt{M-1}+(M-1)\sqrt{\sigma M-1}}\right), \tag{5}$$

$$u = \frac{2\rho}{\sigma M}\left(1+\sqrt{(M-1)(\sigma M-1)}\right), \quad \tau = \frac{2\rho}{\sigma M}\left(1-\sqrt{\frac{\sigma M-1}{M-1}}\right), \tag{6}$$

$\rho_c = \rho_c(\sigma_c)$, σ_c is the root of the equation

$$\log\left(\frac{1-\sqrt{(\sigma M-1)/(M-1)}}{1+\sqrt{(M-1)(\sigma M-1)}}\right) \times \left(\frac{1+\sqrt{(M-1)(\sigma M-1)}}{M} - \frac{\sigma}{2}\sqrt{\frac{M-1}{\sigma M-1}}\log\sigma M\right) = \log\left(1-\sqrt{\frac{\sigma M-1}{M-1}}\right). \tag{7}$$

Proof. Let us present the value $P_{m,n}(M)$ as the sum $P_{m,n}(M) = I_0 + I_1 + \cdots + I_{M-1}$. The sum I_0 consists of the terms that correspond to the indices $s_1, \ldots, s_M$ such that $s_i > \omega_m$, $i = 1, \ldots, M$. The sum I_1 consists of the terms that correspond to the indices such that all $s_i > \omega_m$, $i = 1, \ldots, M$, except one index, etc. Here $\omega_m \to \infty$, $\omega_m/\sqrt{m} \to 0$ as $m \to \infty$. Using the Stirling's formula

$$m! = \sqrt{2\pi m}\, m^m e^{-m+\theta(m)}, \qquad \theta(m) = \theta_m/(12m), \quad 0 < \theta_m < 1,$$

for the indices $s_1, \ldots, s_M$ in the sum I_0, one can obtain that

$$\frac{m!}{s_1! \cdots s_M!} = \frac{1}{(\sqrt{2\pi m})^{M-1}\sqrt{z_1 \cdots z_m}} \exp\left\{m\left(-\sum_i z_i \log z_i\right)\right\} \times \exp\left\{\frac{1}{12m}\left(\theta_m - \frac{\theta_{s_1}}{z_1} - \cdots - \frac{\theta_{s_M}}{z_M}\right)\right\},$$

where $z_i = s_i/m$, $i = 1, \ldots, M$. That is why one can obtain the following estimates for the sum I_0:

$$\sum_{\substack{z_1,\ldots,z_M>0\\ z_1+\cdots+z_M=1}} \frac{1}{(\sqrt{2\pi m})^{M-1}\sqrt{z_1 \cdots z_m}} \exp\left\{\frac{n}{m-1}\left(1+\frac{n-1}{m-1}\right)\frac{\sum_i z_i^2 - 1}{\sum_i z_i^2}\right\} \times \exp\left\{m\left(-\sum_i z_i \log z_i + \rho\log\left(\sum_i z_i^2\right)\right)\right\} \times \exp\left\{-\frac{1}{12m}\left(\frac{1}{z_1} + \cdots + \frac{1}{z_M}\right)\right\}(1+o(1)) \le I_0 \le I_0' \exp\left\{\frac{1}{12m}\right\}(1+o(1)), \tag{8}$$

where

$$I_0' = \sum_{\substack{z_1,\dots,z_M>0 \\ z_1+\dots+z_M=1}} \frac{1}{(\sqrt{2\pi m})^{M-1}\sqrt{z_1\cdots z_m}} \exp\left\{-\rho(1+\rho)\left(\frac{1}{\sum_i z_i^2}-1\right)\right\}$$
$$\times \exp\left\{m\left(-\sum_i z_i\log z_i + \rho\log\left(\sum_i z_i^2\right)\right)\right\}.$$

Let us find the asymptotics of I_0. Using inequality (8), we calculate first the asymptotics of the sum I_0'. The terms that correspond to the vectors $(z_1,\dots,z_M)$ from the neighbourhood of the absolute maximum of the function $\rho f - g$ give the main contribution to the sum I_0', where

$$f = f(z_1,\dots,z_M) = \log\sum_i z_i^2,$$

$$g = g(z_1,\dots,z_M) = \sum_i z_i\log z_i, \qquad z_i \ge 0, \quad z_1+\dots+z_M = 1.$$

Let us set

$$\Phi = \rho f - g + \lambda\left(\sum_i z_i - 1\right),$$

where λ is a constant. The extremum points of the function $\rho f - g$ are the solutions of the following system of equations:

$$\begin{cases} \dfrac{\partial\Phi}{\partial z_j} = \rho\dfrac{2z_j}{\sum_i z_i^2} - \log z_j - 1 + \lambda = 0, \\ \sum_i z_i = 1, \qquad j = 1,\dots,M. \end{cases} \tag{9}$$

Put

$$\sum_i z_i^2 = \sigma, \quad w_j = z_j 2\rho/\sigma, \quad b = e^{\lambda-1}2\rho/\sigma,$$

then system (9) can be written in the form

$$\begin{cases} w_j e^{-w_j} = b, \qquad j = 1,\dots,M, \\ \sum_j w_j = 2\rho/\sigma, \quad \sum_j w_j^2 = 4\rho^2/\sigma. \end{cases} \tag{10}$$

It is known (see (Markushevich, 1967)) that each of the first M equations of system (10) has two roots, if $0 < b < e^{-1}$. These roots we denote by u and $\tau(u)$, $u \ge 1$, $\tau(u) \ge 1$, $\tau(u)$ is an analytic function of u; it is defined for $u \ge 1$ and decreases monotonically from 1 to 0, $d\tau/du\big|_{1+} = -1$. That is why the unknown w_j of system (10) must be equal to u or to $\tau(u)$, the values of u and $\tau(u)$ themselves are defined by b. That is, if the solutions of system (10) exist they must be of the following form:

$$(\underbrace{u,\dots,u}_{k},\underbrace{\tau(u),\dots,\tau(u)}_{M-k}), \qquad 0 \le k \le M. \tag{11}$$

Thus, the solutions of system (10) are the vectors of form (11), which lie in the intersection of the hyperplane $\sum_j w_j = 2\rho/\sigma$ and the hypersphere $\sum_j w_j^2 = 4\rho^2/\sigma$, where $k = 0, \dots, M$, and all the vectors with $M!/(k!(M-k)!)$ possible permutations of these coordinates.

Let us determine solutions of system (9) that correspond to the maximum point of the function $\rho f - g$.

LEMMA. *For $\rho < 1/2$ the point z^0 with coordinates $z_1^0 = \dots = z_M^0 = M^{-1}$ is the maximum point of the function $\rho f - g$. For $\rho > \rho_1$ the maximum points of the function $\rho f - g$ are M points z^i with coordinates*

$$z_i^i = u_1\sigma_1/(2\rho), \quad z_j^i = \tau(u_1)\sigma_1/(2\rho), \qquad j \neq i, \quad i = 1, \dots, M,$$

where $\rho_1 = \rho(\sigma_1^*)$, σ_1^* is the root of the equation $d\rho/d\sigma = 0$, and $\rho(\sigma)$ as a function of parameter σ is defined by formula (5), the value σ_1 is the root of the equation $\rho = \rho(\sigma)$, the values u_1 and $\tau(u_1)$ can be calculated from the value σ_1 by formulas (6).

The proof of Lemma follows from the analysis of the function $\rho f - g$ for different values of the parameter ρ.

The values of the function $\rho f - g$ at points z^0 and z^1 are equal to

$$\begin{aligned} (\rho f - g)(z^0) &= (1-\rho)\log M, \\ (\rho f - g)(z^1) &= (\rho - 1)\log\sigma + \log 2\rho - \frac{\sigma}{2\rho} u \log u - \frac{\sigma}{2\rho}\tau(M-1)\log\tau, \end{aligned} \tag{12}$$

correspondingly, where $u = u_1$, $\sigma = \sigma_1$, $\tau = \tau(u_1)$. The function $\rho f - g$ has the same value at the points z^i, $i = 1, \dots, M$, because it is a symmetric function of its arguments.

Let us define ρ_c by the equation

$$(\rho_c f - g)(z^0) = (\rho_c f - g)(z^1). \tag{13}$$

The value ρ_c can be calculated in the following way. Using formulas (6) for u_1 and $\tau(u_1)$ and formula (5) for ρ, one can transform equation (13) to form (7). It can be proved that the value ρ_c is unique.

Let us estimate the asymptotics of I_0', defined by formula (8), as $m \to \infty$, $n/m \to \rho$, where ρ is a constant. The formula for I_0' can be written in the following form

$$I_0' = \left(\frac{m}{2\pi}\right)^{(M-1)/2} \sum_{\substack{z_1, \dots, z_M > 0 \\ z_1 + \dots + z_M = 1}} L(z_1, \dots, z_M) \exp\{mS(z_1, \dots, z_M)\} m^{-(M-1)},$$

where

$$S(z) = S(z_1, \dots, z_M) = \rho \log \sum_i z_i^2 - \sum_i z_i \log z_i,$$

$$L(z) = L(z_1, \dots, z_M) = (z_1 \dots z_M)^{-1/2} \exp\left\{ -\rho(1+\rho)\left(\left(\sum_i z_i^2\right)^{-1} - 1\right)\right\}.$$

We consider the functions $S(z)$ and $L(z)$ as functions of continuous argument z with the domain of definition

$$D_z = \{z\colon\ z_1 + \cdots + z_M = 1,\ z_i \geq 0,\ i = 1, \ldots, M\}.$$

For calculating the asymptotic value of the sum I_0' we use the relation

$$I_0' \sim \left(\frac{m}{2\pi}\right)^{(M-1)/2} \int_{D_z} L(z) \exp\{mS(z)\}\, dz. \tag{14}$$

According to the Laplace method of asymptotical estimation of integrals containing a large parameter (see Theorem 4.1 of (Fedoryuk, 1977)), the asymptotic value $I(m)$ of integral (14) as $m \to \infty$ is determined by the contribution of a neighbourhood of the maximum point of the function $S(z)$.
Using formula (12) and the equality

$$|\det S''_{zz}|\,\big|_{z_0} = (1 - 2\rho)^{M-1} M^M,$$

one can obtain the asymptotics of $I(m)$ given by formula (3). Here S''_{zz} denotes the matrix

$$\left(\frac{\partial^2 S(z)}{\partial z_i \partial z_j}\right), \qquad i, j = 1, \ldots, M - 1.$$

Using formula (12) and the relation

$$|\det S''_{zz}|\,\big|_{z'} = \left(\frac{2\rho}{\sigma}\right)^{M-1} \left(\frac{1}{\tau} - 1\right)^{M-2}$$
$$\times \left(\frac{u + (M-1)\tau}{u\tau} - M + (M-1)\frac{\sigma}{2\rho^2}(u - \tau)^2\right),$$

one can obtain the value $I(m)$ for $\rho > \rho_c$, given by formula (4).
According to Lemma 1.2 in Chapter II of (Fedoryuk, 1977) and to the relation $\omega_m = o(m^{1/2})$ one can obtain that $I_0' \sim I(m)$. Further, the upper estimate in (8) is equivalent to $I(m)(1 + o(1))$, and the upper and lower estimates in (8) are equivalent to each other, thus $I_0 \sim I(m)$. In the same way one can obtain the asymptotics of the sums I_k, $k = 1, \ldots, M - 1$, and prove the relations $I_{k+1} = o(I_k)$, and further, prove the relation $P_{m,n}(M) \sim I_0$. This completes the proof of Theorem 3.1.
Now let us use the result of Theorem 3.1 to prove Theorem 2.2. From the definition of the small statistical sum it follows that

$$\frac{1}{m} \log \frac{1}{m^n} Z_{m,n}(x) = \frac{1}{m} \log P_{m,n}(x) + \frac{1}{m} \log \frac{1}{m^n} \binom{\binom{m}{2}}{n}.$$

According to Theorem 3.1 one can find that, for $x = M$,

$$\lim_{\substack{m \to \infty \\ n/m \to \rho}} \frac{1}{m} \log \frac{Z_{n,m}(x)}{m^n} = g(x, \rho),$$

where $g(x,\rho) = p(x,\rho) + \rho - \rho \log 2\rho$ and the function $p(x,\rho)$ is defined by formula (12).

As in statistical physics we will proceed from the relation $\rho(x,y) = \rho$ for the densities of graphs $\mathcal{G}_m(x \mid h)$ and $\mathcal{G}_{m,n}(x)$. Using formula (1), we express y as a function of α and obtain that

$$y = \frac{1}{\alpha} \log \frac{1+\alpha(x-1)}{1-\alpha}. \tag{15}$$

Using (15) and (2), we can write formula (2) for the density $\rho(x,y)$ of the graph $\mathcal{G}_m(x \mid h)$ in the form

$$\rho(x,y) = \rho(\alpha) = \frac{1}{2x} \frac{1+\alpha^2(x-1)}{\alpha} \log \frac{1+\alpha(x-1)}{1-\alpha}. \tag{16}$$

Formula (6) can be written as

$$u\frac{\sigma}{2\rho} = \frac{1+\beta(M-1)}{M}, \quad \tau\frac{\sigma}{2\rho} = \frac{1-\beta}{M}, \tag{17}$$

where $\beta = \sqrt{(\sigma M - 1)/(M-1)}$. Since $\beta \in (0,1)$, the variable σ as a function of β has the following expression

$$\sigma = \frac{1+\beta^2(M-1)}{M}. \tag{18}$$

Using formula (5) we represent ρ as

$$\rho = \rho(\beta) = \frac{1}{2M} \frac{1+\beta^2(M-1)}{\beta} \log \frac{1+\beta(M-1)}{1-\beta}. \tag{19}$$

According to Lemma formula (17) corresponds to the domain of definition $\beta \in [\beta_1, 1)$, where $\beta_1 = \sqrt{(\sigma_1 M - 1)/(M-1)}$. In the interval $[\beta_1, 1)$ the function $\rho(\beta)$ increases monotonically from ρ_1 to infinity. Comparing formulas (16) and (19) and using the fact that the function $\rho^{-1}(\beta)$ is single-valued for $\beta \in [\beta_1, 1)$, the function $\rho(\alpha)$ is defined for $\alpha \in (0,1)$, and $\alpha = \alpha(x,y)$ is the largest root of equation (1), we obtain that the variables β and α coincide on the interval $[\beta_1, 1)$, i. e., $\beta = \alpha(x,y)$ for $x = M$. Therefore, formulas (17), (18) can be written in the following form

$$u\frac{\sigma}{2\rho} = \frac{1+\alpha(M-1)}{M}, \quad \tau\frac{\sigma}{2\rho} = \frac{1-\alpha}{M}, \quad \sigma = \frac{1+\alpha^2(M-1)}{M}.$$

Using the expression for $\rho(x,y)$ and (18), we obtain a formula for y as a function of ρ and σ, namely, $y = 2\rho/\sigma$. As a result we obtain the following formula for $u = u_1$, $\tau = \tau(u_1)$ as functions of α and y, which are characteristics of the graph $\mathcal{G}_m(x \mid h)$:

$$\tau = \frac{\alpha y}{e^{\alpha y} - 1}, \quad u = \alpha y \frac{1+\alpha(M-1)}{(1-\alpha)(e^{\alpha y} - 1)}.$$

The value of the function $\rho f - g$ at the point z^1 (formula (12)) can be also expressed in terms of the value $\alpha = \alpha(M, y)$:

$$\begin{aligned} p(M,\rho) &= \max(\rho f - g) \\ &= \rho \log(1 + \alpha^2(M-1)) - \frac{1 + \alpha(M-1)}{M} \log\left(1 + \alpha(M-1)\right) \\ &\quad + \frac{1-\alpha}{M}(M-1)\log(1-\alpha) + (1-\rho)\log M. \end{aligned}$$

This expression for the value $p(M, \rho)$ concurs with the statement of Theorem 3.1, since $\alpha = 0$ for $\rho < \rho_c$ and $\max(\rho f - g) = (\rho f - g)(z^0) = (1-\rho)\log M$ for $\rho > \rho_c$. Using the expression for $p(M, \rho)$ and the relation $\rho(x, y) = \rho$ for fixed x one can obtain the following formula for the thermodynamic potential of the small canonical ensemble:

$$g(x,y) = \log x + \frac{y}{2x}(1-\alpha)^2 - \frac{\alpha^2 y}{2} - \log(1-\alpha) - \rho \log y,$$

where $x = M$. Comparing this expression for $g(x, y)$ and the expression for the thermodynamic potential $\chi(x, y)$ of the graph $\mathcal{G}_m(x \mid h)$, we obtain the assertion of Theorem 2.3.

REFERENCES

Baxter, R. J. (1982). *Exactly Solved Models in Statistical Mechanics.* Academic Press, New York.

Fedoryuk, M. V. (1977). *The Saddle-Point Method.* Nauka, Moscow (in Russian).

Kolchin, V. F. (1986). On the behaviour of a random graph near a critical point. *Theory Probab. Appl.* **31**, 503–515.

Markushevich, A. I. (1967). *Theory of Analytic Functions*, Vol. I. Nauka, Moscow (in Russian).

Minlos, R. A. (1968). Lectures on statistical physics. *Russian Math. Surveys* **23**, 133–190.

Stepanov, V. E. (1970). Phase transitions in random graphs. *Theory Probab. Appl.* **15**, 203–215.

Stepanov, V. E. (1972). Structure of random graphs $\mathcal{G}_m(x|h)$. *Theory Probab. Appl.* **17**, 238–252.

Viennot, G. (1985). Problemes combinatoires poses par la physique statistique. *Seminaire N. Bourbaki* **121–122**, 225–246.

Probabilistic Methods in Discrete Mathematics, pp. 298 – 304
V. F. Kolchin et al. (Eds.)
1993 TVP/VSP

ON SUMS OF VALUES OF THE LEGENDRE SYMBOL OVER COMPONENTS IN A SCHEME OF GENERATION OF RANDOM VECTORS

Yu. I. MAKSIMOV
TVP Laboratories, Krasikov Street 42, 117418 Moscow, Russia

ABSTRACT

A local normal theorem is proved for sums of the Legendre symbol's values over dependent components of a random vector in the group analogue of the classical occupancy problem. The Markov character of the sums is also determined.

1. INTRODUCTION

Let q be a prime exceeding two, R_+ and R_- be the sets of quadratic residues and nonresidues modulo q, and let G be an additive group of residues modulo q. Let $y_1, y_2, \dots$ be a sequence of independent identically distributed random variables on G with the distribution p of the following type: if R_+ or R_- are chosen with probabilities α_+ and α_- ($\alpha_+ + \alpha_- = 1$), then every element $g \in R_+$ (or R_-) is chosen with equal probabilities. In other words,

$$p(g) = \begin{cases} \dfrac{2\alpha_+}{q-1}, & g \in R_+, \\ \dfrac{2\alpha_-}{q-1}, & g \in R_-, \\ 0, & g = 0. \end{cases} \tag{1}$$

The characteristic function $\hat{p}(\gamma)$ has the form

$$\hat{p}(\gamma) = \sum_{g \in R_+} \frac{2\alpha_+}{q-1} e^{2\pi i g\gamma/q} + \sum_{g \in R_-} \frac{2\alpha_-}{q-1} e^{2\pi i g\gamma/q} = \begin{cases} -\dfrac{1}{q-1} + \Delta\varepsilon, & \gamma \in R_+, \\ -\dfrac{1}{q-1} - \Delta\varepsilon, & \gamma \in R_-, \end{cases} \tag{2}$$

where $\gamma \in \Gamma$, Γ is the character group of G, $\gamma \neq 0$ (evidently, $\hat{p}(0) = 1$), $\varepsilon = 1$ for $q \equiv 1 \pmod 4$, $\varepsilon = i$ for $q \equiv 3 \pmod 4$, $\Delta = (\alpha_+ - \alpha_-)q^{1/2}/(q-1)$, and the values of Gaussian sums are used (see, for example, (Davenport, 1967)).

We denote

$$S_n = y_1 + y_2 + \cdots + y_n$$

with the summation in the sense of operation in G; $U_n = \left(\frac{S_n}{q}\right)$, where $\left(\frac{m}{q}\right)$ is the Legendre symbol modulo q. We have

$$\begin{aligned}
\mathbf{E}\, e^{itU_n} &= \frac{1}{q}\sum_{\gamma\in\Gamma}\sum_{g\in G} e^{it\left(\frac{g}{q}\right)} e^{-2\pi i g\gamma/q}\,\mathbf{E}\, e^{2\pi i S_n \gamma/q} \\
&= \frac{1}{q}\sum_{\gamma\in\Gamma}\sum_{g\in G} e^{it\left(\frac{g}{q}\right)} e^{-2\pi i g\gamma/q}(\hat{p}(\gamma))^n \\
&= \frac{1}{q}\sum_{g\in G} e^{it\left(\frac{g}{q}\right)} + \frac{1}{q}\sum_{\gamma\in R_+}\left(-\frac{1}{q-1}+\Delta\varepsilon\right)^n \sum_{\gamma\in\Gamma} e^{it\left(\frac{g}{q}\right)} e^{-2\pi i g\gamma/q} \\
&\quad + \frac{1}{q}\sum_{\gamma\in R_-}\left(-\frac{1}{q-1}-\Delta\varepsilon\right)^n \sum_{g\in G} e^{it\left(\frac{g}{q}\right)} e^{-2\pi i g\gamma/q} \\
&= \frac{1}{q} + \frac{q-1}{2q}\left(-\frac{1}{q-1}+\Delta\varepsilon\right)^n + \frac{q-1}{2q}\left(-\frac{1}{q-1}-\Delta\varepsilon\right)^n \\
&\quad + \frac{q-1}{2q}\left(1+\frac{1}{2}(-1+\bar{\varepsilon}q^{1/2})\left(-\frac{1}{q-1}+\Delta\varepsilon\right)^n\right. \\
&\quad \left. + \frac{1}{2}(-1-\bar{\varepsilon}q^{1/2})\left(-\frac{1}{q-1}-\Delta\varepsilon\right)^n\right) e^{it} \\
&\quad + \frac{q-1}{2q}\left(1+\frac{1}{2}(-1-\bar{\varepsilon}q^{1/2})\left(-\frac{1}{q-1}+\Delta\varepsilon\right)^n\right. \\
&\quad \left. + \frac{1}{2}(-1-\bar{\varepsilon}q^{1/2})\left(-\frac{1}{q-1}-\Delta\varepsilon\right)^n\right) e^{-it}
\end{aligned}$$

(here $\mathbf{E}$ is the expectation symbol; the overline indicates complex conjugation). The constant term and the coefficients of e^{it} and e^{-it} in the right-hand side of the last equality are probabilities that U_n is equal to 0, 1, -1 respectively. The sequence of random variables U_n, $n = 1, 2, \ldots$, is a Markov chain with three states 0, 1, -1. Its transition probability matrix is

$$\begin{pmatrix}
0 & \alpha_+ & \alpha_- \\
\frac{2\alpha_+}{q-1} & \frac{1}{2}-\frac{2\alpha_+}{q-1} & \frac{1}{2} \\
\frac{2\alpha_-}{q-1} & \frac{1}{2} & \frac{1}{2}-\frac{2\alpha_-}{q-1}
\end{pmatrix}$$

if $q \equiv 1 \pmod 4$, and

$$\begin{pmatrix} 0 & \alpha_+ & \alpha_- \\ \dfrac{2\alpha_-}{q-1} & \dfrac{1}{2} - \dfrac{1}{q-1} & \dfrac{1}{2} + \dfrac{1}{q-1} - \dfrac{2\alpha_-}{q-1} \\ \dfrac{2\alpha_+}{q-1} & \dfrac{1}{2} + \dfrac{1}{q-1} - \dfrac{2\alpha_+}{q-1} & \dfrac{1}{2} - \dfrac{1}{q-1} \end{pmatrix}$$

if $q \equiv 3 \pmod 4$.

The Markov property of the sequence U_n follows from the criterion of enlargement of states for Markov chains (Kemeny and Snell, 1959). The structure of the transition probability matrix is obtained with the help of Perron's results on the distribution of quadratic residues and nonresidues over sets $g + R_+$, $g + R_-$ (MacWilliams and Sloane, 1977). It can be easily verified that the eigenvalues of the transition probability matrix of the Markov chain U_n are

$$1, \quad -1/(q-1) + \Delta, \quad -1/(q-1) - \Delta \quad \text{for} \quad q \equiv 1 \pmod 4,$$

$$1, \quad -1/(q-1) + i\Delta, \quad -1/(q-1) - i\Delta \quad \text{for} \quad q \equiv 3 \pmod 4,$$

and coincide with the possible values of the characteristic function $\widehat{p}$ in (2).
It can be also easily verified that the right-hand side eigenvectors of the transition probability matrix are the columns of the matrix

$$\begin{pmatrix} 1 & (q-1)/2 & (q-1)/2 \\ 1 & (-1+\varepsilon q^{1/2})/2 & (-1-\varepsilon q^{1/2})/2 \\ 1 & (-1-\varepsilon q^{1/2})/2 & (-1+\varepsilon q^{1/2})/2 \end{pmatrix}$$

and the left-hand side eigenvectors are the rows of the matrix. Particularly, as supposed, the stationary distribution of the Markov chain U_n is $1/q$, $(q-1)/(2q)$, $(q-1)/(2q)$. In what follows we will consider the sums

$$S_n = Y_1 + Y_2 + \cdots + Y_n$$

of independent identically distributed random vectors with components in G, where the component-wise summation is carried out in the sense of operation in G, and the distribution of Y_i is defined in the following way: each component of Y_i is chosen with equal probabilities $1/N$ (N is the dimension of Y_i), the distribution of the chosen component is p from (1), all other components take value 0 of group G with probability 1.

2. MARKOV PROPERTY

Let, as before, $\left(\frac{m}{q}\right)$ be the Legendre symbol modulo q and χ_0 be the principal character modulo q. Let

$$S_n = (S_{n1}, S_{n2}, \ldots, S_{nN}), \quad U_{n,N}^{(0)} = \sum_{k=1}^{N} \chi_0(S_{nk}), \quad U_{n,N} = \sum_{k=1}^{N} \left(\frac{S_{nk}}{q}\right).$$

The set of possible values (k_0, k) of two-dimensional random variable $(U^{(0)}_{n,N}, U_{n,N})$ is defined in the following way: the first component k_0 takes integer values s, $0 \le s \le N$; if $k_0 = s$, then the second component k takes integer values r, $-s \le r \le s$, $s + r \equiv 0 \pmod 2$. We use the notation K for the set of possible values of $(U^{(0)}_{n,N}, U_{n,N})$. The number of elements of K is $(N+1)(N+2)/2$.
An analogue of the Markov character of U_n mentioned in Introduction is the following one: for fixed N the sequence $(U^{(0)}_{n,N}, U_{n,N})$, $n = 1, 2, \dots, N$, is connected in a Markov chain with the state space K. The last assertion is proved as in the case U_n with the help of criterion of enlargement of states for Markov chains and is based on the distribution of quadratic residues and nonresidues over sets $g + R_+$, $g + R_-$. The eigenvalues of the transition probability matrix of the chain $(U^{(0)}_{n,N}, U_{n,N})$ coincide with the possible values of characteristic function of Y_i in the sum S_n. For this characteristic function we have

$$\widehat{p}_N(\gamma) = \mathbf{E}\, e^{2\pi i (Y_j, \gamma)/q} = \frac{1}{N}\left(N - \frac{q}{q-1}\sum_k \chi_0(\gamma_k) + \Delta\varepsilon \sum_k \left(\frac{\gamma_k}{q}\right)\right).$$

The right-hand side eigenvectors of the transition probability matrix are composed of the coefficients of $e^{itk_0}e^{itk}$, where $(k_0, k) \in K$, in the product of the following factors:

$$\left(1 + \frac{1}{2}(-1 + \varepsilon q^{1/2})e^{it_0}e^{it} + \frac{1}{2}(-1 - \varepsilon q^{1/2})e^{it_0}e^{-it}\right)^{(s+r)/2},$$

$$\left(1 + \frac{1}{2}(-1 - \varepsilon q^{1/2})e^{it_0}e^{it} + \frac{1}{2}(-1 + \varepsilon q^{1/2})e^{it_0}e^{-it}\right)^{(s-r)/2},$$

$$\left(1 + \frac{1}{2}(q-1)e^{it_0}e^{it} + \frac{1}{2}(q-1)e^{it_0}e^{-it}\right)^{N-s}.$$

When (s, r) takes the set of values K, $(N+1)(N+2)/2$ eigenvectors are obtained. The matrix of the left-hand side eigenvectors is obtained as the transposed matrix.

3. GENERATING FUNCTIONS, MOMENTS

We have

$$\begin{aligned}
\mathbf{E}\, e^{it_0 U^{(0)}_{n,N} + itU_{n,N}} &= \frac{1}{q^N}\sum_{\gamma\in\Gamma^N}\sum_{g\in G^N} \exp\Big\{it_0(\chi_0(g_1) + \cdots + \chi_0(g_N)) \\
&\quad + it\Big(\Big(\frac{g_1}{q}\Big) + \cdots + \Big(\frac{g_N}{q}\Big)\Big)\Big\} \\
&\quad \times e^{-2\pi i (g,\gamma)/q}\,\mathbf{E}\, e^{2\pi i (S_n,\gamma)/q} \\
&= \frac{1}{q^N}\sum_{\gamma\in\Gamma^N}\sum_{g\in G^N} \exp\Big\{it_0(\chi_0(g_1) + \cdots + \chi_0(g_N)) \\
&\quad + it\Big(\Big(\frac{g_1}{q}\Big) + \cdots + \Big(\frac{g_N}{q}\Big)\Big)\Big\} \\
&\quad \times e^{-2\pi i (g,\gamma)/q}(\widehat{p}_N(\gamma))^n
\end{aligned}$$

and, for the characteristic function $\widehat{p}_N$,

$$\widehat{p}_N(\gamma) = \frac{1}{N}(N - l_+(\gamma) - l_-(\gamma) + l_+(\gamma)(-1/(q-1) + \Delta\varepsilon) + l_-(\gamma)(-1/(q-1) - \Delta\varepsilon)),$$

where $l_+(\gamma)$, $l_-(\gamma)$ are the numbers of quadratic residues and nonresidues of the vector γ. Therefore,

$$\begin{aligned}
\mathbf{E}\, e^{it_0 U^{(0)}_{n,N} + itU_{n,N}} &= \frac{1}{q^N} \sum_{\gamma\in\Gamma^N} \sum_{g\in G^N} \prod_j e^{it_0\chi_0(g_j)} e^{it(\frac{g_j}{q})} e^{-2\pi i g_j \gamma_j/q} \\
&\quad \times \frac{1}{N}\left(N - l_+(\gamma)\left(-\frac{q}{q-1} - \Delta\varepsilon\right) - l_-(\gamma)\left(-\frac{q}{q-1} + \Delta\varepsilon\right)\right)^n \\
&= \frac{1}{q^N} \sum_{l_-, l_+} \frac{N!}{l_+!\, l_-!\, (N - l_+ - l_-)!} \frac{1}{N^n} \\
&\quad \times \left(N - l_+(\gamma)\left(-\frac{q}{q-1} - \Delta\varepsilon\right) - l_-(\gamma)\left(-\frac{q}{q-1} + \Delta\varepsilon\right)\right)^n \\
&\quad \times a_+^{l_+} a_-^{l_-} a_0^{N-l_+-l_-}, \\
a_+ &= 1 + e^{it_0} e^{it}\left(-\frac{1}{2} + \frac{1}{2}\bar{\varepsilon} q^{1/2}\right) + e^{it_0} e^{-it}\left(-\frac{1}{2} - \frac{1}{2}\bar{\varepsilon} q^{1/2}\right), \\
a_- &= 1 + e^{it_0} e^{it}\left(-\frac{1}{2} - \frac{1}{2}\bar{\varepsilon} q^{1/2}\right) + e^{it_0} e^{-it}\left(-\frac{1}{2} + \frac{1}{2}\bar{\varepsilon} q^{1/2}\right), \\
a_0 &= 1 + \frac{q-1}{2} e^{it_0} e^{it} + \frac{q-1}{2} e^{it_0} e^{-it}.
\end{aligned}$$

Summation over l_+, l_- gives the following result.
The generating function

$$\Phi(z) = \sum_n \frac{N^n z^n}{n!} \mathbf{E}\, e^{it_0 U^{(0)}_{n,N} + itU_{n,N}}$$

can be represented in the form

$$\Phi(z) = \left(A_0(z) + A_+(z) e^{it_0} e^{it} + A_-(z) e^{-t_0} e^{-it}\right)^N,$$

where

$$\begin{aligned}
A_+(z) &= \frac{q-1}{2q} e^z + \frac{q-1}{4q}(-1 + \bar{\varepsilon} q^{1/2}) e^{z(-1/(q-1)+\Delta\varepsilon)} \\
&\quad + \frac{q-1}{4q}(-1 - \bar{\varepsilon} q^{1/2}) e^{z(-1/(q-1)-\Delta\varepsilon)}, \\
A_-(z) &= \frac{q-1}{2q} e^z + \frac{q-1}{4q}(-1 - \bar{\varepsilon} q^{1/2}) e^{z(-1/(q-1)+\Delta\varepsilon)} \\
&\quad + \frac{q-1}{4q}(-1 + \bar{\varepsilon} q^{1/2}) e^{z(-1/(q-1)-\Delta\varepsilon)}, \\
A_0(z) &= \frac{1}{q} e^z + \frac{q-1}{2q} e^{z(-1/(q-1)+\Delta\varepsilon)} + \frac{q-1}{2q} e^{z(-1/(q-1)-\Delta\varepsilon)}.
\end{aligned}$$

The first and second moments of the random vector $(U_{n,N}^{(0)}, U_{n,N})$ can be derived from the generating function as $n, N \to \infty$, $n/N \to \alpha$, in the standard way:

$$\begin{aligned}
\mathbf{E}\, U_{n,N}^{(0)} &= N\frac{q-1}{q}\left(1 - e^{-\alpha q/(q-1)}\,\widehat{\cos}\,\alpha\Delta\right)(1+o(1)),\\
\mathbf{E}\, U_{n,N} &= N\frac{q-1}{q^{1/2}}e^{-\alpha q/(q-1)}\,\widehat{\sin}\,\alpha\Delta(1+o(1)),\\
\operatorname{Var} U_{n,N}^{(0)} &= N\left(\frac{q-1}{q}\frac{1}{q} + \frac{q-1}{q}\frac{q-2}{q}e^{-\alpha q/(q-1)}\,\widehat{\cos}\,\alpha\Delta\right.\\
&\quad - \frac{(q-1)^2}{q^2}e^{-2\alpha q/(q-1)}\,\widehat{\cos}^2\,\alpha\Delta\\
&\quad \left. - \alpha e^{-2\alpha q/(q-1)}\left(\widehat{\cos}\,\alpha\Delta - \frac{q}{q-1}\Delta\,\widehat{\sin}\,\alpha\Delta\right)^2\right)(1+o\,(1)),\\
\operatorname{Var} U_{n,N} &= N\left(\frac{q-1}{q} - \frac{q-1}{q}e^{-\alpha q/(q-1)}\,\widehat{\cos}\,\alpha\Delta\right. \qquad (3)\\
&\quad - \frac{(q-1)^2}{q}e^{-2\alpha q/(q-1)}\,\widehat{\sin}^2\,\alpha\Delta\\
&\quad \left. + \alpha q e^{-2\alpha q/(q-1)}\left(\widehat{\sin}\,\alpha\Delta - \frac{q}{q-1}\Delta\,\widehat{\cos}\,\alpha\Delta\right)^2\right)(1+o\,(1)),\\
\operatorname{Cov}(U_{n,N}^{(0)}, U_{n,N}) &= N\left(\frac{q-1}{q^{3/2}}e^{-\alpha q/(q-1)}\,\widehat{\sin}\,\alpha\Delta + \frac{1}{2}\frac{(q-1)^2}{q^{3/2}}e^{-2\alpha q/(q-1)}\,\widehat{\sin}\,2\alpha\Delta\right.\\
&\quad + \frac{1}{2}q^{1/2}\alpha e^{-2\alpha q/(q-1)}\left(\left(1+\varepsilon^2\left(\frac{q-1}{q}\right)^2\Delta^2\right)\widehat{\sin}\,2\alpha\Delta\right.\\
&\quad \left.\left. - 2\frac{q-1}{q}\Delta\,\widehat{\cos}\,2\alpha\Delta\right)\right)(1+o\,(1)).
\end{aligned}$$

In the above formulas $\widehat{\sin}\,x$, $\widehat{\cos}\,x$ denote the trigonometric functions $\sin x$, $\cos x$ for $q \equiv 3\,(\mathrm{mod}\,4)$, and $\widehat{\sin}\,x$, $\widehat{\cos}\,x$ denote the hyperbolic functions $\sinh x$, $\cosh x$ for $q \equiv 1\,(\mathrm{mod}\,4)$.

4. LOCAL NORMAL THEOREM

A two-dimensional local normal theorem takes place for the vector $(U_{n,N}^{(0)}, U_{n,N})$ with parameters defined in (3) as $n, N \to \infty$, $n/N \to \alpha$. This is proved with the help of the reduction to sums of conditionally independent random variables. Namely, if we have N independent random Poisson-distributed variables $\xi_1, \xi_2, \ldots, \xi_N$ and N independent sequences $\eta_1^{(1)}, \eta_2^{(1)}, \ldots, \eta_1^{(N)}, \eta_2^{(N)}, \ldots$ consisting of independent identically distributed random variables on G with distribution p from (1), then

$$\mathbf{P}\{S_{n1} = g_1, \ldots, S_{nN} = g_N\}$$

$$= \mathbf{P}\{\eta_1^{(1)} + \cdots + \eta_{\xi_1}^{(1)} = g_1, \cdots, \eta_1^{(N)} + \ldots + \eta_{\xi_N}^{(N)} = g_N \mid \xi_1 + \xi_2 + \cdots + \xi_N = n\}.$$

Here $g_1, \dots, g_N$ are elements of G and summation of the random variables $\eta_j^{(k)}$ is carried out in the sense of the operation in G, i. e., modulo q. We define random variables $\zeta^0, \zeta^+, \zeta^-$ with distributions

$$\mathbf{P}\{\zeta^0 = k\} = \mathbf{P}\{\xi_1 = k \mid \eta_1^{(1)} + \cdots + \eta_{\xi_1}^{(1)} = 0\},$$
$$\mathbf{P}\{\zeta^+ = k\} = \mathbf{P}\{\xi_1 = k \mid \eta_1^{(1)} + \cdots + \eta_{\xi_1}^{(1)} \in R_+\},$$
$$\mathbf{P}\{\zeta^- = k\} = \mathbf{P}\{\xi_1 = k \mid \eta_1^{(1)} + \cdots + \eta_{\xi_1}^{(1)} \in R_-\},$$

and obtain

$$\mathbf{P}\{U_{n,N}^{(0)} = k_0, U_{n,N} = k\}$$
$$= \frac{1}{\mathbf{P}\{\xi_1 + \cdots + \xi_N = n\}} \frac{N!}{((k_0 + k)/2)!((k_0 - k)/2)!(N - k_0)!}$$
$$\times \pi_+^{(k_0+k)/2} \pi_-^{(k_0-k)/2} \pi_0^{N-k_0}$$
$$\times \mathbf{P}\left\{\zeta_1^+ + \ldots + \zeta_{(k_0+k)/2}^+ + \zeta_1^- + \cdots + \zeta_{(k_0+k/2}^- + \zeta_1^0 + \cdots + \zeta_{N-k_0}^0 = n\right\}.$$

Here

$$\pi_+ = \mathbf{P}\{\eta_1^{(1)} + \cdots + \eta_{\xi_1}^{(1)} \in R_+\},$$
$$\pi_- = \mathbf{P}\{\eta_1^{(1)} + \cdots + \eta_{\xi_1}^{(1)} \in R_-\},$$
$$\pi_0 = \mathbf{P}\{\eta_1^{(1)} + \cdots + \eta_{\xi_1}^{(1)} = 0\},$$

and $\zeta^0, \zeta^+, \zeta^-$ with subscripts are independent copies of $\zeta^0, \zeta^+, \zeta^-$. Now the local normal theorem for sums of independent summands is applied.

REFERENCES

Davenport, H. (1967). *Multiplicative Number Theory*. Markham, Chicago.

Kemeny, J. and Snell, J. (1960). *Finite Markov Chains*. Van Nostrand, Princeton, New Jersey.

MacWilliams, F. and Sloane, N. (1977). *The Theory of Error-Correcting Codes*. North-Holland, Amsterdam.

Probabilistic Methods in Discrete Mathematics, pp. 305 – 312
V. F. Kolchin et al. (Eds.)
1993 TVP/VSP

ON THE STRONG LAW OF LARGE NUMBERS FOR SUMS OF PAIRWISE INDEPENDENT RANDOM VARIABLES

A. I. MARTIKAINEN
Department of Mathematics and Mechanics, Sankt-Petersburg University, Bibliotechnaya Pl. 2, St. Peterhof, 198904 Sankt-Petersburg, Russia

ABSTRACT

Two versions of the strong law of large numbers will be stated for sums of pairwise independent random variables with a finite moment of order γ, $1 < \gamma < 2$ (Theorem 2) or finite $\mathbf{E}\,|X_1|(\log^+|X_1|)^\tau$, $\tau > 0$ (Theorem 1, see also (Martikainen, 1992)).

1. MAIN RESULTS

Let $\{b_n\}$ be a sequence of positive numbers, $b_n \uparrow \infty$. If $\{X_n\}$ is a sequence of independent random variables, $\mathbf{E}\,X_n = 0$, $S_n = X_1 + \cdots + X_n$, $\sum_{n=1}^{\infty} \mathbf{E}X_n^2/b_n < \infty$, then Kolmogorov's strong law of large numbers states that $S_n/b_n \to 0$ a. s. (as $n \to \infty$, unless stated otherwise). If $\{X_n\}$ is an orthogonal system of random variables, $\mathbf{E}\,X_n = 0$ and

$$\sum_{n=1}^{\infty} \mathbf{E}\,X_n^2(\log n)^2/b_n^2 < \infty \tag{1}$$

then $S_n/b_n \to 0$ a. s. (e. g., see (Loeve, 1963)).
Tandory (Tandory, 1986) proved for $b_n = n$ that the logarithmic multiplier in this theorem could not be changed by other with smaller order of grouth: for any sequence of positive numbers $\{D_n\}$ such that $D_n n^{-1/2}$ does not increase and the condition $\sum_{n=1}^{\infty} D_n(\log n)^2/n^2 = \infty$ holds, there exists an orthogonal system $\{X_n\}$ of random variables with $\mathbf{E}\,X_n = 0$, $\mathbf{E}\,X_n^2 = D_n$ for which $\limsup |S_n|/n = \infty$ a. s. Moreover (cf. (Tandory, 1972)) there exists a sequence $\{X_n\}$ of independent random variables with $\mathbf{E}\,X_n = 0$, $\mathbf{E}\,X_n^2 = D_n$ for which $\limsup |S_n|/n = \infty$ a. s.
Suppose now that X_n are identically distributed. If X_n are mutually independent, then well-known Feller's result states that for any sequence $\{b_n\}$ of positive numbers such that $b_n \uparrow \infty$ and $\sum_{i=n}^{\infty} b_i^{-2} = O(nb_n^{-2})$ the condition $\sum_{n=1}^{\infty} \mathbf{P}\{|X_n| > b_n\} < \infty$ is sufficient for the strong law (Petrov, 1975)

$$(S_n - a_n)/b_n \to 0 \quad \text{a. s.} \tag{2}$$

Here a_n is a median of the random variable S_n. If $b_n/n \downarrow$, then the relation (2) holds with $a_n = \mathbf{E} X_n$. If $b_n/n \uparrow \infty$, then (2) holds with $a_n = 0$ (this is a corollary of Theorem 1 in (Martikainen and Petrov, 1980) and Kroneker lemma, for example).
We know that $S_n/b_n \to 0$ a. s. without assumption of independence of random variables if

$$\sum_{k=n}^{\infty} b_k^{-1} = O\{nb_n^{-1}\} \quad \text{and} \quad \sum_{n=1}^{\infty} \mathbf{P}\{|X_1| > b_n\} < \infty$$

(Martikainen and Petrov, 1980). Suppose now that X_n are pairwise independent and identically distributed. Etemadi (Etemadi, 1981) proved that $S_n/n_n \to \mathbf{E} X_1$ if $\mathbf{E}|X_1| < \infty$. Li Gang (Li Gang, 1988) proved (using the truncation method, orthogonal structure of truncated sequence and Rademaher's inequality) that for $1 < \gamma < 2$ the condition $\mathbf{E}|X|^\gamma(\log^+|X_1|)^\gamma < \infty$ is sufficient for $(S_n - \mathbf{E} S_n)/n^{1/\gamma} \to 0$ a. s. Here $\log^+ x = \log(2 \vee x)$. Note that the implication (2) $\Rightarrow \sum_{n=1}^{\infty} \mathbf{P}\{|X_1| > b_n\} < \infty$ holds for every sequences $\{a_n\}$ and $\{b_n\}$, $b_n \uparrow \infty$ (because the Borel–Cantelli lemma is applicable for sequences of pairwise independent events).
So, in the case $b_n/n \to 0$ there exists a gap between necessary and sufficient conditions for the strong law (necessary conditions are the same as in the case of sums of independent random variables). It is interesting that in the case of identically distributed summands this gap can be diminished (for $b_n = n^{1/\gamma}$, $1 < \gamma < 2$) or dropped (for $b_n = n/(\log n)^\tau, \tau > 0$).

THEOREM 1. *Let $\{X_n\}$ be a sequence of pairwise independent, identically distributed random variables, $\tau > 0$. If*

$$\mathbf{E}|X_1|(\log^+|X_1|)^\tau < \infty \tag{3}$$

then $(S_n - \mathbf{E} S_n)(\log n)^\tau/n \to 0$ a. s.

As was pointed above, condition (3) is necessary for the relation $(S_n - a_n)\times (\log^+ n)^\tau/n \to 0$ with an arbitrary chosen a_n.
The proof of Theorem 1 one can find in (Martikainen, 1992). Nevertheless, lemmas used in the proof of Theorem 2 allow us to give a short proof of Theorem 1 (see below).

THEOREM 2. *Let $\{X_n\}$ be a sequence of pairwise independent identically distributed random variables, $1 < \gamma < 2$, $\tau > 0$, and $\tau > 4\gamma - 6$. If $\mathbf{E}|X_1|^\gamma(\log|X_1|)^\tau < \infty$ then $(S_n - \mathbf{E} S_n)/n^{1/\gamma} \to 0$ a. s.*

The condition $\mathbf{E}|X_1|^\gamma < \infty$ is necessary for the relation $(S_n - a_n)/n^{1/\gamma} \to 0$ a. s. Finally we mention that Bradley (Bradley, 1989) have constructed an example of a sequence of pairwise independent random variables $\{X_n\}$ with $\mathbf{P}\{X_n = -1\} = \mathbf{P}\{X_n = 1\} = \mathbf{P}\{X_n = 0\}/2 = 1/4$ for which the central limit theorem fails. Moreover, $\liminf \mathbf{P}\{S_n = 0\} > 0$ a. s. for this sequence.

2. PROOFS

LEMMA 1. *Let $\alpha, \gamma, \delta, \tau, \omega, A, B$ be constants, $1 < \gamma < 2$ (or $\gamma = 1$, $\tau > 0$), $\omega > \alpha$, $B > A > 0$, $\{c_n\}$ and $\{d_n\}$ be two sequences of constants such that*

$$An^{1/\gamma}(\log n)^{-\omega} \leqslant d_n \leqslant c_n \leqslant Bn^{1/\gamma}(\log n)^{-\alpha}$$

for $n \geqslant 1$. *If* $0 < \beta = (1+\tau+2\delta+\alpha(2-\gamma))^{-1}$, $l_n = e^{n^\beta}$, $\omega(\gamma-1) \leqslant 2\tau+3\delta+(2-\gamma)\alpha$,

$$\mathbf{E}\,|X_1|^\gamma(\log^+|X_1|)^\tau < \infty, \tag{4}$$

then for $Z_{n,i} = X_i \mathrm{I}(d_n \leqslant X_i \leqslant c_n)$, $V_{n,k} = \sum_{i=1}^k Z_{n,i}$ *the relation*

$$\sum_{n=1}^\infty \mathbf{P}\Big\{\max_{l_{n-1}<k\leqslant l_n} |V_{l_n,k} - \mathbf{E}\,V_{l_n,k}| > \varepsilon l_n^{1/\gamma}(\log l_n)^\delta\Big\} < \infty$$

holds for every $\varepsilon > 0$. *Here* $V_{x,k}$ *means* $V_{[x],k}$, *where* $[x]$ *is the integer part of* x.

Remark. For $\gamma > 1$ inequalities $\omega > \alpha$ and

$$\omega \leqslant (2\tau + 3\delta + (2-\gamma)\alpha)/(\gamma - 1)$$

may be simultaneously satisfied if

$$\alpha(\gamma - 1) < 2\tau + 3\delta + (2-\gamma)\alpha. \tag{5}$$

Condition (4) is necessary for (2) with $b_n = n^{1/\gamma}(\log n)^\delta$, $\delta = -\tau/\gamma$ and any a_n. So, when we study sufficient conditions for the strong law, we may restrict our attention to the case $\delta \geqslant -\tau/\gamma$. From this inequality and from $2\alpha\gamma - 3\alpha < 2\tau - 3\tau/\gamma$ we have (5). It proves that inequalities in Lemma 1 are noncontradictory if

$$\delta \geqslant -\tau/\gamma, \quad \delta(2\gamma - 3)(\alpha\gamma - \tau) < 0.$$

Proof of Lemma 1. Put $A_k = \{n\colon Bl_n^{1/\gamma}(\log l_n)^{-\alpha} \geqslant k\}$, $n_k = \inf\{n\colon n \in A_k\}$. Then for $b_n = n^{1/\gamma}(\log n)^\delta$ we have

$$\sum_{n\in A_k} l_n/b_{l_n}^2 = \sum_{n\in A_k} l_n^{1-2/\gamma}(\log l_n)^{-2\delta}$$

$$= \sum_{n\in A_k} e^{n^\beta(\gamma-2)/\gamma} n^{-2\delta\beta} \leqslant \int_{n_k-1}^\infty e^{x^\beta(\gamma-2)/\gamma} x^{-2\delta\beta}\,dx.$$

Since

$$\int_u^\infty e^{x^\beta(\gamma-2)/\gamma} x^{-2\delta\beta}\,dx \sim \frac{\gamma}{\beta(2-\gamma)} u^{1-\beta-2\delta\beta} e^{u^\beta(\gamma-2)/\gamma}$$

as $u \to \infty$, we have

$$\sum_{n\in A_k} l_n/b_{l_n}^2 \leqslant (1+\varphi_k)\frac{\gamma}{\beta(2-\gamma)} n_k^{1-\beta-2\delta\beta} e^{n_k^\beta(\gamma-2)/\gamma},$$

where $\varphi_k \to 0$ as $k \to \infty$. Notice that $0 < \beta \leqslant 1$,

$$Bl_{n_k}^{1/\gamma}(\log l_{n_k})^{-\alpha} \geqslant k > Bl_{n_k-1}^{1/\gamma}(\log l_{n_k-1})^{-\alpha}$$

and $l_{n_k} \asymp l_{n_{k-1}}(u_n \asymp v_n$ means that $0 < \liminf u_n/v_n \leqslant \limsup u_n/v_n < \infty)$.
So,

$$\log k \sim \gamma^{-1} \log l_{n_k} = \gamma^{-1} n_k^{\beta},$$

$$l_{n_k}^{1/\gamma} \asymp k(\log l_{n_k})^{\alpha} \asymp k(\log k)^{\alpha}, \quad l_{n_k}^{1/\gamma} = e^{n_k^{\beta}/\gamma},$$

and

$$\sum_{n \in A_k} l_n/b_{l_n}^2 \leqslant C_1 k^{\gamma-2}(\log k)^{\beta^{-1}-1-2\delta+\alpha(\gamma-2)}.$$

Chebyshev's inequality gives for $P_k = \mathbf{P}\{k-1 \leqslant X_1 < k\}$, $\varepsilon > 0$:

$$\sum_{n=1}^{\infty} \mathbf{P}\{|V_{l_n,l_n} - \mathbf{E}\, V_{l_n,l_n}| > \varepsilon b_{l_n}\} \leqslant \sum_{n=1}^{\infty} l_n \mathbf{E}\, Z_{l_n,1}^2 \varepsilon^{-2} b_{l_n}^{-2}$$

$$\leqslant \sum_{n=1}^{\infty} l_n \varepsilon^{-2} b_{l_n}^{-2} \sum_{k=1}^{\infty} k^2 P_k \mathbf{I}(k \leqslant c_{l_n})$$

$$\leqslant \sum_{n=1}^{\infty} l_n\, \varepsilon^{-2} b_{l_n}^{-2} \sum_{k=1}^{\infty} k^2 P_k \mathbf{I}(k \leqslant B l_n^{1/\gamma}(\log l_n)^{-\alpha})$$

$$\leqslant \sum_{k=1}^{\infty} k^2 P_k \sum_{n \in A_k} l_n \varepsilon^{-2} b_{l_n}^{-2} \leqslant C_2 \sum_{k=1}^{\infty} k^{\gamma}(\log k)^{\beta^{-1}-1-2\delta+\alpha(\gamma-2)} P_k < \infty \tag{6}$$

if $\beta^{-1} - 1 - 2\delta + \alpha(\gamma-2) \leqslant \tau$, i. e., if $\beta \geqslant (\tau + 1 + 2\delta + \alpha(2-\gamma))^{-1}$ (but this is one of conditions of Lemma 1).
In a similar way,

$$\sum_{n=1}^{\infty} \mathbf{P}\{|V_{l_n,l_{n-1}} - \mathbf{E}\, V_{l_n,l_{n-1}}| > \varepsilon b_{l_n}\}$$

$$\leqslant \sum_{n=1}^{\infty} l_{n-1} \varepsilon^{-2} b_{l_n}^{-2} \mathbf{E}\, Z_{l_n,1}^2 \leqslant \sum_{n=1}^{\infty} l_n \varepsilon^{-2} b_{l_n}^{-2} \mathbf{E}\, Z_{l_n,1}^2 < \infty. \tag{7}$$

If $l_{n-1} \leqslant k < l_n$, $\Delta_n = \mathbf{E}\, V_{l_n,l_n} - \mathbf{E}\, V_{l_n,l_{n-1}}$, then the inequalities

$$V_{l_n,l_{n-1}} - \mathbf{E}\, V_{l_n,l_{n-1}} - \Delta_n \leqslant V_{l_n,k} - \mathbf{E}\, V_{l_n,k} \leqslant V_{l_n,l_n} - \mathbf{E}\, V_{l_n,l_n} + \Delta_n$$

hold. So

$$|V_{l_n,k} - \mathbf{E}\, V_{l_n,k}| \leqslant |V_{l_n,l_n} - \mathbf{E}\, V_{l_n,l_n}| + |V_{l_n,l_{n-1}} - \mathbf{E}\, V_{l_n,l_{n-1}}| + \Delta_n.$$

From (4) it follows that for $F(x) = \mathbf{P}\{X_1 < x\}$ the inequality

$$\int_x^{\infty} y\, dF(y) \leqslant x^{1-\gamma}(\log x)^{-\tau} \int_x^{\infty} y^{\gamma}(\log y)^{\tau}\, dF(y)$$

holds with $x > 1$. Therefore,

$$\mathrm{E}\, Z_{n,1} = o\left(d_n^{1-\gamma}(\log d_n)^{-\tau}\right),$$

$$\mathrm{E}\, Z_{n,1} = o\left(n^{(1-\gamma)/\gamma}(\log n)^{-(1-\gamma)\omega-\tau}\right), \qquad n \to \infty.$$

Having in mind the relations $l_n - l_{n-1} \asymp n^{\beta-1} l_n$, $n = (\log l_n)^{1/\beta}$, we obtain

$$\Delta_n = (l_n - l_{n-1})\mathrm{E}\, Z_{l_n,1} = o\left(l_n^{1/\gamma}(\log l_n)^{\beta^{-1}(\beta-1)-(1-\gamma)\omega-\tau}\right).$$

The inequality $\beta^{-1}(\beta-1)-(1-\gamma)\omega-\tau \leqslant \delta$ follows from conditions of our lemma. It means that $\Delta_n = o(b_n)$. Therefore,

$$\mathbf{P}\left\{\max_{l_{n-1}<k\leqslant l_n} |V_{l_n,k} - \mathrm{E}\, V_{l_n,k}| > 3\varepsilon b_{l_n}\right\}$$

$$\leqslant \mathbf{P}\{|V_{l_n,l_n} - \mathrm{E}\, V_{l_n,l_n}| > \varepsilon b_{l_n}\} + \mathbf{P}\{|V_{l_n,l_{n-1}} - \mathrm{E}\, V_{l_n,l_{n-1}}| > \varepsilon b_{l_n}\}$$

for all sufficiently large n. Summing these inequalities over all sufficiently large n and using (6) and (7), we obtain the assertion of Lemma 1.

LEMMA 2. *Let $C, D, \gamma, \delta, \tau, \omega$ be constants,*

$$X_1 \geqslant -C, \quad \omega \geqslant (2-2\delta-\tau)/(2-\gamma), \quad 0<\gamma<2,$$

$$D > 0, \quad b_n = n^{1/\gamma}(\log n)^{\delta},$$

$\{d_n\}$ be a sequence of positive numbers such that $d_n \leqslant D n^{1/\gamma}(\log n)^{-\omega}$. If the condition (4) holds, then

$$\sum_{n=1}^{\infty} \mathbf{P}\left\{\max_{1\leqslant k\leqslant 2^n} |U_{2^n,k} - \mathrm{E}\, U_{2^n,k}| > \varepsilon b_{2^n}\right\} < \infty$$

for every $\varepsilon > 0$ with

$$Y_{n,i} = X_i \mathrm{I}(X_i < d_n), \quad U_{n,k} = \sum_{i=1}^{k} Y_{n,i}.$$

Proof. The pairwise independence of $Y_{n,i}$ for $1 \leqslant i \leqslant n$ yields the orthogonality of system $\{Y_{n,i} - \mathrm{E}\, Y_{n,i};\ 1 \leqslant i \leqslant n\}$. By Rademaher's inequality, we have

$$Q_n = \mathbf{P}\left\{\max_{1\leqslant k\leqslant n} |U_{n,k} - \mathrm{E}\, U_{n,k}| > \varepsilon b_n\right\} \leqslant \varepsilon^{-2} b_n^{-2} \mathrm{E} \max_{1\leqslant k\leqslant n} (U_{n,k} - \mathrm{E} U_{n,k})^2$$

$$\leqslant C_3 (\log n)^2 b_n^{-2} \mathrm{E}\, U_{n,n}^2 = C_3 n^{1-2/\gamma}(\log n)^{2-2\delta} \mathrm{E}\, Y_{n,1}^2$$

$$\leqslant C_3 n^{1-2/\gamma}(\log n)^{2-2\delta} \sum_{k=1}^{h_n} k^2 P_k$$

with $P_n = \mathbf{P}\{n-1 \leqslant X_1 < n\}$, $h_n = Dn^{1/\gamma}(\log n)^{-\omega}$. Put $m_k = \min\{n\colon h_{2^n} \geqslant k\}$. Evidently $m_k \sim \gamma \log k$, $h_{2^n} \sim 2^{1/\gamma} h_{2^{n-1}}$, $2^{m_k/\gamma} m_k^{-\omega} \asymp k$ as $k \to \infty$; $2^{m_k/\gamma} \asymp k(\log k)^{\omega}$ as $k \to \infty$. Hence,

$$\sum_{n=1}^{\infty} Q_{2^n} \leqslant C_3 \sum_{k=1}^{\infty} k^2 P_k \sum_{n=m_k}^{\infty} 2^{n(1-2/\gamma)} n^{2-2\delta}$$

$$\leqslant C_4 \sum_{k=1}^{\infty} k^2 P_k [k(\log k)^{\omega}]^{\gamma-2} (\log k)^{2-2\delta} < \infty.$$

Here we use (4) and $\omega(\gamma-2)+2-2\delta \leqslant \tau$. This follows from conditions of Lemma 2 and proves it.

Proof of Theorem 1. It is sufficient to consider the case $\mathbf{E}X_n = 0$. Put $X_n^+ = X_n \mathbf{I}(X_n \geqslant 0)$, $X_n^- = X_n \mathbf{I}(X_n < 0)$. Notice that $X_n = X_n^+ - X_n^- = (X_n^+ - \mathbf{E}X_n^+) - (X_n^- - \mathbf{E}X_n^-)$ and so we can prove this theorem separately for two sequences of pairwise independent random variables $\{X_n^+ - \mathbf{E}X_n^+\}$ and $\{X_n^- - \mathbf{E}X_n^-\}$. Therefore, we suppose that $X_n \geqslant -C$ for some positive constant C.
Let $b_n = n/(\log n)^{\tau}$, $L_n = \{i\colon 2^{n-1} < l_i \leqslant 2^n\}$ (constants l_i were defined in the statement of Lemma 1), $M_n = L_n$ if $l_t = 2^n$ with $t = \sup L_n$, and $M_n = L_n \cup \{t+1\}$ otherwise. We introduce random variables $Y_{2^n,i}$, $U_{2^n,k}$, $Z_{l_i,j}$ and $V_{l_i,k}$ by the equalities:

$$Y_{2^n,i} = X_i \mathbf{I}(X_i < d_{2^n}), \quad U_{2^n,k} = \sum_{i=1}^{k} Y_{2^n,i},$$

$$Z_{n,l_i,j} \equiv Z_{l_i,j} = X_j \mathbf{I}(d_{2^n} \leqslant X_j \leqslant b_{l_i}), \quad V_{l_i,k} = \sum_{j=1}^{k} Z_{n,l_i,j}$$

(for $l_{i-1} < k \leqslant l_i$, $2^{n-1} < k \leqslant 2^n$). For $\varepsilon > 0$, $m \geqslant 1$ we have

$$\mathbf{P}\{S_n > 4\varepsilon b_n \text{ i.o.}\} \leqslant \mathbf{P}\Big\{ \bigcup_{n=2^{m-1}+1}^{\infty} \{S_n > 4\varepsilon b_n\} \Big\}$$

$$\leqslant \sum_{n=m}^{\infty} \sum_{i \in M_n} \sum_{k=1}^{\infty} \mathbf{P}\{X_k > b_{l_i}\} \mathbf{I}(l_{i-1} < k \leqslant l_i, 2^{n-1} < k \leqslant 2^n)$$

$$+ \sum_{n=m}^{\infty} \sum_{i \in M_n} \mathbf{P}\Big\{ \max_k |V_{l_i,k} - \mathbf{E}V_{l_i,k}| \mathbf{I}(l_{i-1} < k \leqslant l_i, 2^{n-1} < k \leqslant 2^n) > \varepsilon b_{l_i} \Big\}$$

$$+ \sum_{n=m}^{\infty} \mathbf{P}\Big\{ \max_{k \leqslant 2^n} |U_{2^n,k} - \mathbf{E}U_{2^n,k}| > \varepsilon b_{2^n} \Big\}.$$

The first series in the right-hand side is majorized by $\sum_{i=2^{m-1}}^{\infty} \mathbf{P}\{X_1 > b_i\}$. This sum is the remainder of a convergent series by (3). Using Lemma 1 with $\gamma = 1$, $\omega = 2+\tau$, $\alpha = \tau$, $\delta = -\tau$, $\beta = 1$ we find that it is true for the second series too. Finally, using Lemma 2 with the same γ, ω, δ, $D = 1$ we verify that the third series tends to 0 as $m \to \infty$. Thus, $\mathbf{P}\{S_n > 4\varepsilon b_n \text{ i.o.}\} = 0$ for every $\varepsilon > 0$ that completes the proof.

Proof of Theorem 2. By the same reason as in Theorem 1, we can suppose that $X_n \geqslant -C$, $\mathbf{E}X_n = 0$. Define

$$b_n = n^{1/\gamma}, \quad d_n = n^{1/\gamma}(\log n)^{-(2-\tau)/(2-\gamma)}.$$

The application of Lemma 2 with $Y_{n,i} = X_i\mathrm{I}(X_i < d_n)$, $U_{n,k} = \sum_{i=1}^{n} Y_{n,i}$ and Borel–Cantelly lemma proves that

$$\max_{k \leqslant 2^n} |U_{2^n,k} - \mathbf{E}U_{2^n,k}|/b_{2^n} \to 0 \quad \text{a.s.}$$

We will use Lemma 1 with $\delta = 0$ and different values of α, β and ω for several times. Put first

$$\alpha = \tau/\gamma, \quad \omega \equiv \omega_1 = (2+\gamma)\tau\gamma^{-1}(\gamma-1)^{-1}, \quad \beta = (1+\tau)^{-1}.$$

Let random variables $Z_{n,i}^{(1)}$ and $V_{n,k}^{(1)}$ be defined by

$$Z_{n,i}^{(1)} = X_i\mathrm{I}(d_n^{(1)} \leqslant X_i < n^{1/\gamma}(\log n)^{-\alpha})$$

with

$$d_n^{(1)} = n^{1/\gamma}(\log n)^{-\omega_1}, \quad V_{n,k}^{(1)} = \sum_{i=1}^{k} Z_{i=1}^{(1)}$$

From Lemma 1 it follows that

$$\max_{l_{n-1} < k \leqslant l_n} |V_{l_n,k} - \mathbf{E}V_{l_n,k}|/l_n^{1/\gamma} \to 0 \quad \text{a.s.}$$

We define the random variable $Z_{n,i}^{(m)}$ for $m \geqslant 2$ by truncation of X_i with the lower point of truncation $d_n^{(m)} = n^{1/\gamma}(\log n)^{\omega_m}$, where $\omega_m = \omega(\omega_{m-1})$, $\omega(t) = at + b$, $a = (2-\gamma)/(\gamma-1)$, $b = 2\tau/(\gamma-1)$, the upper point of truncation being the lower point of truncation of $Z_{i,i}^{(m-1)}$. Using Lemma 1 with

$$V_{n,k}^{(m)} = \sum_{i=1}^{k} Z_{n,i}^{(m)}, \quad l_n \equiv l_n^{(m)} = e^{n^{\beta_m}}, \quad \beta_m = (1+\tau+\omega_{m-1}(2-\gamma))^{-1},$$

we obtain that

$$\max_{l_{n-1}^{(m)} < k \leqslant l_n^{(m)}} |V_{l_n^{(m)},k}^{(m)} - \mathbf{E}V_{l_n^{(m)},k}^{(m)}|/(l_n^{(m)})^{1/\gamma} \to 0 \quad \text{a.s.}$$

For $\gamma \leqslant 3/2$ we have $a \geqslant 1$. It is clear that $b > 0$. The function $\omega(t)$ is linear, therefore $\omega_m \to \infty$ as $m \to \infty$ and $\omega_m \geqslant (2-\tau)/(2-\gamma)$ for some $m \geqslant 2$. Lemma 1 makes it possible to diminish intervals of truncation if necessary. So, applying this Lemma for a finite number of times we prove that

$$(V_n - \mathbf{E}V_n)/b_n \to 0 \quad \text{a.s.} \tag{8}$$

for $\gamma \leqslant 3/2$. Here $V_n = \sum_{i=1}^{n} Z_i$, $Z_i = X_i \mathrm{I}(X_i < i^{1/\gamma}(\log^+ i)^{-\tau/\gamma}/2)$. By the condition (4),

$$\sum_{n=1}^{\infty} \mathbf{P}\{X_n \neq Z_n\} < \infty, \quad \mathbf{E}V_n = o(b_n)$$

and therefore

$$S_n/b_n \to 0 \quad \text{a.s.} \tag{9}$$

This completes the proof for the case $\gamma \leqslant 3/2$. If $\gamma > 3/2$, the sequence $\{\omega_n\}$ tends to a finite limit which is the only solution of the equation $\omega(t) = t$. This solution equals $2\tau/(2\gamma - 3)$. Noting that this solution is greater than $(2 - \tau)/(2 - \gamma)$ if $\tau > 4\gamma - 6$ we conclude that as in the previous case the application of Lemma 1 for a finite number of times gives (8). Using (9), we complete the proof.

REFERENCES

Bradley, R. (1989). A stationary, pairwise independent, absolutely regular sequence for which the central limit theorem fails. *Probab. Theory Rel. Fields.* **81**, 1–10.

Etemadi, N. (1981). An elementary proof of the strong law of large numbers. *Z. Wahrscheinlichkeitstheor. verw. Geb.* **55**, 119–122.

Li Gang (1988). On the strong convergence of random elements in Banach spaces. *Natur. Sci. J.Sichuan. Univ.* **25**, 381–389.

Loeve, M. (1963). *Probability Theory.* Van Nostrand, New York.

Martikainen, A. I. and Petrov, V. V. (1980). On a theorem of Feller. *Theory Probab. Appl.* **25**, 191–194.

Martikainen, A. I. (1992). A remark on the strong law of large numbers for sums of pairwise independent random variables. *J. Soviet Math., Problems of Theory of Probabilistic Distributions* **12** (in Russian).

Petrov, V. V. (1975). *Sums of Independent Random Variables.* Springer-Verlag, New York.

Tandory, K. (1972). Bemerkung zum Gesetz der grossen Zahlen. *Period. Math. Hungar.* **2**, 33–39.

Tandory, K. (1986). Bemerkung uber die parweise unabhanginen zufalligen Grossen. *Acta Math. Hungar.* **48**, 357–359.

Probabilistic Methods in Discrete Mathematics, pp. 313 – 315
V. F. Kolchin et al. (Eds.)
1993 TVP/VSP

ON A SPECIAL CLASS OF FUNCTIONS ASSOCIATED WITH THE COMPLEXITY OF STRING CORRECTION PROBLEM

S. S. MARTYNOV
TVP Laboratories, Krasikov Street 32, 117418 Moscow, Russia

ABSTRACT

The problem of correction the strings against a language with the strings satisfying some restriction for a set of their substrings is considered. The possibility of classification of the functions specifying the restriction with regard for the complexity of string correction problem is discussed.

Let Σ be a finite alphabet, Σ^* is a set of all finite strings in the alphabet Σ, $L \subseteq \Sigma^*$ is some language.
Consider a set $\Phi = \{\varphi_1, \ldots, \varphi_r\}$ of transformations φ_i: $\Sigma^* \to \Sigma^*$, which will be called below the correction operations. The correction of string $S \in \Sigma^*$ against the language L is a procedure of choice of the shortest sequence of correction operations which transforms S into some string belonging to L.
The intensive research of the complexity of string correction problem (see (Garey and Johnson 1979; Sankoff and Kruskal, 1983)) is stimulated by numerous needs of applications in the fields of the systems of data security, storage and communications, automated systems of text correction and some branches of biochemical and genetic researches.
In the case where the set Φ includes such correction operations as removing a letter, insertion a letter, transposition of a pair of letters or substitution a new letter for a letter, Martynov (Martynov, 1989) have proved the NP-completeness of the string correction problem against the language L that consists of strings satisfying some restriction on a set of their substrings.
Let $\mathcal{F}(\omega_1, \ldots, \omega_l)$ be some map

$$\mathcal{F}\colon \Sigma^l \to \Sigma$$

and let $\sigma \in \Sigma$. The string $S = s_1 \cdots s_n$ $(n \geq l)$ is regarded as satisfying the restriction $\langle \mathcal{F}, \sigma \rangle$ (or $S \in \Sigma^* \langle \mathcal{F}, \sigma \rangle$) if

$$\mathcal{F}(s_i, \ldots, s_{i+l-1}) = \sigma \qquad (i = 1, \ldots, n - l + 1).$$

In other words, $S \in \Sigma^* \langle \mathcal{F}, \sigma \rangle$ if and only if any substring of S with the length l is transformed by $\mathcal{F}$ into σ. We consider the following problem $\mathcal{K}$ of correcting the string against the language $L = \Sigma^*\langle \mathcal{F}, \sigma\rangle$. For a given alphabet Σ, integers k, l, restriction $\langle \mathcal{F}(\omega_1, \dots, \omega_l), \sigma\rangle$ and a string $S \in \Sigma^*$ determine whether there exists a sequence of operations from the set Φ, which transforms S into some string from $\Sigma^*\langle \mathcal{F}, \sigma\rangle$ and the length of which does not exceed k.

Martynov (Martynov, 1989) have obtained some theorems on the complexity of the subproblems $\mathcal{K}_t$ of $\mathcal{K}$, when the function $\mathcal{F}$ depends only on t fixed variables. We may specify

$$\mathcal{F}(x_1, \dots, x_t) = F(x_{j_1}, \dots, x_{j_t}),$$

where $F(x_1, \dots, x_t)\colon \Sigma^t \to \Sigma$ is a fixed function, $1 = j_1 < j_2 < \dots < j_t = l$, and under such an assumption consider a new subproblem

$$\mathcal{K}_F \subseteq \mathcal{K}.$$

The condition of the problem $\mathcal{K}_F$ will contain the integer l, the indexes $j_2 < j_3 < \dots < j_{t-1}$ ($j_1 = 1, j_t = l$), the string $S = s_1 \cdots s_n \in \Sigma^*(n \geq l)$, the integer k. The question is the same as in the problem $\mathcal{K}$.

The proofs of Theorems 2–6 in (Martynov, 1989) contain several examples of the functions F providing the NP-completeness of the problem $\mathcal{K}_F$ under different assumptions concerning the alphabet Σ and the set of correction operations.

Therefore, the problem of classification of functions F by the complexity of the problem $\mathcal{K}_F$ with a certain fixed alphabet Σ and the set of correction operations seems to be posed correctly.

The functions F providing the NP-completeness of the problem $\mathcal{K}_F$ will be called below the $\mathcal{K}$-complete functions, and the class of $\mathcal{K}$-complete functions will be denoted by $[F]_{\mathcal{K}}$.

Now we shall assume that $\Sigma = A \cup \{*\}$, where A is some alphabet, and that the set Φ contains only one operation of substitution letter for letter. We may specify the function F so that $F(x_1, \dots, x_t) = \sigma$ only if $(x_1, \dots, x_t) \dot{\in} A^t$ and demand the integer k in the condition of problem $\mathcal{K}_F$ to be equal to the number of the $*$ in the string S. Then the problem $\mathcal{K}_F$ may be regarded as the problem $\tilde{\mathcal{K}}_F$ of filling the blanks in the string from the language $A^*\langle F, \sigma\rangle$.

It can be easily shown that if $t = 2$, then $\tilde{\mathcal{K}}_F \in P$. This fact follows from the possibility of checking the existence of the way with a given lenght between two vertices of a given graph in polynomial time (see Theorem 13.1 in (Harary, 1969)).

The case of the binary alphabet $A = \{0, 1\}$ may be considered particularly. According to (Shaefer, 1978) the necessary condition for $F \in [F]_{\mathcal{K}}$ can be formulated as follows: F does not belong to each of six special classes of functions specified in (Shaefer, 1978).

But this condition cannot serve as a criterion of $\mathcal{K}$-completeness of the function F. For example, the function

$$f = x_1 x_2 x_3 \oplus x_1 \oplus x_2 \oplus x_3$$

satisfies the condition but the corresponding problem $\tilde{\mathcal{K}}_f$ belongs to P.

Really, let $\varkappa$ be an instance of $\tilde{\mathcal{K}}_f$ with the condition containing an integer l, a string S and the indices of variables $1 = j_1 < j_2 < j_3 = l$. In this case to check the possibility of filling the blanks in the string S so that the result will belong to $A^*\langle f, 1\rangle$ it is sufficient to check the following two conditions

1) $j_{i_1} \not\equiv j_{i_2} \pmod 3$ for $i_1 \neq i_2$;

2) if the integers $1 \leq h_1 < h_2 < \cdots < h_m \leq |S|$ are such that $s_{h_i} = 1$ ($i = 1, \ldots, m$), and $s_g \neq 1$ for $g \notin \{h_1, \ldots, h_m\}$, then $h_i - h_j \equiv 0 \pmod 3$ ($i, j, \in \{1, \ldots, m\}$). Obviously the checking of these conditions may be carried out in the polynomial time.

REFERENCES

Garey, M. R. and Johnson, D. S. (1979). *Computers and Intractability*. Bell Laboratories, Murray Hill, New Jersey.

Harary, F. (1969). *Graph Theory*. Addison-Wesley, Reeding, Massachusets.

Martynov, S. S. (1989). On a complexity of some string correction problems. *Discrete Math. Appl.* 1, 104–112 (in Russian).

Sankoff, D. and Kruskal, J. B. (1983). *Time Warps, String Edits and Macromolecules: The Theory and Practice of Sequence Comparison*. Addison-Wesley, London.

Shaefer, T. J. (1978). The complexity of satisfiability problems. In: *Proceedings of the 10th Annual ACM Symposium on Theory of Computing*, Association for Computing Machinery, New York, pp. 216–226.

Probabilistic Methods in Discrete Mathematics, pp. 316 – 328
V. F. Kolchin et al. (Eds.)
1993 TVP/VSP

OPERATOR MOMENTS OF RANDOM EUCLIDEAN MAPPINGS*

V. M. MAXIMOV

Tver State University, Zhelyabov Street 33, 170013 Tver, Russia

ABSTRACT

In the present paper we introduce operator moments of random variables in the group of Euclidean motions of plane or space. Special properties of the first and second moments, different from the classical case, are indicated. The application to the central limit theorem for the convolution of nonidentical distributions is given.

1. INTRODUCTION

Random Euclidean mappings can be considered as random variables with values in the group of Euclidean motions. For many problems of natural science one needs an assignment of distributions on this group suitable for studying their convolutions. Such problems are, for example, a simulation of random sets in space (stochastic geometry), a description of sufficient statistics of shifts on the group, an evolution of points (nucleous, cells), an evolution of random chains (albumen), a destruction of large systems (polymers) and so on.

The assignment of distributions on groups is always connected with their linear representations. Any linear representation $T(g)$ of a group G, $g \in G$, may be linearly extended to the set of measures on G: $\mu \to T(\mu) = \int T(g)\mu(dg)$. It is evident that this extention preserves the main property of the representation

$$\mu_1 * \mu_2 \to T(\mu_1 * \mu_2) = T(\mu_1)T(\mu_2)$$

and possesses the linear property $T(\lambda_1\mu_1 + \lambda_2\mu_2) = \lambda_1 T(\mu_1) + \lambda_2 T(\mu_2)$. If $T_\gamma(G)$, $\gamma \in \Gamma$, is a complete series of irreducible representations G, then $T_\gamma(\mu) = F_\gamma(\mu)$ may be considered as an ***operator characteristic function*** (o. ch. f.) of the distribution μ. Such o. ch. f. were considered in (Grenander, 1963). Since the representations $T_\gamma(g)$ and $T_\gamma(\mu)$ has usually an infinite dimension, application of the o. ch. f. is difficult. For this reason we consider the operater moments of distributions on the Euclidean motions which are well defined.

*The work was supported by AO CST-Tver, Ltd.

2. PROBABILITY DISTRIBUTIONS ON THE GROUP OF EUCLIDEAN MOTIONS

An Euclidean motion of space is characterized by the parameters z of the shift and α of the rotation. We denote by $g(z,\alpha)$ the motion such that any point x of space at first is transposed to $x+z$ and after that the rotation α of the space is fulfilled. For the sake of simplicity we consider the group of Euclidean motion on the plane $M(2)$, although the main results are valid for the general motions. In this case we set $g(0,\alpha)x = xe^{i\alpha}$. Then the product of any two motions can be written in the form

$$g(z_1,\alpha_1)g(z_2,\alpha_2) = g(z_1 + z_2e^{i\alpha_1}, \alpha_1+\alpha_2). \tag{1}$$

The linear representation T_R, $R \in \mathbf{C}$, of the group $M(2)$ is given on the space $\mathbf{L}_2$ of functions on the unit circle by

$$T_R(g(z,\alpha))f(\psi) = e^{Rr\cos(\psi-\varphi)}f(\psi-\alpha), \tag{2}$$

where $z = re^{i\varphi}$, $f(\psi) \in \mathbf{L}_2$.
It is well known (see (Vilenkin, 1965)), that T_R, $R \neq 0$, is irreducible. If $R = it$, $t \in \mathbf{R}_1$, then the representation T_{it} is a unitary one. For $R = 0$, the representation T_0 corresponds to the homomorphism $g(z,\alpha) \to \alpha$, as is seen from (1), and can be decomposed into a direct sum of one-dimensional representations.
Let μ be a distribution on $M(2)$. Then $p = \int \mu(d\alpha)$ and $q = \int \mu(dz)$ are projections of μ on z and α. The measure μ is called *decomposable* if $\mu(dz,d\alpha) = p(dz)q(d\alpha)$. We introduce $\rho = \frac{1}{2\pi}\int p(d\varphi)$ as the density of the radial distribution of p. If p is invariant with respect to the rotations φ, then a decomposable measure μ is called a *circular distribution* and p is characterized by the radial distribution only. It is easy to check that the convolution p of circular distributions p_1 and p_2 is again circular and the radial density $\rho(r)$ of the distribution p can be expresed in terms of the densities ρ_1 and ρ_2:

$$\rho(r) = \frac{1}{2\pi}\int_0^{2\pi}\int_0^{\infty} \rho_1\left(\sqrt{r^2+r_2^2+2rr_2\cos\psi}\right)\rho_2(r_2)r_2\,dr_2\,d\psi. \tag{3}$$

Let us consider the operator

$$T_{it}(\mu)f(\psi) = \int e^{itr\cos(\psi-\varphi)}f(\psi-\alpha)\,\mu(dr,d\varphi,d\alpha). \tag{4}$$

It is clear that for any measure μ the operator $T_{it}(\mu)$ is defined for all $f(\psi)$ from $\mathbf{L}_2$ and the unique measure μ is defined by an operator $T_{it}(\mu)$, $t \in \mathbf{R}_1$.

DEFINITION 1. The operator $T_{it}(\mu) = F_t(\mu)$, $t \in \mathbf{R}_1$, is called the *operator valued characteristic function of measure μ on $M(2)$.*

It is evident that

$$F_t(\mu_1 * \mu_2) = F_t(\mu_1)F_t(\mu_2), \quad \|F_t(\mu)\| < 1$$

and

$$F_0(\mu)f(\varphi) = \int f(\psi - \alpha)q\,(d\alpha),$$

i.e., $F_0(\mu)$ is not an identical operator. If $\|F_{t_0}(\mu)\| = 1$, $t_0 \neq 0$, then the distribution μ is called the operator-lattice and the operator-lattice property depends on the support of μ only. If the support is finite, then the operator is described completely and generated by elements of quadratic extension of the field of rational numbers (Maximov, 1970).

3. OPERATOR MOMENTS OF DISTRIBUTIONS ON $M(2)$

Let us expand the operator characteristic function $F_t(\mu)$ in powers of t. Then

$$F_t(\mu)f(\psi) = \int e^{itr\cos(\psi-\varphi)} f(\psi - \alpha)\,\mu(dr, d\varphi, d\alpha)$$

$$= \sum_{n=0}^{\infty} \frac{(it)^n}{n!} \int r^n \cos^n(\psi - \varphi) f(\psi - \alpha)\,\mu(dr, d\varphi, d\alpha). \tag{5}$$

DEFINITION 2. The operator M_n on L_2 defined by

$$M_n f(\psi) = \int r^n \cos^n(\psi - \varphi)\, f(\psi - \alpha)\,\mu(dr, d\varphi, d\alpha) \tag{6}$$

is called the nth *operator moment* of the distribution μ. As in the classical case of characteristic functions the convergence of the operator series $\sum \frac{(it)^n}{n!} M_n(\mu)$ and its coincidence with $F_t(\mu)$ are very important. We will use the following sufficient condition

$$\sum_{n=1}^{\infty} \frac{1}{n!} |t|^n \|M_n\| < \infty. \tag{7}$$

Let us consider some useful sets of distributions for which (5) is valid.

a) $\mathfrak{M}_0$ is the set of measures with a bounded support on $M(2)$. Then evidently $\|M_n(\mu)\| < \lambda^n(\mu)$ for some $\lambda(\mu)$ and, consequently, series (7) converges at any t.

b) $\mathfrak{M}_\lambda$, $\lambda > 0$, is the set of measures for which

$$\int e^{\lambda r} \mu(dr, d\varphi, d\alpha) < \infty. \tag{8}$$

For this class one has the following estimate of $\|M_n(\mu)\|$:

$$\|M_n(\mu)\| \le \int e^{-\lambda r} r^n e^{\lambda r} \mu(dg) \le \max_{r \ge 0}(e^{-\lambda r} r^n) c(\mu).$$

Since

$$\max_{r \ge 0}(e^{-\lambda r} r^n) = \left(\frac{n}{\lambda e}\right)^n,$$

where $r = n/\lambda$, then (7) is valid for $\mu \in \mathfrak{M}_\lambda$ for all $|t| < \lambda$.

c) $\mathfrak{M}_\infty$ is the class of measures for which (8) is fulfilled for any λ, $\lambda > 0$.
d) $\mathfrak{M}^\rho$, $\rho > 1$, is the class of measures for which

$$\int e^{\lambda r^\rho} \mu(dr, d\varphi, d\alpha) < \infty \tag{9}$$

for some $\lambda > 0$. For this class, similarly to $\mathfrak{M}_\lambda$, one has

$$\|M_n(\mu)\| \leq \int e^{-\lambda r^\rho} r^n e^{\lambda r^\rho} \mu(dg) \leq \max_{r \geq 0}(e^{-\lambda r^\rho} r^n) c(\mu).$$

Since

$$\max_{r>0}(e^{-\lambda r^\rho} r^n) = \max_{r \geq 0}(e^{-\lambda r} r^{n/\rho}) = \left(\frac{n}{\rho\lambda e}\right)^{n/\rho},$$

for sufficiently large n

$$\frac{|t|^n}{n!}\left(\frac{n}{\rho\lambda e}\right)^{n/\rho} = \sqrt{2\pi n}\frac{c^{n/\rho}}{n^n} n^{n/\rho} < c_2^n(\mu), \qquad |c_2| < 1.$$

Consequently series (7) converges for $\mu \in \mathfrak{M}^\rho$. Further, it is clear that $\mathfrak{M}_0 \subset \mathfrak{M}^\infty \subset \mathfrak{M}_\lambda$.

PROPOSITION 1. *The class of distributions $\mathfrak{M}_\lambda$ is closed with respect to convolution.*

Proof. For any element $g(z, \alpha)$ from $M(2)$ we set $|g| = |z|$. Then

$$|g_1(z_1, \alpha_1) g_2(z_2, \alpha_2)| = |g(z_1 + z_2 e^{i\alpha_1}, \alpha_1 + \alpha_2)| = |z_1 + z_2 e^{i\alpha_1}| \leq |g_1| + |g_2|.$$

Let μ_1, $\mu_2 \in \mathfrak{M}_\lambda$ have the densities $p_1(g)$ and $p_2(g)$ with respect to invariant measure on $M(2)$. Then

$$\int e^{\lambda|g_1|}\left(\int p_1(g_1 g_2^{-1}) p_2(g_2)\, dg_2\right) dg_1 = \int p_2(g_2)\left(\int e^{\lambda|g_1|} p_1(g_1 g_2^{-1})\, dg_1\right) dg_2$$

$$= \int p_2(g_2)\left(\int e^{\lambda|g_1 g_2|} p_1(g_1)\, dg_1\right) dg_2 \leq \int\int e^{\lambda|g_1| + \lambda|g_2|} p_1(g_1) p_2(g_2)\, dg_1\, dg_2$$

$$= \left(\int e^{\lambda|g_1|} p_1(g_1)\, dg_1\right)\left(\int e^{\lambda|g_2|} p_2(g_2)\, dg_2\right) < \infty.$$

Thus the set $\mathfrak{M}_\lambda$ is convex and forms a semigroup with respect to convolution.

PROPOSITION 2. *Let $\{\mu_n\}$ be completely uniform, belong to $\mathfrak{M}_\lambda$, and let $M_k(\mu_n) \to M_k(\mu)$ in norm. Then $\mu_n \to \mu$ completely.*

Really, considering the operator moments on the basis $\{e^{im\psi}\}$ one can obtain the convergence of an integral with respect to μ_n of a function of the kind $\{r^n e^{im\varphi} e^{ik\alpha}\}$.

By virtue of the density of this system in the set of functions on the parameter space (r, φ, α) that question is reduced to the classical case.

4. THE FIRST OPERATOR MOMENT AND ITS PROPERTIES

Since

$$F_t(\mu_1 * \mu_2) = M_0(\mu_1 * \mu_2) + itM_1(\mu_1 * \mu_2) + \frac{(it)^2}{2!} M_2(\mu_1 * \mu_2) + \cdots$$
$$= F_t(\mu_1)F_t(\mu_2) = \left(\sum_k \frac{(it)^k}{k!} M_k(\mu_1)t^k\right) \left(\sum_m \frac{(it)^m}{m!} M_m(\mu_2)t^m\right),$$

the comparison of coefficients at powers t gives the following equalities:

$$\begin{array}{l} 1)\ M_0(\mu_1 * \mu_2) = M_0(\mu_1)M_0(\mu_2), \\ 2)\ M_1(\mu_1 * \mu_2) = M_0(\mu_1)M_1(\mu_2) + M_1(\mu_1)M_0(\mu_2), \\ 3)\ M_2(\mu_1 * \mu_2) = M_0(\mu_1)M_2(\mu_2) + 2M_1(\mu_1)M_1(\mu_2) + M_2(\mu_1)M_0(\mu_2), \\ \dots\dots\dots\dots\dots\dots\dots\dots\dots\dots\dots\dots \end{array} \qquad (10)$$

The operator moment $M_0(\mu)$ coincides with $F_0(\mu)$ and in general is not permutable with the other moments. This follows from the fact that $M_0(\mu)$ is diagonal in the base $\{e^{im\varphi}\}$, but from (5) it follows that other moments are not diagonal. Consider for simplicity a measure μ with a density $p(r, \varphi, \alpha)$ with respect to the invariant measure which can be represented by Fourier series in φ and α:

$$p(r, \varphi, \alpha) = \sum_{k,l} p_{kl}(r)e^{ik\varphi}e^{il\alpha}, \quad p_{kl}(r) = \frac{1}{(2\pi)^2}\int pe^{-ik\varphi}e^{-il\alpha}\, d\varphi\, d\alpha.$$

Then choosing $f(\psi) = e^{in\psi}$ we obtain that

$$\begin{aligned} M_1(\mu)e^{in\psi} &= \int r\cos(\psi - \varphi)e^{in\psi}e^{-in\alpha}\mu(dg) \\ &= \frac{1}{2}\left(\int re^{-i\varphi}e^{-in\alpha}\mu(dg)\right)e^{i(n+1)\psi} \\ &\quad + \frac{1}{2}\left(\int re^{i\varphi}e^{-in\alpha}\mu(dg)\right)e^{i(n-1)\psi} \end{aligned}$$
$$= \frac{1}{2}\left(\int rp_{1,r}(r)\, dr\right)e^{i(n+1)\psi} + \frac{1}{2}\left(\int rp_{-1,n}(r)\, dr\right)e^{i(n-1)\psi}. \qquad (11)$$

Consequently, the operator expectation is equal to zero, if

$$\int rp_{-1,n}(r)\, dr = \int rp_{1,n}(r)\, dr = 0$$

for all integer n.

In contrast to the classical case centering of arbitrary distributions on $M(2)$ is impossible even if $M_1(\mu)$ exists.
Indeed, it is easy to construct measures μ for which $M_1(\mu) = 0$. If we assume the measure μ has a density $p(z, \alpha)$, where z is an element of the plane of shifts and α is a rotation, then in view of (11) the equality

$$\int zp(z,\alpha)\,dz\,d\alpha = 0 \qquad \forall \alpha \in SO(2)$$

is necessary and sufficient for $M_1(\mu) = 0$. It follows from (11) that $M_1(g) \neq 0$ for any degenerate in g distribution, if $|g| \neq 0$, and $M_1(g) = 0$ if $|g| = 0$, i. e., if g is a rotation. Finally, if $M_1(\mu) = 0$, then $M_1(\mu g) \neq 0$ if $|g| \neq 0$. Indeed, if we suppose that $M_1(\mu g) = 0$, then in view of (10)

$$0 = M_1(\mu g) = M_0(\mu)M_1(g) + M_1(\mu)M_0(g) = M_0(\mu)M_1(g).$$

Since $M_1(g) \neq 0$ and the operator M_0 is the operator of multiplication for basic elements $\{e^{in\psi}\}$, we have $M_0(\mu)M_1(g) \neq 0$. Analogously, $M_1(\mu\nu) \neq 0$, if ν is an arbitrary measure with $M_1(\nu) \neq 0$.
However there are many measures which can not be centered by shifts (left or right).

PROPOSITION 3. *Let $M_1(\mu) \neq 0$. Then $M_1(\mu g_0) = 0$ for some $g_0 \in M(2)$ if and only if*

$$z_0 \int e^{-i(n+1)\alpha}\mu(dz,d\alpha) = -\int \bar{z}e^{-in\alpha}\mu(dz,d\alpha),$$

where

$$z_0 = \int ze^{i\alpha}\mu(dz,d\alpha), \quad \bar{z} = re^{-i\varphi}.$$

Proof. If $M_1(\mu g_0) = 0$ for some $g_0 \in M(2)$, then by virtue of (10) $M_0(\mu) \times M_1(g_0) = -M_1(\mu)M_0(g_0)$. Denoting

$$g_0 = (r_0, \varphi_0, \alpha_0), \quad \gamma_n = \int e^{-in\alpha}\mu(dz,d\alpha)$$

and using (11), we obtain

$$M_0(\mu)M_1(g_0) = \frac{1}{2}r_0e^{-i\varphi_0}e^{-in\alpha_0}\gamma_{n+1}e^{i(n+1)\psi} + \frac{1}{2}e^{-in\alpha_0}e^{i\varphi_0}e^{-in\alpha}\gamma_{n-1}e^{i(n-1)\psi}$$

and

$$-M_1(\mu)M_0(g_0) = -\frac{1}{2}e^{-in\alpha_0}\left[\int re^{-i\varphi}e^{-in\alpha}\mu(dz,d\alpha)\right]e^{i(n+1)\psi}$$
$$-\frac{1}{2}e^{-in\alpha_0}\left[\int re^{i\varphi}e^{-in\alpha}\mu(dz,d\alpha)\right]e^{i(n-1)\psi}.$$

Let us compare the coefficients at $e^{i(n+1)\psi}$ which must be equal. Transfering all the terms to one side, substituting $\int e^{-in\alpha}\mu(dz,d\alpha)$ for γ_n, and reducing $\frac{1}{2}e^{-in\alpha_0}$, we obtain that

$$\int (z_0 e^{-i\alpha} + re^{-i\varphi})e^{-in\alpha}\mu(dz,d\alpha) = 0$$

for any integer n. For $n = -1$

$$z_0 = -\int \bar{z}e^{i\alpha}\mu(dz,d\alpha).$$

The simple example of a measure μ noncentered by shifts may be constructed as follows. Let μ has a density $p(z,\alpha)$,

$$p(z,\alpha) = \sum_{k=0}^{N} c_k p_k(z)\cos^{2k}\alpha.$$

The function $p(z,\alpha)$ is a density on $M(2)$ if $p_k(z) \geq 0$, $\int p_k(z)\,dz = 1$ and the constants c_k satisfy the conditions

$$\sum c_k \cos^{2k}\alpha \geq 0, \qquad \alpha \in SO(2),$$

$$\sum c_k \int_0^{2\pi} \cos^{2k}\alpha\,d\alpha = 1.$$

We assume that $\int \bar{z}p_N(z)\,dz \neq 0$.
Then for $n = 2N$

$$\int e^{-i(2N+1)\alpha}p(z,\alpha)\,dz\,d\alpha = 0.$$

On the other hand,

$$\int \bar{z}e^{i2N\alpha}p(z,\alpha)\,dz\,d\alpha = \int \bar{z}e^{i2N\alpha}\left(\sum_{k=0}^{N} c_k p_k(z)\cos^{2k}\alpha\right)dz\,d\alpha =$$

$$= 2\pi c_N \int \bar{z}p_N(z)\,dz \neq 0.$$

Consequently, the constructed measure μ does not satisfy the main condition of Proposition 3.
There are another specific properties of the operator expectation on $M(2)$.

PROPOSITION 4. *If $M_1(\mu) = 0$, then $M_1(\mu^n) = 0$. If $M_1(\mu)$ exists and the projection of μ^n on $SO(2)$ tends weakly to the uniform distribution, then $M_1(\mu^k) \to M$ weakly.*

Indeed, from 1) and 2) of (10) it follows that

$$M_1(\mu^{n+1}) = \sum_{k=0,n} M_0^k(\mu)M_1(\mu)M_0^{n-k}.$$

This implies the first statement. Let $M_1(\mu) \neq 0$. In view of the condition on μ^n the moduli of all coefficients $M_0(\mu)e^{im\psi} = \gamma_m$, $m \neq 0$, are less than 1. Acting by operator $M_1(\mu^{n+1})$ on $e^{im\psi}$, $m \neq 0$, and taking into account (11) and the equality $M_0(\mu^n)e^{im\psi} = \gamma_m^n e^{im\psi}$ it is easy to obtain that $M_1(\mu^{n+1})e^{im\psi} \to 0$ for $m \neq 0$. If $m = 0$, then $M_0(\mu)1 = 1$ and

$$M_1(\mu^{n+1})1 = \sum_{k=0}^{n} M_0^k(\mu)M_1(\mu)1$$

has a limit as $n \to \infty$ which is equal to $c_1 e^{i\psi} + c_2 e^{-i\psi}$, where the constants c_1 and c_2 may be easily calculated.

5. THE SECOND OPERATOR MOMENT AND A NUMERICAL ANALOG OF VARIANCE

By definition of the second operator moment $M_2(\mu)$

$$M_2(\mu)f(\psi) = \int r^2 \cos^2(\psi - \varphi)f(\psi - \alpha)\,\mu(dg).$$

Expressing $\cos^2(\psi - \varphi)$ in terms of the exponents $\{e^{im\psi}\}$, we obtain

$$M_2(\mu)e^{im\psi} = \frac{1}{4}\left(\int r^2 e^{-i2\varphi} e^{-im\alpha}\mu(dg)\right) e^{i(m+2)\psi} + \frac{1}{2}\left(\int r^2 e^{-im\alpha}\mu(dg)\right) e^{im\psi}$$

$$+\frac{1}{4}\left(\int r^2 e^{i2\varphi} e^{-im\alpha}\mu(dg)\right) e^{i(m-2)\psi}.$$

Thus the condition $M_2(\mu) = 0$ is equivalent to the equalities

$$\int r^2 p_{2n}(r)\,dr = \int r^2 p_{-2,n}(r)\,dr = \int r^2 p_{0,n}(r)\,dr = 0$$

for all integer n. In particular, since

$$\int r^2 p_{00}(r)\,dr = 0,$$

the distribution μ is concentrated on the rotations $SO(2)$. Conversely, if μ is concentrated on $SO(2)$, then $M_2(\mu) = 0$.

That fact is in complete agreement with the classical case, since the distribution μ with zero variance is concentrated in one point. This permits us to regard the characteristic $\int r^2\mu(dg) = \sigma^2(\mu)$ as a numerical analogy of variance. This characteristic coincides with the usual variance in the case of the radial distribution. Unfortunately the equality $\sigma^2(\mu_1 * \mu_2) = \sigma^2(\mu_1) + \sigma^2(\mu_2)$ is not true for all distributions.

Definition 3. Distributions μ_1 and μ_2 on $M(2)$ are called ***orthogonal***, if for any $m \neq 0$ in each pair

$$\left(\int e^{im\alpha}\mu_1(d\varphi, d\alpha),\ \int e^{im\varphi}\mu_2(d\varphi, d\alpha)\right),$$

$$\left(\int e^{im\varphi}\mu_1(d\varphi, d\alpha),\ \int e^{im\alpha}\mu_2(d\varphi, d\alpha)\right)$$

one of the values is equal to zero.

PROPOSITION 5. *If distributions μ_1 and μ_2 are orthogonal and*

$$p_{00}^{(i)}(r) = \frac{1}{(2\pi)^2}\int \mu_i(d\varphi, d\alpha)$$

is a density of the radial distributions μ_i, $i = 1, 2$, then the density of the radial distribution of convolution $\mu_1 * \mu_2$ is equal to

$$p_{00}(r) = \frac{1}{2\pi}\int p_{00}^{(1)}\left(\sqrt{r^2 + r_2^2 + 2rr_2\cos\psi}\right)p_{00}^{(2)}(r_2)r_2\,dr_2\,d\psi \tag{12}$$

and the equality $\sigma^2(\mu_1 * \mu_2) = \sigma^2(\mu_1) + \sigma^2(\mu_1)$ is valid.

Proof. For simplicity we consider the case where the distributions μ_1 and μ_2 have densities p_1 and p_2 with respect to the invariant measure on $M(2)$. Then the density p of convolution $\mu_1 * \mu_2$ is

$$p(g_1) = \int p_1(g_1g_2^{-1})p_2(g_2)\,dg_2, \quad p_{00}(r) = \frac{1}{(2\pi)^2}\int p(g)\,\mu(d\varphi, d\alpha). \tag{13}$$

Multiplying the both parts of the left-side equality by the Bessel function $J_0(Rr) = J_0(R|g_1|)$ and integrating with respect to the measure $\mu_1(dg_1)$, we obtain

$$\begin{aligned} J_0(Rr)p_{00}(r)r\,dr &= \int\!\!\int J_0(R|g_1|)p_1(g_1g_2^{-1})p_2(g_2)\,dg_1\,dg_2 \\ &= \int J_0(R|g_1g_2|)p_1(g_1)p_2(g_2)\,dg_1\,dg_2. \end{aligned} \tag{14}$$

Put

$$p_1(g_1) = \sum_{k,l} p_{kl}^{(1)}(r)e^{ik\varphi_1}e^{il\alpha_1}, \quad p_2(g_2) = \sum_{m,n} p_{m,n}^{(2)}(r)e^{im\varphi_2}e^{in\alpha_2}$$

and

$$g_1 = g_1(z_1, \alpha_1), \quad g_2 = g_2(z_2, \alpha_2), \quad g_1g_2 = g(z_1 + z_2e^{i\alpha_1}, \alpha_1 + \alpha_2),$$

$$|g_1| = r, \quad |g_2| = r_2, \quad |g_1g_2| = |z_1 + z_2e^{i\alpha_1}|, \quad \arg z_1 = \varphi_1, \quad \arg z_2e^{i\alpha_1} = \varphi_2 + \alpha_1.$$

Then applying the addition formula of the Bessel functions, we obtain (Vilenkin, 1968) that

$$J_0(R|g_1g_2|) = \sum_{\alpha=-\infty}^{\infty} (-1)^d e^{id(\varphi_2+\alpha_1)} J_d(r)J_d(r_2). \tag{15}$$

Substituting the expansions of $p_1(g_1)$, $p_2(g_2)$ and (15) into (14) and integrating with respect to φ_1, φ_2, α_1 and α_2, we obtain

$$\int J_0(Rr)p_{00}(r)r\,dr$$

$$= \sum_d (-1)^d \int J_d(Rr_1)J_d(Rr_2)p^{(1)}_{0,-d}(r_1)p^{(2)}_{-d,0}(r_2)r_1r_2\,dr_1\,dr_2. \tag{16}$$

By the definition of orthogonal measures, the product $p^{(1)}_{0,-d}(r_1)p^{(2)}_{-d,0}(r_2)$ is always zero if $d \neq 0$. Consequently, (16) yields

$$\int J_0(Rr)p_{00}(r)r\,dr$$

$$= \left(\int J_0(Rr_1)p^{(1)}_{00}(r_1)r_1\,dr_1\right)\left(\int J_0(Rr_2)p^{(2)}_{00}(r_2)r_2\,dr_2\right). \tag{17}$$

The function $\int J(itr)p(r)r\,dr$ is the characteristic function of the circular distribution with the density $p(r) = p(0,r)$ on the plane. Consequently, from (17) it follows that $p_{00}(r)$ is the density of the radial distribution of the sum of two circular distributions on the plane with radial densities $p^{(1)}_{00}(r)$ and $p^{(2)}_{00}(r)$. Then the density is expressed by (3) and $\sigma^2(\mu_1 * \mu_2) = \sigma^2(\mu_1) + \sigma^2(\mu_2)$ is valid.
Thus it remains unsolved the problem of definition of a numerical variance for distributions on $M(2)$ or $M(d)$ satisfying the axioms which were used in the definition of the variance of distributions on compact groups (Maximov, 1970).

6. APPLYING TO THE CENTRAL LIMIT THEOREM

The operator moments can be applied to obtain the central limit theorem for convolution of the nonidentical distributions μ_n from $\mathfrak{M}_\lambda$. This class contains the important particular case where the supports of every distribution are bounded but expand as $n \to \infty$. The central limit theorem is well-known for identically distributed random variables when the second moment for the projections is finite (see (Tutubalin, 1967; Gorostiza, 1973)).

PROPOSITION 6. *Let distributions $\mu_1, \mu_2, \ldots$ from $\mathfrak{M}_\lambda$ on $M(2)$ be given and let $p_1, p_2, \ldots; q_1, q_2, \ldots$ be their projections on the plane and $SO(2)$ respectively. Let $\nu_1, \nu_2, \ldots$ be the circular distributions on the plane obtained by rotation of $p_1, p_2, \ldots$. Then, if*

1) *the central limit theorem is applied to the sums $\frac{1}{\sqrt{n}}(\nu_1 + \cdots + \nu_n)$ and the kth moment of $\frac{1}{\sqrt{n}}(\nu_1 + \cdots + \nu_n)$ converges as $n \to \infty$ for every k;*

2) *the distributions* $q_1 q_2 \cdots q_n = q^{(n)}$ *converge as* $n \to \infty$ *to the uniform distribution on* $SO(2)$ *quickly enough, for example, so that*

$$\sum_{n=0}^{\infty} |e^{im\alpha} q^{(n)}(d\alpha)| < \infty \quad \textit{for all} \quad m,\ m \neq 0,$$

the central limit theorem is applicable to distributions $\frac{1}{\sqrt{n}}\mu_1 * \cdots * \mu_n$.
Clearly, we have no possibility to give here a complete proof, but the principal idea will be demonstrated while estimating the first moments.

1) For $M_0(\mu_1 \cdots \mu_n)$ we have

$$M_0(\mu_1 \cdots \mu_n) e^{im\psi} = \gamma_m^{(1)} \cdots \gamma_m^{(n)} e^{im\psi}, \quad \gamma_m^{(k)} = \int e^{im\alpha} q_k(d\alpha).$$

In view of condition 2

$$\lim_{n\to\infty} \gamma_m^{(1)} \cdots \gamma_m^{(n)} = 0 \quad \text{for } m \neq 0$$

and

$$M_0(\mu_1 \cdots \mu_n) 1 = 1.$$

2) For the first operator moment we have from 1), 2) of (10) that

$$M_1(\mu_1 \cdots \mu_n) = M_0(\mu_1 \cdots \mu_{n-1}) M_1(\mu_n)$$

$$+ M_0(\mu_1 \cdots \mu_{n-2}) M_1(\mu_{n-1}) M_0(\mu_n) + \cdots + M_1(\mu_1) M_0(\mu_2 \cdots \mu_n). \tag{18}$$

Let us show that

$$M_1\left(\frac{1}{\sqrt{n}}\mu_1 \cdots \mu_n\right) = \frac{1}{\sqrt{n}} M_1(\mu_1 \cdots \mu_n) \to 0$$

weakly. From (18)

$$\frac{1}{\sqrt{n}} M_1(\mu_1 \cdots \mu_n) 1 = \frac{1}{\sqrt{n}} \{ M_0(\mu_1 \ldots \mu_{n-1}) M_1(\mu_1) 1$$

$$+ M_0(\mu_1 \cdots \mu_{n-2}) M_1(\mu_{n-1}) M_0(\mu_n) 1 + \cdots + M_1(\mu_1) M_0(\mu_2 \cdots \mu_n) 1 \}.$$

Then

$$M_0(\mu_1 \cdots \mu_n) 1 = 1, \quad M_1(\mu_k) 1 = \sigma_{1,0}^{(k)} e^{i\psi} + \sigma_{-1,0}^{(k)} e^{-i\psi},$$

and

$$M_0(\mu_1 \cdots \mu_{k-1}) M_1(\mu_k) 1 = \sigma_{1,0}^{(k)} \gamma_1^{(1)} \cdots \gamma_1^{(k-1)} e^{i\psi} + \sigma_{-1,0}^{(k)} \gamma_{-1}^{(1)} \cdots \gamma_{-1}^{(k-1)} e^{-i\psi}.$$

By virtue of conditions 1) and 2)

$$\frac{1}{\sqrt{n}}\sigma_{1,0}^{(k)} \to 0, \quad \frac{1}{\sqrt{n}}\sigma_{-1,0}^{(k)} \to 0, \quad \sum_k |\gamma_m^{(1)} \cdots \gamma_m^{(k-1)}| < \infty.$$

Therefore,

$$\left|\frac{1}{\sqrt{n}} M_1(\mu_1 \cdots \mu_n) e^{im\psi}\right| \to 0.$$

We note that for odd moments the expressions $M_{2k+1}(\mu)e^{im\psi}$ do not contain a constant term. Then it can be analogously shown that

$$\left(\frac{1}{\sqrt{n}}\right)^{2k+1} |M_{2k+1}(\mu_1 \cdots \mu_n) e^{im\psi}| \to 0, \qquad m \neq 0.$$

3) For the second operator moment we have from (10)

$$M_2(\mu_1 \cdots \mu_n) = M_0(\mu_1 \cdots \mu_{n-1}) M_2(\mu_n)$$

$$+ M_1(\mu_1 \cdots \mu_{n-1}) M_1(\mu_n) + M_2(\mu_1 \cdots \mu_{n-1}) M_0(\mu_n).$$

If we continue the decomposition of $M_2(\mu_1 \cdots \mu_{n-1})$, we obtain on the nth step

$$M_2(\mu_1 \cdots \mu_n) = M_0(\mu_1 \cdots \mu_{n-1}) M_2(\mu) + \cdots + M_2(\mu_1) M_0(\mu_2 \cdots \mu_n) + S, \quad (19)$$

where S is the sum of all terms containing $M_1(\mu_1 \cdots \mu_k)$. Since

$$\frac{1}{\sqrt{n}} |M_1(\mu_1 \cdots \mu_k) e^{im\psi}| \to 0,$$

we obtain that

$$\frac{1}{n} |S e^{im\psi}| \to 0.$$

The remaining part of the sum is similar to sum (18), where M_1 is changed by M_2.
Consequently, by the same arguments

$$\frac{1}{n} |M_2(\mu_1 \cdots \mu_n) e^{im\psi}| \to 0$$

for $m \neq 0$. For $m = 0$

$$M_2(\mu_1 \cdots \mu_n) 1 = M_0(\mu_1 \cdots \mu_{n-1}) M_2(\mu_n) 1$$

$$+ M_0(\mu_1 \cdots \mu_{n-2}) M_2(\mu_{n-1}) 1 + \cdots + M_2(\mu_1) 1 + S1. \quad (20)$$

For even operator moments the expression $M_{2d}(\mu_k)1$ contains the constant term

$$\frac{1}{2^{2d}}\binom{2d}{d}\int r^{2d}\mu_k(dr,d\varphi,d\alpha)=\frac{1}{2^{2d}}\binom{2d}{d}\frac{1}{(2\pi)^2}\int r^{2d}\rho_k(r)r\,dr\neq 0, \qquad (21)$$

where $\rho_k(r)\,r\,dr$ is the density of the radial distribution of ν_k. The sum of terms (21), $k=1,\ldots,n$, divided by n under condition 1) tends to some positive number. The sum of the other terms divided by n tends to zero. Similar reasonings are applicable to arbitrary operator moments and can be realized by induction. The application of Propositions 1 and 2 completes the proof.

REFERENCES

Gorostiza, L. G. (1973). The central limit theorem for random motions of d-dimensional Euclidean space. *Ann. Probab.* 1, 603–612.

Grenander, U. (1963). *Probabilities on Algebraic Structures.* Wiley, New York–London.

Maximov, V. M. (1970). On the theory of variance for probability distributions on compact groups. *Soviet Math. Dokl.* 192, 732–735 (in Russian).

Maximov, V. M. (1979). Lattice motions of the Euclidean plane. *Math. USSR-Sb.* 109 (151), 275–290 (in Russian).

Tutubalin, V. N. (1967). The central limit theorem for random motions of the Euclidean space. *Moscow Univ. Math. Bull.* 6, 100–108 (in Russian).

Vilenkin, N. Ya. (1968). *Special Functions and the Theory of Group Representation.* AMS, Providence.

Probabilistic Methods in Discrete Mathematics, pp. 329 – 336
V. F. Kolchin et al. (Eds.)
1993 TVP/VSP

PERCOLATION METHOD IN A SYSTEM OF NUMEROUS PARTICLES

M. V. MENSHIKOV and K. D. PELIKH
Moscow State University, Vorobjevy Gory, 119899 Moscow, Russia
Moscow Institute of Steel and Composites, Pervomaiskaya Street 7, 144000 Elektrostal, Russia

ABSTRACT

A discrete time interacting particle system and a certain non-Bernoulli percolation model associated with it are investigated. We develop a new technique to generalize some Bernoulli percolation results concerning critical probability estimates, the uniqueness of the critical probability, exponential decay of the cluster size distribution for subcritical phase. These results are interpreted for the introduced particle system.

1. INTRODUCTION

A discrete time voter model is a conditionally independent Markov process $\xi_t^1(x)$, $x \in \mathbf{Z}$, $t \in \mathbf{Z}_+$, where $\xi_t^1(x)$ takes values 0 and 1. In addition,

$$\mathbf{P}\{\xi_{t+1}^1(x) = 1 \mid \bar{\xi}_t^1\} = \sum_{y\in Q} a_y \xi_t^1(x+y), \tag{1}$$

where $Q = \{y_1, \dots, y_k\} \subset \mathbf{Z}$ is a finite set, $a_y > 0$, $\sum_{y\in Q} a_y = 1$.
Malyshev (Malyshev, 1991) considered the following perturbation $\xi_t^2(x)$ of the model (1) by quadratic monomial:

$$\mathbf{P}\{\xi_{t+1}^2(x) = 1 \mid \bar{\xi}_t^2\} = \sum_{y\in Q} \beta_y \xi_t^2(x+y) + \gamma \xi_t^2(x+y_l)\xi_t^2(x+y_m) + \delta,$$

where $\sum_{y\in Q} \beta_y + \gamma + \delta = 1$, $\beta_y, \gamma, \delta > 0$, y_l, y_m are fixed elements of the set Q. The behaviour of this process for different cases of parameters β_y, γ, δ was investigated. The author has considered in detail the case $\delta > \gamma$ $(\gamma < (1 - \sum \beta_y)/2)$ and the case $\delta = 0$ $(\gamma = 1 - \sum \beta_y)$.
Our aim is to analyze the domain $0 < \delta \leq \gamma$. We investigate the process behaviour under linear variation of the parameters. Let b_y and c are arbitrary but fixed numbers such that $\sum_{y\in Q} b_y + c = 1$. We deal with the following process

$$\mathbf{P}\{\xi_{t+1}(x) = 1 \mid \xi_t\} = \sum_{y\in Q} \varepsilon b_y \xi_t(x+y) + \varepsilon c \xi_t(x+y_l)\xi_t(x+y_m) + 1 - \varepsilon, \quad \varepsilon \in [0,1].$$

It is convenient to analyze interaction processes with the help of dual processes (see (Holley and Stroock, 1979; Griffeath, 1979)). Such dual process for ξ_t is described in (Malyshev et al., 1989), see also (Malyshev, 1991). This process η_τ spreads in "converse time" and can be interpreted in the birth-death particle terminology. Let us consider n particles placed at points $x_1, x_2, \dots, x_n \in \mathbf{Z}$ at the time $\tau = 0$. At each time each of particles can take independently of others one of the following actions:

1) to jump from x_i to $x_i + y$, $y \in Q$, with probability εb_y;
2) to split into two particles placed at points $x_i + y_l$, $x_i + y_m$ with probability εc;
3) to die with probability $1 - \varepsilon$.

Furthermore, if two or more particles come to the same point at a time, they glue together and are considered as a single particle.
The process η is associated with the representation of the correlation function

$$p_t(x_1, \dots, x_n) = \langle \xi_t(x_1)\xi_t(x_2)\dots\xi_t(x_n)\rangle$$

by functions $p_k(\dots)$, where $k < t$.
The duality of $\xi_t(x)$ and η_τ means that

$$\mathbf{P}\{\xi_t^D(x) = 0 \text{ for some } x \in E\} = \mathbf{P}\{\eta_\tau^E(y) = 1 \text{ for some } y \in D\},$$

where ξ_t^D is the process starting with 0 on $D \subset \mathbf{Z}$ and 1 on $\mathbf{Z}\backslash D$ and η_t^E is the process starting with 1 on $E \subset \mathbf{Z}$ and 0 otherwise.
The duality makes it possible to consider the process η_τ instead of ξ_t. It is well known that percolation models are widely used for studying the interaction particle systems (see (Toom, 1968; Harris, 1978; Durrett, 1988)). We introduce a percolation model associated with η_τ.
Consider an oriented graph G with $V \rightleftharpoons \mathbf{Z} \times \mathbf{Z} = \{(x,t)\}$ as the vertex set and $U = \bigcup_{x,t\in\mathbf{Z}} U_{x,t}$ as the set of oriented edges, where $U_{x,t}$ is a set of oriented edges $[a; b\rangle$ with the starting point $a = (x,t)$ and the terminal $b = (x+y, t+1)$, $y \in Q$. Each edge can be open or closed. The stochastic behaviour is determined by the following rules. For each (x,t) independently of other vertices

i) all edges $u \in U_{x,t}$ are closed with probability $1-\varepsilon$;
ii) the edge $[(x,t);(x+y,t+1)\rangle$ is open and other edges from $U_{x,t}$ are closed with probability εb_y, $y \in Q$;
iii) edges $[(x,t);(x+y_l,t+1)\rangle$ and $[(x,t);(x+y_m,t+1)\rangle$ are open and other edges from $U_{x,t}$ are closed with probability εc.

As usual, we introduce the critical values ε_T and ε_H corresponding to the open edge cluster of the origin being finite a. s. for $\varepsilon < \varepsilon_H$, infinite with positive probability for $\varepsilon > \varepsilon_H$, having finite mean size for $\varepsilon < \varepsilon_T$ and infinite mean size for $\varepsilon > \varepsilon_T$.
The sense of the critical values ε_H and ε_T for the process ξ_t is the following. For $\varepsilon < \varepsilon_H$ the process $\xi_t(x)$ tends to the state $\xi_\infty(x) \equiv 1$ as $t \to \infty$ regardless of the initial state $\bar{\xi}_0(x)$. For $\varepsilon > \varepsilon_H$ the process tends to a nontrivial measure μ_∞, which depends on the initial measure. For $\varepsilon < \varepsilon_T$ the process converges quickly to $\xi_\infty(x) \equiv 1$.

Using the percolation technique,

1) we give a method of finding lower bounds for ε_T; in particular we get the estimate $\varepsilon_T > (1+c)^{-1}$;
2) we show that $\varepsilon_H < 1$; for the process ξ_t it means that there exists a nontrivial supercritical phase different from subcritical, in particular, when $\varepsilon > \varepsilon_H$ any initial measure such that

$$\sup_{x_1,\dots,x_n} p_0(x_1,\dots,x_n) \to 0 \qquad (n\to\infty)$$

converges to some nontrivial measure with $p_\infty(x) < 1$;
3) we show that the exponential estimate

$$\mathbf{P}_\varepsilon\{D_n\} < \exp(-an), \tag{2}$$

holds for $\varepsilon < \varepsilon_H$, where D_n is the event

$$D_n = \{\text{There exists a path of open edges starting at the origin with the length not exceeding } n\}. \tag{3}$$

Estimate (2) implies the coincidence of ε_T and ε_H and means an exponential convergence of the correlation functions of the process ξ_t when $\varepsilon < \varepsilon_H$:

$$\left|p_\infty(x_1,\dots,x_n) - p_t(x_1,\dots,x_n)\right| = \left|1 - p_t(x_1,\dots,x_n)\right| < \exp(-at).$$

Proofs of these facts are similar to ones for the Bernoulli case. That is why we briefly sketch them and refer the reader to the literature.

2. PERCOLATION MODEL

Consider the oriented graph G introduced above. We suppose that $Q = \{y_1, y_2, \dots, y_k\}$ and $y_l, y_m \in Q$. We declare the edge states (open or closed) being induced by vertex states. We take for a state of the vertex $v = (x,t)$ a k-vector composed of zeros and units. The vertex state determines the edge states according to the following rule. We set the state of edge $[(x,t);(x+y_i,t+1)\rangle$ to be open (closed) if the ith coordinate of the vertex (x,t) state is 1 (0). There are $k+2$ possible states of the vertex only and each of them can be realized with some probability. Namely,

$$\mu_v\big(\omega(v) = e_i\big) = \varepsilon(v) b_{y_i}, \qquad i = 1,2,\dots,k,$$

where $e_i = (e_i(1),\dots,e_i(k))$, $e_i(j) = \delta_i(j)$ ($\delta_i(j)$ is the Kronecker symbol);

$$\mu_v\big(\omega(v) = e_{ml}\big) = \varepsilon(v)c,$$

where $e_{ml}(m) = 1$, $e_{ml}(l) = 1$, $e_{ml}(j) = 0$ for $j \neq m, l$,

$$\mu_v\big(\omega(v) = e_0\big) = 1 - \varepsilon(v),$$

where $e_0(j) = 0, j = 1, 2, \ldots, k$.
We set $\varepsilon(v) \in [0,1]$, $\sum_{y \in Q} b_y + c = 1$. Thus, we take

$$\Omega_V = \prod_{v \in V} \{\omega(v)\}$$

as the configuration space, where $\{\omega(v)\} = \mathcal{E} = \{e_1, e_2, \ldots, e_k, e_{ml}, e_0\}$ is the set of possible vertex states described above. Configurations of Ω_V are denoted by $\omega = \{\omega(v)\}_{v \in V}$. We take $\mathcal{B}_V$ to be the σ-field generated by the finite cylinder sets of Ω_V. And we define the measure on $(\Omega_V, \mathcal{B}_V)$ as follows

$$\mathbf{P}_V = \prod_{v \in V} \mu_v,$$

where μ_v is given above.
Thus, we have the field of independent random variables which take $k+2$ values. Let $\omega \in \Omega_V$ be a fixed configuration. The cluster of a vertex v, denoted by $W(v)$, is the set of all vertices which can be reached from v by paths of open edges. Introduce the following quantity:

$$\Theta_{\mathbf{P}_V}(v) = \mathbf{P}_V\{\#W(v) = \infty\},$$

where $\#W(v)$ denotes the number of vertices in the cluster $W(v)$.
It is easy to show that $\Theta_{\mathbf{P}_V}(v)$ is a monotone function of $\varepsilon(v_1)$, where v_1 is an arbitrary vertex. In particular, if we set $\varepsilon(v) = \varepsilon$ for each vertex $v \in V$ and denote the corresponding measure by $\mathbf{P}_\varepsilon$, then

$$\Theta(v, \varepsilon) = \mathbf{P}_\varepsilon\{\#W(v) = \infty\}$$

is a monotone function of ε. Furthermore, it is easy to see that $\Theta(v, 0) = 0$ and $\Theta(v, 1) = 1$. It follows that there is a critical value

$$\varepsilon_H = \sup\{\varepsilon \in [0,1]\colon\ \Theta(v, \varepsilon) = 0\}.$$

The value ε_H does not depend on the choice of the vertex v by virtue of the translation invariance of G and the measure $\mathbf{P}_\varepsilon$.
Then we can introduce

$$\varepsilon_T = \sup\{\varepsilon \in [0,1]\colon\ \mathbf{E}_\varepsilon\{\#W(v)\} < \infty\},$$

where $\mathbf{E}_\varepsilon(\cdot)$ denotes the expectation with respect to $\mathbf{P}_\varepsilon$. Like ε_H, the value ε_T does not depend on the choice of the vertex v. It is clear that if $\varepsilon > \varepsilon_H$, that is $\Theta(v, \varepsilon) > 0$, then $\mathbf{E}_\varepsilon\{\#W(v)\} = \infty$. This implies that $\varepsilon_T \le \varepsilon_H$. In reality $\varepsilon_T = \varepsilon_H$, but it is not obvious (see Section 5).

3. LOWER AND UPPER BOUNDS FOR THE CRITICAL POINTS

First we show that

$$0 < \varepsilon_T \leq \varepsilon_H < 1.$$

To prove the inequality $\varepsilon_T > 0$, we use the comparison with a branching process (see (Menshikov et al., 1986, Chapter 3).
The vertices

$$V^{(r)} = \{(x,t)\colon\ x \in \mathbf{Z},\ t = r\}$$

are called r-level vertices.
Let ξ_k be the random value which is equal to the number of k-level vertices of the cluster $W(0)$. We denote the mean number of the k-level vertices $\mathbf{E}_\varepsilon(\xi_k)$ by $M_{k,\varepsilon}$. Let us introduce a discrete time branching process family $\zeta_k(t), k = 1, 2, \ldots, m$. For $\zeta_k(t)$ at each time each of particles can be splitted into j particles with probability $\mathbf{P}_\varepsilon\{\xi_k = j\}$. By comparing the percolation process and the process $\zeta_k(t)$, one can prove the following theorem.

THEOREM 1. *If for ε_0 there exists k such that $M_{k,\varepsilon_0} < 1$, then*

$$\mathbf{E}_{\varepsilon_0}(\#W(0)) < \infty$$

and hence $\varepsilon_0 < \varepsilon_T$.

It is easy to see that

$$M_{1,\varepsilon} = \sum_{y \in Q} \varepsilon b_y + 2\varepsilon c = \varepsilon(1 + c).$$

Considering the quantity $M_{2,\varepsilon}$, one can show that $M_{2,\varepsilon} < (M_{1,\varepsilon})^2$ for $\varepsilon > 0$. This implies the strict inequality $\varepsilon_T > (1 + c)^{-1}$. Generally, Theorem 2 is true (see (Menshikov et al., 1986)).

THEOREM 2. *The equality $\varepsilon_M = \varepsilon_T$ holds, where*

$$\varepsilon_M = \sup\{\varepsilon \in [0,1]\colon\ \inf_{r \in \mathbf{N}} M_{r,\varepsilon} < 1\}.$$

Making use of Theorem 2, we prove the lemma on the process behavior at the critical point.

LEMMA 1. *For $\varepsilon = \varepsilon_T$*

$$\mathbf{E}_\varepsilon\{\#W(0)\} = \infty.$$

Proof. We suppose that the converse statement holds. Then there exists r_0 such that $\mathbf{E}_{\varepsilon_T}(\xi_{r_0}) < 1$. We note that $\mathbf{E}_{\varepsilon_T}(\xi_{r_0})$ is a polynomial of ε_T. Hence there exists $\eta > 0$ such that $\mathbf{E}_{\varepsilon_T+\eta}(\xi_{r_0}) < 1$ and $\varepsilon_T + \eta < \varepsilon_M$. It contradicts to Theorem 1. Lemma is proved.

THEOREM 3. *The inequality $\varepsilon_H < 1$ holds.*

To prove this, we cannot use the standard contour argument because the structure of dual circuits is very complicated. To avoid this, we follow the paper (Harris, 1978) and combine the renormalization method and dual technique.

4. INEQUALITIES FOR PROBABILITIES OF MONOTONE EVENTS

Introduce some notions. We say that a state $e' \in \mathcal{E}$ is greater than a state $e'' \in \mathcal{E}$ ($e' \succ e''$) if and only if $e' \neq e''$ and each of coordinates of the vector e' is more or equal to the corresponding coordinate of the vector e''. A function $f(\omega): \Omega_V \to \mathbf{R}$ is said to be increasing if $\omega_1(v) \succ \omega_2(v)$ for all $v \in V$ implies $f(\omega_1) \geq f(\omega_2)$. An event E is said to be increasing if its indicator function $I_E(\omega)$ is increasing.

Example. The event $D_n(v)$ from (3) is increasing.

The definition of an essential vertex differs from its Bernoulli analogue.

Definition. Let $E \in \mathcal{B}_V$ be an event and $\omega \in \Omega_V$ be a configuration. A vertex v is called essential for (E, ω) if there exists $\omega' \in \Omega_V$ such that ω' differs from ω by the state of the vertex v only and $I_E(\omega) = 1 - I_E(\omega')$.

PROPOSITION (Russo's formula). *Let E be an increasing event which depends on finitely many vertex states. Then*

$$\frac{d}{d\varepsilon}\mathbf{P}_\varepsilon(E) \geq C\mathbf{E}_\varepsilon(N(E)), \tag{4}$$

where $N(E)$ is the number of vertices which are essential for E and C is a constant.

Inequality (4) is used for proving statements of the next section.

The well-known BK-inequality holds for our problem:

$$\mathbf{P}_\varepsilon\{A \square B\} \leq \mathbf{P}_\varepsilon\{A\}\mathbf{P}_\varepsilon\{B\}, \tag{5}$$

where A, B are increasing events and $A\square B$ is the event "A and B occur disjointly". Proof of these theorems based on the original ideas of (Berg and Kesten, 1985; Russo, 1981).

5. AN EXPONENTIAL ESTIMATE FOR SUBCRITICAL PHASE. THE UNIQUENESS OF THE CRITICAL POINT

Our aim is to prove the equality $\varepsilon_T = \varepsilon_H$. Consider $\mathbf{P}_\varepsilon\{D_n(v)\}$, where $D_n(v)$ is the event from (3).

THEOREM 4. *For any $\varepsilon < \varepsilon_H$ there exists $a_1 = a_1(\varepsilon) > 0$ such that for any n*

$$\mathbf{P}_\varepsilon\{D_n(v)\} < \exp\{-a_1 n\}.$$

THEOREM 5. *The equality $\varepsilon_T = \varepsilon_H$ holds for this (non-Bernoulli) percolation model.*

We describe the main ideas of the proof, which are similar to the corresponding ideas for the Bernoulli percolation (see (Menshikov et al., 1986; Grimmett, 1989)).

First, one can deduce from (4) the inequality

$$\frac{\mathbf{P}_{\varepsilon_1}\{D_n(0)\}}{\mathbf{P}_{\varepsilon_0}\{D_n(0)\}} \le \exp\left\{-C_1 \int_0^1 \mathbf{E}_{\varepsilon(t)}\left[N(D_n(0)) \mid D_n(0)\right] dt\right\}, \tag{6}$$

where $0 < \varepsilon_1 < \varepsilon_0 < 1$ are arbitrary and $\varepsilon(t) = t\varepsilon_0 + (1-t)\varepsilon_1$. This inequality is very significant for the proof. The main idea is to show that if $\varepsilon < \varepsilon_H$ and there is a "large" cluster, then there are "many" vertices which are essential for the event $D_n(0)$.

The structure of this cluster is similar to the Bernoulli cluster. We suppose that the event $D_n(0)$ occurs for configuration ω. Let $b_1, b_2, \ldots, b_s$ be the essential vertices for $(D_n(0), \omega)$ arranged according to increasing of their levels. All open paths of the length n pass through these vertices in the order indicated.

Let $\nu_i(\omega)$ be the difference between levels of the essential vertices b_i and b_{i-1}. Using BK-inequality (5), one can prove the following lemma.

LEMMA 2. *There exists $a_1 = a_1(\varepsilon) > 0$ such that for any $n, d \in \mathbf{N}$*

$$\mathbf{P}_\varepsilon\{\nu_k > d \mid D_n(0)\} \le a_1 \mathbf{P}_\varepsilon\{D_d(0)\}.$$

By Lemma 2 we can obtain a lower bound for

$$\mathbf{P}_\varepsilon\{N(D_n(0)) \ge k \mid D_n(0)\}.$$

Using inequality (6), we can estimate the value

$$\mathbf{E}_\varepsilon(N(D_n(0)) \mid D_n(0))$$

(see (Menshikov et al., 1986, Lemmas 6.4–6.9) or (Grimmett, 1989)). In particular, it is proved that there exist $b > 0$ and $\beta > 0$ such that

$$\mathbf{E}_{\varepsilon(t)}(N(D_n(0)) \mid D_n(0)) > bn^\beta$$

for all $\varepsilon(t)$ satisfying $0 < \varepsilon_1 < \varepsilon(t) < \varepsilon_0 < \varepsilon_H$ (ε_1 and ε_0 are fixed). We use inequality (6) once more to obtain the statement of Theorem 4.

To prove Theorem 5, it is sufficient to prove inequality $\varepsilon_T \ge \varepsilon_H$, because the converse inequality is evident by definitions.

If $\varepsilon < \varepsilon_H$, then by Theorem 4

$$\mathbf{P}_\varepsilon\{D_n(0)\} < \exp(-a_1 n).$$

Hence the series

$$\sum_{k=1}^{\infty} \mathbf{P}_\varepsilon\{\#W(0) \ge k\} = \mathbf{E}_\varepsilon(\#W(0))$$

is convergent and it follows by definition of ε_T that $\varepsilon < \varepsilon_T$. Therefore $\varepsilon_T \ge \varepsilon_H$.

Remark. All results are valid for the general model

$$\mathbf{P}\{\xi_{t+1}(x) = 1 \mid \bar{\xi}_t\} = 1 - \varepsilon + \varepsilon \sum_{A \in \mathcal{A}} b_A \prod_{y \in A} \xi_t(x+y),$$

where the summation is carried over a collection $\mathcal{A}$ of nonempty finite subsets A, $\sum_{A \in \mathcal{A}} b_A = 1$. We suppose that there exists $A \in \mathcal{A}$ such that $\# A > 1$.

Acknowledgement
The authors are grateful to R. Durrett for constructive remarks.

REFERENCES

van den Berg, J. and Kesten, H. (1985). Inequalities with applications to percolation and reliability. *J. Appl. Probab.* **22**, 556–569.

Durrett, R. (1988). *Lecture Notes on Particle Systems and Percolation.* Wadsworth and Brooks/Cole, Pacific Grove.

Griffeath, D. (1979). *Additive and Cancellative Interacting Particle Systems.* Springer-Verlag, New York.

Grimmett, G. R. (1989). *Percolation.* Springer-Verlag, New York.

Harris, T. E. (1978). Additive set-valued Markov processes and graphical methods. *Ann. Probab.* **6**, 355–378.

Holley, R. and Stroock, D. (1979). Dual processes and their applications to infinite interacting systems. *Adv. Math.* **2**, 149–174.

Malyshev, V. A. (1991). Breakdown of conservation laws in stochastic cellular automata. *Probl. Inform. Transmission* **27**, No. 2, 3–8 (in Russian).

Malyshev, V. A., Petrova, E. N., and Scacciatelli, E. (1989). *Marginally Closed Processes with a Local Interaction.* Preprint No. 26/89, CARR Reports in Math. Physics.

Menshikov, M. V., Molchanov, S. A., and Sidorenko, A. F. (1986). Percolation theory and some applications. *J. Soviet Math.* **42** (1988), 1776–1810.

Russo, L. (1981). On the critical percolation probabilities. *Z. Wahrschienlichkeitstheor. verw. Geb.* **56**, 229–237.

Toom, A. L. (1968). A family of uniform nets of formal neurons. *Soviet Math. Dokl.* **9**, 1338–1341.

Probabilistic Methods in Discrete Mathematics, pp. 337 - 347
V. F. Kolchin et al. (Eds.)
1993 TVP/VSP

MODELS OF ρ-PERCOLATION

M. V. MENSHIKOV and S. A. ZUEV
Moscow State University, Department of Mechanics and Mathematics,
Vorobjevy Gory, 119899 Moscow, Russia
Moscow Forest Engineering Institute, Mytishchi-1, 141001 Moscow reg., Russia

ABSTRACT

We introduce a new type of percolation models in which a path of a graph is ρ-passable in a given configuration of passable and blocked edges if the concentration of passable edges on it is at least ρ. There exists a ρ-percolation threshold $p_{cr}(\rho)$ corresponding to appearance of an infinite ρ-passable path. We find $p_{cr}(\rho)$ for the Bethe lattice and give upper and lower bounds for hypercubic lattice.

1. INTRODUCTION

Percolation models deal with critical phenomena on infinite graphs corresponding to transition from a disordered state to the formation of an infinite order. First introduced in (Broadbent and Hammersley, 1957) they have become a source of many interesting and important results and are now applied widely in other fields (see, e. g., (Kesten, 1982; Menshikov et al., 1988; Grimmett, 1988)).

Our aim is to focus the researcher's attention on a new class of percolation models which appear naturally in queueing and reliability theories. These models which we call *ρ-percolation models* can be considered as a generalization of the classical percolation obtained by a new concept of an occupied or passable path. In spite of an obvious similarity they possess new interesting properties requiring uncommon approaches.

For convenience we formulate the main results in Section 3 giving the proofs in Section 5. In Section 4 we list some open problems.

2. DEFINITIONS

Let G be an infinite connected graph without loops and multiple edges with the vertex, set V and the edge set E. In addition, we suppose that the degree of a vertex, which is the number of edges meeting at the vertex, is uniformly bounded on V. A *path* on G is a sequence $\pi = \{v_0, e_1, v_2, e_2, \ldots, e_n, v_n\}$, where vertices v_{i-1} and v_i are the endpoints of the edge e_i $(i = 1, 2, \ldots, n)$. The length n of the path is the number of its edges. The path is called *self-avoiding* if $v_i \neq v_j$ for all $i \neq j$. A *distance* between two vertices on the graph is the infimum over the lengths of all the paths connecting these vertices. $R_n(v)$ denotes a ball centered at vertex v, which is a collection of vertices and edges belonging to all paths of

the length not exceeding n starting at v. $S_n(v)$ is a sphere of radius n, i.e., a collection of vertices which are at distance n from the vertex v.
One of the important characteristics of regular infinite graphs is the so-called *connectivity constant* $\alpha(G)$ defined by $\log \alpha(G) = \lim_{n\to\infty}(\log(K_n)/n)$, where K_n is the number of self-avoiding paths of length n on the graph G having the same initial vertex. The existence of this limit was first proved for a wide class of graphs in (Broadbent and Hammersley, 1957).
Each edge $e \in E$ of the graph independently of the others can be "*passable*" with probability p and is assigned the value $\omega(e) = +1$, while with probability $1-p$ it is "*blocked*" and is assigned the value $\omega(e) = -1$. Thus we get a configuration of passable and blocked edges on the graph. Formally, the configuration space is $\{-1,+1\}^E$ and the probability is the product $\mathbf{P}_p = \prod_{e\in E} \mu_e$, where μ_e is the Bernoulli measure associated with the edge e: $\mu_e\{+1\} = \mathbf{P}_p\{\text{edge } e \text{ is passable}\} = p = 1 - \mu_e\{-1\}$. Expectation with respect to $\mathbf{P}_p$ is denoted by $\mathbf{E}_p$.
Let π_n be a self-avoiding path of length n. Fix a configuration $\omega \in \{-1,+1\}^E$ and denote by $\eta(\pi_n)$ the number of passable edges lying on the path π_n. We call π_n ρ-passable for some parameter $0 \leq \rho \leq 1$ in configuration ω if the concentration of passable edges of the path is at least ρ, i.e., $(\eta(\pi_n)/n) \geq \rho$. A vertex w is called *ρ-reachable* from a vertex v if there exists a ρ-passable path between them.

Remark. A path constructed of two ρ-passable paths is ρ-passable in its turn. Obviously, the converse statement is not always valid.

Similarly, we call an infinite self-avoiding path $\pi = \{v_0, e_1, \ldots, e_n, v_n, \ldots\}$ ρ-passable, if $\limsup_{n\to\infty}(\eta(\pi_n)/n) \geq \rho$, where $\pi_n = \{v_0, e_1, \ldots, e_n, v_n\}$ is subpath of the length n of the path π with the same initial vertex.
Denote $\mathcal{U}_N$ the σ-algebra generated by the system of random variables $\{\omega(e), e \notin R_N(v)\}$ and $\mathcal{U}_\infty = \bigcap_{N\geq 1} \mathcal{U}_N$.
Let $A_v(\rho)$ be the event that there exists an infinite ρ-passable path $\pi(v)$ with the initial vertex v. Since $A_v(\rho) \in \mathcal{U}_\infty$, the zero-one law can be applied due to which $\mathbf{P}_p\{A_v(\rho)\}$ can take values 0 and 1 only. In the last case we say that a ρ-percolation occurs.
In fact the definition of a ρ-percolation does not depend on the choice of a vertex v. Since the graph is connected, there always exists a finite path between any pair of vertices. Thus any infinite path, starting at other vertex w and coinciding with the infinite ρ-passable path $\pi(v)$ beyond a finite set of edges is also ρ-passable.
This allows us to define the sets $\mathcal{P}_0$ and $\mathcal{P}_1$ of parameters $(\rho, p) \in [0,1]^2$, for which $\mathbf{P}_p\{A_v(\rho)\} = 0$ and $\mathbf{P}_p\{A_v(\rho)\} = 1$ correspondingly, where the above definition does not depend on the choice of $v \in V$.
Note some immediate properties of $\mathcal{P}_0$ and $\mathcal{P}_1$.
First, $\mathbf{P}_0\{A_v(\rho)\} = 0$ for all $\rho > 0$. Hence $(0,1] \times \{0\} \subseteq \mathcal{P}_0$. Then $\mathbf{P}_1\{A_v(\rho)\} = 1$ for all ρ and $\mathbf{P}_p\{A_v(0)\} = 1$ for all $p > 0$, that gives $\{0\} \times (0,1] \subseteq \mathcal{P}_1$ and $[0,1] \times \{1\} \subseteq \mathcal{P}_1$.
Now we present a very suitable construction which will help us to study properties of the ensemble $\{\omega(e)\}$. It seems to be first successfully used for percolation systems in (Berg and Keane, 1984).
Consider an independent system of random variables $\{X(e),\ e \in E\}$ assigned with the edges of the graph each is uniformly distributed in $[0,1]$. We get a configuration

$\omega = \omega_p \in \{-1,+1\}^E$ setting $\omega(e) = +1$ if $X(e) \leq p$ and $\omega(e) = -1$ otherwise for each edge $e \in E$.

Suppose now that a path π is ρ-passable in the configuration ω_p. Then it is also ρ-passable in configurations $\omega_{p'}$ for all $p' > p$. Together with the fact that each ρ-passable path is also ρ'-passable for all $\rho' < \rho$, this implies that the rectangle $[0,\rho] \times [p,1]$ belongs to $\mathcal{P}_1$ as soon as $(\rho,p) \in \mathcal{P}_1$. Similarly, $(\rho,p) \in \mathcal{P}_0$ implies $[\rho,1] \times [0,p] \in \mathcal{P}_0$. Thus $\mathcal{P}_0$ and $\mathcal{P}_1$ are connected subsets of $[0,1]^2$.

One of the most important objects is the critical set $\mathcal{P}_{cr}$ which is the intersection of the closures of $\mathcal{P}_0$ and $\mathcal{P}_1$ in Euclidean space $[0,1]^2$. Essentially this article develops methods of estimating the critical set for various types of graphs.

3. RESULTS

3.1. *Characterization of the critical set*

The following theorem establishes the main properties of the critical set.

THEOREM 3.1 1) *There exists a strictly increasing continuous function $p_{cr}(\rho)$, $0 \leq \rho < 1$, such that its graph represents $\mathcal{P}_{cr}$ in the set $[0,1) \times [0,1)$.*
2) *For all $\rho \in [0,1)$, $\varepsilon > 0$, and $\rho + \varepsilon < 1$*

$$0 < P_1(\varepsilon) \leq p_{cr}(\rho+\varepsilon) - p_{cr}(\rho) \leq \frac{\varepsilon(1-p_{cr}(\rho))}{1-\rho},$$

where $P_1(\rho)$ is the implicit function defined by $H(p,\rho) - \log\alpha(G) = 0$, $H(p,\rho) = \rho\log(\rho/p) + (1-\rho)\log((1-\rho)/(1-p))$ is the entropy function, $\alpha(G)$ is the connectivity constant of the graph G.

Due to Theorem 3.1 the function $p_{cr}(\rho)$ can be defined as

$$\begin{aligned} p_{cr}(\rho) &= \sup\{p\colon\ \mathbf{P}_p\{A_v(\rho)\} = 0 \text{ for all } v \in V\} \\ &= \inf\{p\colon\ \mathbf{P}_p\{A_v(\rho)\} > 0 \text{ for some } v \in V\}. \end{aligned}$$

Equivalently, the inverse function

$$\begin{aligned} \rho_{cr}(p) &= \inf\{\rho\colon\ \mathbf{P}_p\{A_v(\rho)\} = 0 \text{ for all } v \in V\} \\ &= \sup\{\rho\colon\ \mathbf{P}_p\{A_v(\rho)\} > 0 \text{ for some } v \in V\}. \end{aligned}$$

These two definitions reflect two approaches to ρ-percolation: one can fix a percolating level varying the probability of a pass p, or fix a value of p manipulating ρ. Each has its own advantages as we show in the article.

3.2. *ρ-percolation on the Bethe lattice*

One of the most frequently used class of infinite graphs is the *Bethe lattice* B_m, i. e., an infinite graph without cycles (tree) with each vertex having exactly $m+1$ neighbors ($m \geq 2$).

THEOREM 3.2. 1) $p_{cr}(\rho)$ *for* B_m *is the implicit function defined by* $H(p,\rho) - \log m = 0$, *where* $H(p,\rho) = \rho\log(\rho/p) + (1-\rho)\log((1-\rho)/(1-p))$ *is the entropy of the system.*
2) $\mathbf{E}_p\{\xi_\rho(v)\} < \infty$ *for all* $p < p_{cr}(\rho)$ *and all* $v \in V$, *where* $\xi_\rho(v)$ *is the number of* ρ*-reachable vertices from the vertex* v.
3) *For all* $p < p_{cr}(\rho)$ *and* $\delta = H(p,\rho) - \log m > 0$

$$\sqrt{\frac{\pi}{\delta}} C_1 \operatorname{erf c}(\sqrt{\delta}) \le \mathbf{E}_p\Big\{\xi_\rho(v)\Big\} \le C_2\left(e^{-\delta} + \sqrt{\frac{\pi}{\delta}} \operatorname{erf c}(\sqrt{\delta})\right),$$

where

$$C_1 = \frac{m}{m+1}\frac{1}{\sqrt{8\rho(1-\rho)}},$$

$$C_2 = \frac{m}{m+1}\frac{\rho(1-p)}{\sqrt{2\pi\rho(1-\rho)}(\rho(1-p) - p(1-\rho))},$$

$$\rho(1-p) - p(1-\rho) > 0,$$

$$\operatorname{erf c}(x) = \frac{2}{\sqrt{\pi}}\int_x^\infty e^{-u^2}\,du.$$

4) $p_{cr}(\rho) = \rho\exp\{-(1/\rho)\log m - 1\}(1 + o(1))$ as $\rho \downarrow 0$; $p_{cr}(1-0) = 1/m$.
5) $\mathbf{E}_p\{\xi_\rho(v)\} = O(\delta^{-1/2})$ as $\delta \downarrow 0$.

3.3. *ρ-percolation on the hypercubic lattice*

As we see the Bethe lattice gives an example of explicitly solvable model. We are not too enthusiastic to expect an exact solution for general graphs. But some estimates of the critical probability can be obtained. In this section, we deal with a hypercubic lattice, but the choice of the lattice does not play a key role. One can easily fit the proof to a wide class of graphs periodically imbedded in $\mathbf{R}^d$.
Let G_d be a *hypercubic lattice* in $\mathbf{R}^d$, i.e., the graph with the vertex set $V = \mathbf{Z}^d$ in which two vertices $\mathbf{m} = (m_1,\dots,m_d)$ and $\mathbf{n} = (n_1,\dots,n_d)$ share an edge iff $\sum_{i=1}^d |m_i - n_i| = |m - n| = 1$. The norm introduced defines the distance between any pair of vertices of the graph as the minimum length of a path between them. As we mentioned in Secton 2 the number K_n of self-avoiding paths of the length n starting from the origin is asymptotically α^n, where α is the connectivity constant of the graph G_d.

THEOREM 3.3 *Let* $p_{cr}(G_d)$ *be the percolation threshold for the lattice* G_d. *Then*
1) $P_1(\rho) \le p_{cr}(\rho) \le \min\{\rho p_{cr}(G_d),\ P_2(\rho)\}$, *where* $P_1(\rho)$ *is the implicit function* $H(p,\rho) - \log\alpha = 0$, α *is the connectivity constant of* G_d,

$$H(p,\rho) = \rho\log\frac{\rho}{p} + (1-\rho)\log\frac{1-\rho}{1-p}$$

is the entropy of the system,

$$P_2(\rho) = 1 - \Big(1 - p_{cr}(G_d)\Big)^{[1/2pd]^{-d}} 1\{\rho < d/4\} + 1\{\rho \ge d/4\},$$

2) $P_2(\rho) = (2\rho d)^d \log(1 - p_{cr}(G_d))(1 + o(1))$ as $\rho \downarrow 0$.

As will be clear from the proof, the connectivity constant here can be replaced by any upper bound of α. For example, in total there are $2d(2d-1)^{n-1} \overset{l}{\sim} (2d-1)^n$ paths of the length n starting at the origin. Thus $\alpha \leq 2d - 1$ for the graph G_d. Obviously, the estimate $\alpha < M - 1$, where M is the maximum number (over the vertices) of edges incident to a vertex, holds for any graph which has this number uniformly bounded and for which connectivity constant exists.

4. OPEN PROBLEMS

In percolation theory there are defined few critical probabilities. One ($p_H = p_{cr}$) corresponds to the appearance of an infinite passable path (i. e., the path containing passable edges only) and another one (p_T) corresponding to finite/infinite expectation of the number of vertices, which are reachable by passable paths from a fixed vertex. By the analogy with the classical percolation theory we denote this threshold by

$$p_T(\rho) = \sup\{p\colon\ \mathbf{E}_p\{\xi_\rho(v)\} < \infty \text{ for all } v \in V\}.$$

It is easy to see that $p_{cr}(\rho) \geq p_T(\rho)$ always. But it has been proved in (Menshikov, 1986) that $p_{cr} = p_T$ under rather weak conditions. Statement 2 of the theorem immediately implies that $p_{cr}(\rho) = p_T(\rho)$ for all ρ for the Bethe lattice.

PROBLEM 1. *Is $p_{cr}(\rho) = p_T(\rho)$ for other graphs?*

Another interesting case which we did not touch here is 1-percolation in our definition. Clearly it differs from the ordinary percolation for infinite paths. However $p_{cr}(1) \leq p_{cr}$ since every infinite passable path is also 1-passable.

PROBLEM 2. *Is $p_{cr}(1) = p_{cr}$?*

In other words, can there exist infinite paths with concentration of passable edges equal to 1 below the percolation threshold? As statement 4) of Theorem 3.2 shows it is not the case for the Bethe lattices.

Let $F_n(\rho)$ be the event {there exists a ρ-passable path of the length not exceeding n starting from the origin}. The proof of Lemma 5.1 implies that the estimate $\mathbf{P}_p\{F_n(\rho)\} \leq C\exp(-An)$ is valid for all $p < P_1(\rho)$ and some $C, A > 0$.

PROBLEM 3. *Is $\mathbf{P}_p\{F_n(\rho)\} \leq C\exp(-An)$ for all $p < p_{cr}(\rho)$?*

Theorem 3.2 solves this problem positively for the Bethe lattices.

Relation to first-passage percolation (as an introduction to the field see, e. g., the survey (Kesten, 1986)).

Consider Bernoulli measure assigning to edges of the graph the passable time equal to 0 and 1 with probability p and $q = 1 - p$ correspondingly. Then the event {path π_n of the length n has the passing time $t(\pi_n) < (1-\rho)n$} coincides with the event {path π_n is ρ-passable}. Let $\beta(x)$ denote the time constant in the xth direction (see (Cox and Durrett, 1981) for details). It can be easily proved that $\rho_{cr}(p) \geq 1 - \beta$, where $\beta = \inf_x \beta(x)$ $(= \beta(\{1,1\})$ for $\mathbf{R}^2)$.

PROBLEM 4. *Is $\rho_{cr}(p) = 1 - \beta$?*

We have to mention here that for oriented hypercubic lattices these two models are equivalent.
This problem is bound up with the previous one since some exponential bounds exist for the passing time (see (Kesten, 1980) and (Kesten, 1986)).

5. PROOFS

In this section we prove the statements of Section 3.
First of all we need the following lemma.

LEMMA 5.1. *Let $M = M(G)$ be such that*

$$\limsup_{n\to\infty} \frac{\log K_n(v)}{\log M^n} \le 1$$

for all $v \in V$, where $K_n(v)$ is the number of self-avoiding paths of the length n starting from the vertex v. Then
1) $p_{cr}(\rho) \ge P_1(\rho) > 0$ for all $\rho > 0$, where $P_1(\rho)$ as in Theorem 3.3,
2) $\mathbf{E}_p\{\xi_\rho(v)\} < \infty$ for all $p < p_{cr}(\rho)$ and all $v \in V$, where $\xi_\rho(v)$ is the number of ρ-reachable vertices from the vertex v.

Proof. Let $\eta(\pi_n)$ denote the number of passable edges on a path π_n of length n. Then

$$\begin{aligned}\mathbf{P}_p\{\eta(\pi_n) \ge \rho n\} &\le \frac{\rho(1-p)\exp\{-nH(p,\rho)\}}{\sqrt{2\pi n\rho(1-\rho)}(\rho(1-p) - p(1-\rho))} \\ &= \frac{A_2}{\sqrt{n}}\exp\{-nH(p,\rho)\}\end{aligned}$$

for all $\rho < p$. This entropy estimate is known in the theory of information (see, e. g., (Gallager, 1968)).
For any $\beta > 1$ there exists N such that $K_n < (\beta M)^n$ for all $n > N$. Thus,

$$\begin{aligned}\mathbf{E}_p\{\xi_\rho(v)\} &= \sum_{w\in R_N(v)} \mathbf{P}_p\{w \text{ is reachable from } v\} \\ &\quad + \sum_{w\notin R_N(v)} \mathbf{P}_p\{w \text{ is reachable from } v\} \\ &\le C(p,N) + A_2 \sum_{n=N+1}^{\infty} \frac{(\beta M)^n}{\sqrt{n}} \exp\{-nH(p,\rho)\} \qquad (5.1) \\ &= C(p,N) + A_2 \sum_{n=N+1}^{\infty} \frac{1}{\sqrt{n}} \exp\Big\{-n(H(p,\rho) - \log(\beta M))\Big\} < \infty,\end{aligned}$$

if $H(p,\rho) - \log(\beta M) > 0$.
Take $(\rho, p) \in \mathcal{P}_1$. Then with probability 1 for each $\varepsilon > 0$ there exists an infinite self-avoiding path $\pi = \pi(\varepsilon)$ starting from v such that $\eta(\pi_n)/n \ge \rho - \varepsilon$ infinitely often, that implies $\mathbf{P}_p\{\xi_{\rho-\varepsilon}(v) = \infty\} = 1$ for all $\varepsilon > 0$. As (5.1) shows it is

necessary that $H(p, \rho - \varepsilon) - \log(\beta M) \leq 0$. Using the continuity of the function H we get $H(p, \rho) - \log(\beta M) \leq 0$ for all $(\rho, p) \in \mathcal{P}_1$ and hence $H(p, \rho) - \log(\beta M) > 0$ yields $\mathbf{E}_p\{\xi_\rho(v)\} < \infty$ and $(\rho, p) \in \mathcal{P}_0$, or equivalently $p_{cr}(\rho) \geq P_1(\rho)$ for $P_1(\rho)$ defined by $H(p, \rho) - \log(\beta M) = 0$.

Proof of Theorem 3.1. To get the first statement of the theorem it suffices to prove that $\mathcal{P}_{cr}$ contains neither horizontal nor vertical segments in $(0, 1)^2$.
Let $\{X(e), e \in E\}$ be the system of independent uniformly distributed in $(0, 1)$ variables, as described in Section 2. Paint an edge e of the graph black if $0 \leq X(e) \leq p - \Delta$, and red if $p - \Delta < X(e) \leq p$ for some $\Delta > 0$ to be specified later. Denote by $\eta_b(\pi_n)$ and $\eta_r(\pi_n)$ the number of black and red edges on a given finite path π_n.
Suppose $(\rho, p) \in \mathcal{P}_1$ and ε $(0 < \varepsilon < \rho)$ is fixed. Then there exists an infinite path π such that

$$\limsup_{n \to \infty} \frac{1}{n}\Big(\eta_b(\pi_n) + \eta_r(\pi_n)\Big) \geq \rho.$$

Choose $\Delta = \Delta(\varepsilon)$ to have $(\varepsilon, \Delta) \in \mathcal{P}_0$. Such Δ exists due to Lemma 5.1 and can be chosen arbitrarily in the interval $(0, P_1(\varepsilon))$. Note that the concentration of red edges is Δ and so $\limsup_{n \to \infty}(\eta_r(\pi_n)/n) < \varepsilon$ and consequently $\limsup_{n \to \infty}(\eta_b(\pi_n)/n) \geq \rho - \varepsilon$.
Thus we have showed that for any $\varepsilon > 0$ there exists $\Delta > 0$ such that $(\rho, p) \in \mathcal{P}_1$ implies $(\rho - \varepsilon, p - \Delta) \in \mathcal{P}_1$. Consequently $\mathcal{P}_{cr}$ cannot contain horizontal segments and even more, if $(\rho_1, p_1), (\rho_2, p_2) \in \mathcal{P}_{cr}$ for $\rho_1 < \rho_2$, then $p_1 - p_2 \geq \Delta(\rho_2 - \rho_1)$ for all $0 < \Delta(\rho_2 - \rho_1) < P_1(\rho_2 - \rho_1)$ and hence $p_2 - p_1 \geq P_1(\rho_2 - \rho_1)$. This proves the first inequality in 2).
To get another one choose $(\rho_1, p_1) \in \mathcal{P}_{cr}$. As it was just proved, for all $\delta > 0$ and $\varepsilon < \min(\rho_1, 1 - \rho_1)$ we have $(\rho_1 - \delta, p_1) \in \mathcal{P}_1$ while $(\rho_1 + \delta, p_1) \in \mathcal{P}_0$.
Again paint an edge e with $X(e) \leq p_1$ black and denote by $\eta_{\bar{b}}(\pi_n)$ the number of edges of π_n which are not black. Let π be an infinite $(\rho_1 - \delta)$-passable path. Then

$$\rho_1 - \delta \leq \limsup_{n \to \infty} \frac{1}{n}\eta_b(\pi_n) \leq \rho_1 + \delta.$$

Thus

$$\frac{1}{n}\eta_{\bar{b}}(\pi_n) \geq 1 - (\rho_1 + \delta)$$

for all sufficiently large n and all paths π_n.
Now paint the edges e red for which $p_1 < X(e) \leq p_2$ for some $p_2 < 1$. Then the law of large numbers guarantees the existence of $\lim_{n \to \infty} \eta_r(\pi_n)/\eta_{\bar{b}}(\pi_n) = (p_2 - p_1)/(1 - p_1)$ for π_n – *a* subpaths of π. Consequently,

$$\limsup_{n \to \infty} \frac{1}{n}\eta_r(\pi_n) \geq (1 - \rho_1 - \delta)\frac{p_2 - p_1}{1 - p_1}$$

and

$$\limsup_{n \to \infty} \frac{1}{n}\Big(\eta_b(\pi_n) + \eta_r(\pi_n)\Big) \geq \rho_1 - \delta + (1 - \rho_1 - \delta)\frac{p_2 - p_1}{1 - p_1}.$$

Tending δ to 0 we find that the choice of $\rho_2 > \rho_1 + (1-\rho_1)(p_2-p_1)/(1-p_1)$ gives $(\rho_2, p_2) \in \mathcal{P}_1$. Finally,

$$p_{cr}(\rho_2) - p_{cr}(\rho_1) \le p_2 - p_1 \le \frac{(\rho_2-\rho_1)(1-p_1)}{1-\rho_1}$$

and we obtain the last inequality in 2) by putting $\rho = \rho_1$ and $\varepsilon = \rho_2 - \rho_1$.

Proof of Theorem 3.2. Since the number of paths of the length n on the Bethe lattice is $(m+1)m^{n-1}$, it follows that Lemma 5.1 can be applied with $M(G) = m$, giving $p_{cr}(\rho) \ge P_1(\rho)$ and $\mathbf{E}_p\{\xi_\rho(v)\} < \infty$ for all $p < p_{cr}(\rho)$. To get first two statements it suffices now to prove that $p_{cr}(\rho) \le P_1(\rho)$.

Let e be an edge incident to a vertex $v \in V$. Denote by $T_n(v)$ the set of vertices w, for which there exists a path $\pi(v,w)$ between v and w of the length n not passing through the edge e.

Fix an occupancy configuration on the edge set E. First we assign to the vertex v the generation index 0. Assign to a vertex $w \in T_n(v)$ the generation index 1 if the path $\pi(v,w)$ (a single!) is ρ-passable. Define the vertices of the further generations by induction. If a vertex w' has been assigned the generation index k, then there exists a ρ-passable path $\pi(v',w')$ of the length n connecting the vertex w' and some vertex v' which have the generation index $k-1$. Let e' be the edge of the path $\pi(v',w')$ with the endpoint w'. Assign the generation index $k+1$ to those unmarked vertices $u \in V$, for which there exists a ρ-passable path between w' and u of the length n not passing through the edge e'.

Thus we have defined a branching process $\zeta_n(k)$, $k = 0,1,2,\ldots$, where $\zeta_n(k)$ is the number of vertices with the generation index k. For each such a vertex there exists a single path with the length kn, which is ρ-passable by virtue of Remark. Consequently the relation $\zeta_n(k) > 0$ for all $k > 0$ guarantees the existence of an infinite ρ-passable path.

It is well known in the theory of branching processes (see, e. g., (Feller, 1970)) that the probability of the event $\{\zeta_n(k) > 0 \text{ for all } k > 0\}$ is positive if $\mathbf{E}_p\{\zeta_n(1)\} > 1$. Using the following entropy estimate

$$\begin{aligned}\mathbf{P}_p\{\eta(\pi_n) \ge \rho n\} &\ge \frac{1}{\sqrt{8n\rho(1-\rho)}} \exp\{-nH(p,\rho)\} \\ &= \frac{A_1}{\sqrt{n}} \exp\{-nH(p,\rho)\}\end{aligned}$$

for any path π_n of the length n (this estimate is known in the theory of information (Gallager, 1968)) we get

$$\begin{aligned}\mathbf{E}_p\{\zeta_n(1)\} &= \sum_{w\in T_n(v)} \mathbf{P}_p\{\text{path } \pi(v,w) \text{ is } \rho\text{-passable}\} \\ &= m^n \mathbf{P}_p\{\eta(\pi_n) \ge \rho n\} \\ &\ge \frac{A_1}{\sqrt{n}} \exp\Big\{-n(H(p,\rho) - \log m)\Big\}.\end{aligned}$$

Thus, if $H(p,\rho) - \log m < 0$, then there exists n such that $\mathbf{E}_p\{\zeta_n(1)\} > 1$ and consequently with a positive probability there exists an infinite ρ-passable path containing the vertex v.

This yields inequality $p_{cr}(\rho) \leq P_1(\rho)$ and hence $p_{cr}(\rho) = P_1(\rho)$.
For further consideration we need an estimate analogous to (5.1)

$$\mathbf{E}_p\{\xi_\rho(v)\} = \sum_{n=1}^{\infty}(m+1)m^{n-1}\mathbf{P}_p\{\eta(\pi_n) \geq \rho n\}$$
$$\leq \frac{m+1}{m}A_2\sum_{n=1}^{\infty}\frac{1}{\sqrt{n}}\exp\Big\{-n(H(p,\rho)-\log m)\Big\}. \tag{5.2}$$

Let $\delta = H(p,\rho) - \log m$. If $\delta > 0$, then the sequence $f(n) = n^{-1/2}\exp\{-\delta n\}$ is decreasing and we can apply the integral bounds

$$\int_1^\infty f(x)\,dx \leq \sum_{n=1}^{\infty} f(n) \leq f(1) + \int_1^\infty f(x)\,dx.$$

Since

$$\int_1^\infty \frac{e^{-\delta x}}{\sqrt{x}}dx = \frac{2}{\sqrt{\delta}}\int_{\sqrt{\delta}}^\infty e^{-u^2}du,$$

statements 3) and 5) of the theorem follow immediately from (5.2).
Now the only thing left is statement 4).
As 1) shows $\rho \downarrow 0$ yields $p_{cr}(\rho) \downarrow 0$. Consequently, as $\rho \downarrow 0$

$$\begin{aligned}\log m &\equiv H(p_{cr}(\rho),\rho)\\ &= \rho\log\frac{\rho}{p_{cr}(\rho)} + (1-\rho)\log\left(1+\frac{p_{cr}(\rho)-\rho}{1-p_{cr}(\rho)}\right)\\ &= \rho\log\frac{\rho}{p_{cr}(\rho)} + p_{cr}(\rho) - \rho + o(\rho).\end{aligned}$$

Dividing both sides of this equality by ρ we get

$$\log\frac{\rho}{p_{cr}(\rho)} + \frac{p_{cr}(\rho)}{\rho} - 1 = \frac{1}{\rho}\log m + o(1).$$

Substitution $\rho = 0$ in Theorem 3.1 gives $p_{cr}(\varepsilon)/\varepsilon \leq 1$ for all $\varepsilon > 0$. Since the right-hand side of the last equation goes to infinity as $\rho \downarrow 0$, it follows that the relation $p_{cr}(\rho) = o(\rho)$ is necessary. This gives

$$\frac{\rho}{p_{cr}(\rho)}\exp\left\{\frac{p_{cr}(\rho)}{\rho}\right\} = \exp\left\{\frac{1}{\rho}\log m + 1\right\}(1+o(1)).$$

Finally, since $p_{cr}(\rho)/\rho = o(1)$,

$$\exp\left\{\frac{p_{cr}(\rho)}{\rho}\right\} = 1 + o(1)$$

and

$$p_{cr}(\rho) = \rho\exp\left\{-\frac{1}{\rho}\log m - 1\right\}(1+o(1))$$

as $\rho \downarrow 0$.
The proof is complete.

Proof of Theorem 3.3. The proof of the lower bound on $p_{cr}(\rho)$ can be readily obtained from Lemma 5.1 with $M(G) = \alpha$.
To prove the upper bound we use a renormalization approach.
For a natural $r \geq 2$ consider in $\mathbf{R}^d$ the set of cells $D(\mathbf{k}) = D(k_1,\dots,k_d) = [rk_1, r(k_1+1)) \times \dots \times [rk_d, r(k_d+1))$, $k_i \in \mathbf{Z}$, $i = 1,\dots,d$, each containing r^d edges of the lattice G_d. Let $\tilde{G}_d$ be another copy of the hypercubic lattice with the vertex set $\mathbf{Z}^d$. Each occupancy configuration on the edge set E of the graph G_d corresponds an occupancy configuration on $\mathbf{Z}^d$ under the following rule. We declare a vertex $\mathbf{k} \in \mathbf{Z}^d$ occupied if there exists at least one passable edge of the graph G_d in the cell $D(\mathbf{k})$ and vacant otherwise. Since different cells are disjoint, vertices of the graph $\tilde{G}_d$ are occupied or blocked independently. The probability that a fixed vertex is occupied equals $\tilde{p} = 1-(1-p)^{r^d}$ obviously.
Set $\tilde{p} > p_{cr}(\tilde{G}_d)$. Then there exists an infinite occupied path $\tilde{\pi} = \{\mathbf{n}_0, \mathbf{n}_1, \dots\}$ on the graph $\tilde{G}_d$ with the following property: $\mathbf{n}_i$ and $\mathbf{n}_{i+j}$ share an edge of $\tilde{G}_d$ iff $j = 1$. Suppose $\mathbf{n}_i$ and $\mathbf{n}_{i+1}$ are two neighboring vertices of the path $\tilde{\pi}$. Then there exist at least one passable vertex in each cell $D(\mathbf{n}_i)$ and $D(\mathbf{n}_{i+1})$. If there are few such edges in a cell we simply choose one of them at random. It can be shown that in case $r \geq 2$ one can connect the chosen edges by a self-avoiding path π on G_d possessing such a property: between passable edges of neighboring cells $D(\mathbf{n}_i)$ and $D(\mathbf{n}_{i+1})$ there lies not more than $2rd$ edges. Since $\tilde{\pi}$ is infinite, π is infinite too.
So the concentration of passable edges on the path π is at least $1/(2rd)$. Thus, for $\rho < 1/(4d)$, choosing $2 \leq r < 1/(2\rho d)$, one can infer that there exists an infinite ρ-passable path on G_d if $(1-p)^{r^d} < 1 - p_{cr}(G_d)$. This gives

$$(1-p)^{[1/(2\rho d)]^d} < 1 - p_{cr}(G_d)$$

or

$$p > 1 - \Big(1 - p_{cr}(G_d)\Big)^{[1/(2\rho d)]^{-d}}$$

and finally

$$p_{cr}(\rho) \leq 1 - \Big(1 - p_{cr}(G_d)\Big)^{[1/(2\rho d)]^{-d}} = p_2(\rho).$$

To get the inequality $p_{cr}(\rho) \leq \rho p_{cr}(G_d)$ we use the Kesten's proof.
Let $p > p_{cr}(G_d)$. Then with positive probability there exists an infinite path π starting from the origin and containing passable edges only. Paint the edges e for which $X(e) < \rho p$ black (see Section 2 for the definition of X's). Then concentration of black edges on the infinite path π is ρ and hence $p_{cr}(\rho) \leq \rho p$ for all $p > p_{cr}(G_d)$ that completes the proof.

Acknowledgments
Authors are grateful to Vadim Malyshev for setting the problem and to Harry Kesten for very useful discussions.

REFERENCES

van den Berg, J. and Keane, M. (1984). On the continuity of percolation probability function. *Contemp. Math.* **26**, 61–65.

Broadbent, S. R. and Hammersley, J. M. (1957). Percolation processes. *Proc. Cambridge Phil. Soc.* **53**, 629–645.

Cox, J. T. and Durrett, R. (1981). Some limit theorems for percolation processes with necessary and sufficient conditions. *Ann. Probab.* **9**, 583–603.

Feller, W. (1970). *An Introduction to Probability Theory and Its Applications*, Vol. I. John Wiley, New York.

Gallager, R. G. (1968). *Information Theory and Reliable Communication.* John Wiley, New York.

Grimmett, G. (1989). *Percolation.* Springer–Verlag, New York.

Kesten, H. (1980). On the time constant and path length of first passage percolation. *Adv. Appl. Probab.* **12**, 848–863.

Kesten, H. (1982). *Percolation Theory for Mathematicians.* Birkhäuser, Boston.

Kesten, H. (1986). Aspects of first-passage percolation. *Lect. Notes Math.* **1180**, 123–264.

Menshikov, M. V. (1986). Coincidence of critical points in percolation problems. *Soviet Math. Dokl.* **33**, 856–859.

Menshikov, M. V., Molchanov, S. A., and Sidorenko, A. F. (1988). Percolation theory and some applications. *J. Soviet Math.* **42**, 1766–1810.

Probabilistic Methods in Discrete Mathematics, pp. 348 – 352
V. F. Kolchin et al. (Eds.)
1993 TVP/VSP

ON ESTIMATION OF THE COEFFICIENT OF ASYMPTOTIC RELATIVE EFFICIENCY OF DISCRETE–ANALOGUE ALGORITHMS TO DETECT SIGNALS IN A NON-GAUSSIAN NOISE

T. Yu. MOROZOVA and A. N. TRUNOV
Scientific Research Institute of Automation, Botanicheskaya Street 25,
127106 Moscow, Russia

ABSTRACT

Estimates of the asymptotic efficiency of discrete-analogue algorithms to detect signals in a stationary additive noise can be reduced to estimates of Fisher's information of the marginal density of the noise. Various approaches to estimation methods of Fisher's information depending on a prior information about the density are considered.

The theory of the optimum discrete-analogue algorithms to detect a nonrandom signal in a stationary additive Gaussian noise is well developed and became classic. This theory begins with Wiener's and Kotel'nikov's works and in principle cal solve the problem of synthesis of the optimum discrete-analogue algorithm to detect a signal in a communication channel with a Gaussian noise. In contrast to the classic theory, the theory of the optimum algorithms to detect a signal in a stationary additive non-Gaussian noise is in an initial stage of development and is fragmentary now.

The necessity of developing this theory is caused by practical problems with essentially non-Gaussian noises. As examples of such noises we can consider active noises generated by modulation of parameters of high-frequency oscillation by noise voltage, impulse noises, atmospheric noises etc. (Levin, 1989).

If we use methods of the classic theory which give the optimal decision for Gaussian noises in this situation we obtain a nonoptimal result. From the practical point of view a numerical measure of loss in optimality is needed. This measure could give a possibility to estimate the efficiency of the algorithm used. This measure has a particular importance in problems dealing with weak signals, which can be detected only by the optimum algorithms. In statistical radio-engineering the coefficient of asymptotic relative efficiency is used as such a measure, i. e.,

$$e(p) = \sigma^2 i(p),$$

where σ^2. is the variance of the noise, $i(p)$ is the nonzero Fisher's information of the marginal continuous density of the noise $p(x)$.

For any differentiable density

$$e(p) \geq 1,$$

where the equality is attained only if the density $p(x)$ is Gaussian. This fact follows from the Rao–Cramer inequality and Hogg's and Craig's theorem (Zacks, 1971). Let $f(x)$ be the density of the standardized noise with variance equal to 1. It follows from the equality $f(x) = \sigma p(\sigma x)$ that $e(p) = i(f)$.

Thus the estimation of the coefficient of asymptotic relative efficiency is reduced to the estimation of Fisher's information

$$i(f) = i(F) = \int_{-\infty}^{\infty} \frac{(f'(x))^2}{f(x)}\, dx = \int_{-\infty}^{\infty} \left(\frac{f'(x)}{f(x)}\right)^2 dF(x) \qquad (1)$$

for the density $f(x)$ with distribution function $F(x)$.

In the present paper we consider various approaches to the estimation of $i(f)$ or $i(F)$, which depend on a prior information about the density $f(x)$ or the distribution function $F(x)$. If an explicit expression of the density $p(x)$ or $f(x)$ is known the estimation of $i(f)$ is the estimation of integral (1). If this integral is not expressed in elementary functions, then it can be calculated by numerical methods.This case corresponds to maximum information about the density and creates no difficulties, but, unfortunately, it is very seldom in practice.

It is more common situation where the density $f(x)$ of a standardized noise is not known and an investigator can only use observed values of the noise. The problem is to construct a consistent estimate for $i(f)$ on the base of these data. If the problem is parametric and an explicit expression for $f(x)$ is known up to the value of a parameter θ, $f(x) = f_\theta(x)$, and the function $i(f)$ is continuous with respect to the parameter θ, then a consistent estimate $i^*(f)$ of $i(f)$ can be obtained by substitution a consistent estimate of θ^* for the unknown θ:

$$i^*(f) = i(f_\theta^*).$$

An estimate of Fisher's information for the density of the generalized exponential distribution is a typical example of such a problem. Here

$$f(x) = f_\theta(x) = \frac{1}{2\theta C(\theta)\Gamma(\theta)} \exp\left\{\left(\frac{|x|}{C(\theta)}\right)^{1/\theta}\right\}, \qquad (2)$$

where $C(\theta) = (\Gamma(\theta)/\Gamma(3\theta))^{1/2}$ and θ, $0 < \theta < 2$, is an unknown parameter. The normal and Laplace distributions are particular cases of (2), where $\theta = 1/2$, and $\theta = 1$, respectively.

If θ^* is a consistent estimate of the parameter θ of density (2), then the estimate

$$i^*(f) = \frac{\Gamma(3\theta^*)\Gamma(2-\theta^*)}{(\theta^*)^2\Gamma^2(\theta^*)}$$

is consistent for Fisher's information.

The estimation problem of Fisher's information for a density $f(x)$ becomes essentially more complicated, if the density $f(x)$ cannot be parameterized. The smoothed empirical functional $F^*(x)$ of the distribution function $F(x)$ is a consistent estimate for $F(x)$ but it may be not suitable for the construction of a consistent estimate of Fisher's information in the form

$$i^*(F) = i(F^*),$$

because of the complicated dependence of the functional $i(F)$ on the distribution function F. In our opinion, in this situation the confidence interval estimation should be used in preference to construction of a point estimation of Fisher's information. The following representation of the functional $i(F)$ is given in (Huber, 1984).

For a given distribution function F, consider the Hilbert space $L_2(F)$ of differentiable functions ψ satisfying $\int_{-\infty}^{\infty} \psi^2 dF < \infty$.

Let us determine a linear functional A on the set $L_2(F)$ by the equality

$$A\psi = -\int_{-\infty}^{\infty} \psi' \, dF.$$

Then the representation

$$i(F) = \|A\|^2 = \left(\sup_{\psi} \left(\int_{-\infty}^{\infty} \psi' \, dF\right)^2\right) \left(\int_{-\infty}^{\infty} \psi^2 \, dF\right)^{-1} \tag{3}$$

is valid. Representation (3) obtained by Huber (Huber, 1984) makes it possible to prove convexity of the nonlinear functional $i(F)$ with respect to F and gives a possibility to estimate the degree of nonlinearity of $i(F)$ since the functions $\int_{-\infty}^{\infty} \psi' \, dF$ and $\int_{-\infty}^{\infty} \psi^2 \, dF$ from (3) are linear functions in F.

The convexity of the nonlinear functional $i(F)$ permits us to find the distribution F_0 (or the density f_0), which minimizes Fisher's information $i(F)$.

Thus if a prior information about the distribution F or f consists of belonging F to the class of distributions for which we can find $i(F_0)$, then the value $i(F_0)$ can be chosen as a bound for $i(F)$.

For example, let us consider a lower estimate of Fisher's information for the ε-normal and littered ε-normal distribution, $0 < \varepsilon < 1$ (Huber, 1984).

For the littered ε-normal distribution the density is equal to

$$f_0(x) = \begin{cases} (1-\varepsilon)\varphi(x), & \text{if } |x| \le k, \\ \frac{1-\varepsilon}{(2\pi)^{1/2}} \exp\{-k|x| + k^2/2\}, & \text{if } |x| > k, \end{cases} \tag{4}$$

where φ is the density of standard normal distribution $\Phi(x)$ and k and ε are related by the equality

$$(2\varphi(k)/k) - 2\Phi(-k) = \varepsilon/(1-\varepsilon). \tag{5}$$

From (4) and (5) it follows that

$$i(F_0) = 1 + (4\varphi(k)(1-\varepsilon)/k^3)((1-k\varphi(k))(1-\varepsilon)) - (2\varphi(k)(1-\varepsilon))/k. \tag{6}$$

For the ε-normal distribution, $0 < \varepsilon < 0.03$, the density is equal to

$$f_0(x) = \begin{cases} \varphi(a)(\cos(ca/2))^{-2}(\cos(cx/2))^2, & 0 \le x \le a, \\ \varphi(x), & a \le x \le b, \\ \varphi(b)\exp\{-b(x-b)\}, & x \ge b, \end{cases} \tag{7}$$

where the constants a, b, c are defined by equalities

$$\begin{gathered} c\tan(ca/2) = a, \qquad 0 \le ca \le \pi, \\ (ac^2\varphi(a))/(2\cos^2(ca/2)) = c^2(\Phi(a) - 1/2 - \varepsilon) - a, \\ \varphi(b)/b + \Phi(b) = 1 + \varepsilon. \end{gathered} \tag{8}$$

From (7) and (8) it follows that

$$\begin{aligned} i(F_0) = 4\big(c^2\Phi(a) - c^2/2 - c^2\varepsilon - a + \Phi(b) - \Phi(a)\big)\big(\varphi(a)a^3/(6\cos^2(ca/2)) \\ -\varphi(a)a/(c^2\cos^2(ca/2)) + 2\varphi(a)a/c^4\big)\big((a^2c^2/2 - 1) + 2\varphi(a)a/c^2 \\ +\Phi(b) - \Phi(a) + a\varphi(a) + 4\varphi(b)/b^3 + 2\varphi(b)/b\big). \end{aligned}$$

From the practical point of view, it is important to have a possibility to estimate the precision of the lower bound $i(F_0)$ for $i(F)$; if the class of distributions is normed, then such an estimate may be obtained by the formula of finite increments for the functionals (Kolmogorov and Fomin, 1972):

$$i(F) - i(F_0) \le \sup_{0\le t\le 1} \|i'_c(F_t)\|\|F - F_0\|,$$

where $F_t = (1-t)F_0 + tF$, $0 \le t \le 1$, and $i'_c(F_t)$ is the weak derivative (Gato's derivative) of the functional $i(F_t)$.

The formula of finite increments for the functionals permit to estimate (by Gato's derivative) the distance between $i(F_1)$ and $i(F_2)$ with the help of the distance between the distribution functions F_1 and F_2.

In conclusion we consider a case which is often met in practice. Let a prior information about the density $p(x)$ or $f(x)$ imply that the first several moments (the coefficients γ_1 and γ_2 of asymmetry and excess) are given and the density is close in some sense to the normal density.

THEOREM. *Let $f(x)$ have a finite Fisher's information, $\varphi(x)$ be the standard normal density and*

$$\rho = \int_{\mathbf{R}} \frac{(f(x) - \varphi(x))^2}{\varphi(x)}\, dx.$$

If $f(x)/\varphi(x)$ is uniformly bounded and $\rho \to 0$, then

$$i(f) - 1 = \sum_{n\ge 3} n!\, nc_n^2 + O(\sqrt{\rho})$$

where

$$c_n = \frac{1}{n!}\int_{\mathbf{R}} f(x)H_n\, dx,$$

and $H_n = (-1)^n \varphi^{(n)}(x)/\varphi(x)$ is the Hermit polynomial of the nth order, $H_0 = 1$.

A proof of the theorem under more strong conditions is given in (Morozova, 1991). The coefficients c_n can be expressed in terms of the moments μ_n of the distribution, in particular,

$$c_3 = \mu_3/6, \quad c_4 = (\mu_4 - 2)/24.$$

The theorem enables an estimate of $i(f)$ to be obtained by means of the estimates of the moments. For example, if we can obtain estimates of the first four moments, then the statistic $\hat{i}(f) = 1 + T_n/n$, where $T_n = n\hat{\gamma}_1^2/2 + n\hat{\gamma}_2^2/6$, and $\hat{\gamma}_1, \hat{\gamma}_2$ are the sampled coefficients of asymmetry and excess constructed on the base of independent observations, can be used as an estimate of $i(f)$.
In the case of normal density φ, following (D'Agostino and Pearson, 1973) we can prove that the statistic T_n has in limit as $n \to \infty$ the noncentral χ^2-distribution with two degrees of freedom and parameter of noncentrality $\lambda = 5$. Thus for normal density φ and any $\varepsilon > 0$

$$\mathbf{P}\{|\hat{i}(\varphi) - i(\varphi)| \le \varepsilon\} = \mathbf{P}\{T_n \le n\varepsilon\}. \tag{9}$$

If we assume that for a density f sufficiently close to φ the behaviour of the statistic T_n differs inessentially, then we can substitute f for φ in (9) and obtain an approximate estimate of the size n needed to attain a required accuracy ε.

REFERENCES

D'Agostino R. B. and Pearson, E. S. (1972). Test for departure from normality. Empirical results for the distributions of b_2 and $(b_1)^{1/2}$. *Biometrika* 60, 613–622.

Huber, J. (1984). *Robust Statistics*. Wiley, New York.

Huber, J. (1964). Robust estimation of a location parameter. *Ann. Math. Statist.* 35, 73–101.

Kolmogorov, A. N. and Fomin, S. V. (1972). *Elements of the Theory of Functions and Functional Analysis*. Nauka, Moscow (in Russian).

Levin, B. R. (1989). *Theoretical Base of Statistical Radio-Engineering*. Radio and Communication, Moscow (in Russian).

Morozova, T. Yu. (1991). On estimates of the coefficient of asymptotic relative efficiency in the problem of optimum detection of signals in noise. In: *Stochastic Processes and Their Applications*, MIEE, Moscow pp. 27–32.

Zacks, S. (1971). *The Theory of Statistical Inference*. Wiley, New York.

Probabilistic Methods in Discrete Mathematics, pp. 353 – 365
V. F. Kolchin et al. (Eds.)
1993 TVP/VSP

RANDOM MAPPINGS AND TREES: RELATIONS BETWEEN ASYMPTOTICS OF MOMENTS OF CERTAIN CHARACTERISTICS*

L. R. MUTAFCHIEV

Institute of Mathematics with Computer Center, Bulgarian Academy of Sciences, Akad. G. Bonchev Street 8, 1113 Sofia, Bulgaria

ABSTRACT

Mappings of an n-element set into itself may be represented by directed graphs whose each vertex has an out-degree one. Assuming that $n \to \infty$, we prove a limit theorem which allows to transfer asymptotic results for rooted trees to mappings. Some applications concerning the size of the tree to which a fixed vertex belongs are also given.

1. INTRODUCTION

Let $\mathfrak{M}_n$ denote the set of all single-valued mappings of the n-element set $U_n = \{1, 2, \ldots, n\}$ into itself. It is clear that there are n^n such different mappings, i.e., $|\mathfrak{M}_n| = n^n$ (throughout the paper we shall write $|N|$ for the cardinality of any finite set N). Each mapping $M \in \mathfrak{M}_n$ may be represented by a directed graph G_M with vertex-set U_n and arc-set $\{(d, M(d)) : d \in U_n\}$. Thus, every vertex of G_M has the out-degree one and G_M consists of a number of disjoint components with exactly one cycle in each. The vertices of G_M may be considered as lying in directed trees whose roots coincide with the cyclic vertices; the arcs of each tree are directed towards the root of the tree.

We shall also consider the set $\mathfrak{J}_n$ of all rooted labelled trees on n vertices. It is well-known by Cayley's formula that $|\mathfrak{J}_n| = n^{n-1}$. One may consider $\mathfrak{M}_n$ and $\mathfrak{J}_n$ as probability spaces equipping each of these sets with a uniform probability measure. In other words, one may regard each $M \in \mathfrak{M}_n$ or $T \in \mathfrak{J}_n$ as elementary events realized with probabilities n^{-n} or n^{-n+1}, respectively. In this paper the terms "random mapping" and "random tree" will be used for any elementary event from the probability space defined above. We shall use the notation $\mathbf{P}\{\cdot\}$ simultaneously for both probability measures on $\mathfrak{M}_n$ and $\mathfrak{J}_n$. Thus, conceivable numerical characteristics of random mappings and random trees become random variables in probability spaces $\mathfrak{M}_n$ and $\mathfrak{J}_n$, respectively. There are fairly many results characterizing mappings from $\mathfrak{M}_n$, which study distributions of such variables (see, for example, a survey by Mutafchiev (Mutafchiev, 1984)). The main problems in random mapping theory concern the case $n \to \infty$. Various limit theorems for such

*This research is partially supported by Contract MM17-Ministry of Science and Education, Bulgaria

random variables are contained in the monograph (Kolchin, 1986). It turns out that in most cases the expressions found for fixed values of n are rather more complicated than those obtained after the passage to the limit. Moreover, there are many examples in random mapping theory in which direct combinatorial methods do not work effectively. In this paper a similar situation will be considered. The method proposed here can successfully be applied in the case of mappings from $\mathfrak{M}_n$ where direct combinatorial reasonings are not avialable. Our results below give relations between the asymptotics of variables characterizing an arbitrary property of random trees from $\mathfrak{T}_n$ and the asymptotics of the corresponding variables characterizing the same property of large trees of mappings from $\mathfrak{M}_n$. Since the structure of trees is simpler than the structure of mappings, in many cases it is more convenient to study first the class $\mathfrak{T}_n$ before the class $\mathfrak{M}_n$. Then, by the relations obtained here one can immediately transfer an asymptotic result for trees to mappings. A similar problem concerning relations between asymptotics of characteristics of connected and disconnected mappings has been considered by Mutafchiev (Mutafchiev, 1992).

In Section 2 we begin with some notation and the statement of the main result whose proof is given in Section 4. Section 3 contains some preliminaries. Finally, in Section 5 we apply our main result to the study of probability distribution of the size of the tree to which a fixed vertex belongs under the condition that the vertex obeys a fixed property (as, for example, it is an endvertex, etc.). It appears that the limiting probability depends only on the order of normalization of the number of vertices possessing this property. More precisely, if this order is n^β, then the limiting distribution has a density function of the form $h(\alpha) = (\beta - 1/2)\alpha^{\beta-3/2}$, $0 < \alpha < 1$, $1/2 < \beta \le 1$.

2. NOTATION AND STATEMENT OF THE MAIN RESULT

Let R denote an arbitrary property of the vertices of a graph G, admissible in both cases $G \in \mathfrak{M}_n$ and $G \in \mathfrak{T}_n$. For a random mapping from $\mathfrak{M}_n$ or random tree from $\mathfrak{T}_n$ define the random subset

$$A_n(R) = \{d \in U_n\colon\ d \text{ possesses property } R\}.$$

Moreover, for mappings from $\mathfrak{M}_n$, let

$$B_n(s) = \{d \in U_n\colon\ d \text{ belongs to a tree with more than } s \text{ vertices}\}, \quad s = 0, 1, \ldots$$

We consider two families of random variables depending on R. The first family is defined on $\mathfrak{T}_n$ as $X_n^{(1)}(R) = |A_n(R)|$. The second one is defined on $\mathfrak{M}_n$ as $X_n(R; s) = |A_n(R) \bigcap B_n(s)|$.

Throughout this paper we use the common notation $\mathbf{E}(\cdot)$ for expectations taken for variables on both sets $\mathfrak{T}_n$ and $\mathfrak{M}_n$.

We denote by $\Gamma(z)$ the Euler gamma-function. For a natural number j define

$$K_j = \{(k_1, \ldots, k_j)\colon\ k_l \in \{0, 1, \ldots\}, \quad l = 1, \ldots, j, \quad k_1 + 2k_2 + \cdots + jk_j = j\}. \tag{2.1}$$

Throughout the paper the notation $\sum_{K_j}$ denotes a sum taken over all $(k_1, \ldots, k_j) \in K_j$. Moreover, set $k = k_1 + \cdots + k_j$ and consider the following array of k arbitrary numbers:

$$\begin{array}{ccc} a_{1,1}, & \ldots, & a_{1,k_1} \\ a_{2,k_1+1}, & \ldots, & a_{2,k_1+k_2} \\ \ldots & \ldots & \ldots \\ a_{j,k_1+\cdots+k_{j-1}+1}, & \ldots, & a_{j,k}. \end{array} \tag{2.2}$$

It is clear that the number of rows in it does not exceed j. Some of the rows in array (2.2) may disappear. Thus, if $k_1 = 0$, the first row is empty and the array contains no more than $j-1$ rows; if $k_1 + k_2 = 0$, the first two rows are empty, etc. We need a notation for the product of the numbers in array (2.2). Setting $\prod_1^0 = 1$, we denote for brevity by $\prod_{(k,K_j)} a_{l,p}$ the product

$$\prod_{p=1}^{k_1} a_{1,p} \prod_{p=k_1+1}^{k_1+k_2} a_{2,p} \cdots \prod_{p=P_{j-1}}^{k} a_{j,p} = \prod_{l=1}^{j} \prod_{p=P_{l-1}}^{P_l} a_{l,p}$$

where $P_s = \sum_{\nu=1}^{s} k_\nu + 1, s = 2, \ldots, l-1, P_l = k_1 + \cdots + k_l$.
Finally, we need a notation for some simplexes in the k-dimensional Euclidean space which are the domains of integration of the integrals given below. Thus, let

$$D_k(\alpha) = \left\{ x_p > \alpha,\ p = 1, \ldots, k : \sum_{p=1}^{k} x_p < 1 \right\},$$

where $0 < \alpha < 1$. The following notation will be used for a given number β :

$$I_0(\alpha, \beta) = 1,$$

$$I_k(\alpha, \beta) = \int \cdots \int_{D_k(\alpha)} \left(1 - \sum_{p=1}^{k} x_p\right)^{(k-1)/2} \prod_{(k,K_j)} x_p^{l\beta-3/2} dx_p, \tag{2.3}$$

(see array (2.2) with $a_{l,p} = x_p^{l\beta-3/2} dx_p$).
We now state our main result which is proved in Section 4.

THEOREM. *Let $j \geq 1$, $b > 0$ and $1/2 < \beta \leq 1$. Suppose that*

$$\lim_{n\to\infty} \mathbf{E}\,[X_N^{(1)}(R)]^l / (b^l n^{l\beta}) = m_l(R), \qquad l = 1, \ldots, j, \tag{2.4}$$

where $m_1(R), \ldots, m_j(R)$ are positive constants. Then

$$\lim_{n\to\infty} \mathbf{E}\,[X_n(R;0)]^j / (b^j n^{j\beta}) = \mu_j(R;\beta) \tag{2.5}$$

and, moreover, if $s \sim \alpha n$ as $n \to \infty$, $0 < \alpha < 1$, then

$$\lim_{n\to\infty} \mathbf{E}[X_n(R;s)]^j / (b^j n^{j\beta}) = \mu_i(R;\alpha,\beta), \tag{2.6}$$

where $\mu_j(R;\beta)$ and $\mu_j(R;\alpha,\beta)$ are given by

$$\mu_j(R;\beta) = \frac{\pi^{1/2}}{\Gamma(j\beta+1/2)} \sum_{K_j} \frac{j!k!(2\pi^{1/2})^{-k}}{k_1!\cdots k_j!} \left(\frac{m_1(R)\Gamma(\beta-1/2)}{1!}\right)^{k_1} \cdots \left(\frac{m_j(R)\Gamma(j\beta-1/2)}{j!}\right)^{k_j}, \tag{2.7}$$

$$\mu_j(R;\alpha,\beta) = \pi^{1/2} \sum_{K_j} \frac{j!k!(2\pi^{1/2})^{-k} I_k(\alpha,\beta)}{k_1!\cdots k_j!\Gamma((k+1)/2)} \left[\frac{m_1(R)}{1!}\right]^{k_1} \cdots \left[\frac{m_j(R)}{j!}\right]^{k_j}. \tag{2.8}$$

Remark. Since the method of proofs presented here cannot be applied, if $0 < \beta \le 1/2$, we intend to study this case in a subsequent publication.

3. PRELIMINARY RESULTS

Denote by Y_l the number of trees with l vertices of a random mapping $M \in \mathfrak{M}_n$ and introduce the following probability generating functions:

$$f_0 = 1, \quad f_n(x;R) = \mathbf{E}[x^{X_n^{(1)}(R)}] = \sum_{l=0}^{n} \mathbf{P}\{X_n^{(1)}(R) = l\}x^l, \qquad n \ge 1, \tag{3.1}$$

$$\begin{aligned} \varphi_0 = 1, \quad \varphi_n(x_1,\ldots,x_n) &= \mathbf{E}\left[x_1^{Y_1}\cdots x_n^{Y_n}\right] \\ &= \sum_{K_n} \mathbf{P}\{Y_1 = k_1,\ldots,Y_n = k_n\} x_1^{k_1}\cdots x_n^{k_n}, \qquad n \ge 1, \end{aligned} \tag{3.2}$$

$$\begin{aligned} \psi_{n,s}(x;R) &= \mathbf{E}\left[x^{X_n(R;s)}\right] \\ &= \sum_{l=0}^{n} \mathbf{P}\{X_n(R;s) = l\}x^l, \qquad n \ge 1, \quad s = 0,1,\ldots,n-1, \end{aligned} \tag{3.3}$$

where $x, x_1, \ldots, x_n$ are arbitrary variables in the unit disk.
The next lemma establishes some relations between these generating functions.

LEMMA 1. *The following relations hold*

$$\psi_{n,0}(x;R) = \varphi_n(f_1(x;R),\ldots,f_n(x;R)),$$

$$\psi_{n,s}(x;R) = \varphi_n(\underbrace{1,\ldots,1}_{s \text{ times}}, f_{s+1}(x;R),\ldots,f_n(x;R)), \qquad 1 \le s \le n-1.$$

Proof. By the construction of the graphs G_M, $M \in \mathfrak{M}_n$, and the definitions of $X_n^{(1)}(R)$, Y_l and $X_n(R;s)$ it is easily checked that

$$X_n(R;s) = \zeta_{Y_l} = \sum_{l=s+1}^{n} \sum_{p=1}^{Y_l} \xi_{l,p}^{(1)}(R),$$

where $\{\xi_{l,p}^{(1)}(R),\ p=1,\dots,Y_l,\ l=1,\dots,n\}$ are independent random variables such that $\{\xi_{l,p}^{(1)}(R),\ p=1,\dots,Y_l\}$ have the distribution function of $X_l^{(1)}(R)$. This representation holds also when $s=0$. Now, using conditional expectations in (3.3) and the second equality of (3.2), we get

$$\begin{aligned}\psi_{n,s}(x;R) &= \mathbf{E}\{\mathbf{E}[x^{X_n(R;s)}|Y_1,\dots,Y_n]\}\\ &= \sum_{K_n}\mathbf{P}\{Y_1=k_1,\dots,Y_n=k_n\}\mathbf{E}\left[x^{\zeta_{k_l}}\right]\\ &= \sum_{K_n}\mathbf{P}\{Y_1=k_1,\dots,Y_n=k_n\}f_{s+1}^{k_s+1}(x;R)\cdots f_n^{k_n}(x;R)\\ &= \begin{cases}\varphi_n(f_1(x;R),\dots,f_n(x;R)), & \text{if } s=0;\\ \varphi_n(\underbrace{1,\dots,1}_{s \text{ times}},f_{s+1}(x;R),\dots,f_n(x;R)), & \text{if } 1\le s\le n-1.\end{cases}\end{aligned}$$

Further, we need the exponential generating function of the sequence $\{n^{n-1}\}$. Thus, let

$$\theta(z)=\sum_{n=1}^{\infty}n^{n-1}\frac{z^n}{n!}. \tag{3.4}$$

We also need the following representation of $\psi_{n,s}(x;R)$ as a contour integral.

LEMMA 2. *The following equality is valid*

$$\psi_{n,s}(x;R)=\frac{n!}{2\pi i n^n}\int_\gamma\left\{1-\theta(z)-\sum_{l=s+1}^{\infty}\frac{l^{l-1}}{l!}z^l\left[f_l(x;R)-1\right]\right\}^{-1}\frac{dz}{z^{n+1}},$$

$$s=\ 0,1,\dots,n-1,$$

where γ is a counter clockwise oriented contour around the origin such that $\gamma\subset\{z:\ |z|\le e^{-1}\}$ and $|x|\le 1$.

Proof. We shall use Stepanov's formula (see (Stepanov, 1969))

$$\sum_{n=0}^{\infty}n^n\varphi_n(x_1,\dots,x_n)z^n/n! = \left(1-\sum_{l=1}^{\infty}z^l x_l\frac{l^{l-1}}{l!}\right)^{-1}.$$

Setting $x_1=\dots=x_s=1$, if $s\ge 1$, and $x_l=f_l(x;R)$, if $l\ge s+1$ (see (3.1)), and using the definition of $\psi_{n,s}(x;R)$ we immediately get

$$\sum_{n=0}^{\infty}n^n\psi_{n,s}(x;R)z^n/n! = \Big\{1-\theta(z) - \sum_{l=s+1}^{\infty}\frac{l^{l-1}}{l!}z^l\left[f_l(x;R)-1\right]\Big\}^{-1}.$$

To obtain the required relation it suffices to apply the integral Cauchy formula taking into account that $\theta(z)$ has the radius of convergence equal to e^{-1} (see 3.4)).

Now we turn our attention to an expansion of $\theta(z)$ near $z = r_n e^{iu}$, where

$$r_n = (1 - n^{-1/2})e^{-1+n^{-1/2}} \tag{3.5}$$

and $-\infty < u < \infty$.

LEMMA 3. (Kolchin, 1986; Chapter I, Lemma 12.3). *The following representation holds*

$$\left[1 - \theta(r_n e^{iu})\right]^2 = \frac{1}{n} - 2iu + \varepsilon_1(u) + \varepsilon_2(n,u),$$

where $\varepsilon_1(u)/u \to 0$ *uniformly with respect to* n *and*

$$|\varepsilon_2(n,u)| \le 2|u|n^{-1/2}.$$

Further, for the sake of simplicity let

$$m_l^{(n)}(R) = \mathbf{E}\left[X_n^{(1)}(R)\right]^l, \qquad l = 1, 2, \ldots$$

Consider the sums

$$\sigma_{n,s}^{(l)} = \sum_{p=s+1}^{\infty} \frac{p^{p-1}}{p!} r_n^p e^{ivp/n} m_l^{(p)}(R), \tag{3.6}$$

where r_n is given by (3.5). The asymptotic behaviour of $\sigma_{n,s}^{(l)}$ is given by the following lemma.

LEMMA 4. *Suppose that the limits (2.4) exist for a certain natural number j and proper constants $b > 0$ and $\beta \in (1/2, 1]$.*

(i) *If $s \sim \alpha n$ as $n \to \infty$ and $0 < \alpha < 1$, then*

$$\sigma_{n,s}^{(l)} = \frac{1}{\sqrt{2\pi}} b^l m_l(R) n^{l\beta - 1/2} \int_\alpha^\infty e^{-y(1/2 - iv)} y^{l\beta - 3/2} dy\left[1 + o(1)\right], \qquad l = 1, \ldots, j,$$

(ii) $\sigma_{n,0}^{(l)} = \dfrac{b^l m_l(R)}{(2\pi)^{1/2}} \Gamma(l\beta - 1/2)\left(\dfrac{n}{1/2 - iv}\right)^{l\beta - 1/2} [1 + o(1)], \quad l = 1, \ldots, j,$ as $n \to \infty$.

Proof. (i) Straightforward calculations yield (see (3.5)) that for $n, p \to \infty$

$$(r_n e)^p = (1 - n^{-1/2})^p e^{pn^{1/2}} = e^{-p/(2n)}\left[1 + \Delta(n,p)\right], \tag{3.7}$$

where

$$\Delta(n,p) = O(pn^{-3/2}). \tag{3.8}$$

On the other hand, using Stirling's formula, we get

$$\frac{p^{p-1}}{p!} = e^p/(2\pi p^3)^{1/2}(1 + o(1)), \qquad p \to \infty; \tag{3.9}$$

(3.7) and (3.9) imply that

$$\frac{p^{p-1}r_n^p}{p!} = \left[e^{-p/(2n)}/(2\pi p^3)^{1/2}\right]\left[1+\Delta(n,p)\right], \qquad n,\ p\to\infty. \tag{3.10}$$

Moreover, the condition of the lemma yields the following asymptotic relation for $m_l^{(p)}(R)$

$$m_l^{(p)}(R) = m_l(R)(bp^\beta)^l\left[1+o(1)\right], \qquad p\to\infty. \tag{3.11}$$

Replacing $p^{p-1}r_n^p/p!$ by the right-hand side of (3.10) and $m_l^{(p)}(R)$ by the right-hand side of (3.11), from (3.6) we obtain that

$$\sigma_{n,s}^{(l)} = (2\pi)^{-1/2}m_l(R)b^l\sum_{p>s}e^{-(1/2-iv)p/n}p^{l\beta-3/2}\left[1+\Delta(n,p)\right] \times\left[1+o(1)\right], \qquad n,\ s\to\infty.$$

Now for $\sigma_{n,s}^{(l)}$ we can write

$$\sigma_{n,s}^{(l)} = \sum\nolimits_1 + \sum\nolimits_2, \tag{3.12}$$

where

$$\sum\nolimits_1 = \frac{1}{\sqrt{2\pi}}m_l(R)b^l\sum_{p>s}e^{-(1/2-iv)p/n}p^{l\beta-3/2}\left[1+o(1)\right],$$

$$\sum\nolimits_2 = \frac{1}{\sqrt{2\pi}}m_l(R)b^l\sum_{p>s}e^{-(1/2-iv)p/n}p^{l\beta-3/2}\Delta(n,p)\left[1+o(1)\right].$$

Using the definition of Riemann's integral and the asymptotic of s, it is easy to check that as $n\to\infty$

$$\sum\nolimits_1 = (2\pi)^{-1/2}m_l(R)b^l n^{l\beta-1/2}\int_\alpha^\infty e^{-y(1/2-iv)}y^{l\beta-3/2}dy\left[1+o(1)\right], \qquad n\to\infty. \tag{3.13}$$

Further, by (3.8) as $n\to\infty$

$$\left|\sum\nolimits_2\right| \le C_0\sum_{p>s}e^{-p/(2n)}p^{l\beta-3/2}pn^{-3/2}\left[1+o(1)\right] \tag{3.14}$$

$$= C_0 n^{l\beta-1}\int_\alpha^\infty e^{-y/2}y^{l\beta-1/2}dy = O(n^{l\beta-1}) = o(n^{l\beta-1/2}),$$

where C_0 is a positive constant. Now, part (i) of the lemma follows immediately from (3.12)–(3.14).

(ii) To obtain the asymptotic of $\sigma_{n,0}^{(l)}$ we need to investigate the behaviour of

$$\varepsilon_n = \sum_{1\le p\le n^{1/2}}\frac{p^{p-1}}{p!}r_n^p e^{ivp/n}m_l^{(p)}(R).$$

From (3.5) it follows that $r_n < e^{-1}$, $n = 1, 2, \ldots$ Therefore

$$|\varepsilon_n| < \sum_{1 \le p \le n^{1/2}} \frac{p^{p-1}}{p!} e^{-p} m_l^{(p)}(R). \tag{3.15}$$

With the aid of (2.4) it is easy to see that there exists a constant $C_1 > 0$ such that

$$m_l^{(p)}(R) \le C_1 p^{l\beta} = O(n^{l\beta/2}), \qquad 1 \le p \le n^{1/2}. \tag{3.16}$$

Thus, from (3.9), (3.15) and (3.16) we obtain the estimate

$$|\varepsilon_n| = O(n^{l\beta/2}) \sum_{1 \le p \le n^{1/2}} p^{-3/2} = O(n^{l\beta/2}), \qquad n \to \infty. \tag{3.17}$$

Since

$$\sigma_{n,0}^{(l)} = \varepsilon_n + \sigma_{n,[n^{1/2}]+1}, \tag{3.18}$$

where $[n^{1/2}]$ denotes the integer part of $n^{1/2}$, the remaining task is to find the asymptotic of $\sigma_{n,[n^{1/2}]+1}$ as $n \to \infty$. It can be done similarly to the proof of (i). We have

$$\begin{aligned} \sigma_{n,[n^{1/2}]+1}^{(l)} &= \frac{1}{\sqrt{2\pi}} m_l(R) b^l n^{l\beta-1/2} \int_0^\infty e^{-y(1/2-iv)} y^{l\beta-3/2} dy \left[1 + o(1)\right] \\ &= \frac{1}{\sqrt{2\pi}} m_l(R) b^l \Gamma(l\beta - 1/2) \left(\frac{n}{1/2 - iv}\right)^{l\beta-1/2} (1 + o(1)), \quad n \to \infty. \end{aligned}$$

Now the combination of (3.17)–(3.19) completes the proof of (ii).

4. PROOF OF THE MAIN RESULT

We begin with the choice of the contour γ of Lemma 2. To prove both relations (2.5) and (2.6) we set

$$\gamma = \gamma_n = \{z : \ z = r_n e^{iu}, \ -\pi \le u \le \pi\}$$

(see also (3.5)). Further, we may also set $x = e^{it}$, $-\infty < t < \infty$, in the relation of Lemma 2 and differentiate it at $t = 0$. Using Bruno's formula (see, e.g., (Riordan, 1958)), after simple manipulations one can obtain that

$$\mathbf{E}\left[X_n(R; s)\right]^j = \sum_{K_j} \frac{j!k! J_{n,s}(k_1, \ldots, k_j)}{k_1! \cdots k_j! (1!)^{k_1} \cdots (j!)^{k_j}}, \tag{4.1}$$

where

$$\begin{aligned} J_{n,s}(k_1, \ldots, k_j) &= \frac{n!}{2\pi i n^n} \int_{\gamma_n} \prod_{l=1}^{j} \left(\sum_{p=s+1}^{\infty} \frac{p^{p-1}}{p!} z^p m_l^{(p)}(R) \right)^{k_l} \frac{dz}{(1 - \theta(z))^{k+1} z^{n+1}} \\ &= \frac{n!}{2\pi (n r_n)^n} \int_{-\pi}^{\pi} \prod_{l=1}^{j} \left(\sum_{p=s+1}^{\infty} \frac{p^{p-1}}{p!} r_n^p e^{ipu} m_l^{(p)}(R) \right)^{k_l} \\ &\quad \times e^{-inu} \frac{du}{(1 - \theta(r_n e^{iu}))^{k+1}}. \end{aligned} \tag{4.2}$$

First, by Stirling's formula

$$\frac{n!}{2\pi(nr_n)^u} = \sqrt{\frac{ue}{2\pi}}(1 + O(n^{-1/2})), \qquad n \to \infty.$$

Then, setting $u = v/n$ in the last integral of (4.2), we get

$$J_{n,s}(k_1, \ldots, k_j) = \int_{-u\pi}^{u\pi} g_n(v)dv,$$

where (see (3.6))

$$g_n(v) = \left(\frac{e}{2\pi n}\right)^{1/2} (1 + O(n^{-1/2}))(1 - \theta(r_n e^{iv/n}))^{-k-1} e^{-iv} \prod_{l=1}^{j} (\sigma_{n,s}^{(l)})^{k_l}.$$

Lemma 3 implies that for fixed v and $n \to \infty$

$$\left(\frac{e}{2\pi n}\right)^{1/2} \left(1 - \theta(r_n e^{iv/n})\right)^{-k-1} = \frac{1}{2}\left(\frac{n^k e}{2^k \pi}\right)^{1/2} \left(\frac{1}{2} - iv\right)^{-(k+1)/2} \left(1 + o(1)\right).$$

Therefore, applying both results of Lemma 4 and taking into account that $(k_1, \ldots, k_j) \in K_j$ (see (2.1)), we obtain for such v that if $s = 0$, then

$$\begin{aligned} \lim_{n\to\infty} g_n(v)/(bn^\beta)^j &= h_1(k_1, \ldots, k_j; v) \\ &= \frac{1}{2\pi^{1/2}} e^{1/2-iv} \left(\frac{1}{2} - iv\right)^{-j\beta-1/2} \prod_{l=1}^{j} \left[\frac{m_l(R)}{2\pi^{1/2}} \Gamma(l\beta - 1/2)\right]^{k_l}, \end{aligned}$$

and if $s \sim \alpha n,\ 0 < \alpha < 1$, then

$$\begin{aligned} \lim_{n\to\infty} g_n(v) &= h_2(k_1, \ldots, k_j; v) \\ &= \frac{1}{2\pi^{1/2}} e^{1/2-iv} \left(\frac{1}{2} - iv\right)^{-(k+1)/2} \\ &\quad \times \prod_{l=1}^{j} \left[\frac{m_l(R)}{2\pi^{1/2}} \int_\alpha^\infty e^{-y(1/2-iv)} y^{l\beta-3/2} dy\right]^{k_l}. \end{aligned}$$

Now, a slight modification of the Lebesgue dominated convergence theorem yields

$$\lim_{n\to\infty} J_{n,s}(k_1, \ldots, k_j)/(bu^\beta)^j$$

$$= \begin{cases} \int_{-\infty}^{\infty} h_1(k_1, \ldots, k_j; v)dv, & \text{if} \quad s = 0; \\ \int_{-\infty}^{\infty} h_2(k_1, \ldots, k_j; v)dv, & \text{if} \quad s \sim \alpha n, \quad 0 < \alpha < 1. \end{cases} \tag{4.3}$$

To obtain the final expressions (2.7) and (2.8) for $\mu_j(R;\beta)$ and $\mu_j(R;\alpha,\beta)$, respectively, it suffices to get the values of the above integrals and to substitute them into (4.1). Starting with the expression for $h_1(k_1,\ldots,k_j;v)$ we set in the first integral of the right-hand side of (4.3)

$$\omega = \frac{1}{2} - iv. \tag{4.4}$$

Then, obviously,

$$\frac{1}{2\pi^{1/2}}\int_{-\infty}^{\infty} e^{1/2-iv}\left(\frac{1}{2}-iv\right)^{-j\beta-1/2}dv = \frac{\pi^{1/2}}{2\pi i}\int_{1/2-i\infty}^{1/2+i\infty} e^{\omega}\omega^{-j\beta-1/2}d\omega. \tag{4.5}$$

Taking into account that $\omega^{-j\beta-1/2}$ is the Laplace transform of $y^{j\beta-1/2}/\Gamma(j\beta+1/2)$ (see, e. g., (Doetsch, 1967)) and that the integral of the right-hand side of (4.5) represents the inverse Laplace transform at the point $y=1$, i. e., the value of the original at the same point, we immediately obtain

$$\int_{-\infty}^{\infty} h_1(k_1,\ldots,k_j;v)dv = \frac{\pi^{1/2}}{\Gamma(j\beta+1/2)}\prod_{l=1}^{j}\left(\frac{m_l(R)}{2\pi^{1/2}}\Gamma(l\beta-1/2)\right)^{k_l} \tag{4.6}$$

Now, by (4.1), the first line of (4.3) and (4.6) we obtain (2.5) and (2.7).
Further, for the second integral of the right-hand side of (4.3) the substitution (4.4) yields

$$\int_{-\infty}^{\infty} h_2(k_1,\ldots,k_j;v)dv$$

$$= \frac{\pi^{1/2}}{2\pi i}\int_{-\infty}^{\infty}\omega^{-(k+1)/2}e^{\omega}\prod_{l=1}^{j}\left(\frac{m_l(R)}{2\pi^{1/2}}Q(\alpha,l\beta-1/2;\omega)\right)^{k_l}d\omega, \tag{4.7}$$

where $Q(\alpha,a,\omega)=\int_{\alpha}^{\infty}e^{-y\omega}y^{a-1}dy$. Obviously, $Q(\alpha,a,\omega)$ represents the Laplace transform of

$$q(\alpha,a,y)=\begin{cases} y^{a-1}, & \text{if } y\ge\alpha,\\ 0, & \text{if } y<\alpha.\end{cases}$$

Therefore the product $\prod_{l=1}^{j}[(Q(\alpha,l\beta-1/2,\omega)]^{k_l}$ is the Laplace transform of the corresponding convolution

$$\begin{aligned}&\rho(\alpha,\beta,k_1,\ldots,k_j;y)=\\ &=\underbrace{q(\alpha,\beta-1/2;y)*\cdots*q(\alpha,\beta-1/2;y)}_{k_1\ \text{times}}\\ &*\underbrace{q(\alpha,2\beta-1/2;y)*\cdots*q(\alpha,2\beta-1/2;y)}_{k_2\ \text{times}}\\ &*\cdots*\underbrace{q(\alpha,j\beta-1/2;y)*\cdots*q(\alpha,j\beta-1/2;y)}_{k_j\ \text{times}}.\end{aligned}$$

Moreover, $\omega^{-(k+1)/2}$ is the Laplace transform of $y^{(k-1)/2}/\Gamma((k+1)/2)$ (see again (Doetsch, 1967)). Thus, similarly to the previous reasoning we conclude that the integral of the right-hand side of (4.7) is the inverse Laplace transform at the point $y = 1$. Taking into account the expressions for the originals, we get

$$\int_{-\infty}^{\infty} h_2(k_1, \ldots, k_j; v)dv \tag{4.8}$$

$$= \pi^{1/2} \prod_{l=1}^{j} \left(\frac{m_l(R)}{2\pi^{1/2}}\right)^{k_l} \left(\frac{y^{(k-1)/2}}{\Gamma((k+1)/2)} * \rho(\alpha, \beta, k_1, \ldots, k_j; y)\right)_{y=1}.$$

Using the conventions (2.3), one can check by simple manipulations that

$$\left(y^{(k-1)/2} * \rho(\alpha, \beta, k_1, \ldots, k_j; y)\right)_{y=1} = I_k(\alpha, \beta). \tag{4.9}$$

Now, it is obvious that (4.1), the second line of (4.3), (4.8) and (4.9) yield (2.6) and (2.8).

5. APPLICATIONS AND COMMENTS

First, we apply Theorem in case of expectations. Thus, let $j = 1$. Definition (2.1) of the set K_1 implies that $\sum_{K_1}$ in (2.7) and (2.8) contains exactly one summand and that

$$\begin{aligned} \mu_1(R; \beta) &= \Gamma(\beta - 1/2) m_1(R)/(2\Gamma(\beta + 1/2)) \\ &= m_1(R)/(2(\beta - 1/2)), \end{aligned} \tag{5.1}$$

$$\begin{aligned} \mu_1(R; \alpha, \beta) &= \left[m_1(R)/2\right] I_1(\alpha, \beta) = \frac{m_1(R)}{2} \int_{\alpha}^{1} x^{\beta-3/2} dx \\ &= \frac{m_1(R)}{2(\beta - 1/2)}(1 - \alpha^{\beta-1/2}). \end{aligned} \tag{5.2}$$

Now, consider the following experiment. Suppose that a mapping $M \in \mathfrak{M}_n$ is picked at random with probability n^{-n} and afterwards a vertex from the vertex-set of the corresponding graph G_M is chosen independently from the choice of M with probability $1/n$. Let Z_n denote the size (the number of vertices) of the tree to which the chosen vertex belongs. The following corollary presents a limit theorem for the probability that $Z_n/n > \alpha$, $0 < \alpha < 1$, under the condition that this vertex possesses a property R (or, equivalently, the vertex $\in A_n(R)$).

COROLLARY. *Suppose that* (2.4) *holds for* $j = 1$. *Then*

$$\lim_{n\to\infty} \mathbf{P}\{Z_n/n > \alpha \mid \text{the chosen vertex belongs to } A_n(R)\} = 1 - \alpha^{\beta-1/2},$$

where $0 < \alpha < 1$.

Proof. By the formula of total probability, for any fixed $s \geq 1$,

$$\mathbf{P}\{Z_n > s \mid \text{the chosen vertex } \in A_n(R)\}$$

$$= \sum_l \frac{l}{n} \mathbf{P}\{X_n(R; s) = l\} / \sum_l \frac{l}{n} \mathbf{P}\{X_n(R; 0) = l\}$$

$$= \mathbf{E}[X_n(R; s)] / \mathbf{E}[X_n(R; 0)].$$

This together with $s \sim \alpha n$, $0 < \alpha < 1$, (2.5), (2.6), (5.1), and (5.2) implies the required result.

Example 1. Let R_0 denote the property "endvertex". Then, $A_n(R_0)$ is the set of all endvertices of a tree $T \in \mathcal{J}_n$ (mapping M belongs to $\mathfrak{M}_n$). Rényi (Rényi, 1959) proved that

$$\mathbf{E}[X_n^{(1)}(R_0)]/n \to m_1(R_0) = e^{-1} \quad \text{as} \quad n \to \infty.$$

Hence, $b = \beta = 1$, and Corollary implies that

$$\lim_{n \to \infty} \mathbf{P}\{Z_n/n > \alpha \mid \text{an endvertex is chosen}\} = 1 - \alpha^{1/2}. \tag{5.3}$$

Example 2. From Corollary we can also immediately find the limit distribution of Z_n/n. Thus, let R_1 denote a property which does not restrict the vertices of the mappings (trees), as, for instance, "the vertex belongs to U_n". Hence, $X_n^{(1)}(R_1) = n$, $b = \beta = 1$ and from Corollary it follows that Z_n obeys the same limit law (5.3).

Finally, we turn our attention to possible simplifications of (2.8) if

$$\frac{1}{2} \leq \alpha < 1. \tag{5.4}$$

Actually, from (2.3) and the definition of the simplexes $D_k(\alpha)$ it follows that $I_k(\alpha, \beta) = 0$ if $\alpha \geq 1/k$. Hence, by (5.4) we conclude that $I_k(\alpha, \beta) = 0$ for $k \geq 2$. Moreover, definition (2.1) implies again that the sum of the right-hand side of (2.8) has one term. Therefore (2.8) becomes

$$\mu_j(R; \alpha, \beta) = [m_j(R)/2] I_1(\alpha, \beta)$$
$$= [m_j(R)/2] \int_\alpha^1 x^{j\beta - 3/2} dx = \frac{m_j(R)}{2j\beta - 1}(1 - \alpha^{i\beta - 1/2}).$$

Similar reasoning can be also used to study the cases $1/3 \leq \alpha < 1/2$, $1/4 \leq \alpha < 1/3$, etc.

REFERENCES

Doetsch, G. (1967). *Auleitung zum Praktischen Gebranch der Laplace-Transformation und der Z-Transformation*. R. Oldenbourg, München.

Kolchin, V. F. (1986). *Random Mappings*. Optimization Softwere, New York.

Mutafchiev, L. (1984). On some stochastic problems of discrete mathematics. In: *Mathematics and Mathematical Education* (Sunny Beach, 1984). Bulgaria Akad. Nauk. Sofia, pp. 57–80.

Mutafchiev, L. (1992). On random mappings of finite sets. In: *Random Graphs'89*, John Wiley, New York.

Rényi, A. (1959). Some remarks on the theory of trees. *Publ. Math. Inst. Hungarian Acad. Sci.*, **4**, 73–85.

Riordan, J. (1958). *An Introduction to Combinatorial Analysis*. John Wiley, New York.

Stepanov, V. E. (1969). Limit distributions of certain characteristics of random mappings. *Theory Probab. Appl.* 14, 612–626.

Probabilistic Methods in Discrete Mathematics, pp. 366 – 375
V. F. Kolchin et al. (Eds.)
1993 TVP/VSP

ON A NONLINEAR GENERALIZATION OF THE OCCUPANCY PROBLEM

G. V. PROSKURIN
TVP Laboratories, Krasikov Street 32, 117418 Moscow, Russia

ABSTRACT

Let $\{s(i)\}$ be a sequence of random independent identically distributed two-elements subsets of $\{0,1,\dots,N\}$, $s(t)=\{s_1(t),s_2(t)\}, s_j(t)\in\{0,1,\dots,N\}$, $s_1(t)\neq s_2(t)$, and let the graph

$$G=(\{0,1,\dots,N\},\ \{(i,j):\ \mathbf{P}\{s(t)=\{i,j\}\}>0\})$$

be connected.

We consider the Markov chain $l(t)=(l_0(t),l_1(t),\dots,l_N(t))$, $t=0,1,2,\dots$, where $l(0)=(0,0,\dots,0)$, $l_i(t+1)=l_i(t)$, if $i\in s(t+1)$, and $l_i(t+1)=1+\max\{l_i(t),l_j(t)\}$, $i\in s(t+1)=\{i,j\}$. The asymptotic behaviour of characteristics of the Markov chain $l(t)$, including the behaviour of the range $r(t)=\max\{l_i(t)-l_j(t)\colon\ i,j=0,\dots,N\}$, is studied. In particular, it is shown that, if $\mathbf{P}\{s(t)=(i,j)\}=2/(N(N+1))$ and $(4e/b)^{-b}\geq e_0>e$, then

$$\lim_{N\to\infty}\lim_{t\to\infty}\mathbf{P}\{r(t)<b\ \log N\}=1.$$

This model we can interpret as a certain nonlinear generalization of the occupancy problem (Kolchin et al., 1978).

1. Let $l=(l_0,l_1,\dots,l_N)$, $l(0)=(0,0,\dots,0)$, let $\{s(t)\}$ be a sequence of random independent identically distributed two-element subsets of $\{0,\dots,N\}$, and let

$$\mathbf{P}\{s(t)=\{i,j\}\}=p_{\{i,j\}}=p_{i,j},$$

$$f_d(i,j,l)=\begin{cases} l_d, & \text{if } d\neq i, d\neq j,\\ 1+\max\{l_i,l_j\}, & \text{if } d=i.\end{cases} \tag{1.1}$$

$$F_{i,j}l=\big(f_0(i,j,l),f_1(i,j,l),\dots,f_N(i,j,l)\big).$$

In the present paper we consider the asymptotic behaviour of characteristics of the Markov chain

$$l(t)=\big(l_0(t),l_1(t),\dots,l_N(t)\big)=F(t)l(0)=F_{s(t)}F_{s(t-1)}\cdots F_{s(l)}l(0), \tag{1.2}$$

and first of all the behaviour of the range

$$r(t)=\max_{i,j}\{l_i(t)-l_j(t)\},$$

as $t \to \infty$.
Since $\max_i l_i(t) \to \infty$ as $t \to \infty$, it is more convenient to study the Markov chain $m(t) = (m_0(t), m_1(t), m_2(t), \dots, m_N(t)) = Ll(t)$, $m_i(t) = l_i(t) - l_0(t)$. We denote by $p_t\{m \to M'\} = \mathbf{P}\{LF(t)m \in M'\}$ the t-step transition probability of the Markov chain $m(t)$ from the state $m \in M = \{0, 1, 2, \dots\}^N$ into the set $M' \subseteq M$. The probability that the Markov chain at time t is in a state m denote by $p_t(m)$. Taking into accoint that $m_0(t) = 0$, sometimes we use the brief notation $m(t) = (m_1(t), m_2(t), \dots, m_N(t))$.
We suppose that the graph G with the set of vertices $I = \{0, 1, \dots, N\}$ and the set of edges $\Gamma = \{(i, j)\colon \mathbf{P}\{s(t) = \{i, j\}\} > 0$ is connected.

THEOREM 1.1. *The Markov chain $m(t)$ has a single class on non-null recurrent states with the period equal to one and its stationary distribution*

$$p(m) = \lim_{t \to \infty} p_t(m), \qquad m \in M$$

is a unique solution of the system of equations

$$p(m) = \sum_{m' \in M} p(m') p_1\{m' \to m\}, \qquad m \in M\colon \quad \sum_{m \in M} p(m) = 1. \tag{1.3}$$

Proof. To prove the statement of the theorem, it is sufficient to show that there exists a state m^* satisfying the conditions:
1) for every state m there exists a sequence $s_1, s_2, \dots, s_t$, $s_i \in \Gamma$, such that

$$m^* = LF_{s_t} \cdots F_{s_2} F_{s_1} m; \tag{1.4}$$

2) there exists $T_0 > 0$ such that for every $t \geq T_0$ and for every m a sequence $s_1, s_2, \dots, s_t$, for which (1.4) is valid, can be found.
From condition 1) it follows that the Markov chain has a single class of recurrent states, and from condition 2) it follows that this class is aperiodic. The reccurent states are non-null because

$$p_{T+t}(m^*) = \sum_{m \in M} p_t(m) p_T\{m \to m^*\}$$

$$\geq \sum_{m \in M} p_t(m) \min\{p_{s_1} p_{s_2} \cdots p_{s_{T_0}}\colon s_1, s_2, \dots, s_{T_0}\} \geq (\min\{p_s\colon s \in \Gamma\})^{T_0} > 0.$$

The existence of the stationary distribution satisfying the system of equations (1.3) follows from the existence of a single class of non-null recurrent aperiodic states and Theorems 6.4 and 7.1 in (Chung, 1960).
Let us go to the proof of the theorem. In the graph G choose a certain fixed tree $G' = (\{0, 1, \dots, N\}, \Gamma')$, $\Gamma' \subseteq \Gamma$. All vertices of the tree are devided into the strata corresponding to the length of the path in G' to the vertex 0: a vertex i belongs to the stratum labelled with h, if the lenght $h(i)$ of the path from the vertex i to the vertex 0 is equal to h, $h(0) = 0$, $h(\Gamma') = \max\{h(i)\}$.

Introduce the notation

$$\Gamma^+(i) = \{(i,j) \in \Gamma' : h(i) > h(j)\}, \quad \Gamma^-(i) = \{(i,j) \in \Gamma' : h(i) < h(j)\},$$

$$\prod_{i=1}^{r}{}^{+} a_i = a_1 a_2 \cdots a_r, \quad \prod_{i=1}^{r}{}^{-} a_i = a_1 a_{r-1} \cdots a_1,$$

$L^* = \{l = (l_0, l_1, \dots, l_N) \colon l_i \le l_j\}$ for any $(i,j) \in \Gamma'$ such that $h(i) > h(j)\}$, $a_h^1, a_h^2, \dots, a_h^{k_h}$ are the vertices of the stratum h of the graph G'.
From any state l we can reach the set L^* in N steps, since

$$\prod_{h=1}^{h(\Gamma')}{}^{+} \Big(\prod_{k=1}^{k_h}{}^{+} \Big(\prod_{s \in \Gamma^+(a_h^k)} F_s\Big)\Big) L^*.$$

Consider the set

$$L' = \prod_{h=1}^{h(\Gamma')}{}^{-} \Big(\prod_{k=1}^{k_h}{}^{-} \Big(\prod_{s \in \Gamma^-(a_h^k)} F_s\Big)\Big) L^*.$$

It is easy to show that L' consists of the elements of the form $(l_0, l_0 + l_1^*, \dots, l_0 + l_N^*)$, where the values $l_1^*, \dots, l_N^*$ are fixed. Therefore the set L' of the states of the Markov chain $l(t)$ corresponds to the single state $m^* = (l_1, \dots, l_N^*)$ of the Markov chain $m(t)$. Thus, for any m

$$L\Big(\prod_{h=1}^{h(\Gamma')}{}^{-} \Big(\prod_{k=1}^{k_h}{}^{-} \Big(\prod_{s \in \Gamma^-(a_h^k)} F_S\Big)\Big)\Big)\Big(\prod_{h=1}^{h(\Gamma')}{}^{+} \Big(\prod_{k=1}^{k_h}{}^{+} \Big(\prod_{s \in \Gamma^+(a_h^k)} F_s\Big)\Big)\Big) m = m^*.$$

This verifies condition 1). The validity of condition 2) follows from the fact that $F_{0,a}\, L^* \subseteq L^*$ for any $t > 0$. The theorem is proved.
In fact we have described a simple method for constructing a sequence $s(1), s(2), \dots, s(t)$ such that if it occurs, then for any state l of the Markov chain a transition from l to the state of the type $(l_0, l_0 + l_1^*, \dots, l_0 + l_N^*)$ takes place. With the help of this method for any given graph G we can estimate the probability $p_T\{m \to m^*\}$ of transition in T steps from a fixed state m to the state m^*. For example, for the graph $\Gamma' = \{(0,1),\ (1,2), \dots, (N-1, N)\}$ the mapping

$$F_{0,1} F_{1,2} \dots F_{N-1,N},$$

transforms any state to a state $l = (l_0, l_1, \dots, l_N)$ such that $l_0 \ge l_1 \ge \cdots \ge l_N$, and then the mapping $F_{N-1,N} F_{N-2,N-1} \cdots F_{0,1}$ transforms this state to the state $(l_0, l_0 + 1, \dots, l_0 + N - 1)$ corresponding to the state $m = (0, 1, \dots, N-2, N-1, N-1)$. The probability that $s(1), s(2), \dots, s(2N)$ is the sequence

$$(N-1, N), (N-2, N-1), \dots, (0,1), (0,1), (1,2), \dots, (N-1, N)$$

is equal to $(p_{N-1,N}, p_{N-2,N-1} \cdots p_{0,1})^2$. Therefore in this case

$$p_{2N}\{m \to m^*\} > (p_{N-1,N} p_{N-2,N-1} \cdots p_{0,1})^2.$$

With the help of Theorem 1.1 we can estimate the stationary distribution of the range of $l(t)$ as $t \to \infty$.

THEOREM 1.2. *For $N > 1$ and sufficiently large t the inequalities*

$$c_1 a_1^k \le \mathbf{P}\{r(t) \ge k\} \le c_2 a_2^k, \qquad k \ge k_0,$$

are valid, where the constants k_0, $c_1 > 0$, $c_2 > 0$, $a_1 > 0$, $a_2 < 1$ depend on the probabilities $\{p_{i,j}\}$.

Proof. From Theorem 1.1 it follows that there exist a state $m^* = (m_1^*, \dots, m_N^*)$ and a sequence $s_1, s_2, \dots, s_t$ such that for any m

$$LF_{s_t} F_{s_{t-1}} \cdots F_{s_1} m = m^*.$$

Let $r'(m) = \max_{i,j}\{m_i - m_j\}$, $r(m^*) = u$. If $s_1, s_2, \dots, s_t$ occurs in the sequence

$$s(T-t+u-k+2),\ s(T-t+u-k+3), \dots, s(T), \quad T > t-u+k-1, \quad u < k,$$

then the sequence $m(T-t+u-k+2)$, $m(T-t+u-k+3), \dots, m(T)$ contains the state m^*. Suppose that $m^* = m(d)$, $T - m + u < d < T + 1$. Taking into account that $r'(m(t)) < r'(m(t-1)) + 2$, in this case we have

$$u = r'(m^*) = r'(m(d)) \ge r'(m(d+1)) - 1 \ge \cdots \ge r'(m(T)) - (T-d).$$

As a result we obtain that $r'(m(T)) \le u + T - d < k$. Thus, if the sequence $s(T-t+u-k+2)$, $s(T-t+u-k+3), \dots, s(T)$ contains $s_1, s_2, \dots, s_t$, then $r(T) < k$. Hence, the probability $\mathbf{P}\{r(T) \ge k\}$ is not greater than the probability that the sequence

$$s(T-t+u-k+2),\ s(T-t+u-k+3), \dots, s(T)$$

does not contain $s_1, s_2, \dots, s_t$. This probability can be estimated from above by the value

$$(1 - p_{s_1} p_{s_2} \cdots p_{s_t}^{[(t+k-u-1)/t]} \le (1 - p_{s_1} p_{s_2} \cdots p_{s_t})^{1+[(k-u-1)/t]}.$$

From the last inequality it is easy to see that $\lim_{t\to\infty} \mathbf{P}\{r(t) \ge k\} \le c_2 a_2^k$.
Let us prove the second inequality. It is easy to obtain that for every fixed value of $m(t-k) = m = m(m_1(T-k), \dots, m_N(T-k))$ we can find i and j such that

$$\max\{m_s(T-k)\colon s = 1, \dots, N\} = m_i(T-k) = m_j(T-k)$$

$$> \min\{m_s(T-k)\colon s = 1, \dots, N\}$$

and $p_{ij} > 0$. If the sequence $s(T-k+1), \dots, s(T)$ consists of pairs (i,j) only, then

$$r'\big(m(T-k)\big) = r'\big(m(T-k+1)\big) - 1 = \cdots = r'\big(m(T)\big) - k.$$

Hence,

$$\mathbf{P}\{r'(m(T)) \ge k \,|\, m(T-k) = m\} \ge (p_{ij})^k,$$

$$\mathbf{P}\{r'(m(T)) \geq k \mid m(T-k) = m\} \geq \min\{(p_{ij})^k : (i,j) \in \Gamma\},$$

i.e., there exist positive a_1 and c_1 such that $\lim_{t\to\infty} \mathbf{P}\{r(t) \geq k\} \geq c_1 a_1^k$. The proof of the theorem is completed.

From Theorem 1.2 it follows that the stationary distribution of the range $r(t)$ has a finite mean and a finite variance. It is necessary to notice that the constants a_1 and a_2 in the theorem (if we calculate them according to the proof of the theorem) differ considerably from each other: the value a_1 is close to 0, and the value a_2 is close to 1. At the same time we can considerably improve the upper bound of the limit of the probability $\mathbf{P}\{r(t) \geq k\}$.

We say that a sequence $s_1, \ldots, s_t$, $s_d \in \Gamma$, connects i and j, if $i \in s_1$ and $j \in s_t$, $s_d \cap s_{d+1} \neq \emptyset$, $d = 1, 2, \ldots, t-1$.

Let $B_{i,j}(t,u)$ be the event which occurs if the sequence $s(1), s(2), \ldots, s(t)$ contains a certain subsequence $s(d_1), s(d_2), \ldots, s(d_u)$ connecting i with j. Let further $A_{i,j}(t,u)$ be the complement of $B_{i,j}(t,u)$,

$$A_i(t) = \bigcup_{j\neq i} \bigcap_{u=1}^{t} A_{i,j}(t,u), \quad B(t,u) = \bigcup_{i,j=0}^{N} B_{i,j}(t,u).$$

THEOREM 1.3. *The inequality*

$$\lim_{v\to\infty} \mathbf{P}\{r(v) \geq k\} \leq \max\{\mathbf{P}\{A_i(t)\} : i \in I\} + \mathbf{P}\{B(t,k)\}, \tag{1.5}$$

holds. Here $I = \{i\colon \lim_{v\to\infty} \mathbf{P}\{\max_j\{l_j(v)\} = l_i(v)\} > 0\}$,

$$\mathbf{P}\{B(t,k)\} < \binom{t}{k} \sum_{\substack{s_1,\ldots,s_k \\ s_i \cap s_{i+1} \neq \emptyset,\ s_i \in \Gamma}} p_{s_1} p_{s_2} \cdots p_{s_k}, \tag{1.6}$$

$$\mathbf{P}\{A_i(t)\} = \sum_{j_1,\ldots,j_N} \sum_{u=t+1}^{N} \operatorname{Coef}_{z^u} \frac{z p_u(j_1, j_2, \ldots, j_N)}{1 - z q_u(j_1, j_2, \ldots, j_N)}, \tag{1.7}$$

where the summation is carried out over all permutations of the set $\{0, 1, \ldots, N\} \backslash i$, $p_u(j_1, j_2, \ldots, j_N) = \sum_{0 \leq a < u} p_{j_a j_u}$,

$$q_u(j_1, j_2, \ldots, j_N) = 1 - \sum_{0 \leq a < u} p_{j_a j_u}, \qquad j_0 = i.$$

Proof. It is clear that

$$\lim_{v\to\infty} \mathbf{P}\{r(v) \geq k\} = \sum_{m \in M} p(m) \mathbf{P}\{r'(F(t)m) \geq k\}$$

$$\leq \{\mathbf{P}\{r'(F(t)m) \geq k\}\colon p(m) > 0\}. \tag{1.8}$$

Consider the value

$$\mathbf{P}\{r'(F(t)m) \geq k\}, \qquad m = (m_0, m_1, \ldots, m_N), \quad m_i = \max_j\{m_j\}.$$

Evidently,

$$\mathbf{P}\{r'(F(t)m) \geq k\} = \mathbf{P}\{r'(F(t)m) \geq k, A_i(t)\} + \mathbf{P}\{r'(F(t)m) \geq k, \ \overline{A}_i(t)\}$$
$$\leq \mathbf{P}\{A_i(t)\} + \mathbf{P}\{r'(F(t)m) \geq k, \ \overline{A}_i(t)\} \tag{1.9}$$

The enent $\overline{A}_i(t)$ means that for every j there exists a subsequence connecting i and j, and consequently all coordinates of the vector $l(t) = F(t)m = (l_0(t), l_1(t), \ldots, l_N(t))$ are not less that m_i. Thus, the event $\{r'(F(t)m) \geq k, \overline{A}_i(t)\}$ implies the event

$$\left\{\max_j l_j(t) - \min_j l_j(t) \geq k, \ \min_j l_j(t) \geq m_i\right\} \subseteq \left\{\max_j l_j(t) \geq m_i + k\right\}.$$

Therefore

$$\mathbf{P}\{r'(F(t)m) \geq k, \ \overline{A}_i(t)\} \leq \mathbf{P}\left\{\max_j\{l_j(t)\} \geq m_i + k\right\}.$$

This probability attains the maximum value at $m = m(0) = (0, 0, \ldots, 0)$. In this case the event $\{\max_j\{l_j(t)\} \geq k\}$ implies the event $B(t,k)$. Therefore

$$\mathbf{P}\{r'(F(t)m) \geq k, \ \overline{A}_i(t)\} \leq \mathbf{P}\{B(t,k)\}. \tag{1.10}$$

Substituting (1.9) and (1.10) into (1.8) we obtain inequality (1.5).
Let us go over to the estimation of the summands of (1.5). The probability $\mathbf{P}\{B(t,k)\}$ can be estimated from above by the average number of subsequences $s(d_1), s(d_2), \ldots, s(d_k)$, connecting some two points; this is done in (1.6). The event $A_i(t)$ can be represented as a union of mutually disjoint events $A_i(j_1, j_2, \ldots, j_N, t_1, t_2, \ldots, t_N)$:

$$A_i(t) = \bigcup_{j_1,\ldots,j_N} \bigcup_{u=t+1}^{\infty} \bigcup_{\sum_{v=1}^{n} t_v = u} A_i(j_1, j_2, \ldots, j_N, t_1, t_2, \ldots, t_N),$$

where the event $A_i(j_1, j_2, \ldots, j_N, t_1, t_2, \ldots, t_N)$ means that for the vertex j_u, $u = 1, \ldots, N$, the sequence $s(1)$, $s(2), \ldots, s(d)$ contains a subsequence connecting i with j_u, if $d = t_1 + t_2 + \cdots + t_u$, and does not contain such a subsequence, if $d < t_1 + t_2 + \cdots + t_u$. Therefore,

$$\mathbf{P}\{A_i(t)\} = \sum_{j_1,\ldots,j_N} \sum_{u=t+1}^{\infty} \sum_{t_1+\cdots+t_N=u} A_i(j_1, \ldots, j_N, t_1, \ldots, t_N). \tag{1.11}$$

The probability of the event $A_i(j_1, \ldots, j_N, t_1, \ldots, t_N)$, evidently, is equal to

$$\prod_{u=1}^{N} p_u(j_1, \ldots, j_N)\Big(q_u(j_1, \ldots, j_N)\Big)^{t_u - 1}$$

Substituting this expression into (1.11), by simple transformations we obtain (1.7). It seems that the theorem gives nontrivial estimates of the distribution of the random variable r.

2. In this section we consider the stationary distribution of the vector $m(t)$ and the stationary distribution of its range $r(t)$ as $t \to \infty$ for certain graphs G.

THEOREM 2.1. *Let*

$$\mathbf{P}\{s(t) = (0,j)\} = p_j > 0, \qquad \sum_{j=1}^{N} p_j = 1.$$

Then

$$\lim_{t\to\infty} \mathbf{P}\{m(t) = (m_1,\dots,m_N)\} = g(-m_1,\dots,-m_N),$$

where $g(-m_1,\dots,-m_N)$ is the probability that in a sequence of multinomial trials with the probability of the jth outcome equal to p_j; $j = 1,2,\dots,N$, the first realization of the ith outcome occurs at the (m_{i+1})th trial.

Proof. Observe that $l_0(t+1) = l_0(t) + 1 = t+1$ in the Markov chain under consideration. Therefore the difference $l_0(t) - l_j(t)$ is equal to $t - \max\{d\colon d = 1,\dots,t,\ s(d) = (0,i)\}$. In other words, $-m_i(t)$ is equal to $d-1$, where d is determined as the number of the first term of the sequence

$$s(t),\ s(t-1),\dots,s(1),\ (0,i) \tag{2.1}$$

which coincides with $(0,i)$.
Sequence (2.1) can be considered as a sequence of multinomial trias with the probability of the jth outcome $p_j = \mathbf{P}\{s(t) = (0,i)\} > 0$, $j = 1,\dots,N$. This implies the assertion of the theorem. It is interesting to note that in this case $\mathbf{P}\{r \geq k\} = \mathbf{P}\{A_0(k)\}$.
The theorem reduces the investigation of the Markov chain under consideration to the well-studied problem (see, for example, (Kolchin et al., 1978)). In particular,

$$\lim_{t\to\infty} \mathbf{P}\{m_i(t) = -m_i\} = p_i(1-p_i)^m, \qquad i > 0,$$

$$\lim_{t\to\infty} \mathbf{P}\{r(t) \leq k\} = \mathbf{P}\{\mu_0(k+1,N) = 0\},$$

where $\mu_0(k+1,N)$ is the number of empty cells in the scheme of random allocation of k particles into N cells, where each particle occupies the jth cell with probability p_j. If $p_i = 1/N$, $N \to \infty$, then uniformly with respect to a such that $|a| < C$ and aN are integers $\lim_{t\to\infty} N\mathbf{P}\{m_i(t) = -aN\}e^{-a}(1+o(1))$, $\lim_{t\to\infty} \mathbf{P}\{r(t) \leq N\log N + aN\} = \exp(-e^{-a})(1+o(1))$.

THEOREM 2.2. *Let $p_{i,j} = 2/(N(N+1))$, $N \to \infty$ and $(4e/b)^{-b} \geq e_0 > e$, where e_0 is a constant. Then*

$$\lim_{t\to\infty} \mathbf{P}\{r(t) < b\log N\} = 1 + o(1).$$

In particular, this statement is valid if $b = 4e^2$.

Proof. Put in (1.5)

$$k = b\log N, \quad t = N(\log N + w),$$

where w tends to infinity arbitrary slowly. In this case (1.6) implies that

$$\mathbf{P}\{B(t,k)\} < \binom{t}{k}\binom{N+1}{2}^{1-k}(2N-1)^{k-1}$$

$$= \binom{t}{k}\left(\frac{4}{N+1}\right)^{k-1}\left(1-\frac{1}{2N}\right)^{k-1}.$$

Applying Stirling's formula, we obtain that

$$\mathbf{P}\{B(t,k)\} < C(4t/N)^k N/k! = CN\left(\frac{4te}{Nk}\right)^k/\sqrt{k}(1+o(1))$$

$$= CN(4e/b)^{b\log N+w}/\sqrt{\log N}(1+o(1)). \tag{2.2}$$

Let us estimate the probability $\mathbf{P}\{A_i(t)\}$ with the help of (1.7). Remark that

$$q_u(j_1,\dots,j_N) = 1 - 2u(N+1-u)/(N(N+1)) = q_u,$$

$$p_u(j_1,\dots,j_N) = 2u/(N(N+1)) = (1-q_u)/(N+1-u).$$

Thus,

$$\mathbf{P}\{A_i(t)\} = \sum_{j_1,\dots,j_N}\sum_{v=t+1}^{\infty}\mathrm{Coef}_{z^v}\prod_{u=1}^{N} z(1-q_u)/((N+1-u)(1-q_u z))$$

$$= \sum_{v=t+1}^{\infty}\mathrm{Coef}_{z^v}\prod_{u=1}^{N} z(1-q_u)/(1-q_u z). \tag{2.3}$$

Expression (2.4) shows that $\mathbf{P}\{A_i(t)\} = \mathbf{P}\{d_i > t\}$, where d_i is a random variable which can be represented as a sum of N independent random variables with geometrical distributions:

$$d_i = d_i^{(1)} + d_i^{(2)} + \cdots + d_i^{(N)},$$

$$\mathbf{P}\{d_i^{(u)} = a\} = q_u^{a-1}(1-q_u), \quad \mathbf{E}\,d_i^{(u)} = (1-q_u)^{-1}, \quad \mathrm{Var}\,d_i^{(u)} = q_u(1-q_u)^{-2}.$$

The mean $\mathbf{E}\,d_i$ and the variance $\mathrm{Var}\,d_i$ are given by

$$\mathbf{E}\,d_i = \sum_{u=1}^{n}\mathbf{E}\,d_i^{(u)} = \sum_{u=1}^{N}(1-q_u)^{-1} = N\sum_{u=1}^{N}\frac{1}{u} \sim N\log N,$$

$$\operatorname{Var} d_i = \sum_{u=1}^{N} \operatorname{Var} d_i^{(u)} < CN^2.$$

Using Chebyshev's inequality, we get that if $N \to \infty$, then

$$\mathbf{P}\{A_i(t)\} = \mathbf{P}\{d_i > t\} = \mathbf{P}\{d_i - \mathbf{E}\, d_i > t - \mathbf{E}\, d_i\}$$

$$\leq \mathbf{P}\{|d_i - \mathbf{E}\, d_i| > t - \mathbf{E}\, d_i\} \operatorname{Var} d_i/(t - \mathbf{E}\, d_i)^2 < C/w^2 = o(1).$$

Thus we obtained that $\mathbf{P}\{B(t,k)\} = o(1)$, $\mathbf{P}\{A_i(t)\} = o(1)$. Substituting these estimates into (1.5), we obtain the assertion of the theorem.
The stationary distribution of the vector $m(t)$ for the Markov chain under consideration can be calculated for $N = 3$.

THEOREM 2.3. *Let $N = 3$, $p_{i,j} - 1/6$. Then the stationary distribution of the vector $m(t)$ is given by the following formulas:*
1) $p(0,0,0) = g(1)/5 - v$;
2) *if $b > 0$, then*

$$p(0,b,b) = p(b,0,b) = p(b,b,0) = p(-b,-b,0)$$

$$= p(-b,0,-b) = p(0,-b,-b) = f_b/6;$$

3) *if $0 < a < b$, then*

$$p(a,b,b) = p(b,a,b) = p(b,b,a) = p(a-b,-b,0) = p(a-b,0,-b) = p(0,a-b,-b)$$

$$= p(-b,a-b,0) = p(-b,0,a,-b) = p(0,-b,a-b) = p(-a,b-a,b-a)$$

$$= p(b-a,-a,b-a) = p(b-a,b-a,a) = (1/6)^{b-a-1} g_a/12,$$

where $v = 3 - 2\sqrt{2}$,

$$f(z) = \sum_{b=1}^{\infty} f_b z^b = 6g(1)z/((5-v)(1-vz)) - g(1)z/(1-z/6),$$

$$g(z) = \sum_{b=1}^{\infty} g_b z^b = 2f(z)(1-z/6)/(3(1-z/2)) + g(1)z/(3(1-z/2)),$$

$g(1) = 2 - \sqrt{2}$;
4) *in other cases the stationary probabilities are equal to* 0.

For $N = 2$ the stationary distribution of the vector $m(t)$ can be found for arbitrary probabilities $p_{i,j}$.

THEOREM 2.4. *Let $N = 2$. Then the stationary distribution of the vector $m(t)$ is given by the following formulas:*
1) *if $b > 0$, then*

$$p(b,b) = p_{1,2}^b(1-p_{1,2}), \quad p(0,-b) = p_{0,1}^b(1-p_{0,1}), \quad p(-b,0) = p_{0,2}^b(1-p_{0,2});$$

2) *in other cases* $P(a, b) = 0$.

From the last theorem it follows that

$$\mathbf{P}\{r = 0\} = 0,$$

$$\mathbf{P}\{r = s\} = p_{1,2}^{s}(1 - p_{1,2}) + p_{0,1}^{s}(1 - p_{0,1}) + p_{0,2}^{s}(1 - p_{0,2}), \qquad s > 0.$$

To prove Theorems 2.3 and 2.4 it is sufficient to verify that the formulas for stationary distributions satisfy (1.3).

REFERENCES

Kolchin, V. F., Sevastyanov, B. A., and Chistyakov, V. P. (1978). *Random Allocations.* John Wiley, New York.

Chung, K. L. (1960). *Markov Chains with Stationary Transition Probabilities.* Springer, Berlin.

Probabilistic Methods in Discrete Mathematics, pp. 376 – 385
V. F. Kolchin et al. (Eds.)
1993 TVP/VSP

WEAK CONVERGENCE OF CERTAIN MULTIPLE SUMS OF DEPENDENT RANDOM VARIABLES

I. RAKHIMOV
Institute of Mathematics, Academy of Sciences of the Republic of Uzbekistan, Hodgaev Street 29, 700143 Tashkent, Uzbekistan

ABSTRACT

In this article a new scheme to prove limit theorems for multiple sums of random number of random variables is proposed. This scheme is based on an approximation of such sums by sums of the form

$$V_n = \sum_{i=1}^{n} \nu_i(n) X_i(n),$$

where $X_i(n)$ are (may be dependent) random variables and $\nu_i(n)$ are indicator random variables with values 0 or 1.

We consider a family of random variables $\{\xi_i^n; i, n \in \mathbf{N}\}$, $\mathbf{N} = \{1, 2, \ldots\}$ defined on a probability space $\{\Omega, \mathcal{F}, \mathbf{P}\}$ and a family $\{\mathcal{F}_i^n,\ i \in \mathbf{N}_0,\ n \in \mathbf{N}\}$, $\mathbf{N}_0 = \{0\} \cup \mathbf{N}$, of sub-$\sigma$-algebras of $\mathcal{F}$, such that ξ_i^n is measurable with respect to $\mathcal{F}_i^n$, $\mathcal{F}_i^n \subseteq \mathcal{F}_{i+1}^n$, $i \in \mathbf{N}_0$, $n \in \mathbf{N}$.
The conditions for the weak convergence of sums

$$S_n = \sum_{i=1}^{k_n} \xi_i^n \tag{1}$$

of independent random variables to the given infinitely divisible distribution are well known (see, for example, (Petrov, 1987)). In a series of papers analogous conditions were obtained for sums of dependent ξ_i^n. For example, B. Brown in 1971 (see also (Dvoretzky, 1972)) has obtained conditions for the convergence to the normal distribution, B. Brown and G. Eagleson in (Brown and Eagleson, 1971) have investigated the case where the limiting infinitely divisible distribution has a finite variance. A. Klopotowski (see (Klopotowski, 1977)) found such conditions without any assumptions on finiteness of moments.
The conditions obtained in these papers have the following form. Let a set of conditions be sufficient for the weak convergence of sums of independent random variables to the given infinitely divisible distribution. If we replace all probabilities, expectations etc. occurring in these conditions by conditional values with respect

to some σ-algebras and if this random quantities converge in probability to some constants, then the weak convergence of sums holds also in a dependent case.
More detailed information about such results may be obtained from (Jacod and Shiryaev, 1987). R. Liptzer and A. N. Shiryaev (see Liptzer and Shiryaev, 1986) have found a method of proof of such results by means of "stochastic exponents" for semimartingales. They demonstrated also the possibility to deduce the functional central limit theorem for the sum of dependent random variables from the theorem on convergence of semimartingales.
It seems that for the first time the weak convergence of sums of dependent random variables to a mixture of infinitely divisible distributions was proved in (Eagleson, 1975) (he considered the sum of martingale differences with finite variances). Similar results were obtained in (Klopotowski, 1980) without conditions on finiteness of moments.
Several papers (see (Gaenssler et al., 1978; Rychlik, 1979; Beska et al., 1982; Kubaski, 1983)) were devoted to generalizations of theorems on convergence to given infinitely divisible distributions or their mixtures for sums of random number of random variables. In these papers the number of summands was assumed to be a stopping time with respect to $\{\mathcal{F}_i^n,\ i \in \mathbf{N}_0\}$. The scheme of proofs is the same as in the case of nonrandom number of summands.
In this article we suggest a simple proof of limit theorems for sums of random number of random variables. Our method is based on an approximation of distribution of a random sum by the distribution of

$$V_n = \sum_{i=1}^{m_n} \nu_i(n) X_i(n), \tag{2}$$

where random variables $X_i(n)$ are measurable with respect to $\mathcal{F}_i^n$ for any pair (i, n) and $\nu_i(n)$ are indicator random variables with values 0 or 1 such that the variable $\nu_i(n)$ is measurable with respect to $\mathcal{F}_{i-1}^n$ for any pair (i, n).
It is essential for us that $\mathbf{P}\{\nu_i(n) = 0 \text{ or } \nu_i(n) = 1\} = 1$. The case where $\nu_i(n)$ are integer-valued random variables may be reduced to our simple case.
In Section 2 this scheme will be used in proofs of limit theorems for sums

$$S_n^{(r)} = \sum_{i_1=1}^{n} \sum_{i_2=1}^{\eta_{i_1}(n)} \cdots \sum_{i_r=1}^{\eta_{i_1 \cdots i_{r-1}}(n)} \xi_{i_1 \cdots i_r}(n), \tag{3}$$

where $\eta_{i_1}(n), \ldots, \eta_{i_1 \cdots i_{r-1}}(n)$ are integer-valued random variables, $\xi_{i_1 \cdots i_r}(n)$, $(i_1, \ldots, i_r) \in \mathbf{N}^r$, are some (possibly dependent) random variables. The sums $S_n^{(2)}$ were previously investigated in (Rakhimov, 1987; Rakhimov and Sirazhdinov, 1988) in connection with applications to the theory of branching random processes. There it was assumed that variables $\xi_{i_1 i_2}^{(n)}$ are independent and that the distribution of $\xi_{i_1 i_2}^{(n)}$ does not depend on i_2.
Under the conditions analogous to (14) limit theorems for $S_n^{(2)}$ were proved in (Rakhimov, 1991). These results were applied there to two problems: (a) an investigation of branching processes with immigration allowing dependence of reproduction and immigration processes; (b) a study of sampling sums from finite

population of dependent random variables. But the proof in (Rakhimov, 1991) involved a rather complicated method of "stochastic exponents" for semimartingales.

1. SUMS OF RANDOM NUMBER OF RANDOM SUMMANDS

In this section we show that it is possible to deduce limit theorems for sums of random number of random variables from a limit theorem for sums of nonrandom number of dependent random variables. To this aim we prove a theorem on a convergence to the mixture of infinitely divisible distributions.
Let random variables ξ_i^n be the same as in the sum (1),

$$S_n^* = \sum_{i=1}^{\zeta_n} \xi_i^n, \tag{4}$$

where $\{\zeta_n,\ n \in \mathbf{N}\}$ is a sequence of integer-valued random variables such that ζ_n is a stopping time with respect to $\{\mathcal{F}_i^n,\ i \in \mathbf{N}_0\}$, that is $\{\zeta_n \leq i\} \in \mathcal{F}_i^n,\ i \in \mathbf{N}_0$, for any n. We put $\mathcal{F}_0 = \cap \mathcal{F}_0^n$. Let us denote by $K = K(x, \omega)$ the mapping from $\mathbf{R} \times \Omega$ into $\mathbf{R} = (-\infty, \infty)$, such that it is
a) an $\mathcal{F}_0$-measurable random variable for fixed x,
b) bounded nondecreasing function of x for almost all fixed ω,
c) $\mathbf{P}\{K(-\infty, \omega) = 0\} = 1$.
Setting

$$h(t,x) = \begin{cases} \left(e^{itx} - 1 - \frac{itx}{1+x^2}\right)\frac{1+x^2}{x^2}, & x \neq 0, \\ -\frac{t^2}{2}, & x = 0, \end{cases}$$

we define the function

$$\psi(t) = \exp\left\{it\gamma + \int_{-\infty}^{\infty} h(t,x)\, dK(x)\right\}, \tag{5}$$

where γ is an $\mathcal{F}_0$-measurable random variable. Then for almost all fixed $\omega \in \Omega$, function $\psi(t)$ is the characteristic function of infinitely divisible distribution (see (Petrov, 1987)).
Let

$$g_1(x) = \frac{x}{1+x^2}, \quad g_2(x,y) = x g_1(x) \chi(x < y), \tag{6}$$

where $\chi(A)$ is the indicator function of an event A; $\mathbf{E}_{n,i}[\xi] = \mathbf{E}[\xi \mid \mathcal{F}_i^n]$, and τ is a positive constant. We introduce the notation

$$A_{ni} = \mathbf{E}_{n,i-1}\left[\xi_i^n \chi(|\xi_i^n| \leq \tau)\right], \quad Y_i^n = \xi_i^n - A_{ni},$$

$$B_{ni} = \mathbf{E}_{n,i-1} g_1(Y_i^n), \quad D_{n,i}(x) = \mathbf{E}_{n,i-1} g_2(Y_i^n, x).$$

The following statement was proved in (Klopotowski, 1977).

STATEMENT 1. *Let the following conditions be satisfied.*

C1. *For every point $x \in \mathbf{R} \cup \{+\infty\}$*

$$\sum_{i=1}^{k_n} D_{n,i}(x) \xrightarrow{\mathbf{P}} K(x), \qquad n \to \infty, \tag{7}$$

for some $\mathcal{F}_0$-measurable function $K(x)$.

C2. *For some $\mathcal{F}_0$-measurable random variable γ*

$$\sum_{i=1}^{k_n} \{A_{n,i} + B_{n,i}\} \xrightarrow{\mathbf{P}} \gamma, \qquad n \to \infty. \tag{8}$$

C3. *For any fixed $t \in \mathbf{R}$*

$$\sum_{j=1}^{k_n} \left|\mathbf{E}_{n,j-1}(e^{i\ tY_j^n} - 1)\right|^2 \xrightarrow{\mathbf{P}} 0, \qquad n \to \infty. \tag{9}$$

Then $S_n \xrightarrow{D} W$ as $n \to \infty$, where $\mathbf{E}\, e^{itW} = \mathbf{E}\, \psi(t)$.

In order to generalize Statement 1 for sums (4) we choose the sequence of integers $\{k_n,\ n \in \mathbf{N}\}$ such that $\mathbf{P}\{\zeta_n > k_n\} \to 0$ as $n \to \infty$. Letting $\nu_i(n) = \chi(i \le \zeta_n)$ we obtain

$$\chi(\zeta_n \le k_n) \sum_{i=1}^{\zeta_n} \xi_i^n = \chi(\zeta_n \le k_n) \sum_{i=1}^{k_n} \nu_i(n) \xi_i^n, \tag{10}$$

where $\chi(A)$ is the indicator function of an event A.

Since $\{i \le \zeta_n\} \in \mathcal{F}_{i-1}^n$, $i, n \in \mathbf{N}$, Statement 1 is applicable to the right-hand side of (10). Hence if the conditions C1–C3 with $\nu_i(n)\xi_i^n$ in place of ξ_i^n are satisfied, then

$$\sum_{i=1}^{k_n} \nu_i(n) \xi_i^n \xrightarrow{D} W, \qquad \mathbf{E}\, e^{itW} = \mathbf{E}\, \psi(t). \tag{11}$$

In view of the relations

$$g_1\big(\nu_i(n)\xi_i^n\big) = \nu_i(n) g_1(\xi_i^n), \quad g_2\big(\nu_i(n)\xi_i^n,\ x\big) = \nu_i(n) g_2(\xi_i^n, x),$$

$$\mathbf{E}_{n,j-1} e^{it\xi_j^n \nu_j(n)} = 1 - \nu_j(n) + \nu_j(n) \mathbf{E}_{n,j-1} e^{it\xi_j^n}$$

conditions C1–C3 are equivalent to the same ones with the sums $\sum_{i=1}^{k_n} \nu_i(n)\xi_i^n$ instead of $\sum_{i=1}^{k_n} \xi_i^n$.

It should be noted that relation (10) is true for any sequence $\{\xi_i^n;\ i, n \in \mathbf{N}\}$. Therefore the conditions C1–C3 with $\nu_i(n)\xi_i^n$ instead of ξ_i^n are equivalent to C1–C3 with ζ_n in place of k_n.

Thus (11) is true under the conditions C1–C3 where variables k_n are replaced by the variables ζ_n.

From (10), (11) and the inequality

$$\big|\mathbf{P}\{\xi\chi < x\} - \mathbf{P}\{\xi < x\}\big| \le \mathbf{P}\{\chi = 0\}, \tag{12}$$

where χ is the indicator function, ξ is an arbitrary random variable, and $x \in \mathbf{R}$, we obtain that

$$S_n^* = \sum_{i=1}^{\zeta_n} \xi_i^n \xrightarrow{D} W. \tag{13}$$

Thus we have proved the following theorem.

THEOREM 1. *Assume the following conditions are satisfied.*
C1$'$. *For every point $x \in \mathbf{R} \cup \{+\infty\}$*

$$\sum_{i=1}^{\zeta_n} D_{ni}(x) \xrightarrow{\mathbf{P}} K(x), \qquad n \to \infty,$$

for some $\mathcal{F}_0$-measurable function $K(x)$.
C2$'$. *For some $\mathcal{F}_0$-measurable random variable γ*

$$\sum_{i=1}^{\zeta_n} \{A_{ni} - B_{ni}\} \xrightarrow{\mathbf{P}} \gamma, \qquad n \to \infty.$$

C3$'$. *For any fixed $t \in \mathbf{R}$*

$$\sum_{i=1}^{\zeta_n} \left|\mathbf{E}_{nj-1}(e^{itY_j^n} - 1)\right|^2 \xrightarrow{\mathbf{P}} 0, \qquad n \to \infty.$$

Then $S_n^* \xrightarrow{D} W$ as $n \to \infty$, where $\mathbf{E}\, e^{itW} = \mathbf{E}\, \psi(t)$ and the function $\psi(t)$ is defined by (5).

Analogous results were obtained in (Gaenssler et al., 1978; Rychlik, 1979; Beska et al., 1982; Kubaski, 1983) by more tedious arguments.

2. LIMIT THEOREMS FOR $S_N^{(R)}$

Let $J = (i_1, \dots, i_r)$, $r \geq 1$, be an ordered r-tuple, $i_1, \dots, i_r \in \mathbf{N}^r$, $\{\xi_J(n),\ J \in \mathbf{N}^r\}$ be the family of random variables defined on a probability space $\{\Omega,\ \mathcal{F},\ \mathbf{P}\}$ and let $\{\mathfrak{F}_J(n),\ J \in \mathbf{N}^r\}$ be a family of sub-σ-algebras of $\mathcal{F}$ such that variable $\xi_J(n)$ is measurable with respect to $\mathfrak{F}_J(n)$ for any $\bar{J} \in \mathbf{N}^r$.
Below we shall use the direct product of random number of σ-algebras in the usual sense (see (Jacod and Shiryaev, 1987; Liptzer and Shiryaev, 1986)):

$$\prod_{i=1}^{\eta} \mathfrak{F}_i(n) = \left\{ A \in \mathcal{F} \colon A \cap \{\eta = j\} \in \prod_{i=1}^{j} \mathfrak{F}_i(n) \right\}.$$

We consider sums (3) defined above. Suppose that

$$\{i_2 \leq \eta_{i_1}(n),\ i_3 \leq \eta_{i_1 i_2}(n), \dots, i_r \leq \eta_{i_1 \cdots i_{r-1}}(n)\} \in \mathcal{F}_{J'}(n) \tag{14}$$

for any $(i_1, \ldots, i_r) \in \mathbf{N}^r$, where

$$\mathcal{F}_J(n) = \mathfrak{F}_0(n) \times \prod_{j_r=1}^{i_r} \mathfrak{F}_{i_1 \cdots i_{r-1} j_r}(n)$$

$$\times \prod_{l=1}^{r-1} \prod_{j_l=1}^{i_l - 1} \prod_{j_{l+1}=1}^{\eta_{i_1 \cdots i_{l-1} j_l}(n)} \cdots \prod_{j_r=1}^{\eta_{i_1 \cdots i_{l-1} j_l \cdots j_{r-1}}} \mathfrak{F}_{i_1 \cdots i_{l-1} j_l \cdots j_r}(n),$$

$J' = (i_1, \ldots, i_{r-1}, i_r - 1)$, $\mathfrak{F}_0(n)$ is some sub-σ-algebra of $\mathcal{F}$ and

$$\prod_{j_l=1}^{0} \prod_{j_{l+1}=1}^{\eta_{i_1 \cdots i_{l-1} j_l}(n)} \cdots \prod_{j_r=1}^{\eta_{i_1 \cdots i_{l-1} j_l \cdots j_{r-1}}(n)} \mathfrak{F}_{i_1 \cdots i_{l-1} j_l \cdots j_r}(n) = \{\varnothing, \Omega\}$$

for $l = 1, 2, \ldots, r$. We shall denote

$$\mathbf{E}_J^n [\xi] = \mathbf{E} [\xi \mid \mathcal{F}_J(n)].$$

Let τ be some positive constant, and functions $g_1(x)$, $g_2(x, y)$ be defined by (6). We put

$$\theta_J(n) = \mathbf{E}_{J'}^n \left[\xi_J(n) \chi \left(|\xi_J(n)| \le \tau \right) \right], \quad Y_J(n) = \xi_J(n) - \theta_J(n),$$

$$\varphi_J^{(n)}(t) = \mathbf{E}_{J'}^n \left[e^{itY_J(n)} \right], \quad B_J(n) = \mathbf{E}_{J'}^n \left[g_1 \left(Y_J(n) \right) \right],$$

$$D_J(n, x) = \mathbf{E}_{J'}^n \left[g_2 \left(Y_J(n), x \right) \right].$$

Let $\mathcal{F}_0$ be the σ-algebra such that $\mathcal{F}_0 \subseteq \mathfrak{F}_0(n)$ for any $n \in \mathbf{N}$ and $\sum'(\cdot)$ denote the sum of the form

$$\sum_{i_1=1}^{n} \sum_{i_2=1}^{\eta_{i_1}(n)} \cdots \sum_{i_r=1}^{\eta_{i_1 \cdots i_{r-1}}(n)} (\cdot).$$

We introduce the following conditions:

C4. *For every point $x \in \mathbf{R} \cup \{+\infty\}$*

$$\sum\nolimits'_J D_J(n, x) \xrightarrow{\mathbf{P}} K(x) \tag{15}$$

for some $\mathcal{F}_0$-measurable function $K(x)$;

C5. *For some $\mathcal{F}_0$-measurable random variables γ*

$$\sum\nolimits'_J \left[\theta_J(n) + B_J(n) \right] \xrightarrow{\mathbf{P}} \gamma; \tag{16}$$

C6. *For any fixed $t \in \mathbf{R}$*

$$\sum\nolimits'_J \left| 1 - \varphi_J^{(n)}(t) \right|^2 \xrightarrow{\mathbf{P}} 0. \tag{17}$$

THEOREM 2. *If conditions* (14)–(17) *are satisfied, then* $S_n^{(r)} \xrightarrow{D} W$ *as* $n \to \infty$, *the distribution of the random variable* W *is a mixture of infinitely divisible laws with the characteristic function* $\mathbf{E}\,\psi(t)$, *where function* $\psi(t)$ *is defined by* (5).

Proof. First of all we show that the distribution of $S_n^{(r)}$ can be approximated by the distribution of sum (2). We put

$$C(m,n) = \{m,\ m+1,\dots,n\}, \qquad m \le n.$$

Let $N_n^{(0)} = n,\ N_n^{(1)},\dots,N_n^{(r-1)}$ be sequences of natural numbers such that

$$\mathbf{P}\Big\{\max_{(i_1,\cdots,i_k)\in\Delta_{nk}} \eta_{i_1\cdots i_k}(n) > N_n^{(k)}\Big\} \to 0 \tag{18}$$

as $n \to \infty$, where

$$\Delta_{nk} = C(1,n)C(1,N_n^{(1)})\cdots C(1,N_n^{(k-1)}), \qquad k = 1,\dots,r-1.$$

For any r-tuple $J = (i_1,\dots,i_r) \in \Delta_{nr}$ we define a sequence $\{j_n\}$, $n \ge 1$, by the relation

$$j_n = (i_1-1)N_n^{(1)}\cdots N_n^{(r-1)} + (i_2-1)N_n^{(2)}\cdots N_n^{(r-1)} + \cdots + (i_{r-1}-1)N_n^{(r-1)} + i_r.$$

Then letting

$$\nu_{j_n}(n) = \chi\big(i_2 \le \eta_{i_1}(n),\ i_3 \le \eta_{i_1i_2}(n),\dots,\ i_r \le \eta_{i_1\cdots i_{r-1}}(n)\big), \tag{19}$$

$$X_{j_n}(n) = \xi_{i_1\cdots i_r}(n), \tag{20}$$

we obtain $T_{nr}S_n^{(r)} = T_{nr}V_n$, where V_n is defined by (2), $m_n = \prod_{i=0}^{r-1} N_n^{(i)}$, and

$$T_{nr} = \chi\Big\{\bigcap_{k=1}^{r-1}\Big\{\max_{(i_1,\dots,i_k)\in\Delta_{nk}} \eta_{i_1\dots i_k}(n) \le N_n^{(k)}\Big\}\Big\}.$$

Then, using inequality (12) we find that

$$\big|\mathbf{P}\{S_n^{(r)} < x\} - \mathbf{P}\{V_n < x\}\big| < 2\mathbf{P}\{T_{nr} = 0\}. \tag{21}$$

Since the probability in the right-hand side of (21) is smaller than

$$\sum_{k=1}^{r-1}\mathbf{P}\Big\{\max_{(i_1,\dots,i_k)\in\Delta_{nk}} \eta_{i_1\dots i_k}(n) > N_n^{(k)}\Big\},$$

in view of the choice of sequences $N_n^{(k)}$, $k = 0,\dots,r-1$, we obtain

$$\mathbf{P}\{S_n^{(r)} < x\} = \mathbf{P}\{V_n < x\} + o(1), \qquad n \to \infty. \tag{22}$$

We consider now the sum V_n and introduce an additional notation. Let τ^* be some positive constant, functions $g_1(x)$, $g_2(x,y)$ be defined by (6), and let

$$\mathbf{E}_{n,i}[\xi] = \mathbf{E}[\xi \mid \mathcal{F}_i^n], \quad \theta_i^*(n) = \mathbf{E}_{n,i-1}\Big[X_i(n)\chi\big(|X_i(n)| \le \tau^*\big)\Big],$$

$$Y_i^*(n) = X_i(n) - \theta_i^*(n), \quad \varphi_j^{*(n)}(t) = \mathbf{E}_{n,j-1}\, e^{itY_j^*(n)}, \quad \mathcal{F}_0 = \bigcap_{n=1}^{\infty} \mathcal{F}_0^n,$$

$$B_i^*(n) = \mathbf{E}_{n,i-1}\, g_1\big(Y_i^*(n)\big), \quad D_i^*(n,x) = \mathbf{E}_{n,i-1}\, g_2\big(Y_i^*(n), x\big).$$

STATEMENT 2. *Let the following conditions are satisfied.*

C4′. *For every point $x \in \mathbf{R} \cup \{+\infty\}$*

$$\sum_{i=1}^{m_n} \nu_i(n) D_i^*(n,x) \xrightarrow{\mathbf{P}} K(x), \qquad n \to \infty, \tag{23}$$

for some $\mathcal{F}_0$-measurable function $K(x)$.

C5′. *For some $\mathcal{F}_0$-measurable random variable γ*

$$\sum_{i=1}^{m_n} \nu_i(n)\big\{\theta_i^*(n) + B_i^*(n)\big\} \xrightarrow{\mathbf{P}} \gamma, \qquad n \to \infty. \tag{24}$$

C6′. *For any fixed $t \in \mathbf{R}$*

$$\sum_{i=1}^{m_n} \nu_i(n) \big|1 - \varphi_i^{*(n)}(t)\big|^2 \xrightarrow{\mathbf{P}} 0, \qquad n \to \infty. \tag{25}$$

Then $V_n \xrightarrow{D} W$, $\mathbf{E}\, e^{itW} = \mathbf{E}\, \psi(t)$.

Proof of Statement 2 follows from Statement 1 by the arguments used in the proof of Theorem 1.

In order to get the proof of Theorem 2 it is sufficient to apply Statement 2 to the relation (22).

We turn to the case of finite variances.

Assume that

$$\mathbf{E}_{J'}^n \left[\xi_J^2(n)\right] < \infty \tag{26}$$

almost everywhere for any tuple $J \in \mathbf{N}^r$, $n \in \mathbf{N}$. Let

$$\theta_J(n) = \mathbf{E}_{J'}^n \left[\xi_J(n)\right], \quad Y_J(n) = \xi_J(n) - \theta_J(n)$$

$$\sigma_J^2(n) = \mathbf{E}_{J'}^n \left[Y_j^2(n)\right], \quad g(\xi, x) = \xi^2 \chi(\xi < x),$$

$$L(t,x) = \begin{cases} (e^{itx} - 1 - itx)x^{-2}, & x \neq 0, \\ -\frac{t^2}{2}, & x = 0, \end{cases}$$

$$\pi(t) = \exp\left\{it\gamma + \int_{-\infty}^{\infty} L(t,x)\, dK(x)\right\},$$

where γ is some $\mathcal{F}_0$-measurable random variable, $K(x)$ is the function from (5).

THEOREM 3. *If condition (14) is satisfied and there exists a finite $\mathcal{F}_0$-measurable random variable T such that for some positive constants B_n*

$$\lim_{n\to\infty} \mathbf{P}\left\{ \frac{1}{B_n^2} \sum_J{}' \sigma_J^2(n) > T \right\} = 0, \tag{27}$$

$$\frac{1}{B_n^2} \max_{i_1 \in C(1,n)} \max_{i_2 \in C(1,\eta_{i_1}(n))} \cdots \max_{i_r \in C(1,\eta_{i_1 \cdots i_{r-1}}(n))} \sigma_J^2 \to 0, \qquad n \to \infty, \tag{28}$$

$$\frac{1}{B_n} \sum_J{}' \theta_J(n) \xrightarrow{\mathbf{P}} \gamma, \qquad n \to \infty, \tag{29}$$

for some $\mathcal{F}_0$-measurable random variable γ,

$$\sum_J{}' \mathbf{E}_{J'} \left[g\left(\frac{1}{B_n} Y_J(n), x \right) \right] \xrightarrow{\mathbf{P}} K(x), \qquad n \to \infty, \tag{30}$$

for $\mathcal{F}_0$-measurable function $K(x)$ and for any $x \in \mathbf{R}$, then $S_n^{(r)}/B_n \xrightarrow{D} w$ as $n \to \infty$, where the characteristic function of w is $\mathbf{E}\,\pi(t)$.

Proof of Theorem 3 follows from (22) and the following statement. Let

$$\theta_i(n) = \mathbf{E}_{n,i-1}[X_i(n)], \quad Y_i^*(n) = X_i(n) - \theta_i(n), \quad \sigma_i^2(n) = \mathbf{E}_{n,i-1}[Y_i^{*2}(n)].$$

STATEMENT 3. *Let T be a finite $\mathcal{F}_0$-measurable random variable such that for some positive constants B_n*

$$\lim_{n\to\infty} \mathbf{P}\left\{ \frac{1}{B_n^2} \sum_{i=1}^{m_n} \nu_i(n)\sigma_i^2(n) > T \right\} = 0, \tag{31}$$

$$B_n^{-2} \max_{1\le i\le m_n} \nu_i(n)\sigma_i^2(n) \xrightarrow{\mathbf{P}} 0, \qquad n \to \infty. \tag{32}$$

If for some $\mathcal{F}_0$-measurable random variable γ

$$B_n^{-1} \sum_{i=1}^{m_n} \nu_i(n)\theta_i(n) \xrightarrow{\mathbf{P}} \gamma, \qquad n \to \infty, \tag{33}$$

and for $\mathcal{F}_0$-measurable function $K(x)$

$$B_n^{-1} \sum_{i=1}^{m_n} \nu_i(n) \mathbf{E}_{ni-1} \left[g\left(\frac{1}{B_n} Y_i^*(n), x \right) \right] \xrightarrow{\mathbf{P}} K(x), \qquad n \to \infty, \quad x \in \mathbf{R},$$

then $V_n/B_n \xrightarrow{D} w$, $\mathbf{E}\, e^{itw} = \mathbf{E}\,\pi(t)$.
Statement 3 follows from Theorems 3 of (Eagleson, 1975).

REFERENCES

Beska, M., Klopotowski, A., and Slominski, L. (1982). Limit theorems for random sums of dependent d-dimensional random vectors. *Z. Wahrscheinlichkeitstheor. verw. Geb.* **61**, 43–57.

Brown, B. (1971). Martingale central limit theorems. *Ann. Math. Statist.* **42**, 59–62.

Brown, B. and Eagleson, G. (1971). Martingale convergence to infinitely divisible laws with finite variances. *Trans. Amer. Math. Soc.* **162**, 449–453.

Dvoretzky, A. (1972). Asymptotic normality for sums of dependent random variables. In: *Proceedings of the 6th Berkeley Symposium on Mathematical Statistics and Probability* **2**, pp. 513–536.

Eagleson, G. (1975). Martingale convergence to mixtures of infinitely divisible laws. *Ann. Probab.* **3**, 557–562.

Gaenssler, P., Strobel, J., and Stute, W. (1978). On central limit theorems for martingale triangular arrays. *Acta Math. Acad. Sci. Hungar.* **31**, 205–216.

Jacod, J. and Shiryaev, A. (1987). *Limit Theorems for Stochastic Processes.* Springer-Verlag, New York

Klopotowski, A. (1977). Limit theorems for sums of dependent random vectors in $\mathbf{R}^d$. In: *Dissert. Math. C.L.I.* , PWN, Warszawa, pp. 1–62.

Klopotowski, A. (1980). Mixtures of infinitely divisible distributions as limit laws for sums of dependent random variables. *Z. Wahrscheinlichkeitstheor. verw. Geb.* **51**, 101–113.

Kubaski, K. (1983). On a random sums limit problem. In: *Proceedings of the 4th Pannonian Symposium on Mathematical Statistics*, pp. 231–263.

Liptzer, R. and Shiryaev, A. (1986). *Theory of Martingales.* Nauka, Moscow (in Russian).

Petrov, V. (1987). *Limit Theorems for Sums of Independent Random Variables.* Nauka, Moscow, (in Russian).

Rakhimov, I. (1987). On an approximation of the distribution of a sum of random number of random summands. *Dokl. Akad. Nauk UzSSR* **1**, 5–7 (in Russian).

Rakhimov, I. (1991). Sampling sums of dependent random variables, mixtures of infinitely divisible laws, and branching random processes. *Discrete Math.* **3**, 128–147.

Rakhimov, I. and Sirazhdinov, S. (1988). Approximation of the distribution of a sum in a certain scheme for summation of independent random variables. *Dokl. Akad. Nauk SSSR* **301**, 31–34, (in Russian).

Rychlik, Z. (1979). Martingale random central limit theorems. *Acta Math. Acad. Sci. Hungar.* **34**, 129–139.

Probabilistic Methods in Discrete Mathematics, pp. 386 – 394
V. F. Kolchin et al. (Eds.)
1993 TVP/VSP

ON PROBLEMS OF RENEWAL SYSTEM THEORY FOR FINITE TIME INTERVAL

A. A. ROGOV and V. I. TCHERNETSKII
Petrozavodsk State University, 185610 Petrozavodsk, Russia

ABSTRACT

In this paper we present some results of renewal theory.

1. INTRODUCTION

At the present time problems of renewal system theory have been sufficiently deeply elaborated only for the infinitely growing time interval of system functioning ($t \to \infty$), and this is an essential modelling assumption but is not always adequate to real conditions of system functioning (Cox, 1963; Bayhelt and Franken, 1988; Barlow and Proshan, 1965; Barzilovich, 1982; Barzilovich et al., 1983). New problems of renewal theory for systems functioning on a finite time interval ($t \in [0, T]$) are of increasing interest at recent time because of practical requirements (Barlow and Proshan, 1965; Barzilovich, 1982).

We do not pretend in full account, and the authors present the following list of renewal system problems on finite time intervals.

1. Identification both of reliability function $F(t) = \mathbf{P}\{\gamma < t\}$ and renewal time distribution function $G(t) = \mathbf{P}\{\psi < t\}$ from statistical data.

2. Prediction identification of parameters both of the reliability function and of the renewal time distribution function on the basis of the information about supposed system functioning regimes and technical modernization of separate system units and tehnological function schemes.

3. Characterization of a class of renewal functions $B(t) = \mathbf{E}(N(t))$, $t \in [0, T]$, where $\mathbf{E}(N(t))$ is an expected value of renewal $N(t)$ on time interval $[0, t], t \in [0, T]$ for given $F(t), G(t)$.

4. Algorithmization of solution of the main integral equation of the renewal theory

$$B(t) = F(t) * G(t) + \int_0^t B(t-x) d(F(x) * G(x)),$$

for any probability distribution functions $F(t), G(t)$ concentrated on $[0, \infty)$ which describe absolutely continuous random variables; $F(t) * G(t)$ is the convolution of $F(t)$ and $G(t)$.

5. Identification of renewal function $B(t) = \mathbf{E}(N(t))$ on the basis of statistical data from the system functioning.
6. Synthesis of mathematical models of a functioning of renewal system which are adequate to chosen optimization criterion and to real functioning conditions for accepted model assumptions.
7. Usage of a computer in a solution of optimization problems for the renewal theory.
8. Investigation of the complexity of these problems and the stability of its optimal solutions. Estimates of the number of quasioptimal solutions.
9. Computing of industrial, economical, technical, and diagnostical characteristics of renewal systems. Estimation of economical and industrial effectiveness of a functioning of renewal system on a finite time interval.

2. RESULTS OF INVESTIGATION OF RENEWAL SYSTEMS ON FINITE TIME INTERVAL. IDENTIFICATION AND CHARACTERIZATION PROBLEMS

In this section only the results obtained by the authors will be briefly presented. First of all we note that in course of practical solution of optimization problems of the renewal system theory the identification of the reliability function $F(t)$ and of the renewal time distribution function $G(t)$ are auxiliary problems, and we have solved them applying traditional methods of mathematical statistics to arrays of real data for investigated renewal systems.
A specific nature of functions $F(t)$ and $G(t)$ leads to assumptions on the possible analytic structure of these functions which approximate corresponding empirical distribution functions.
The problems of prediction identification of $F(t)$ and $G(t)$ (or $\widehat{F} = F(t) * G(t)$) are new. It requires to develop some special mathematical models. Our results on this subject will be published in the nearest future.
The characterization problem for a class of renewal functions is to indicate necessary and sufficient conditions for $B(t)$ to be a solution of integral renewal equation

$$B(t) = F(t) + \int_0^t B(t-x)dF(x), \tag{1}$$

where $F(t)$ is a distribution function of an absolutely continuous random variable concentrated on $[0, \infty)$.
Let $b^*(p) = \int_0^\infty e^{-pt}b(t)dt$ be the Laplace transform of the renewal density function $b(t) = dB/dt$. The characteristic function of the random variable with distribution function $F(t)$,

$$g(\omega) = \int_0^\infty e^{i\omega t}dF(t), \qquad \omega \in (-\infty, +\infty), \quad i^2 = -1,$$

can be represented as the Laplace image

$$g^*(p) = \int_0^\infty e^{-pt}f(t)dt$$

of probability density function $f(t) = dF/dt$, namely, $g(\omega) = g^*(-i\omega)$.
It follows from (1) that

$$g^*(p) = \frac{b^*(p)}{1+b^*(p)}, \quad \text{i. e., } g(\omega) = \frac{b^*(-i\omega)}{1+b^*(-i\omega)}. \tag{2}$$

THEOREM 1 (Tchernetskii and Rogov, 1991). *Function $b(t)\,(b(t) \equiv 0,\ t < 0)$ is the intensity renewal function if and only if the following conditions are satisfied:*
a)

$$1 = g(0) = \lim_{\omega \to 0} \frac{b^*(-i\omega)}{1+b^*(-i\omega)}; \tag{3}$$

b) $g(\omega)$ is continuous for any $\omega \in (-\infty, +\infty)$;
c)

$$\sum_{s=1}^{N}\sum_{k=1}^{N} \frac{b^*(-i(\omega_s - \omega_k))}{1+b^*(-i(\omega_s - \omega_k))} z_s \bar{z}_k \geq 0, \tag{4}$$

for any integer $N > 0$, any real ω_s, ω_k, and any complex $z_s, \bar{z}_k$, where $\bar{z}_k$ is complex conjugate to z_k.

All known properties (Barlow and Proshan, 1963; Bayhelt and Franken, 1988; Barzilovich, 1982) of reneval function $B(t)$ follow from conditions a)–c) presented above.
The characterization problem of a class of renewal functions became practically important due to the needs in justifying the approximations for renewal functions used in effective algorithms of numerical solution of the integral renewal equation. Moreover, the characterization problem of renewal functions also requires more careful consideration because Tchernetskii's idea to identify the renewal function directly from statistical data without preliminary identification of functions $F(t), G(t)$ (or its convolution $\widehat{F}(t) = F(t) * G(t)$) followed by the solution of the integral renewal equation.
The following theorems of the renewal theory may be very useful for the solution of applied problems.

THEOREM 2 (Tchernetskii and Rogov, 1991). *If $B(t)$, $t \in [0, \infty)$, is a twice continuously differentiable nondecreasing function such that*

$$B(0) = 0, \quad B'(0) = 0 \quad \text{and} \quad \lim_{t \to \infty} B(t)/t = c, \tag{5}$$

then the solution $F(t)$ of the integral equation

$$F(t) = B(t) - \int_0^t B'(t-x)F(x)dx, \tag{6}$$

exists and is a continuous distribution function such that:
1) $F(0) = 0$;
2) *$F(t)$ is twice continuously differentiable and $F'(0) = 0$.*

The reverse statement is also true. If $F(t)$ is a continuous distribution function of some random variable ξ taking values from $[0, \infty)$ and conditions 1)–2) are

satisfied, then the solution $B(t)$ of integral equation (6) is a renewal function satisfying conditions (5).
Note that $B'(0) = 0$ for a wide class of distributions encountered in reliability theory.
The simplest approximation for a renewal function which takes in account the strain aging and wear process, is a quadratic function. The known limit property of renewal functions shows that it could not be quadratic in the whole domain $[0, +\infty)$. However, it may be approximated by a quadratic form on a finite interval $[0, T]$. Since a quadratic function does not satisfy the conditions of the preceding theorem we need to develop separately characteristic theorems for such functions.

THEOREM 3. *The renewal function $B(t)$ has the form*

$$B(t) = \frac{b^2 - c^2}{2} t^2 + 2bt, \tag{7}$$

on the interval $[0, T]$, where $T \leq \log((b + c)/(b - c))/(2c)$, $b > 0$, $0 < c < b$, if and only if the generalized reliability function $F(t)$ connected with the renewal function by equation (6) has the form

$$F(t) = 1 - e^{-bt}(\cosh(ct) - (b/c)\sinh(ct)), \qquad t \in [0, T].$$

THEOREM 4. *The renewal function $B(t)$ has the form*

$$B(t) = \frac{c^2 + b^2}{2} t^2 + 2bt,$$

on interval $[0, T]$, where $T \leq (1/c)\arctan(c/b), c > 0, 0 \leq b < c$, if and only if the generalized reliability function $F(t)$ connected with renewal function by equation (6) has the form

$$F(t) = 1 - e^{-bt}(\cos(ct) - (b/c)\sin(ct)), \qquad t \in [0, T].$$

3. ALGORITHMIZATIONS OF SOLUTION OF THE MAIN INTEGRAL EQUATION

A statistical processing of experimental data on refusals and repairings of a system shows that Poisson distribution with the probability density function

$$f(t) = \lambda \exp(-\lambda t), \qquad \lambda > 0, \quad t \geq 0$$

could not be used for the approximation of empirical distributions. Therefore corresponding renewal function $B(t) = \lambda t$ is used for theoretical considerations only.

The Erlang distribution with the probability density function

$$f(t) = \frac{\lambda^m t^{m-1}}{(m-1)!} \exp(-\lambda t), \qquad \lambda > 0, \quad t \geq 0, \quad m \in \{1, 2, \ldots\},$$

can be used only in partial cases. For this distribution

$$\frac{dB(t)}{dt} = b(t) = \sum_{k=1}^{m} \frac{1}{m}(s_k \lambda) \exp(-s_k t), \qquad s_k = -a + a/\sqrt{m}\theta_k, \tag{8}$$

$$\theta_k = \exp\left(i\frac{2\pi k}{m}\right), \qquad k = 0, 1, \ldots, m.$$

THEOREM 5 (Tchernetskii). *If*

$$f(t) = \sum_{k=1}^{n} p_k \lambda_k e^{-\lambda_k t}, \tag{9}$$

where $p_k > 0$, $\sum_{k=1}^{n} p_k = 1, \lambda_k > 0, \lambda_k \neq \lambda_s\,(k \neq s)$, *then*

$$b(t) = C_0 + \sum_{j=1}^{n} C_j \exp(-\beta_j t), \tag{10}$$

$$C_0 = Q_1(0)/P'(0), \quad C_j = Q_1(\beta_j)/P'(\beta_j), \qquad Q_1(p) = \sum_{k=1}^{n} \frac{p_k \lambda_k Q_2(p)}{p + \lambda_k},$$

$$P(p) = Q_2(p) - Q_1(p), \qquad Q_2(p) = \prod_{k=1}^{n} (p + \lambda_k),$$

$\beta_k > 0, k = 1, 2, \ldots, n-1$, β_k *are roots of polynomial* $P(p)$.

A Scetch of Proof

Let us show that $p = 0$ is a simple root of polynomial $P(p)$. Clearly,

$$P(0) = Q_2(0) - Q_1(0)$$

$$= \lambda_1 \lambda_2 \cdots \lambda_n - \sum_{k=1}^{n} p_k \lambda_1 \lambda_2 \cdots \lambda_n = \lambda_1 \lambda_2 \cdots \lambda_n \left(1 - \sum_{k=1}^{n} p_k\right) = 0,$$

because $\sum_{k=1}^{n} p_k = 1$. Calculating $P'(0)$, we obtain

$$P'(0) = Q_2'(0) - Q_1'(0) + Q_2(0) \sum_{k=1}^{n} \frac{p_k}{\lambda_k} > 0$$

since

$$Q_2(0) = \lambda_1\lambda_2\cdots\lambda_n > 0 \quad \text{and} \quad \sum_{k=1}^{n} p_k/\lambda_k > 0,$$

so zero is a simple root.
Applying the residue theorem and taking into account that all roots $-\beta_1, -\beta_2, \ldots, -\beta_{n-1}$ of $P(p)$ are simple, we obtain that

$$b(t) = \frac{Q_1(0)}{P'(0)} + \sum_{j=1}^{n-1} \frac{Q_1(\beta_k)}{P'(\beta_k)} \exp(-\beta_k t). \tag{11}$$

The proof of the fact that all values β_k are positive and different is very complicated.

Remark 1. A general formula for $b(t)$ can be similarly obtained for a mixture of Erlang distributions. In this case some polynomials of t will replace constants in the formula analogous to (11).

Remark 2. The real distribution obtained by an experimental data fitting may be approximated rather well by a mixture of Poisson and Erlang distributions.
These remarks make it possible to construct an algorithm which could be used on a computer for the solution of renewal theory problems on finite time intervals with real statistical data.

4. SYNTHESIS OF AN OPTIMIZATION MATHEMATICAL MODEL OF THE FUNCTIONING OF RENEWAL SYSTEMS

Real practice leads to a series of optimization problems for functioning of renewal systems. As an example of approach to such problems we present results of development and investigation of optimization models for planning of prophylactical measures (PPM) for a complex system on finite time interval $[0, T]$.
Let us accept the following assumptions:
– a complex system has only one repair team which could make a prophylactic repair only for one machine at a time;
– a planning is carried out for a given finite time interval T (planning period) and it is assumed that at the initial time all the machines are in the working state;
– time of renewal after a damage of every machine does not depend on the number and times of previous damages;
– duration of a prophylactic repair is a constant depending on type of every individual machine;
– all characteristics are completely renewed after a repair of the machine.
The authors model is based on the following strategy of PPM: in case of a refusal of any machine its repairing begins immediately by means of the staff without the repair team and without change of time of next the prophylactic repair which will be carried out at a fixed and optimal time.
As an optimality criterion we have chosen the expected value of the total profit with an account of losses caused by the idle time to repair machines and expenses on both the repair and PPM.

The mathematical description of the chosen optimality criterion and limitations of economical and technical nature are stated in a form of the following optimization problem: find a vector x^* and plan-plot $y(x)^*$ maximizing the mean total income of object function R, i. e.,

$$R(x^*, y(x)^*) = \max\Big(\sum_{i=1}^{n}(C(i)T - (C_{pr}(i) + C(i))x(i)r(i)) \\ - \sum_{i=1}^{n}\sum_{s=0}^{x(i)}(C(i) + C_a(i))\bar{r}_a(i)B_i\big(u_s(i) - m(s)r(i)\big)\Big) \tag{12}$$

under the conditions

$$\sum_{i=1}^{n} x(i)r(i) \le T, \qquad x(i) \ge 0, \quad x(i) \text{ is integer}, \quad i = 1, \dots, n; \tag{13}$$

$$u_0(i) \ge 0, \quad u_s(i) \ge r(i) > 0, \qquad i = 1, \dots, n, \quad s = 1, \dots, x(i); \tag{14}$$

$$\sum_{s=0}^{x(i)} u_s(i) = T, \qquad i = 1, \dots, n; \tag{15}$$

$$\sum_{s=0}^{p} u_s(i)(2z_{pq}^{ij} - 1) + r(i)z_{pq}^{ij} + \sum_{s=0}^{q} u_s(j)(1 - 2z_{pq}^{ij}) + r(j)(1 - z_{pq}^{ij}) \le 0, \tag{16}$$

$$z_{pq}^{ij} \in \{0, 1\}, \qquad i \ne j, \quad i = 1, \dots, n, \quad j = 1, \dots, n, \\ p = 0, \dots, x(i) - 1, \quad q = 0, \dots, x(j) - 1,$$

where n is the number of machines; $r(i)$, $i = 1, \dots, n$, is the necessary time for prophylactic repair of ith machine (in hours): $x(i)$, $i = 1, \dots, n$, is the number of planned prophylactic repairs for ith machine;
$C(i), i = 1, \dots, n$, is the cost of a production made by the ith machine per one hour;
$C_a(i), i = 1, \dots, n$, is the mean cost of an emergency repair of ith machine per one hour;
$C_{pr}(i), i = 1, \dots, n$, is the cost of a prophylactic repair of ith machine per one hour;
$\bar{r}_a(i), i = 1, \dots, n$, is the mean time of an emergency repair of the ith machine (in hours);
$B_i(t)$, $i = 1, \dots, n$, is the renewal function for the ith machine;

$$m(s) = \begin{cases} 0, & \text{if } s = 0, \\ 1 & \text{otherwise}; \end{cases}$$

z_{pq}^{ij} $(i \ne j, i = 1, \dots, n, j = 1, \dots, n,\ p = 0, \dots, x(i) - 1, q = 0, \dots, x(j) - 1)$ are auxiliary variables.
The time interval between the beginnings of $(s-1)$th and sth prophylactic repairs of the ith machine we denote by its sth work cycle ($s = -1$ corresponds to the beginning of planning period at $t = 0$, $s = x(i)$ corresponds to its end). Let

$u_s(i), i = 1, \ldots, n, s = 0, \ldots, x(i)$, be the duration of the sth work cycle of the ith machine.

A vector $x = (x(1), x(2), \ldots, x(n))$ consists of the number of prophylactic repairs for every machine from the system. A set $y(x) = \{(u_s(i), z_{pq}^{ij}): i \neq j,\ i = 1, \ldots, n,\ j = 1, \ldots, n,\ p = 0, \ldots, x(i) - 1,\ q = 0, \ldots, x(j) - 1,\ s = 0, \ldots, x(i)\}$ is a plot of PPM corresponding to vector x.

Inequalities (13)–(16) mean that the total duration of all PPMs for all machines of the system could not exceed the planning period T; intervals between the beginnings of repairs for the same machine could not be less than the prophylactic repair time for this machine; total duration of all work cycles for every individual machine is equal to the planning period T; a simultaneous prophylactic repair of two or more machines is impossible.

The objective functional is nonlinear; it depends on the following variables: $x(i)$ are non-negative integers, variables $u_s(i)$ are non-negative real numbers and values z_{pq}^{ij} are Boolean. According to the modern classification an optimization problem is a mixed multidimensional problem of linear programming with constraints of type "or-or".

To estimate the complexity of problem we have found the number z of pure continuous subproblems, which should be considered to find the optimal solution, i. e., the number of different values of the set of discrete variables:

$$\left([T/r] + 1\right)^n 2^{(n(n-1))^2([T/r])^2} \leq z \leq \left([T/p] + 1\right)^n 2^{(n(n-1))^2([T/p])^2},$$

where $r = \max_{1 \leq i \leq n} r(i),\ p = \min_{1 \leq i \leq n} r(i)$.

The following theorem on the existence of solutions of the optimization problem has been proved in (Rogov, 1991 b).

THEOREM 6. *If functioning of each machine of the system is described by renewal process with continuous renewal function, then the solution of problem* (12)–(16) *exists.*

Papers (Rogov, 1991 a,b; Rogov and Tchernetskii, 1990, 1991) contain further information on investigations of optimization problem (12)–(16) and its particular cases in which analytical or numerical solutions can be found.

REFERENCES

Bayhelt, F. and Franken, P. (1988). *Reliability and Technical Service. Mathematical Approach.* Radio and Svyaz, Moscow (in Russian).

Barlow, R. E. and Proshan, F. (1965). *Mathematical Theory of Reliability.* John Wiley, New York.

Barzilovich, E. Yu. (1982). *Technical Service Models.* Vyschaya Shkola, Moscow (in Russian).

Barzilovich, E. Yu., Belyaev, Yu. K., and Kashtanov V. A. (1983). *Problems of Mathematical Theory of Reliability.* Radio and Svyaz, Moscow (in Russian).

Cox, D. R. (1963). *Renewal Theory.* Wiley, New York.

Rogov, A. A. (1991 a). *On a Problem of Optimal Service of n Service Stations with Fixed Number of Repairs.* Dep. in VINITI, No. 3477–B91 (in Russian).

Rogov, A. A. (1991 b). *On Some Cases of Solving Constructing Problem for Schedule of Planning Repair of n Machines*. Dep. in VINITI, No. 3477-B91 (in Russian).

Rogov, A. A. and Tchernetski V. I. (1990). On a method of calculating optimal schedule for prophylactic repair of paper producing machine. In: *Mathematical Modelling of Economic Processes*, Petrozavodsk, pp. 42–56.

Tchernetskii, V. I. and Rogov, A. A. (1991). Optimization model of prophylactic repair of machine system. In: *Actual Problems of Fundamental Sciences*, Moscow State Technical University, Moscow (in Russian).

Probabilistic Methods in Discrete Mathematics, pp. 395 – 402
V. F. Kolchin et al. (Eds.)
1993 TVP/VSP

ON THE GENERATING FUNCTIONS OF COLOURED PARTITIONS AND LIMIT THEOREMS FOR SUMS OF INDICATORS

M. N. ROKHLIN
Automation Research Institute, Botanicheskaya Street 25, 127276 Moscow, Russia

ABSTRACT

This paper deals with estimates for the characteristic functions of the sums of indicators based on the calculation of their binomial moments. The conditions that give a possibility to apply this method of proving limit theorems are formulated and a new result about the distribution of the number of vertices of given degree in random graphs is obtained.

There are many methods that may be used in studying limit distributions of sums of indicators. Some of them are based on investigations of characteristic or distribution functions (see, for example, (Sachkov, 1978)), or on investigations of some conditional distributions (Kolchin, 1986; Holst, 1979). Some limit distributions may be obtained by estimation of central moments (Mikhailov, 1988; Ronzhin, 1979). In (Sevastyanov, 1972; Zubkov, 1979; Ambrosimov, 1984; Rokhlin, 1991) binomial moments are used to find limit distributions of sums of indicators. The preference of the latter approach is explained by the fact that binomial moments are the most easily calculated characteristics of the sums. In this paper we use this approach and develop the method suggested in (Rokhlin, 1991). Let us remind some details.

Let $\xi = \xi(n)$, $n = 1, 2, \ldots$, be a sequence of random variables with binomial moments

$$B_k = \mathbf{E}\binom{\xi}{k}, \qquad k = 1, 2, \ldots$$

The characteristic functions of the random variables ξ may be formally represented as

$$\mathbf{E}\, e^{it\xi} = \mathbf{E}\,(e^{it} - 1 + 1)^{\xi} = \mathbf{E}\sum_k \binom{\xi}{k}(e^{it} - 1)^k = \sum_k B_k(e^{it} - 1)^k.$$

Let us suppose that the binomial moments of ξ can be given as $B_k = \lambda^k F(k)/k!$, where $\lambda = \lambda(n) \approx \mathbf{E}\,\xi$, $F(k) = F(k, n)$ and $F(k)$ may be replaced by the convergent series

$$F(k) \approx \sum_l f_l k^l \approx \sum_l f_l(n) k^l$$

with the sufficient accuracy. Moreover, let us suppose that the power k^l in last series may be replaced by the factorial power $(k)_l = k(k-1)\cdots(k-l+1)$. In this case we would have

$$\mathbf{E}\, e^{it\xi} \approx \sum_k (\lambda(e^{it}-1))^k/k! \sum_l f_l(k)_l \approx \sum_l f_l c^l \sum_l c^l/l! \approx \exp(c)F(c),$$

where $c = \lambda(e^{it}-1)$.

This idea was used in (Rokhlin, 1991) for proving a limit theorem for sums of indicators. One of the results, proved in that paper, is that if $B_k = \lambda^k F(k)/k!$, $F(k)$ satisfies some conditions connected with the accuracy of the representation of $F(k)$ by the convergent series, and $F(k) = f(k)(1+o(1))$ as $n \to \infty$ uniformly for $k = O(\lambda^{1/2})$, then

$$\varphi(t) = \mathbf{E}\exp\left(it(\xi-\lambda)/\lambda^{1/2}\right) \to \exp(-t^2/2) f\left(\lambda(\exp(it/\lambda^{1/2})-1)\right).$$

The class of functions $F(k)$ that allow the direct use of this result is also described in (Rokhlin, 1991). In particular, the functions $F(k) = (n)_k/n^k$, $F(k) = \exp(ak)(1-k/n)^{an-bk}$ and their productions belong to this class.

A typical problem that may be solved by this method is the obtaining of the limit distribution of the number of isolated trees in large random graphs, or of the random variables μ_r in the classical occupancy problem. The essential property of such problems is the absence of deep dependencies between indicators. This implies that the binomial moments of such sums are of very specific kind. In the present paper we show that the main ideas of (Rokhlin, 1991) may be useful in more general situations. Before formulating the main theorem, let us consider a well-known example of U-statistic.

Let $\xi_1, \xi_2, \ldots, \xi_N$ be a sequence of outcomes in the equiprobable multinomial scheme with N trials and n outcomes. We denote by $\varkappa$ the sum

$$\varkappa = \sum_{1 \le i < j \le N} \theta_{ij},$$

where

$$\theta_{ij} = \begin{cases} 1, & \text{if } \xi_i = \xi_j, \\ 0 & \text{otherwise.} \end{cases}$$

It is well known that binomial moments of $\varkappa$ may be represented as

$$\mathbf{B}_k(\varkappa) = \mathbf{E}\binom{\varkappa}{k} = \frac{1}{k!} \sum_{(i_1,j_1),\ldots,(j_k,j_k)} \mathbf{P}\{\theta_{i_1 j_1} = \cdots = \theta_{i_k j_k} = 1\}, \qquad (1)$$

where the summation is carried over all different pairs $(i_p, j_p) \neq (i_q, j_q)$.

To find the last sum, let us consider the graph $\mathbf{H}$ formed by the vertices $i_1, j_1, \ldots, i_k, j_k$ and the edges $(i_1, j_1), \ldots, (i_k, j_k)$. Suppose that there are k_s^r connected components of the graph $\mathbf{H}$ with s vertices and r arcs. The set of $\{k_s^r\}$ we shall call the structure of $\mathbf{H}$. Notice that every structure of $\mathbf{H}$ satisfies the equality

$$\sum_{r,s} r k_s^r = k. \qquad (2)$$

The probabilities in (1) obviously depend on the structure of **H** only, and does not depend on the values of indices. Moreover, each graph **H** with the same structure adds equal contribution

$$\mathbf{p}\{k_s^r\} = \prod_r \prod_s \left(\frac{1}{n}\right)^{(s-1)k_s^r}$$

to the sum. This formula is correct, if we suppose that $0^0 = 1$. The change of summation over the indices i, j by the summation over the structures of graphs formed by indices in (1) leads us to the formula

$$\mathbf{B}_k = \mathbf{B}_k(\varkappa) = \sum_{\{k_s^r\}} N(N-1)\cdots(N-M+1) \prod_{r,s} \frac{(g(r,s)n^{-s+1}/s!)^{k_s^r}}{k_s^r!}$$

$$= \sum_{\mathbf{K}=\{k_s^r\}} \prod_r \prod_s \frac{a(r,s)^{k_s^r}}{k_s^r!} \frac{(N)_M}{N^M},$$

where $g(r,s)$ is the number of connected graphs with s vertices and r edges,

$$a(r,s) = \frac{N^s g(r,s)}{n^{s-1}s!}, \quad M = \sum_s s k_s^r.$$

The summation in the last formula is taken over all structures satisfying (2). Let us analyze the variables $a(r,s)$. It is well known from many works on U-statistics (see, for example, (Ronzhin, 1979; Mikhailov, 1988)), that the limit distribution (as $n, N \to \infty$) of $\varkappa$ is determined by its first two moments that depend only on $a(1,s)$, $a(2,s)$. Since there exists exactly one graph with one edge and exactly three graphs with two edges, we have $\lambda = a(1,2) = N^2/n$, $2\mu = 2a(2,3) = N^3/n^2$. Given $\nu = \sqrt{\lambda + 2\mu}$, let us estimate $a(r,s)$ for $r > 2$. Using the rough inequality $\sum_s g(r,s) < s^{2r}/s!$ it is easy to show that

$$\sum_s a(r,s) = o(\nu^r), \tag{3}$$

and this estimation is correct for all $r = O(\nu)$ as $n, N \to \infty$.

Let us consider the characteristic function of the random variable $\zeta = (\varkappa - \lambda)/\nu$. Since the random variable $\varkappa$ has a finite number of positive integer outcomes, only a finite number of their binomial moments may be nonzero. This fact, as well as in (Rokhlin, 1991), permits to define the characteristic function of ζ as

$$\psi(t) = \mathbf{E}\exp\{it\zeta\} = \exp\{-it\lambda/\nu\} \sum_k \mathbf{B}_k(\exp\{it/\nu\} - 1)^k. \tag{4}$$

Using the expression obtained for $\mathbf{B}_k$ and equality (2), including the multiplier $c = c(t,\nu) = \exp(it/\nu) - 1$ into the inner sum, and denoting $\mathbf{F}\{k\} = (N)_M/N^M$, we obtain that

$$\psi(t) = \exp\{-it\lambda/\nu\} \sum_{\mathbf{K}} \sum_{\{k_s^r\}} \prod_r \prod_s \frac{(c^r a(r,s))^{k_s^r}}{k_s^r!} \mathbf{F}\{k\}.$$

This equality and relation (3) show that the main contribution to the characteristic function is made by the summands that are independent of the structure elements with indices $r > 2$. As in (Sevastyanov, 1972), these relation we will call exceptional. Here we temporarily interrupt the investigation of this example and formulate general definitions and the main theorem.
Let $s_1, s_2, \dots, s_k, \dots$ be a sequence of positive integers. Any non-negative integer solution $\{k_s^r\}$ of the equation

$$\sum_{r=1}^{\infty}\sum_{s=1}^{s_r} r k_s^r = k$$

is called the coloured partition of order k. If $k_s^r = l$, we say that the partition has l blocks (or components) of rank r and colour s. All components of the solution with ranks $r > 2$ we consider as exceptional and denote by $\bar{k}$. The value

$$V = \sum_{r>2}\sum_{s} r k_s^r$$

is called the volume of exceptional components. The blocks of rank $r = 1, 2$ are called the main blocks and denoted by $\underline{k}$. All coloured partitions of order k we denote by $\mathbf{K}$. Let $\mathbf{F}\{k_s^r\} = \mathbf{F}_n\{k_s^r\}$, $n = 1, 2, \dots$, be a sequence of functionals which are determined on the set of the coloured partitions and takes real values. Let the set $a(r,s) = a_n(r,s)$, $r = 1, 2, \dots$, $s = 1, \dots, s_r$, of real non-negative numbers be determined for each $n = 1, 2, \dots$ We define the sum

$$\varphi_n(k) = \sum_{\mathbf{K}=\{k_s^r\}} \prod_r \prod_s \frac{a(r,s)^{k_s^r}}{k_s^r!} \mathbf{F}\{k_s^r\} \tag{5}$$

as the generating function of the coloured partition.
Let us consider the sequence of double generating functions

$$\psi(t) = \psi(c_n(t)) = \sum_k c_n^k(t)\varphi_n(k)$$

with complex $c = c_n(t)$. Imposing constraints upon $c_n(t)$ and the functionals $\mathbf{F}\{k_s^r\}$ we find the limit of $\psi(t)$ as $n \to \infty$. Let $\lambda = \lambda_n = \sum a_n(1,s)$, $\mu = \mu_n = \sum a_n(2,s)$ and $\nu = \sqrt{\lambda + 2\mu}$. It is sufficient for our aims to consider the arguments c, that are of the form $c = c_n(t) = \exp(it/\nu) - 1$ only. The essential condition in (Rokhlin, 1991) that allowed us to find the limit of $\psi(t)$ was the possibility of decomposing the functional $\mathbf{F}$ into an absolute convergent series. In our case this condition is formulated as follows. We say that the functional $\mathbf{F}$ has the kernel $\mathbf{f} = \mathbf{f}_n\{k_s^r\}$ in the domain λ/ν if there exist functionals $\mathbf{g} = \mathbf{g}_n\{k_s^r\}$ such that $\mathbf{F} = \mathbf{f}(1 + \mathbf{g})$ and the following conditions are satisfied as $n \to \infty$.
(1) For any non-negative constant h and any partitions of order $k < h\lambda/\nu$ functional $\mathbf{f}$ may be represented as an absolutely convergent series by the powers of main components:

$$\mathbf{f} = \sum_l f_l\{\bar{k}\}\underline{k}^l, \tag{6}$$

where $l = (l_1^1, l_2^1, \ldots, l_{s_1}^1, l_1^2, l_2^2, \ldots, l_{s_2}^2)$ is a multi-index,

$$\underline{k}^l = \prod_{r=1}^{2} \prod_{s=1}^{s_r} (k_s^r)^{l_s^r},$$

and $f_l\{\bar{k}\}$ is a sequence of functionals determined on the exceptional components.
(2) For any non-negative constants h, v and any partitions of order k

$$\max_{2<k<v\lambda/\nu} \mathrm{g}\{k_s^r\} \exp(h\lambda/\nu) = o(1).$$

The kernel $\mathbf{f}$ of the functional $\mathbf{F}$ is called linearly bounded, if, for any constant h,

$$\sum_l |f_l\{\bar{k}\}| \underline{k}^l = O(1) \quad \text{as} \quad n \to \infty$$

uniformly with respect to $k < h\nu$.
If condition (1) is fulfilled, it is possible to extend the domain of the functional $\mathbf{F}$ into the complex plane, putting for $u = (u_1^1, \ldots, u_{s_1}^1, u_1^2, \ldots, u_{s_2}^2)$ from the convergence domain of the series (6)

$$\mathbf{f}(u) = \mathbf{f}(u, \bar{k}) = \sum f_l(\bar{k}) u^l.$$

In what follows the values of $\mathbf{F}$ on the partitions without exceptional components are of special interest for us. Using the symbol $\bar{0}$ for such partitions, we determine

$$\mathbf{f}_0(u) = \sum f_l(\bar{0}) u^l.$$

In contrast to (Rokhlin, 1991) we will not study the general conditions of convergence of the double generating function $\psi(t)$, but confine ourselves to a consequence useful in studying the limit behaviour of the characteristic function of the indicator sum $\eta = \eta(n) = \sum \theta_i(n)$, $i = 1, \ldots, N(n)$, as $n \to \infty$.

THEOREM 1. *If the binomial moments of the sequence of random variables $\eta(n)$ are of the form of a generating function of coloured partitions and, as $n \to \infty$, the following conditions are satisfied:*
(1) $\nu = \sqrt{\lambda + 2\mu} \to \infty$;
(2) *the functional $\mathbf{F}$ in the domain λ/ν has a linearly bounded kernel $\mathbf{f}$;*
(3) *for any positive constants B, h*

$$\limsup_{n\to\infty} \max_{2<r<h\lambda/\nu} \left(\left(\frac{B}{\nu}\right)^r \sum_s \left|a(r,s)\right| \right)^{1/r} = 0;$$

(4) *for* $u(t) = (c_n(t)a(1,1), \ldots, c_n(t)a(1,s_1), c_n^2(t)a(2,1), \ldots, c_n^2(t)a(2,s_2))$

$$\lim_{n\to\infty} \mathbf{f}_0(u) = \chi(t);$$

then for any fixed t, $-\infty < t < \infty$,

$$\lim_{n\to\infty} \psi(t) = \lim_{n\to\infty} \mathbf{E}\exp(it(\eta-\lambda)/\nu) = \exp(-t^2/2)\chi(t).$$

Proof. Since the random variables η for any n have only a finite number of different values, their characteristic function $\psi(t)$ is of form (4). Substitute into this formula expression (5) for $\mathbf{B}_k$, and group the addends that correspond to the coloured partitions with no more than $H\lambda/\nu$ main blocks of each rank and colour, and with values of exceptional components not greater than $H\lambda/\nu$. Formally, the set $\mathbf{L}$ of all such addends is determined by the system of inequalities

$$\mathbf{L} = \mathbf{L}\{k_s^r\} = \begin{cases} k_s^r < H\lambda/\nu, & r=1,2, \quad s=1,\dots,s_r, \\ \sum\limits_{r>2}\sum\limits_s rk_s^r < H\lambda/\nu. \end{cases} \tag{7}$$

Let us first suppose that there exists a constant $A > 0$ such that

$$\mathbf{P}\{\eta < A\lambda\} = 1. \tag{8}$$

It is easy to show that in this case $\mathbf{B}_k(\eta) \le (A\lambda)^k/k!$, and therefore for any $\varepsilon > 0$ it is possible to find a constant H large enough such that

$$\left|\psi(t) - \sum_{k<h\lambda/\nu} c^k\mathbf{B}_k(\eta)\right| < \varepsilon.$$

If at least one of inequalities (7) is wrong, the order $k = \sum\sum rk_s^r$ of the partitions, which the considered blocks correspond to, is not less than $H\lambda/\nu$. It means that only the addends corresponding to the domain $\mathbf{L}$ make the main contribution to the sum for $\psi(t)$. Moreover, since the functional $\mathbf{F}$ has the kernel $\mathbf{f}$, we can replace $\mathbf{F}$ by $\mathbf{f}$:

$$\left|\psi(t) - \sum_{\mathbf{L}}\prod_r\prod_s \frac{(c^r a(r,s))^{k_s^r}}{k_s^r}\mathbf{f}\{k_s^r\}\right| < \varepsilon.$$

The only difficulty in summation in the last formula is that (in obvious notation) $\exp(|u(t)|) \to \infty$. But

$$|\exp(u(t))| = \left|\prod_{r=1}^{2}\prod_{s=1}^{s_r} \exp(c^r(t)a_n(r,s))\right|$$

is bounded, and therefore, to find the sum over $\mathbf{L}$, we group addends

$$\sum_{\mathbf{L}_2}\prod_r\prod_s \frac{(c^r a(r,s))^{k_s^r}}{k_s^r!}\left(\sum_{\mathbf{L}_1}\prod_r\prod_s \frac{(c^r a(r,s))^{k_s^r}}{k_s^r!}\mathbf{f}\{k_s^r\}\right),$$

introducing all members corresponding to the main blocks into the inner sum:

$$\mathbf{L}_1 = \{0 \le k_s^r < H\lambda/\nu\colon\ r=1,2;\ s=1,\dots,s_r\},$$

and leaving in the external sum all members corresponding to the exceptional components:

$$\mathbf{L}_2 = \left\{k_s^r\colon\ 0 \le \sum\sum r k_s^r < H\lambda/\nu,\ r = 3,4,\ldots;\ s = 1,\ldots,s_r\right\}.$$

We can estimate the sum over the domain $\mathbf{L}_1$ by using Theorem 1 from (Rokhlin, 1991) and taking into account that $\mathbf{f}$ is linearly bounded. We obtain that

$$\left|\psi(t) - \exp(-it\lambda/\nu)\exp(u(t))\mathbf{f}_0(u)\right| \le \varepsilon + \tau, \tag{9}$$

where

$$\tau = O\left(\exp(u(t))\sum_{\mathbf{L}_3}\prod_r\prod_s \frac{|c^r a(r,s)|^{k_s^r}}{k_s^r!}|\mathbf{f}(u,\bar{k})|\right),$$

and $\mathbf{L}_3$ differs from $\mathbf{L}_2$ only by absence of addends corresponding to zero value of the exceptional components. It is easy to show, using condition 3 of the theorem, that τ does not exceed ε. Turning to the limit in (9) and replacing $\mathbf{f}_0(u)$ by $\chi(t)$ we complete the proof, if (8) is valid. If $\eta > A\lambda$ with a positive probability, then as in (Rokhlin, 1991) we can use truncation and instead of the random variable η consider the variable $\xi = \mathbf{I}\eta$, where $\mathbf{I}$ is an indicator of the event $\{\eta < (s_1 + s_2)\lambda\}$. From conditions (1)–(3) of the theorem one can get that limit distributions of $(\eta - \lambda)/\nu$ and$(\xi - \lambda)/\nu$ are identical, if they exist. But the random variable ζ satisfies (8) and, that is easy to show, the difference between the moments of η and ξ may be neglected. The proof is complete.

One of the main difficulties in verifying the conditions of the theorem is the investigation of the functional $\mathbf{F}$. But in some combinatorial problems the functionals frequently appear that may be investigated by the methods described in (Rokhlin, 1991). Particularly, turning to the example above, it is easy to verify that $\mathbf{F}\{k_s^r\}$ in the domain λ/ν has a bounded kernel $\mathbf{F}_0 = \exp(-(k_2^1 + k_3^2)/2N)(1 + o(1))$, and this estimate for any fixed $H > 0$ is uniform with respect to $k_s^r < H\lambda/\nu$. Using the theorem proved above, we obtain that if $n, N \to \infty$ in such a way that $\lambda/\nu \to \infty$ and $N/n \to \alpha$, then

$$\psi(t) \to \exp\left(-\frac{t^2}{2(1+\alpha)}\right).$$

More substantial results can be obtained with the help of this theorem, for example, for the distribution of the random variables $\tau = \tau_t$, the number of vertices of degree t in the random graph $G_{n,p}$. That result develops the investigation of τ_t, that was begun in (Ivchenko, 1973; Palka, 1984).

Theorem 2. *If $n \to \infty$ and $np \to \lambda > 0$, then*

$$\lim_{n\to\infty} \exp\left(iu(\tau - na)/n^{1/2}\right) = \exp\left(-u^2\sigma^2/2\right),$$

where $a = \lambda^t e^{-\lambda}/t!$, $\sigma^2 = a(1 - a(1 - (\lambda - t)^2/\lambda))$.

REFERENCES

Ambrosimov, A. S. (1984). Asymptotic normality of sums of dependent random variables considered by B. A. Sevastyanov. *Theory Probab. Appl.* **21**, 183–189.

Holst, L. (1979). A unified approach to limit theorems for urn models. *J. Appl. Probab.* **16**, 154–162.

Ivchenko, G. I. (1973). On the asymptotic behavior of degrees of vertices in a random graph. *Theory Probab. Appl.* **18**, 188–196.

Kolchin, V. F. (1986). *Random Mappings*. Springer, New York.

Mikhailov, V. G. (1988). On the asymptotic normality of U-statistics with non-negative kernels. *Proc. Steklov Inst. Math.* **177**, 111–120.

Palka, Z. (1984). On number of vertices of given degree in a random graph. *J. Graph Theory* **8**, 167–170.

Rokhlin, M. N. (1991). Binomial moments and exponential generating functions. *Discrete Math. Appl.* **3**, 114–132 (in Russian).

Ronzhin, A. F. (1979). On limit theorems for sums of dependent random variables. *Theory Probab. Appl.* **24**, 549–559.

Sachkov, V. N. (1978). *Probabilistic Methods in Combinatorial Analysis*. Nauka, Moscow (in Russian).

Sevastyanov, B. A. (1972). Poisson limit law for a scheme of sums of dependent random variables. *Theory Probab. Appl.* **17**, 695–700.

Zubkov, A. M. (1979). Inequalities for transition probabilities and their applications. *Math. USSR Sb.* **109**, 491–532 (in Russian).

Probabilistic Methods in Discrete Mathematics, pp. 403 – 412
V. F. Kolchin et al. (Eds.)
1993 TVP/VSP

PROBABILISTIC METHODS IN THE THEORY OF APPROXIMATION OF DISCRETE FUNCTIONS

B. V. RYAZANOV

TVP Laboratories, Krasikov Street 32, 117418 Moscow, Russia

ABSTRACT

In this paper the limit theorems of probability theory are applied to the analysis of the approximation of discrete functions (i. e., functions mapping a finite set of cardinality N into a finite set of cardinality M). The following three objects are investigated:

the Hamming distance between a random discrete function of arbitrary nature and an arbitrary equidistant set (the distance between every two functions from this set is $N(M-1)/M$);

the code distance of a random linear code;

the number of nonzero Walsh coefficients (Walsh spectral complexity) of a random Boolean function.

On the set of original functions (a linear code also is represented as a discrete function generated by a system of vectors) a uniform measure is defined, so the objects under consideration become random variables (r. v.). We investigate the asymptotic distributions of these r. v.'s as $N \to \infty$ by means of new multidimensional integral and local limit theorems for large deviations of sums of random vectors.

1. AN APPROXIMATION WITH EQUIDISTANT SETS

1.1. The statement of the problem and the main result

1.1.1. Let $\mathcal{X}$ and $\mathcal{Y}$ be arbitrary finite sets of cardinalities $N = |\mathcal{X}|$ and $M = |\mathcal{Y}|$, $\mathcal{B}_{N,M} = B(\mathcal{X}, \mathcal{Y})$ be the set of all mappings from $\mathcal{X}$ into $\mathcal{Y}$ and let

$$\rho(g,h) = \sum_{x \in \mathcal{X}} \mathrm{I}\{g(x) \neq h(x)\}$$

be the Hamming distance between $g, h \in \mathcal{B}_{N,M}$ ($\mathrm{I}\{\cdot\}$ is the indicator function of the event specified in brackets).

We assume that N/M is integer. Let $\mathcal{B}$ be an arbitrary equidistant subset of the set $\mathcal{B}_{N,M}$ such that

$$\rho(g,h) = N(M-1)/M \qquad \forall g, h \in \mathcal{B}; \tag{1}$$

$$\rho(f, \mathcal{B}) = \min_{g \in \mathcal{B}} \rho(f, g)$$

be the distance between the function $f \in \mathcal{B}_{N,M}$ and the set B.
Let us define a uniform measure $\mathbf{P}$ on $\mathcal{B}_{N,M}$ and pick f at random according to this measure; we will call such f a *random discrete function* (r. d. f.) from $\mathcal{B}_{N,M}$. In this case the distance $\rho(f, B)$ becomes a random variable (r. v.), and we will be interested in the distribution of this r. v.

1.1.2. The main result of this section is the following assertion.

THEOREM 1. *Let f be a r. d. f. and B be an arbitrary equidistant set of cardinality $L = |B|$ satisfying (1). Let*

$$\tilde{u}^2 = 2\log L - \log\log L + 2x - \log 4\pi + o(1), \qquad L \to \infty, \quad M = o(N), \tag{2}$$

where x is an arbitrary fixed number, $-\infty < x < \infty$. Then

$$\lim_{L\to\infty} \mathbf{P}\{\rho(f, B) < Nq - \tilde{u}\sqrt{Npq}\} = 1 - e^{-e^{-x}}, \tag{3}$$

where $p = M^{-1}$, $q = 1 - p$.

Example. Let M be a prime number ($M \geq 2$), n be a positive integer, $N = M^n$,

$$\mathcal{Y} = \{0, 1, \dots, M-1\}, \quad \mathcal{X} = \mathcal{Y}^n.$$

We define the operation $\oplus$ of coordinate-wise addition (mod M) on $\mathcal{X}$ and on $\mathcal{Y}$, and for every $g = (g_1, \dots, g_n)$ and $x = (x_1, \dots, x_n)$ from $\mathcal{X}$ we set

$$g(x) = (g, x) = g_1x_1 \oplus \cdots \oplus g_nx_n.$$

It is easy to see that g is a linear function, and every linear function from $\mathcal{B}(\mathcal{X}, \mathcal{Y})$ has form (3). We will demonstrate below that the set B of all linear functions from $\mathcal{B}(\mathcal{X}, \mathcal{Y})$ is equidistant.
In this example $L = N = M^n$, and the "typical" value ρ_{typ} of $\rho(f, B)$, i. e., the value around of which the distribution of distance $\rho(f, B)$ is concentrated, has the form, as $n \to \infty$,

$$\rho_{\text{typ}} \approx N(M-1)/M - \tilde{u}\sqrt{N(M-1)/M^2}$$

or

$$\rho_{\text{typ}} \approx N(M-1)/M - M^{-1}u\sqrt{2N(M-1)n\log M}.$$

In particular, for $M = 2$, i. e., if $\mathcal{B}_{N,2}$ is the set of Boolean functions of n variables and B is the set of linear Boolean functions

$$\rho_{\text{typ}} \approx 2^{n-1} - \sqrt{2^{n-1}n\log 2}.$$

1.2. Proofs

1.2.1. Let us show that in the example under consideration the set $\mathcal{B}$ of all linear functions from $\mathcal{B}(\mathcal{X},\mathcal{Y})$ is equidistant. Notice that for every integer k

$$M^{-1}\sum_{m=0}^{M-1} e^{2\pi imk/M} = \begin{cases} 1 & \text{if } k \equiv 0 \,(\operatorname{mod} M), \\ 0 & \text{if } k \not\equiv 0 \,(\operatorname{mod} M), \end{cases}$$

so for every distinct $g, h \in \mathcal{B}$ and $m \not\equiv 0 \,(\operatorname{mod} M)$

$$\sum_{x\in\mathcal{X}} \exp\left\{\frac{2\pi i}{M} m(g-h,x)\right\} = 0.$$

Using these two equalities, we obtain the equality

$$\begin{aligned} N - \rho(g,h) &= \sum_{x\in\mathcal{X}} \mathbf{I}\{g(x) = h(x)\} \\ &= \sum_{x\in\mathcal{X}} M^{-1} \sum_{m=0}^{M-1} \exp\left\{\frac{2\pi i}{M} m(g(x) - h(x))\right\} \\ &= M^{-1} \sum_{m=0}^{M-1} \sum_{x\in\mathcal{X}} \exp\left\{\frac{2\pi i}{M} m(g-h,x)\right\} = N/M, \end{aligned}$$

which yields

$$\rho(g,h) = N(M-1)/M.$$

1.2.2. To prove Theorem 1, we put

$$\xi_g(x) = \mathbf{I}\{f(x) = g(x)\}, \quad \xi_g = \sum_{x\in\mathcal{X}} \xi_g(x), \quad \xi = \max_{g\in\mathcal{B}} \xi_g.$$

It is clear that

$$\rho(f,\mathcal{B}) = N - \xi.$$

Further we put

$$u = Np + \tilde{u}\sqrt{Npq}, \quad \eta_g = \mathbf{I}\{\xi_g \ge u\}, \quad \eta = \sum_{g\in\mathcal{B}} \eta_g$$

and notice that

$$\mathbf{P}\{\xi < u\} = \mathbf{P}\{\eta = 0\}. \tag{4}$$

Let us use the Sevastyanov theorem (see (Kolchin et al., 1978)), according to which

$$\lim_{L\to\infty} \mathbf{P}\{\eta = 0\} = e^{-\lambda} \tag{5}$$

(where λ, $0 < \lambda < \infty$, is a fixed number) provided

$$\sum_{g\in\mathcal{B}} \mathbf{P}\{\eta_g = 1\} \to \lambda \tag{6}$$

and for every fixed integer r, $r \geq 1$,

$$\frac{\mathbf{P}\{\eta_{g_1} = \cdots = \eta_{g_r} = 1\}}{\mathbf{P}\{\eta_{g_1} = 1\} \cdots \mathbf{P}\{\eta_{g_r} = 1\}} = 1 + o(1) \tag{7}$$

uniformly over

$$\mathbf{g} = (g_1, \ldots, g_r) \in \mathcal{B}^{(r)},$$

where $\mathcal{B}^{(r)}$ is the set of all possible r-samples from $\mathcal{B}$ without replacement.
To verify conditions (6) and (7), we may assume, without loss of generality, that

$$\mathcal{X} = \{1, 2, \ldots, N\}, \quad \mathcal{Y} = \{0, 1, \ldots, M-1\}.$$

To every function $g \in \mathcal{B}_{N,M}$ we assign the vector

$$\hat{\mathbf{g}} = \left\{ e^{2\pi i m g(x)/M} \colon m = 1, \ldots, M-1;\ x = 1, \ldots, N \right\}$$

of dimension $N(M-1)$ and notice that for every $g, h \in \mathcal{B}$ the corresponding vectors $\hat{\mathbf{g}}$ and $\hat{\mathbf{h}}$ are orthogonal, because due to (1) and the evident equality

$$\sum_{m=1}^{M-1} e^{2\pi i m y/M} = \begin{cases} M-1 & \text{for } y = 0, \\ -1 & \text{for } y \neq 0 \end{cases}$$

(y is integer, $|y| < M$) we have

$$\sum_{m=1}^{M-1} \sum_{x=1}^{N} \exp\left\{ \frac{2\pi i}{M} m(g(x) - h(x)) \right\} = (M-1)(N - \rho(g,h)) - \rho(g,h) = 0.$$

So

$$L = |\mathcal{B}| \leq N(M-1) < N^2$$

and

$$\tilde{u} \sim \sqrt{2 \log L} = O(\log N) = o(N^{1/6}).$$

In this case, setting

$$\lambda = e^{-x}, \quad \tilde{\xi}_g = \frac{\xi_g - Np}{\sqrt{Npq}},$$

we can check condition (6) by means of the Cramer theorem (Feller, 1971):

$$\begin{aligned} L\mathbf{P}\{\eta_g = 1\} &= L\mathbf{P}\{\xi_g \geq u\} = L\mathbf{P}\{\tilde{\xi} \geq \tilde{u}\} \\ &\sim L\Phi(-\tilde{u}) = \frac{L}{\tilde{u}\sqrt{2\pi}} \int_{\tilde{u}}^{\infty} e^{-v^2/2}\, dv \\ &\sim \frac{L}{\tilde{u}\sqrt{2\pi}} e^{-\tilde{u}^2/2} \sim \lambda. \end{aligned}$$

Further we use the following assertion.

THEOREM 2. *Let r, $r \geq 1$, be an arbitrary fixed integer, $L \to \infty$, $M = o(N)$, $v \geq 0$, and $v = o(N^{1/6})$. Then*

$$\mathbf{P}\{\tilde{\xi}_{g_1} \geq v, \ldots, \tilde{\xi}_{g_r} \geq v\} = \Phi^r(-v)\big(1 + o(1)\big) \tag{8}$$

uniformly in $\mathbf{g} = (g_1, \ldots, g_r) \in B^{(r)}$.

Condition (7) is a direct consequence of (8) for $v = \tilde{u}$.
The proof of Theorem 2 is too cumbersome (it is done by the scheme used in (von Bahr, 1967)) and therefore is omitted. To complete the proof of Theorem 1, it is sufficient to return to equalities (4), (5) and notice that

$$\mathbf{P}\{\rho(f, \mathcal{B}) < Nq - \tilde{u}\sqrt{Npq}\} = \mathbf{P}\{\xi \geq u\} = 1 - \mathbf{P}\{\eta = 0\}.$$

2. THE CODE DISTANCE OF A RANDOM LINEAR CODE

2.1. The main result

2.1.1. Let V_n be the set of all n-dimensional binary vectors with coordinate-wise addition $\oplus\,(\text{mod}\,2)$; $\mathbf{v} = (v^{(1)}, \ldots, v^{(m)})$ be an m-sample of elements $v^{(i)} \in V_n$, V_n^m be the set of all m-samples from V_n with replacement, and let $V_n^{(m)}$ be the set of all m-samples from V_n without replacement, $V_n^{[m]}$ be the set of all m-samples $\vec{v}$ of rank m, $m \leq n$. Let

$$V = V(\mathbf{v}) = \{x_1 v^{(1)} \oplus \cdots \oplus x_m v^{(m)}\colon\ x_i \in V_1,\ i = 1, \ldots, m\}$$

be a linear code, i. e., a subspace of the space V_n generated by the system of vectors $\mathbf{v}$, and let

$$\|v\| = \sum_{i=1}^{n} v_i$$

be the Hamming weight of vector $v = (v_1, \ldots, v_n) \in V_n$;

$$d(V) = \min\big\{\|v\|\colon v \in V \setminus \{0\}\big\}$$

be the code distance of code V.
We may interpret the code $V = V(\mathbf{v})$ as a discrete function

$$f_{\mathbf{v}}(x) = x_1 v^{(1)} \oplus \cdots \oplus x_m v^{(m)},$$

which maps $\mathcal{X} = V_m$ into $\mathcal{Y} = V_n$ ($x = (x_1, \ldots, x_m) \in V_m$), and in this case the sets V_n^m, $V_n^{(m)}$ and $V_n^{[m]}$ may be considered as subsets of the set $\mathcal{B}(\mathcal{X}, \mathcal{Y})$.

THEOREM 3. *Let*

1) *a uniform measure* $\mathbf{P}$ *on* V_n^m ($V_n^{(m)}$ or $V_n^{[m]}$) *be determined and* $V = V(\mathbf{v})$ *be the linear code generated by the system of vectors* $\mathbf{v}$ *picked from* V_n^m *(correspondingly, from* $V_n^{(m)}$ *or* $V_n^{[m]}$*) according to this measure;*

2) $n \to \infty, \quad m \to \infty, \quad m = o(n^{1/6});$ (9)

3) $L = 2^m - 1$, x *be an arbitrary fixed number*, $-\infty < x < \infty$,

$$\tilde{u}^2 = 2\log L - \log\log L + 2x - \log 4\pi + o(1). \tag{10}$$

Then

$$\lim_{\substack{n\to\infty \\ m\to\infty}} \mathbf{P}\Big\{d(V) < \frac{n}{2} - \frac{\tilde{u}\sqrt{n}}{2}\Big\} = 1 - e^{-e^{-x}}. \tag{11}$$

2.1.2. From (11) it follows that the "typical" value around which the distribution of distance $d(V)$ of random code V is concentrated (as $n, m \to \infty$) behaves in the following way:

$$d_{\text{typ}} \approx \frac{n}{2} - \frac{\tilde{u}\sqrt{n}}{2} \approx \frac{n}{2} - \sqrt{\frac{nm}{2}\log 2}\,. \tag{12}$$

In particular it means that the overwhelming bulk of linear codes have the code distance less than $n/2$ for large n and m. In this connection it is pertinent to note that the problem of constructing (for given n and m) the linear code V of dimension m with the maximal possible code distance

$$d(n,m) = \max_V \{d(V)\colon \dim V = m\}$$

is one of the central problems of the coding theory, and in general case it is unsolved. The well-known Griesmer estimate (see (MacWilliams and Sloane, 1977))

$$n \ge \sum_{i=0}^{m-2} \,]d(n,m)/2^i[$$

is attainable for some linear codes, for example, for the simplex codes with parameters $n = 2^m - 1$ and $d(n,m) = 2^{m-1}$. Notice that for these parameters

$$d_{\text{typ}} \approx 2^{m-1} - \sqrt{m\,2^{m-1}\log 2}\,.$$

2.2. Proof of Theorem 3

2.2.1. Let us consider the case where $\mathbf{v}$ is picked from V_n^m. We put

$$V_m^* = V_m \setminus \{0\}; \quad g = (g_1, \dots, g_m) \in V_m;$$

$$v^{(s)} = (v_1^{(s)}, \dots, v_m^{(s)}), \qquad v^{(s)} \in V_m, \quad s = 1, \dots, m;$$

$$\xi_g(i) = (-1)^{g_1 v_i^{(1)} \oplus \cdots \oplus g_m v_i^{(m)}};$$

$$\xi_g = n - 2\|g_1 v_i^{(1)} \oplus \cdots \oplus g_m v_i^{(m)}\| = \sum_{i=1}^{n} \xi_g(i); \qquad \xi = \max_{g\in V_m^*} \xi_g.$$

The remaining proof replicates the scheme of Section 1.2.2. We set

$$\eta_g = \mathbf{I}\{\xi_g \geq \tilde{u}\sqrt{n}\}, \quad \eta = \sum_{g \in V_m^*} \eta_g$$

and estimate the probability

$$\mathbf{P}\{\xi < \tilde{u}\sqrt{n}\} = \mathbf{P}\{\eta = 0\}$$

by the same theorem due to Sevastyanov. In doing so condition (6) is verified with the help of the same theorem due to Cramer, and condition (7) is verified with the help of theorem from (von Bahr, 1967).

2.2.2. The case where $\mathbf{v}$ is picked from $V_n^{(m)}$ or $V_n^{[m]}$ can be reduced to the previous case due to relations

$$|V_n^{[m]}| < |V_n^{(m)}| < |V_n^m|, \quad |V_n^{[m]}| = |V_n^m|(1 + o(1)), \qquad n \to \infty, \quad m = o(n^{1/6}).$$

3. THE SPECTRAL COMPLEXITY OF BOOLEAN FUNCTIONS

3.1. Definitions and statement of results

3.1.1. Let B_n be the set of all Boolean functions (B. f.) of n variables; let

$$(x, y) = x_1 y_1 \oplus \cdots \oplus x_n y_n$$

be the scalar product of vectors $x = (x_1, \ldots, x_n)$ and $y = (y_1, \ldots, y_n)$ from V_n; and let

$$f(x) = \sum_{a \in V_n} \hat{f}(a)(-1)^{(a,x)} \tag{13}$$

be the expansion of B. f. f with respect to Walsh system. The set of Walsh coefficients

$$\hat{f} = \left\{ \hat{f}(a) = 2^{-n} \sum_{x \in V_n} f(x)(-1)^{(a,x)}, \ a \in V_n \right\}$$

is called the Walsh spectrum of B. f. f, and the number $s_n(f)$ of nonzero elements of this spectrum is called the spectral complexity of B. f. f. The functional $s_n(f)$, evidently, characterizes the complexity of realization of B. f. f by the scheme based on expansion (13).

Denote by $\xi_n = \xi_n(f) = 2^n - s_n(f)$ the number of zeros in the spectrum $\hat{f}$, and by ξ a r. v. with distribution

$$\mathbf{P}\{\xi = 0\} = \mathbf{P}\{\xi = 1\} = \frac{1}{2}.$$

THEOREM 4. *Let f be picked at random according to the uniform measure on B_n. Then for every fixed $r \geq 1$ and $n \to \infty$*

$$\mathbf{E}\xi_n^r = \frac{1}{2}\left(\frac{2^{n+3}}{\pi}\right)^{r/2}(1 + o(1)), \tag{14}$$

i. e., the sequence of r. v.'s $\tilde{\xi}_n = (2^{n+3}/\pi)^{-1/2}\xi_n$ *converges in distribution to the r. v.* ξ *as* $n \to \infty$.

3.1.2. The form of the limit law of ξ_n, in particular, means that for large n the set B_n is partitioned into asymptotically equiprobable classes depending on the behaviour of $s_n(f)$: the first class contains those B. f. for which

$$s_n(f) = o(2^{n/2}),$$

and the second class contains those B. f. for which the value $s_n(f)$ is of order $2^{n/2}$. The Walsh system has the evident property that every B. f. $f \in B_n$ with odd weight does not have zero Walsh coefficients, and the set of B. f. with odd weight constitutes exactly one half of the set B_n. Therefore, from Theorem 1 it follows that the first class is asymptotically exhausted by B. f.'s with odd weights, and the second class, by B. f.'s with even weights. So, further we consider only B. f.'s with even weights.

On the set $\widehat{B}_n$ of all the B. f.'s from B_n with even weights we introduce the uniform measure and denote by $\hat{\xi}_n$ the number of nonzero Walsh coefficients of a random equiprobable B. f. from B_n.

THEOREM 5. *For every fixed* $r \geq 1$ *and* $n \to \infty$

$$\mathbf{E}\,\hat{\xi}_n^r = \left(\frac{2^{n+3}}{\pi}\right)^{r/2}\left(1 + o(1)\right); \tag{15}$$

hence, the sequence $\tilde{\xi}_n = (2^{n+3}/\pi)^{-1/2}\hat{\xi}_n$ *converges in probability to* 1. *The variance of* $\hat{\xi}_n$ *is asymptotically* $(2^{n+3}/\pi)^{1/2}(1 + o(1))$.

The relationship between the expectation $\mathbf{E}\,\hat{\xi}_n$ and the variance $\mathrm{Var}\,\hat{\xi}_n$, and also the representation of $\hat{\xi}_n$ as a sum of weakly dependent r. v.'s (see the proof) make it possible to suppose that the r. v.

$$\bar{\xi}_n = (\hat{\xi}_n - \mathbf{E}\,\hat{\xi}_n)/\sqrt{\mathrm{Var}\,\hat{\xi}_n}$$

asymptotically has the standard normal distribution. We have not got the proof of this assumption. However, just the values of moments $\mathbf{E}\,\hat{\xi}_n^r$ allow us to obtain sufficiently good probabilistic estimates for the distribution function of r. v. $\hat{\xi}_n$. For example, Theorem 5 and the Markov inequality imply the evident assertion.

COROLLARY. *Let* a *be an arbitrary positive real number and* r *be an arbitrary positive integer. Then*

$$\mathbf{P}\{\hat{\xi}_n \geq a\,2^{n/2}\} \leq \left(\frac{8}{3\pi a^2}\right)^{r/2}\left(1 + \delta_r(n)\right), \tag{16}$$

where $\delta_r(n) \to 0$, $n \to \infty$, *for each fixed* r.

For the number $s_n(f) = 2^n - \hat{\xi}_n$ of nonzero Walsh coefficients of a random B. f. from $\widehat{B}_n$ inequality (16) implies the estimate

$$\mathbf{P}\{s_n(f) \geq 2^n - a\,2^{n/2}\} \leq \left(\frac{8}{3\pi a^2}\right)^{r/2}\left(1 + \delta_r(n)\right).$$

Substituting $a = \varepsilon_n 2^{n/2}$ into this inequality, where ε_n is a sequence tending to zero sufficiently slowly (in such a way that $a \to \infty$) we obtain

$$\lim_{n\to\infty} \mathbf{P}\{s_n(f) \geq 2^n(1-\varepsilon_n)\} = 1,$$

i. e., the overwhelming bulk of B. f.'s from $\widehat{\mathcal{B}}_n$ have the spectral complexity $s_n(f)$ close to 2^n.

3.2. Proofs

3.2.1. To prove Theorem 4, we fix an integer $r \geq 1$ and for a sample $\bar{a} = (a_1, \dots, a_r)$ of vectors $a_k \in V_n$ denote

$$\mathbf{P}\{\bar{a}\} = \mathbf{P}\{\hat{f}(a_1) = \cdots = \hat{f}(a_r) = 0\}.$$

In addition, let U_k be the set of all r-samples $\bar{a} \in V_n$ with replacement such that the rank of the system $a_1, \dots, a_r$ equals k. Then

$$\mathbf{E}\xi_n^r = \sum_{k=1}^{r} \sum_{\bar{a}\in U_k} \mathbf{P}\{\bar{a}\}. \tag{17}$$

THEOREM 6. *For every fixed r, $r \geq 1$, and $n \to \infty$*

$$\mathbf{P}\{\bar{a}\} = \frac{1}{2}(2^{n-3}\pi)^{-r/2}(1+o(1)) \tag{18}$$

uniformly in $\bar{a} \in U_r$.

The proof of this theorem is omitted because of its length.
From (18) it follows that for every k, $1 \leq k \leq r$, uniformly over $\bar{a} \in U_k$

$$\mathbf{P}\{\bar{a}\} = O(2^{-nk/2}), \qquad n \to \infty. \tag{19}$$

In addition, obviously,

$$|U_r| = 2^{nr} + O(2^{n(r-1)}), \quad |U_k| = O(2^{nk}), \qquad k = 1, \dots, r-1. \tag{20}$$

Relations (18)–(20) make it easy to evaluate the sums $T_k = \sum_{\bar{a}\in U_k} \mathbf{P}\{\bar{a}\}$:

$$T_r = \frac{1}{2}\left(\frac{2^{n+3}}{\pi}\right)^{r/2}(1+o(1)), \qquad n \to \infty,$$

and

$$T_k = O(2^{nk/2}), \qquad k = 1, \dots, r-1.$$

Thus, T_r constitutes the main part of sum (17), and $\sum_{k=1}^{r-1} T_k$ is of lower order.

3.2.2. The proof of Theorem 5 is based on the following argument. Let $\tau = \tau(f)$ be a functional defined on $\mathcal{B}_n$, and $\mathcal{B}$ be an arbitrary subset of $\mathcal{B}_n$. Let τ generate two r. v.'s τ_1 and τ_2 by the following rule: $\tau_1 = \tau(g)$ and $\tau_2 = \tau(h)$, where g and h are random functions with uniform distributions on $\mathcal{B}_n$ and $\mathcal{B}$, correspondingly. Then

$$\mathbf{P}\{\tau_2 < x\} = \mathbf{P}\{\tau_1 < x \mid g \in \mathcal{B}\}.$$

To prove (15) it is sufficient to use this equality, the formula of full probability and (14):

$$\mathbf{E}\xi_n^r = \frac{1}{2}\mathbf{E}\{\xi_n^r \mid f \in \widehat{\mathcal{B}}_n\} + \frac{1}{2}\mathbf{E}\{\xi_n^r \mid f \notin \widehat{\mathcal{B}}_n\} = \frac{1}{2}\mathbf{E}\,\hat{\xi}_n^r.$$

REFERENCES

von Bahr, B. (1967). Multi-dimensional integral limit theorems. *Arkiv Math.* 6, 71–88.

Feller, W. (1971). *An Introduction to Probability Theory and Its Applications.* Vol. 2. John Wiley, New York.

Kolchin, V. F., Sevastyanov, B. A., and Chistyakov, V. P. (1978). *Random Allocations.* John Wiley, New York.

MacWilliams, F. and Sloane, N. (1977). *The Theory of Error-Correcting Codes.* North-Holland, Amsterdam.

Probabilistic Methods in Discrete Mathematics, pp. 413 – 427
V. F. Kolchin et al. (Eds.)
1993 TVP/VSP

CHERNOFF'S INEQUALITY FOR SUM OF INDEPENDENT RANDOM VECTORS

N. P. SALIKHOV
TVP Laboratories, Krasikov Street 32, 117418 Moscow, Russia

ABSTRACT

An upper bound is obtained for the probability P of fulfilment of the system of inequalities $\sum_{t=1}^{n} X_{ti} \geq 0$, $i = 1, \dots, k$, where $\mathbf{X}_t = (X_{t1}, \dots, X_{tk})$, $t = 1, \dots, n$, are independent trials in polynomial schemes with outcomes $(a_{j1}, \dots, a_{jk})$, $j = 1, \dots, m$. For identically distributed $\mathbf{X}_t$, $t = 1, \dots, n$ and $k = 1$, the estimate coincides with a well-known Chernoff's estimate. The bound depends upon parameters $c_1, \dots, c_k$. Under special choice of parameters, when c_i can be of only two different values and $\mathbf{X}_t$ are identically distributed a simple method for minimization of the bound for P is suggested. An algorithm is given for minimization of the bound for P in the general case.

1. Let $\mathbf{X}_1, \dots, \mathbf{X}_n$ be independent random vectors,

$$\mathbf{X}_t = (X_{t1}, \dots, X_{tk}) \quad \text{for} \quad t = 1, \dots, n,$$

$$\mathbf{P}\{\mathbf{X}_t = \mathbf{a}_j\} = p_{tj} \geq 0 \quad \text{for} \quad j = 1, \dots, m, \quad t = 1, \dots, n,$$

$$\mathbf{a}_j = (a_{j1}, \dots, a_{jk}) \quad \text{for} \quad j = 1, \dots, m,$$

$\sum_{j=1}^{m} p_{tj} = 1$ for $t = 1, \dots, n$. We obtain an upper bound for the probability $P = \mathbf{P}\left\{\sum_{t=1}^{n} \mathbf{X}_t \geq \mathbf{0}\right\}$. Here and in the sequel the vector inequality $\mathbf{x} \geq \mathbf{y}$ for $\mathbf{x} = (x_1, \dots, x_k)$, $\mathbf{y} = (y_1, \dots, y_k)$ is equivalent to the system of k inequalities

$$x_1 \geq y_1, \dots, x_k \geq y_k.$$

Suppose that for every $i = 1, \dots, k$ there exists $j \in \{1, \dots, m\}$ such that $a_{ji} < 0$. (If $I = \{i \in \{1, \dots, k\} : a_{ji} \geq 0 \text{ for } j = 1, \dots, m\} \neq \varnothing$, then $P = \mathbf{P}\left\{\sum_{t=1}^{n} X_{ti} \geq 0 \text{ for } i = 1, \dots, k, \ i \notin I\right\}$ and instead of vectors $\mathbf{a}_j, \mathbf{X}_t$ we may use their $(k - |I|)$-dimensional subvectors in which only the coordinates with $i \notin I$ are considered.)
If $\mathbf{X}_1, \dots, \mathbf{X}_n$ are identically distributed, i. e.,

$$p_{tj} = p_j > 0 \quad \text{for} \quad j = 1, \dots, m, \quad t = 1, \dots, n, \tag{1}$$

then

$$P = \mathbf{P}\left\{ \sum_{j=1}^{m} a_{ji} n_j \geq 0, \quad i = 1, \dots, k \right\},$$

where n_j is a frequency of jth outcome in n independent trials corresponding to a polynomial scheme with probabilities $p_1, \ldots, p_m$. In particular, for $m = k$, $\mathbf{a}_j = \mathbf{e}_j - \mathbf{u}$, where $\mathbf{e}_j$ is the jth ort in k-dimensional space for $j = 1, \ldots, k$,

$$\mathbf{u} = (u_1, \ldots, u_k) \geq 0, \qquad 0 < \sum_{i=1}^{k} u_i = u \leq 1,$$

probability P equals

$$Q = \mathbf{P}\{n_1 \geq nu_1, \ldots, n_k \geq nu_k\}.$$

Olkin and Sobel (1965) have obtained an exact expression for Q in a form of r-dimensional integral of the polynomial $q(v_1, \ldots, v_r)$ of degree $(n - r)$ where r is the number of positive components in the vector $\mathbf{u}$.

2. We estimate the probability P by means of the well-known method, see (Wozencraft and Reiffen, 1961, pp. 139–140).
Let

$$f(\mathbf{y}) = \exp\left(\sum_{i=1}^{k} c_i y_i\right) \text{ for } \mathbf{y} = (y_1, \ldots, y_k),$$

where $\mathbf{c} = (c_1, \ldots, c_k) \geq 0$ is a fixed vector such that $\sum_{i=1}^{k} c_i = 1$. Then, for any $h \geq 0$,

$$P \leq \mathbf{E} f\left(h \sum_{t=1}^{n} \mathbf{X}_t\right) = \prod_{t=1}^{n} \mathbf{E} f(h\mathbf{X}_t)$$
$$= \prod_{t=1}^{n} \sum_{j=1}^{m} p_{tj} \exp\left(h \sum_{i=1}^{k} c_i a_{ji}\right).$$

Let

$$\sum_{i=1}^{k} c_i a_{ji} = b_j \quad \text{for } j = 1, \ldots, m,$$

$$g_t(h) = \sum_{j=1}^{m} p_{tj} \exp(hb_j) \quad \text{for } t = 1, \ldots, n.$$

Then

$$P \leq \prod_{t=1}^{n} g_t(h) = G(h)$$

for any $h \geq 0$.

THEOREM 1. *Let*

$$I_t = \{j \in \{1, \ldots, m\} : p_{tj} > 0\}, \quad \beta_t = \max_{\lambda \in I_t} b_\lambda,$$

$$J_t = \{j \in I : b_j = \beta_t\} \quad \text{for } t = 1, \ldots, n, \ b = \sum_{t=1}^{n} \beta_t.$$

Then the best estimates of the form $P \le G(h)$ with $h \ge 0$ are

$$P = 0 = \lim_{h\to\infty} G(h) = G(\infty) \quad \text{if} \quad b < 0, \tag{2}$$

$$P \le \prod_{t=1}^{n} \sum_{j\in J_t} p_{tj} = \lim_{h\to\infty} G(h) = G(\infty) \quad \text{if} \quad b = 0, \tag{3}$$

$$P \le G(h_0) < 1 \quad \text{if} \quad \sum_{t=1}^{n}\sum_{j=1}^{m} p_{tj} b_j < 0 < b,$$

where h_0 is the unique positive solution of the equation $\sum_{t=1}^{n} g_t'(h)(g_t(h))^{-1} = 0$,

$$P \le 1 = G(0) \quad \text{if} \quad \sum_{t=1}^{n}\sum_{j=1}^{m} p_{tj} b_j \ge 0.$$

Proof. Estimates (2) and (3) follow from the equality

$$G(h) = \exp(hb) \prod_{t=1}^{n} \sum_{j\in J_t} p_{tj} \exp(h(b_j - \beta_t)).$$

Let $\varphi(h) = \log G(h) = \sum_{t=1}^{n} \log g_t(h)$. Then

$$\varphi(0) = 0,$$

$$\varphi'(h) = \sum_{t=1}^{n} g_t'(h)(g_t(h))^{-1},$$

$$\varphi''(h) = \sum_{t=1}^{n} (g_t(h))^{-2}(g_t''(h)g_t(h) - (g_t'(h))^2).$$

Using Cauchy inequality we find that

$$\begin{aligned}(g_t'(h))^2 &= \left(\sum_{j=1}^{m} p_{tj} b_j \exp(hb_j)\right)^2 \\ &\le \left(\sum_{j=1}^{m} p_{tj} b_j^2 \exp(hb_j)\right)\left(\sum_{j=1}^{m} p_{tj} \exp(hb_j)\right) \\ &= g_t''(h) g_t(h).\end{aligned}$$

Therefore, $\varphi''(h) \ge 0$ for $h \ge 0$. If $\varphi'(0) = \sum_{t=1}^{n}\sum_{j=1}^{m} p_{tj} b_j \ge 0$, then $\varphi(h) \ge \varphi(0)$ for $h \ge 0$, $\min_{h\ge 0} G(h) = G(0) = 1$. If $\varphi'(0) < 0 < b$, then

$$\begin{aligned}\lim_{h\to\infty} \varphi'(h) &= \lim_{h\to\infty} \sum_{t=1}^{n} \sum_{j\in I_t} p_{tj} b_j \exp(hb_j) \left(\sum_{\lambda\in I_t} p_{t\lambda} \exp(hb_\lambda)\right)^{-1} \\ &= \sum_{t=1}^{n} \sum_{j\in J_t} p_{tj} b_j \left(\sum_{\lambda\in J_t} p_{t\lambda}\right)^{-1} \\ &= \sum_{t=1}^{n} \beta_t = b > 0,\end{aligned}$$

and

$$\min_{h\geq 0}\varphi(h)=\varphi(h_0),$$

where h_0 is the unique positive solution of the equation $\varphi'(h) = 0$. Therefore, $\min_{h\geq 0} G(h) = G(h_0) < 1$.

COROLLARY 1. *Let $\beta = \max_{\lambda\in\{1,\dots,m\}} b_\lambda$. If condition (1) is satisfied, then the best estimates of the form $P \leq G(h)$, $h \geq 0$, are*

$$P = 0 \quad \textit{if} \quad \beta < 0,$$

$$P \leq p^n, \quad \text{where} \quad p=\sum_{j\in J} p_j,\ J=\{j\in\{1,\dots,m\}: b_j=\beta\} \quad \textit{if} \quad \beta=0,$$

$$P \leq G(h_0) = (g(h_0))^n < 1,$$

where h_0 is the unique solution of the equation $g'(h) = 0$,

$$g(h)=\sum_{j=1}^{m} p_j \exp(hb_j) \quad \textit{if} \quad \sum_{j=1}^{m} p_j b_j < 0 < \beta,$$

$$P \leq 1 \quad \textit{if} \quad \sum_{j=1}^{m} p_j b_j \geq 0.$$

For $k = 1$ the inequality $P \leq (g(h_0))^n$ from Corollary 1 is known as Chernoff's inequality, see (Wozencraft and Reiffen, 1961, p. 140).

3. Consider the case where the solution h_0 can be obtained in an explicit form. Let

$$\begin{aligned} b_j &= \sum_{i=1}^{k} c_i a_{ji} = A \quad \text{for} \quad j\in R,\\ b_j &= \sum_{i=1}^{k} c_i a_{ji} = a \quad \text{for} \quad j\in S=\{1,\dots,m\}\setminus R, \quad R\subset\{1,\dots,m\},\\ &\qquad R\neq\varnothing, \quad A>a. \end{aligned} \tag{4}$$

Let

$$\sum_{j\in R} p_{tj} = p, \quad \sum_{j\in S} p_{tj} = q = 1-p \quad \text{for} \quad t=1,\dots,n, \tag{5}$$

where $0 < p < 1$. In this case we have $b = nA$,

$$\sum_{t=1}^{n}\sum_{j=1}^{m} p_{tj} b_{tj} = (Ap + aq)n,$$

$$g_t(h) = p\exp(hA) + q\exp(ha) \quad \text{for} \quad t=1,\dots,n.$$

Theorem 1 may be formulated now in the following way.

COROLLARY 2. *Let* $\lambda = -a(A-a)^{-1}$. *If conditions* (4), (5) *are satisfied, then the best estimates of the form* $P \le G(h),\ h \ge 0$, *are*

$$P = 0 \quad \text{if} \quad \lambda > 1;$$
$$P \le \prod_{t=1}^{n} \sum_{j \in J_t} p_{tj} \quad \text{if} \quad \lambda = 1;$$
$$P \le G(h_0) = \left(\left(\frac{p}{\lambda}\right)^{\lambda} \left(\frac{1-p}{1-\lambda}\right)^{1-\lambda} \right)^{n},$$

where $h_0 = (A-a)^{-1} \log(-qa/(pA))$ *if* $p < \lambda < 1$; $P \le 1$ *if* $\lambda \le p$.

Remark 1. The estimate of P does not increase with λ. For every $t = 1, \dots, n$ we have $J_t \subseteq R$, and if $\lambda = 1$, then

$$P \le \prod_{t=1}^{n} \sum_{j \in J_t} p_{tj} \le p^n.$$

The inequality (Borovkov, 1986, p. 131)

$$\left(\frac{p}{\lambda}\right)^{\lambda} \left(\frac{1-p}{1-\lambda}\right)^{1-\lambda} \le \exp\left(-2(\lambda-p)^2\right)$$

permits us to obtain the following estimates in the case $p < \lambda < 1$:

$$P \le \left(\left(\frac{p}{\lambda}\right)^{\lambda} \left(\frac{1-p}{1-\lambda}\right)^{1-\lambda} \right)^{n} \le \exp\left(-2(\lambda-p)^2 n\right).$$

Remark 2. Conditions (5) are satisfied, if, for example, for every $t = 1, \dots, n$ arrays $\{p_{tj},\ j \in R\}$ and $\{p_{tj},\ j \in S\}$ are permutations of arrays $\{p'_1, \dots, p'_{|R|}\}$ and $\{p''_1, \dots, p''_{|S|}\}$ respectively and

$$\sum_{i=1}^{|R|} p'_i = p, \quad \sum_{i=1}^{|S|} p''_i = 1-p, \qquad 0 < p < 1.$$

Conditions (5) follow also from (1).

Remark 3. Consider the problem of existence of c satisfying conditions (4) for some fixed R. Notice that there exists a correspondence between the solutions $(c_1, \dots, c_k, A, a)$ of the system

$$\sum_{i=1}^{k} c_i a_{ji} - A = 0, \qquad j \in R, \qquad \sum_{i=1}^{k} c_i a_{ji} - a = 0, \qquad j \in S,$$

$$\sum_{i=1}^{k} c_i = 1, \qquad c_1 \geq 0, \ldots, c_k \geq 0,$$

and the solutions $(d_1, \ldots, d_k, x, y)$ of the system

$$\begin{aligned} &\sum_{i=1}^{k} d_i(a_{ji} + g) - x = 0, \qquad j \in R, \\ &\sum_{i=1}^{k} d_i(a_{ji} + g) - y = 0, \qquad j \in S, \\ &\sum_{i=1}^{k} d_i = 1, \qquad d_1 \geq 0, \ldots, d_k \geq 0 \end{aligned} \tag{6}$$

(g is fixed), namely:

$$(c_1, \ldots, c_k, A, a) = (d_1, \ldots, d_k, x - g, y - g).$$

If $g \geq -\min_{i,j} a_{ji}$, then for every solution of system (6) we have $x \geq 0$, $y \geq 0$. Using the test given in (Chernikov, 1968, p. 78) we may ascertain the existence or non-existence of a positive solution of system (6), and we may find all positive solutions of system (6) by means of the algorithm in (Chernikov, 1968, pp. 258–260). Only positive solutions with $x > y$ may be used in Corollary 2. The best estimate for P corresponds to the case where $a/(a - A) = (y - g)/(y - x)$ assumes its maximum value.

4. Let us use Corollary 2 to estimate the probability Q. In this case conditions (5) are satisfied for any R. To verify conditions (4) we may rewrite them in the following way: $b_j = c_j - v = A$ for $j \in R$, $b_j = c_j - v = a$ for $j \in S$, $A > a$, where

$$v = \sum_{i=1}^{k} c_i u_i = (v + A)\sigma + (v + a)(u - \sigma) = vu + au + (A - a)\sigma,$$

$$\sigma = \sum_{j \in R} u_j, \quad u = \sum_{j=1}^{k} u_j.$$

Taking into account that $\sum_{i=1}^{k} c_i = 1$, $c_1 \geq 0, \ldots, c_k \geq 0$, we obtain the following system for quantities $A - a$, a, v:

$$(A - a)\sigma + au = v(1 - u), \quad (A - a)\rho + ak = 1 - vk,$$

$$A - a > 0, \quad a \geq -v, \quad v \geq 0, \quad \text{where} \quad \rho = |R| \in \{1, \ldots, k - 1\}.$$

From this system we find that:
(1) if $\Delta = \sigma k - \rho u \neq 0$, then

$$a = \big(\sigma(1 - vk) - \rho v(1 - u)\big)\Delta^{-1}, \quad A = a + (vk - u)\Delta^{-1},$$

where $v \in (u/k, \sigma/\rho]$ for $\Delta > 0$, and $v \in [\sigma/\rho, u/k)$ for $\Delta < 0$;
(2) if

$$\frac{\sigma}{\rho} = \frac{u}{k} = \Theta > 0,$$

then

$$\begin{aligned} &a \in \big(-\Theta(1-u+\Theta k)^{-1}, (1-u)(1-u+\Theta k)^{-1}k^{-1}\big),\\ &A = a + \rho^{-1}\big((1-u)(1-u+\Theta k)^{-1} - ak\big),\\ &v = \Theta(1-u+\Theta k)^{-1}. \end{aligned}$$

Consider two cases.
1. Let $\sigma = 1$. Then $u = 1$, $\Delta = k - \rho > 0$, $a = (1-vk)/(k-\rho)$, $A = 0$, $\lambda = a/(a-A) = 1$ for any $v \in (1/k, 1/\rho]$. According to Corollaries 1, 2 we obtain in this case the estimate $P \leq p^n$, where $p = \sum_{j \in R} p_j$. The best estimate can be obtained, if $R = \{i \in \{1,\ldots,k\}: u_i > 0\}$.
2. Let $0 \leq \sigma < 1$. Then

$$\begin{aligned} \lambda &= \frac{a}{a-A} = \frac{v(\rho(1-u)+\sigma k)-\sigma}{vk-u} = \lambda(v) \quad \text{if} \quad \Delta \neq 0,\\ \lambda &= \rho a(ka - (1-u)(1-u+\Theta k)^{-1})^{-1} = \lambda(a),\\ \lambda(a) &\leq \lambda(-\Theta(1-u+\Theta k)^{-1}) = \sigma \quad \text{if} \quad \Delta = 0. \end{aligned}$$

So, the best estimates of our type for a fixed R are:

$$\begin{aligned} &Q \leq p^n \quad \text{if} \quad p < \sigma = 1,\\ &Q \leq \left(\left(\frac{p}{\sigma}\right)^{\sigma}\left(\frac{1-p}{1-\sigma}\right)^{1-\sigma}\right)^n \quad \text{if} \quad p < \sigma < 1,\\ &Q \leq 1 \quad \text{if} \quad \sigma \leq p. \end{aligned}$$

The following theorem reduces the problem of construction of an optimal R minimizing the quantity

$$Q(\sigma, p) = \begin{cases} (p/\sigma)^{\sigma}\left((1-p)/(1-\sigma)\right)^{1-\sigma}, & \text{if } p < \sigma < 1,\\ p, & \text{if } p < \sigma = 1, \end{cases}$$

to the search of minimum among no more than k values of function $Q(\sigma, p)$.

THEOREM 2. *Let $i_1,\ldots,i_k$ be a permutation of numbers $1,\ldots,k$ such that $\lambda_1 \geq \lambda_2 \geq \cdots \geq \lambda_k$, $\lambda_1 > 1$, $\lambda_k \leq 1$, where $\lambda_\nu = u_{i_\nu} p_{i_\nu}^{-1}$ for $\nu = 1,\ldots,k$. Let for any set $R \subseteq \{1,\ldots,k\}$*

$$\sigma = \sigma(R) = \sum_{i \in R} u_i, \quad p = p(R) = \sum_{i \in R} p_i.$$

For sets R satisfying the condition $u \geq \sigma(R) > p(R) > 0$ the quantity $Q(\sigma,p)$ assumes its minimum value, if $R = R_v = \{i_1,\ldots,i_v\}$, where $v \in \{1,\ldots,k-1\}$ is such that

$$\lambda_v > \frac{z_{2v} - z_{1v}}{\log z_{2v} - \log z_{1v}} \geq \lambda_{v+1}, \quad z_{2v} = \frac{\sigma(R_v)}{p(R_v)} > 1, \quad z_{1v} = \frac{1-\sigma(R_v)}{1-p(R_v)} < 1,$$

or if $R = R_0 = \{i \in \{1,\ldots,k\}: u_i > 0\}$.

Remark 4. If $u < 1$, then the set R_0 cannot be optimal. If $\lambda_k > 1$, then $u = \sum_{i=1}^k u_i > \sum_{i=1}^k p_i = 1$, that contradicts the assumption $0 < u \leq 1$. If $\lambda_1 \leq 1$, then for any R we have $\sigma \leq p$ and hence the estimate $Q \leq 1$ holds.

Remark 5. The following example shows that the necessary condition for optimality may be satisfied for more than one set R_v. Let $k = 3$, $p_1 = p_2 = \varepsilon$, $p_3 = 1 - 2\varepsilon$, $u_1 = 4\varepsilon$, $u_2 = 2\varepsilon$, $u_3 = 1 - 7\varepsilon$, where $0 < \varepsilon < 1/7$. Then for sufficiently small ε

$$\lambda_1 = 4 > \frac{4 - (1 - 4\varepsilon)(1 - \varepsilon)^{-1}}{\log 4 - \log((1 - 4\varepsilon)(1 - \varepsilon)^{-1})} > 2 = \lambda_2,$$
$$\lambda_2 = 2 > \frac{3 - (1 - 6\varepsilon)(1 - 2\varepsilon)^{-1}}{\log 3 - \log((1 - 6\varepsilon)(1 - 2\varepsilon)^{-1})} > \frac{1 - 7\varepsilon}{1 - 2\varepsilon} = \lambda_3,$$

i. e., the necessary condition is satisfied for R_1 and R_2.

Proof. Let $R \subset \{1, \ldots, k\}$ be such that $0 < p = p(R) < \sigma(R) = \sigma < 1$, $R \neq \emptyset$. We introduce the function

$$g(x, y) = \log Q(\sigma + x, p + y) = (\sigma + x) \log \frac{p + y}{\sigma + x} + (1 - \sigma - x) \log \frac{1 - p - y}{1 - \sigma - x},$$

if $0 < p + y < 1, 0 < \sigma + x < 1$, $y \neq 0$, and

$$g(1 - \sigma, y) = \lim_{x \to 1 - \sigma - 0} g(x, y) = \log(p + y),$$

if $0 < p + y < 1$, $y \neq 0$. Now we find conditions for $g(x, y) < g(0, 0)$ to be true. Let first $x < 1 - \sigma$. Denote $H(t) = g(xt, yt)$, $t \in [0, 1]$. Then $g(x, y) - g(0, 0) = H(1) - H(0) = H'(0) + \frac{1}{2} H''(t)$ for some $t \in (0, 1)$. Let us prove that $H''(t) < 0$. We have

$$H'(t) = \frac{\partial}{\partial z} g(z, yt) \Big|_{z=xt} x + \frac{\partial}{\partial z} g(xt, z) \Big|_{z=yt} y,$$
$$H''(t) = -y^2(Az^2 + 2Bz + C), \quad \text{where} \quad z = xy^{-1},$$
$$A = -\frac{\partial^2 g(z, yt)}{\partial z^2} \Big|_{z=xt} = \frac{1}{b} + \frac{1}{1 - b}, \qquad b = \sigma + xt,$$
$$B = -\frac{\partial^2 g(z, w)}{\partial z \partial w} \Big|_{\substack{z=xt \\ w=yt}} = -\left(\frac{1}{a} + \frac{1}{1 - a}\right), \qquad a = p + yt,$$
$$C = -\frac{\partial^2 g(xt, w)}{\partial w^2} \Big|_{w=yt} = \frac{b}{a^2} + \frac{1 - b}{(1 - a)^2}.$$

The inequality $B^2 - AC < 0$ is equivalent to $(b - a)^2 > 0$, so $H''(t) < 0$ for $a \neq b$. If $a = b$, then $A = C = -B > 0$, $H''(t) = -y^2 A(z - 1)^2$. If $H''(t) = 0$, then $z = 1$, $x = y$, $b = \sigma + xt = p + yt = a$, $\sigma = p$, that contradicts the conditions of the theorem. Therefore $H''(t) < 0$ in the case $a = b$. Hence we obtain $g(x, y) < g(0, 0)$, if $H'(0) \leq 0$, i. e., if $x(\log z_2 - \log z_1) - y(z_2 - z_1) \geq 0$, where

$$z_2 = \sigma / p > 1, \quad z_1 = (1 - \sigma)/(1 - p) < 1.$$

If $x > 1 - \sigma - 0$, then the inequality

$$g(x,y) - g(0,0) - H'(0) < 0$$

transforms into

$$g(1-\sigma,y) - g(0,0) - (-(1-\sigma)(\log z_2 - \log z_1) + y(z_2 + z_1)) \leq 0.$$

Therefore, $g(1-\sigma,y) < g(0,0)$, if

$$(1-\sigma)(\log z_2 - \log z_1) - y(z_2 - z_1) > 0.$$

Letting $x = u_i$, $y = p_i$ for $i \in S$, and $x = u_i$, $y = -p_i$ for $i \in R$ we obtain the following necessary condition for R to minimize $Q(\sigma,p)$:

$$\frac{u_i}{p_i} \leq \frac{z_2 - z_1}{\log z_2 - \log z_1} \quad \text{for} \quad i \in S, \quad \text{if} \quad p + p_i < 1, \quad \sigma + u_i \leq 1,$$

$$\frac{u_i}{p_i} > \frac{z_2 - z_1}{\log z_2 - \log z_1} \quad \text{for} \quad i \in R, \quad \text{if} \quad p - p_i > 0, \quad \sigma - u_i > 0.$$

Let us show that the necessary condition is satisfied in some special cases. We notice first of all that according to the Cauchy theorem

$$\frac{z_2 - z_1}{\log z_2 - \log z_1} = z, \qquad z_1 < z < z_2.$$

Let $i \in S$, $p + p_i = 1$. Then $S = \{i\}$, $\sigma + u_i = u$, and

$$\frac{u_i}{p_i} = \frac{u - \sigma}{1 - p} \leq \frac{1 - \sigma}{1 - p} = z_1 < z.$$

Let $i \in R$, $p - p_i = 0$. Then $R = \{i\}$, $\sigma = u_i$, and

$$\frac{u_i}{p_i} = \frac{\sigma}{p} = z_2 > z.$$

Let $i \in R$, $\sigma - u_i = 0$. Then

$$\frac{u_i}{p_i} \geq \frac{\sigma}{p} = z_2 > z.$$

Remark 6. The quantity $-2(\sigma - p)^2$ is an upper bound for $\log Q(\sigma,p)$ and attains its minimum for $R = R_w$, where w is uniquely determined by the conditions

$$w \in \{1, \dots, k-1\}, \qquad \lambda_w > 1 \geq \lambda_{w+1}.$$

It is easy to show that in some cases this set R_w does not minimize the quantity $Q(\sigma,p)$. Let $k = 3$, $p_1 = p_2 = \varepsilon$, $p_3 = 1 - 2\varepsilon$, $u_1 = 6\varepsilon$, $u_2 = 2\varepsilon$, $u_3 = 1 - 9\varepsilon$, where $0 < \varepsilon < 1/9$. Then $\lambda_1 = 6 > \lambda_2 = 2 > 1 > \lambda_3 = (1 - 9\varepsilon)/(1 - 2\varepsilon)$. Therefore $w = 2$. For sufficiently small ε the necessary condition of optimality is fulfilled for R_1 and is not fulfilled for R_2:

$$\lambda_1 = 6 > \frac{6 - (1 - 6\varepsilon)(1 - \varepsilon)^{-1}}{\log 6 - \log\left((1 - 6\varepsilon)(1 - \varepsilon)^{-1}\right)} > 2 = \lambda_2,$$

$$\frac{4 - (1 - 8\varepsilon)(1 - 2\varepsilon)^{-1}}{\log 4 - \log\left((1 - 8\varepsilon)(1 - 2\varepsilon)^{-1}\right)} > 2 = \lambda_2.$$

Therefore,

$$Q\big(\sigma(R_1), p(R_1)\big) < Q\big(\sigma(R_2), p(R_2)\big),$$

if ε is small enough.

5. Taking into account ideas from (Fano, 1961, pp. 296–302) we consider a more general approach to obtain an estimate for the probability P. Put

$$G(h,\mathbf{c}) = \prod_{t=1}^{n} \sum_{j=1}^{m} p_{tj} \exp\left(h \sum_{i=1}^{k} c_i a_{ji}\right),$$

where $h \geq 0$, $\mathbf{c} \in C = \{\mathbf{c} = (c_1,\ldots,c_k): c_1 \geq 0,\ldots,c_k \geq 0,\ \sum_{i=1}^{k} c_i = 1\}$. Instead of the estimate

$$P \leq \min_{h \geq 0} G(h,\mathbf{c})$$

we may consider the estimate

$$P \leq G = \min_{h \geq 0,\, \mathbf{c} \in C} G(h,\mathbf{c}).$$

Let us consider some special cases.
1. If $\sum_{t=1}^{n}\sum_{j=1}^{m} p_{tj} a_{ji} \geq 0$ for $i = 1,\ldots,k$, then for any $\mathbf{c} \in C$ we have $\sum_{t=1}^{n}\sum_{j=1}^{m} p_{tj} \sum_{i=1}^{k} a_{ji} \geq 0$ and $\min_{h \geq 0} G(h,\mathbf{c}) = G(0,\mathbf{c}) = 1$ by Theorem 1. Therefore $G = 1$.
Introduce the sets

$$C_0 = \left\{\mathbf{c}: \mathbf{c} \in C,\ \sum_{t=1}^{n} \max_{j \in I_t} \sum_{i=1}^{k} c_i a_{ji} = 0\right\},$$

$$C_1 = \left\{\mathbf{c}: \mathbf{c} \in C,\ \sum_{t=1}^{n} \max_{j \in I_t} \sum_{i=1}^{k} c_i a_{ji} > 0\right\}.$$

2. If $(C_0 \cup C_1) \subset C$, then for some $\mathbf{c} \in C$ we have

$$\sum_{t=1}^{n} \max_{j \in I_t} \sum_{i=1}^{k} c_i a_{ji} < 0$$

and $\min_{h \geq 0} G(h,\mathbf{c}) = 0$ by Theorem 1. Therefore $G = 0$.
Let $C = C_0 \cup C_1$. For $\mathbf{c} \in C_0$, according to Theorem 1 we have

$$\min_{h \geq 0} G(h,\mathbf{c}) = \prod_{t=1}^{n} \sum_{j \in J_t(\mathbf{c})} p_{tj},$$

where

$$J_t(\mathbf{c}) = \left\{j: j \in I_t, \sum_{i=1}^{k} c_i a_{ji} = \max_{\nu \in I_t} \sum_{i=1}^{k} c_i a_{\nu i}\right\}.$$

If

$$C_{10} = C_1 \cap \left\{\mathbf{c}: \mathbf{c} \in C, \sum_{t=1}^{n}\sum_{j=1}^{m} p_{tj} \sum_{i=1}^{k} c_i a_{ji} \geq 0\right\},$$

$$C_{11} = C_1 \cap \left\{ \mathbf{c}\colon \mathbf{c} \in C, \sum_{t=1}^{n} \sum_{j=1}^{m} p_{tj} \sum_{i=1}^{k} c_i a_{ji} < 0 \right\},$$

then $C_1 = C_{10} \cup C_{11}$ and $\min_{\mathbf{c} \in C_{10}} G(h, \mathbf{c}) = 1$ by Theorem 1. Therefore $G = \min(1, G_0, G_1)$, where

$$G_0 = \begin{cases} \min\limits_{\mathbf{c} \in C_0} \prod\limits_{t=1}^{n} \sum\limits_{j \in J_t(\mathbf{c})} p_{tj}, & \text{if } C_0 \neq \emptyset, \\ 1 & \text{otherwise,} \end{cases}$$

$$G_1 = \begin{cases} \inf\limits_{h>0,\, \mathbf{c} \in C_{11}} G(h, \mathbf{c}), & \text{if } \inf\limits_{h>0,\, \mathbf{c} \in C_{11}} G(h, \mathbf{c}) = G(h^*, \mathbf{c}^*),\ h^* > 0,\ \mathbf{c}^* \in C_{11}, \\ 1 & \text{otherwise.} \end{cases}$$

Therefore the problem is reduced to minimizing the function

$$G(\mathbf{x}) = \prod_{t=1}^{n} \sum_{j=1}^{m} p_{tj} \exp\left(\sum_{i=1}^{k} x_i a_{ji} \right),$$

or the function

$$g(\mathbf{x}) = \log G(\mathbf{x}) = \sum_{t=1}^{n} \log \sum_{j=1}^{m} p_{tj} \exp\left(\sum_{i=1}^{k} x_i a_{ji} \right),$$

on the set

$$M = \left\{ \mathbf{x} = (x_1, \ldots, x_k)\colon x_1 \geq 0, \ldots, x_k \geq 0,\ \sum_{i=1}^{k} x_i > 0, \right.$$
$$\left. \sum_{t=1}^{n} \sum_{j=1}^{m} p_{tj} \sum_{i=1}^{k} x_i a_{ji} < 0 < \sum_{t=1}^{n} \max_{j \in I_t} \sum_{i=1}^{k} x_i a_{ji} \right\}.$$

Let $g(\mathbf{x})$ attains its minimum on M at a point $(x_1^0, \ldots, x_k^0)$ such that $x_i^0 > 0$ for $i \in I \subseteq \{1, \ldots, k\}$, $I \neq \emptyset$, $x_i^0 = 0$ for $i \in \{1, \ldots, k\} \setminus I$. Then the quantities x_i^0, $i \in I$, should be a solution of the system of equations

$$\frac{\partial g(\mathbf{x}, I)}{\partial x_r} = 0, \qquad r \in I, \tag{7}$$

where $g(\mathbf{x}, I) = g(\mathbf{y})$, $\mathbf{y} = (y_1, \ldots, y_k)$, $y_i = x_i$ for $i \in I$, $y_i = 0$ for $i \in \{1, \ldots, k\} \setminus I$.

Let us prove that the value of function $g(\mathbf{x}, I)$ is the same for all points $\mathbf{x}$ satisfying system (7). Assume for simplicity that $I = \{1, \ldots, k\}$, $g(\mathbf{x}, I) = g(\mathbf{x})$. (For $I \subset \{1, \ldots, k\}$ we use $\sum_{i \in I}$ instead of $\sum_{i=1}^{k}$). We shall use the following formulas for derivatives:

$$\frac{\partial g(\mathbf{x})}{\partial x_r} = \sum_{t=1}^{n} \sum_{j=1}^{m} q_{tj}(\mathbf{x}) a_j,$$

where

$$q_{tj}(\mathbf{x}) = p_{tj}\exp\left(\sum_{i=1}^{k} x_i a_{ji}\right)(g_t(\mathbf{x}))^{-1},$$

$$g_t(\mathbf{x}) = \sum_{j=1}^{m} p_{tj}\exp\left(\sum_{i=1}^{k} x_i a_{ji}\right), \qquad j = 1,\dots,m, \quad t = 1,\dots,n, \quad r = 1,\dots,k,$$

$$\begin{aligned}\frac{\partial^2 g(\mathbf{x})}{\partial x_r \partial x_s} &= \sum_{t=1}^{n}\sum_{j=1}^{m} a_{jr}\frac{\partial q_{tj}(\mathbf{x})}{\partial x_s}\\ &= \sum_{t=1}^{n}\sum_{j=1}^{m} q_{tj}(\mathbf{x})a_{jr}\left(a_{js} - \sum_{j=1}^{m} q_{tj}(\mathbf{x})a_{js}\right)\\ &= \sum_{t=1}^{n}\sum_{j=1}^{m} q_{tj}(\mathbf{x})\left(a_{jr} - \sum_{j=1}^{m} q_{tj}(\mathbf{x})a_{jr}\right)\left(a_{js} - \sum_{j=1}^{m} q_{tj}(\mathbf{x})a_{js}\right),\\ &\qquad r = 1,\dots,k, \quad s = 1,\dots,k.\end{aligned}$$

THEOREM 3. *If* $\mathbf{y} = (y_1,\dots,y_k)$, $\mathbf{z} = (z_1,\dots,z_k)$ *are two solutions of the system*

$$\frac{\partial g(\mathbf{x})}{\partial x_r} = d_r, \qquad r = 1,\dots,k,$$

then

$$\sum_{i=1}^{k} a_{ji}(z_i - y_i) = f_t, \qquad j \in I_t, \quad t = 1,\dots,n,$$

where the constants f_t *are such that*

$$\sum_{t=1}^{n} f_t = \sum_{i=1}^{k} d_i(z_i - y_i).$$

Proof. Consider the function

$$\mu_r(v) = \frac{\partial g(\mathbf{y} + v(\mathbf{z} - \mathbf{y}))}{\partial x_r}$$

for $v \in [0,1]$. By definition $\mu_r(1) = \mu_r(0) = d_r$, so

$$\begin{aligned}0 = \mu_r(1) - \mu_r(0) &= \int_0^1 \mu_r'(v)\,dv\\ &= \int_0^1 \sum_{s=1}^{k} \frac{\partial^2 g(\mathbf{y} + v(\mathbf{z} - \mathbf{y}))}{\partial x_r \partial x_s}(z_s - y_s)\,dv, \qquad r = 1,\dots,k,\end{aligned}$$

$$\sum_{r=1}^{k}(z_r - y_r)\int_0^1 \mu_r'(v)\,dv$$
$$= \int_0^1 \sum_{r=1}^{k}\sum_{s=1}^{k} \frac{\partial^2 g(\mathbf{y}+v(\mathbf{z}-\mathbf{y}))}{\partial x_r \partial x_s}(z_r - y_r)(z_s - y_s)\,dv = 0.$$

The estimate

$$\sum_{r=1}^{k}\sum_{s=1}^{k}\frac{\partial^2 g(\mathbf{x})}{\partial x_r \partial x_s}u_r u_s$$
$$= \sum_{t=1}^{n}\sum_{j=1}^{m} q_{tj}(\mathbf{x})\sum_{r=1}^{k}\sum_{s=1}^{k}\left(a_{jr} - \sum_{j=1}^{m} q_{tj}(\mathbf{x})a_{jr}\right)\left(a_{js} - \sum_{j=1}^{m} q_{tj}(\mathbf{x})a_{js}\right)u_r u_s$$
$$= \sum_{t=1}^{n}\sum_{j=1}^{m} q_{tj}(\mathbf{x})\left(\sum_{v=1}^{k} u_v\left(a_{jv} - \sum_{j=1}^{m} q_{tj}(\mathbf{x})a_{jv}\right)\right)^2 \geq 0$$

shows that we integrate a continuous and non-negative function over $[0,1]$ and that all stationary points of function $g(x)$ are the points of its local minima. So

$$\sum_{r=1}^{k}\sum_{s=1}^{k}\frac{\partial^2 g(\mathbf{y}+v(\mathbf{z}-\mathbf{y}))}{\partial x_r \partial x_s}(z_r - y_r)(z_s - y_s)dv \equiv 0.$$

Hence for every $t = 1, \ldots, n$ and $j \in I_t$

$$\sum_{v=1}^{k}(z_v - y_v)\left(a_{jv} - \sum_{j=1}^{m} a_{jv}q_{tj}(\mathbf{y} + x(\mathbf{z}-\mathbf{y}))\right) \equiv 0,$$

$$\sum_{v=1}^{k} a_{jv}(z_v - y_v) \equiv \sum_{j=1}^{m} q_{tj}(\mathbf{y} + x(\mathbf{z}-\mathbf{y}))\sum_{v=1}^{k} a_{jv}(z_v - y_v).$$

Therefore,

$$\sum_{v=1}^{k}(z_v - y_v)a_{jv} = f_t$$

for every $t = 1, \ldots, n$ and $j \in I_t$. Further we have

$$\sum_{r=1}^{k} d_r(z_r - y_r) = \sum_{r=1}^{k}\frac{\partial g(\mathbf{z})}{\partial x_r}(z_r - y_r)$$
$$= \sum_{t=1}^{n}\sum_{j\in I_t} q_{tj}(\mathbf{z})\sum_{r=1}^{k} a_{jr}(z_r - y_r)$$
$$= \sum_{t=1}^{n} f_t \sum_{j\in I_t} q_{tj}(\mathbf{z}) = \sum_{t=1}^{n} f_t.$$

COROLLARY 3. *If* $\mathbf{y},\mathbf{z}$ *are two stationary points of the function* $g(\mathbf{x})$, *then* $g(\mathbf{y}) = g(\mathbf{z})$,

$$\sum_{t=1}^{n}\sum_{j=1}^{m} p_{tj} \sum_{i=1}^{k} z_i a_{ji} = \sum_{t=1}^{n}\sum_{j=1}^{m} p_{tj} \sum_{i=1}^{k} y_i a_{ji},$$
$$\sum_{t=1}^{n} \max_{j\in I_t} \sum_{i=1}^{k} z_i a_{ji} = \sum_{t=1}^{m} \max_{j\in I_t} \sum_{i=1}^{k} y_i a_{ji}. \tag{8}$$

In this case we have

$$\sum_{i=1}^{k} z_i a_{ji} = \sum_{i=1}^{k} y_i a_{ji} + f_t$$

for every $t = 1,\dots,n$ and $j \in I_t$, where $\sum_{t=1}^{k} f_t = 0$. Then

$$g(\mathbf{z}) = \sum_{t=1}^{n} \log \sum_{j\in I_t} p_{tj} \exp\left(\sum_{i=1}^{k} z_i a_{ji}\right)$$
$$= \sum_{t=1}^{n}\left(\log \sum_{j\in I_t} p_{tj} \exp\left(\sum_{i=1}^{k} y_i a_{ji}\right) + f_t\right) = g(\mathbf{y}).$$

Equalities (8) are proved similarly.
Thus, we obtain the following method of searching minima of the function $g(\mathbf{x})$ on M. For every function $g(\mathbf{x}, I)$, $I \subseteq \{1,\dots,k\}$, $I \neq \varnothing$, we find one of the solutions $\mathbf{y} = (y_1,\dots,y_k)$ of system (7) in the region $x_i > 0$ for $i \in I$, if system (7) is solvable. We accept a solution $\mathbf{y}$, if it does not belong to the set M. Notice that due to (8) either all or none of the solutions of system (7) belong to M. For all other solutions $\mathbf{y}$ (their total number does not exceed $2^k - 1$) we compute $g(\mathbf{y})$ and choose the solution with minimum value of $g(\mathbf{y})$.
If $I \subseteq \{1,\dots,k\}$, $I \neq \varnothing$, is such that

$$M(I) = M \cap \{\mathbf{x}: x_i > 0 \text{ for } i \in I,\ x_i = 0 \text{ for } i \in \{1,\dots,k\}\backslash I\} = \varnothing,$$

then we have not to solve the system (7) for $g(\mathbf{x}, I)$. Notice that condition $M(I) \neq \varnothing$ is equivalent to the condition: at least one of $\prod_{t=1}^{n} |I_t|$ systems of homogeneous linear inequalities:

$$x_i > 0 \quad \text{for} \quad i \in I,$$

$$\sum_{i\in I} x_i \sum_{t=1}^{n}\sum_{j=1}^{m} p_{tj} a_{ji} < 0, \quad \sum_{i\in I} x_i a_{j_t i} \geq \sum_{i\in I} x_i a_{ji} \quad \text{for} \quad j \in I_t \backslash j_t, \quad t = 1,\dots,n,$$

$$\sum_{i\in I} x_i \sum_{t=1}^{n} a_{j_t i} > 0,$$

is solvable. Here $j_1 \in I,\dots,j_n \in I_n$ are fixed. The following conditions are sufficient for the equality $M(I) \neq \varnothing$:

1. $\sum_{t=1}^{n} \sum_{j=1}^{m} p_{tj} a_{ji} \geq 0$ for every $i \in I$;
2. $a_{ji} \leq 0$ for every $j = 1, \ldots, m$ and every $i \in I$.

If all systems (7) have no solutions on the set M, then $\inf_{h>0, \mathbf{c} \in C_{11}} G(h, \mathbf{c})$ is not attained in the domain $h > 0$, $\mathbf{c} \in C_{11}$.

REFERENCES

Borovkov, A. A. (1986). *Probability Theory.* Nauka, Moscow (in Russian).

Chernikov, S. N. (1968). *Linear Inequalities.* Nauka, Moscow (in Russian).

Fano, R. M. (1961). *Transmission of Information. A Statistical Theory of Communications.* John Wiley, New York.

Kullback, S. (1959). *Information Theory and Statistics.* John Wiley, New York.

Olkin, I. and Sobel, M. (1965). Integral expressions for tail probabilities of the multinomial and negative multinomial distributions. *Biometrika* 52, 167–169.

Wozencraft, J. M. and Reiffen, B. (1961). *Sequential Decoding.* John Wiley, New York.

Probabilistic Methods in Discrete Mathematics, pp. 428 – 436
V. F. Kolchin et al. (Eds.)
1993 TVP/VSP

SOME LIMIT THEOREMS IN BATCH ALLOCATION SCHEME WITH RANDOM LEVELS DEFINED BY AN ALLOCATION SCHEME

M. SAVCHUK
Cybernetic Center, Ukrainian Academy of Sciences,
Glushkov pr. 40, Kiev, Ukraine

ABSTRACT

A scheme is considered, where ball batches are allocated in N cells having random levels defined by allocating other ball batches in these cells. Factorial moments are derived, a Poisson limit theorem is obtained for random variables related to the number of occupied cells, weak convergence of multidimensional random processes constructed from this scheme to Gaussian diffusion processes is investigated.

We consider a scheme where ball batches are allocated in N cells with random levels defined by an allocation of other batches in these cells. We derive factorial moments and obtain a Poisson limit theorem for random variables related to the nimber of occupied cells; we investigate a weak convergence of multidimensional random processes corresponding to this scheme to Gaussian diffusion processes. For the first time allocation problems with random levels have been considered in (Ivanov, 1981), a detailed analysis of an allocation scheme with random levels, where balls are allocated one by one may be found in (Ivanov, 1982).
Assertions proved in this paper have been stated in (Savchuk, 1986), where the proofs were outlined only. (Kovalenko et al., 1986) contains a detailed exposition of the technique used in (Savchuk, 1983) to prove weak convergence of multidimensional random processes related to allocation problems, by means of general limit theorems of the theory of random processes. This technque is used in this paper to derive a multidimensinal functional limit theorem. A Poisson limit theorem is proved by the method of moments via asymptotic analysis of an explicit formula for factorial moments. Section 1 contains a summary of results and Section 2 provides their proofs.

1. STATEMENT OF THE PROBLEM AND SUMMARY OF RESULTS

Let h_j $(j = 1, \dots, N)$ be a number of balls hitting into the jth cell after equiprobable independent allocation of l batches in N cells, each batch having m balls. The balls of a batch are allocated no more than one in a cell and all $C_N^m = \binom{N}{m}$ possible allocations of this kind are equiprobable; the allocation of different batches are independent. We shall treat h_j, $j = 1, \ldots, N$, as random levels for corresponding cells. Let us allocate in the same cells additinal n bathces of size m equiprobably and independently both of each other and of the preceding l batches. Let us denote by $\bar{\xi} = \bar{\xi}(N, m, l, n)$ a random variable which is equal to the number of $j \in \{1, \ldots, N\}$ such that jth cell contains no more than h_j balls of the last n batches; let $\bar{\xi} = \bar{\xi}(N, m, l, n) = N - \xi$. We shall introduce the following notation: $[x]$ is the integral part of a real number x; $x^{[k]} = x(x-1)\cdots(x-k+1)$, $x^{[0]} = 1$, δ_{ij} is the Kronecker delta, $\bar{\delta}_{ij} = 1 - \delta_{ij}$;

$$b_r(n, \lambda) = \binom{n}{r} \lambda^r \lambda^{n-r}, \qquad 0 \leq \lambda \leq 1,$$

$$R = \sum_{h=0}^{l} \sum_{r=0}^{\min(h,n)} b_h(l, \lambda) b_r(n, \lambda).$$

THEOREM 1. *For any integers N, m, l, $n \geq 0$, $N \geq m$,*

$$\mathbf{E}\,\xi^{[k]} = l!n! \left(N^{[k]}\right)^{1-l-n}$$

$$\times \sum_{l_\omega, n_{\omega'}} \left(\prod_{\omega} l_\omega! \prod_{\omega'} n_{\omega'}!\right)^{-1} \prod_{i=0}^{k} \left(m^{[i]}(N-m)^{[k-i]}\right)^{\Sigma_i} \tag{1}$$

where $\Sigma_i = \sum_{|\omega|=i} l_\omega + \sum_{|\omega'|=i} n_{\omega'}$, indices $\omega = (\omega_1, \ldots, \omega_k)$ and $\omega' = (\omega'_1, \ldots, \omega'_k)$ run independently over the set of Boolean k-dimensional vectors, $|\omega|$ is the number of nonzero coordinates of the vector and the summation is taken over all unodered sets of integers $l_\omega \geq 0$ and $n_{\omega'} \geq 0$ satsfying the system

$$\begin{cases} \sum_{\omega} l_\omega = l, \quad \sum_{\omega'} n_{\omega'} = n, \\ \sum_{\omega \in \Omega(j)} l_\omega \geq \sum_{\omega' \in \Omega(j)} n_{\omega'}, \qquad j = 1, \ldots, k. \end{cases} \tag{2}$$

Here $\Omega(j)$ is the set of Boolean k-dimensional vectors with nonzero jth coordinate.

COROLLARY 1. *For any integer N, m, l, $n \geq 0$*

$$\mathbf{E}\xi = NR, \quad \mathbf{E}\bar{\xi} = N(1-R), \tag{3}$$

where λ in expression for R should be equal to m/N.

In what follows we shall consider the case where the allocation scheme parameters N, m, l, n, vary in such a way that $N \to \infty$ and quantities l and n are bounded. Depending on the behavior of the parameter $\lambda = m/N$ four cases may be classified.

COROLLARY 2.

1) *Let* $N \to \infty$, $\lambda \to 1$, $l < n = O(1)$. *Then*

$$\mathbf{E}\xi = N\binom{n}{l}(1-\lambda)^{n-l}(1+O(1-\lambda)),$$
$$\mathbf{E}\bar{\xi} = N\left(1-\binom{n}{l}(1-\lambda)^{n-l}+O((1-\lambda)^{n-l+1})\right). \tag{4}$$

2) *Let* $N \to \infty$, $\lambda \to 1$, $n \leq l = O(1)$. *Then*

$$\mathbf{E}\xi = N\left(1-\binom{l}{n-1}(1-\lambda)^{l-n+1}+O((1-\lambda)^{l-n+2})\right),$$
$$\mathbf{E}\bar{\xi} = N\binom{l}{n-1}(1-\lambda)^{l-n+1}(1+O(1-\lambda)). \tag{5}$$

3) *Let* $N \to \infty$, $\lambda \to 0$, $l, n = O(1)$. *Then*

$$\mathbf{E}\xi = N(1-n\lambda+O(\lambda^2\bar{\delta}_{0n})),$$
$$\mathbf{E}\bar{\xi} = \bar{\delta}_{0n}Nn\lambda(1+O(\lambda)), \tag{6}$$

4) *Let* $N \to \infty$, $0 < \delta \leq \lambda \leq 1-\delta$, $l, n = O(1)$, $\delta = \text{const}$. *Then*

$$\mathbf{E}\xi = RN, \quad \mathbf{E}\bar{\xi} = (1-R)N. \tag{7}$$

THEOREM 2. *If* l *and* n *are bounded provided* $m, N-m \to \infty$ *and*

$$N\binom{n}{l}(1-m/N)^{n-l} \to \gamma < \infty \qquad \left(N\binom{l}{n-1}(1-m/N)^{l-n+1} \to \gamma < \infty\right),$$

then the distribution of random variable ξ ($\bar{\xi}$) *converges to the Poisson distribution with parameter* γ.

Remark 1. The conditions of Theorem 2 are not contradictory and may hold either in the case 1) with $n \geq l+2$ or in the case 2) with $n \leq l+2$.

Remark 2. If in the case 3) the expectation $\mathbf{E}\bar{\xi}$ is bounded, then it is clear that m is bounded tooo and random variable ξ will be concentrated at the point nm. Theorem 2 involves all cases of the Poisson limit distribution of random variables ξ and $\bar{\xi}$ as $N \to \infty$ and l, n are bounded; here $m/N \to 1$ is a necessary condition for the Poisson convergence.

COROLLARY 3. *If* $m, N-m \to \infty$, $l, n = O(1)$ *and*

$$N\binom{n}{l}(1-m/N)^{n-l} \to \gamma < \infty \qquad \left(N\binom{l}{n-1}(1-m/N)^{l-n+1} \to \gamma < \infty\right),$$

then the number of different allocations of $l+n$ batches such that the jth cell contains more than h_j (at most h_j) balls, $j = 1, \ldots, N$, is asymptotically equivalent to $e^{-\gamma}\binom{N}{m}^{l+n}$.

In the case 4) of Corollary 2 the distribution of random variables ξ and $\bar{\xi}$ is asymptotically normal. There a functional limit theorem holds also. Let us denote random processes on the segment $[0,1]$: $Z_N^i(t)$ is the number of balls from the ith batch hitting the first $[Nt]$ cells; $\varkappa_N(t) = N^{-1/2}(Z_N^i(t) - \mathbf{E}\, Z_N^i(t))$, $i = 1, \ldots, l+n$, $Y(t)$ is the number of cells among the first $[Nt]$ cells which after allocation of all the $l+n$ batches contain no more balls from the first l components than from the last n components; $\xi_N(t) = N^{-1/2}(Y_N(t) - \mathbf{E}\, Y_N(t))$. Let us denote by $\mathbf{D}^{l+n+1}$ the $(l+n+1)$-dimensional space of functions without unremovable discontinuities equipped with the uniform topology and by $w(t)$ the standard $(l+n+1)$-dimensional Wiener process.

THEOREM 3. *If $l \geq 0$, $n \geq 1$, $l+n \geq 2$ are fixed, and $N \to \infty$, $m/N \to \lambda$, $0 < \lambda < 1$, then the sequence of vector-valued random processes*

$$\mu_N(t) = (\varkappa_N^1(t), \ldots, \varkappa_N^{l+n}(t), \xi_N(t)) \tag{8}$$

converges weakly on the segment $[0,1]$ in the space $\mathbf{D}^{l+n+1}$ to the Gaussian diffusion process with zero mean and correlation matrix

$$K(t,s) = \|k_{ij}(t,s)\|_1^{l+n+1}, \tag{9}$$

where for $t \geq s$

$$k_{ij}(t,s) = \lambda(1-\lambda)s(1-t)\delta_{ij}, \qquad i,j = 1, \ldots, l+n,$$

$$k_{l+n+1,i}(t,s) = k_{i,l+n+1}(t,s) = \lambda(R_1 - R)s(1-t), \qquad i = 1, \ldots, l,$$

$$k_{l+n+1,j}(t,s) = k_{j,l+n+1}(t,s) = \lambda(R_2 - R)s(1-t), \qquad j = l+1, \ldots, l+n,$$

$$k_{l+n+1,l+n+1}(t,s) = sR(1-R) + \frac{\lambda st}{1-\lambda}\left[l(R_1 - R)^2 + n(R_2 - R)^2\right],$$

$$R_1 = \sum_{h=0}^{l-1} b_h(l-1,\lambda) \sum_{r=0}^{h+1} b_r(n,\lambda),$$

$$R_2 = \sum_{h=0}^{l} \sum_{r=0}^{h-1} b_h(l,\lambda) b_r(n-1,\lambda).$$

The limit process can be represented in the form

$$\mu(t) = \int_0^t \left(I + A\frac{s-t}{1-s}\right) B\, dw(s), \tag{10}$$

where I is the unit $(l+n+1)$ matrix;

$$A = \|a_{ij}\|_1^{l+n+1}, \qquad a_{ii} = 1, \quad i = 1, \ldots, l+n,$$

$$a_{l+n+1,i} = (R_1 - R)/(1-\lambda), \qquad i = 1, \dots, l,$$

$$a_{l+n+1,j} = (R_2 - R)/(1-\lambda), \qquad j = l+1, \dots, l+n;$$

otherwise $a_{ij} = 0$ *and the drift vector of the diffusion process* $\mu(t)$ *is equal to* $-A(1-t)^{-1}x$; $B^2 = K(0,1)$ *is the diffusion operator of* $\mu(t)$.

COROLLARY 4. *If* $N \to \infty$, $0 < \delta < \lambda = m/N \le 1 - \delta$ ($\delta = \text{const}$), $l, n = O(1)$, *then random variables* ξ *and* $\bar{\xi}$ *have asymptotically normal distribution with parameters*

$$\mathbf{E}\,\xi = RN, \quad \mathbf{E}\,\bar{\xi} = (1-R)N,$$
$$\operatorname{Var}\xi = \operatorname{Var}\bar{\xi} = \left\{R(1-R) + \frac{\lambda}{1-\lambda}\left[l(R_1 - R)^2 + n(R_2 - R)^2\right]\right\}N, \tag{11}$$

where λ *in expressions for* R, R_1, *and* R_2 *is equal to* m/N.

2. PROOFS OF ASSERTIONS OF SECTION 1

Proof of Theorem 1. Let a Boolean k-dimensional vector ω, (ω') correspond to each of the first l (last n) batches in the following manner: jth coordinate of the vector ω, (ω') is equal to 1 if the jth cell contains a ball from the batch and to zero otherwise. Let us denote by l the number of batches which correspond to an arbitrary vector ω and by $n_{\omega'}$ the number of batches corresponding to vector ω'. If random variable ξ is represented as a sum of indicator functions, $\xi = \sum_{i=1}^{N} \theta_i$, then

$$\mathbf{E}\,\xi^{[k]} = \mathbf{E} \sum_{i_1,\dots,i_k} \theta_{i_1} \cdots \theta_{i_k},$$

where $\theta_i = 1$, if the ith cell contains no more than h_i balls from the last n batches, otherwise $\theta_i = 0$; the summation is taken over all ordered sets $(i_1, \dots, i_k)$ with pairwise unequal indices. This implies that

$$\mathbf{E}\,\xi^{[k]} = N^{[k]}\mathbf{P}\left\{\theta_1 = \theta_2 = \dots = \theta_k = 1\right\}. \tag{12}$$

Using the classical probability definition we find

$$\begin{aligned}\mathbf{E}\,\xi^{[k]} = N^{[k]} &\sum_{l_\omega, n_{\omega'}} l!n! \left(\prod_{\omega} l_\omega! \prod_{\omega'} n_{\omega'}!\right)^{-1} \\ &\times \prod_{\omega} \left(\frac{m^{[|\omega|]}(N-m)^{[k-|\omega|]}}{N^{[k]}}\right)^{l_\omega} \\ &\times \prod_{\omega'} \left(\frac{m^{[|\omega'|]}(N-m)^{[k-|\omega'|]}}{N^{[k]}}\right)^{n_{\omega'}}\end{aligned} \tag{13}$$

where the summation is taken over all unordered sets of integers $l_\omega \ge 0$ and $n_{\omega'} \ge 0$ satisfying system (2) and indices ω and ω' run independently over the set of Boolean k-dimensional vectors. Elementary transformations of (13) yield the assertion of Theorem 1 in the form of relations (1), (2).

Corollary 1 follows immediately from Theorem 1 with $k = 1$. So we can easily derive asymptotic formulas (3)–(7) for the mathematical expectations of ξ and $\bar{\xi}$ in Corollary 2.

Proof of Theorem 2. Let $a_{ij} = 1$ if the jth cell contains a ball from the ith batch, otherwise $a_{ij} = 0$. Then to each realisation of allocation there corresponds a rectangular Boolean matrix $A = \|a_{ij}\|$ of size $(l+n)N$ satisfying conditions $\sum_{j=1}^{N} a_{ij} = m$, $i = 1, \dots, l+n$. The probability on the right-hand side of (12) is equal to the probability that in any column $j = 1, \dots, k$ of matrix A the first l positions contain at least the same number of units as the last n positions. For any fixed k and $l, n = O(1)$, m, $N - m \to \infty$

$$\begin{aligned}\mathbf{P}\{\theta_1 = \cdots = \theta_k = 1\} &= \mathbf{P}\{\theta_1 = 1\} \prod_{i=2}^{k} \mathbf{P}\{\theta_i = 1 | \theta_1 = \cdots = \theta_{i-1} = 1\} \\ &= R^k \left(1 + O\left(\frac{1}{m} + \frac{1}{N-m}\right)\right).\end{aligned}$$

Thus under the same conditions

$$\begin{aligned}\mathbf{E}\,\xi^{[k]} &= N^{[k]} R^k \left(1 + O\left(\frac{1}{m} + \frac{1}{N-m}\right)\right) \\ &= (NR)^k \left(1 + O\left(\frac{1}{m} + \frac{1}{N-m}\right)\right).\end{aligned}$$

If, moreover, $N\binom{n}{l}(1-\lambda)^{n-l} \to \gamma < \infty$, then $\mathbf{E}\,\xi = NR \to \gamma < \infty$ and for any fixed k

$$\begin{aligned}\mathbf{E}\,\xi^{[k]} &= (\mathbf{E}\,\xi)^k \left(1 + O\left(\frac{1}{m} + \frac{1}{N-m}\right)\right) \\ &= \gamma^k (1 + o(1)).\end{aligned} \tag{14}$$

In view of (14) the method of moments implies the convergence of distribution of ξ as $m, N - m \to \infty$, $N\binom{n}{l}(1-\lambda)^{n-l} \to \gamma < \infty$, $l, n = O(1)$ to the Poisson distribution with the parameter γ. In a similar way the second part of Theorem 2 can be proved.

Theorem 2 implies immediately Corollary 3.

Proof of Theorem 3 proceeds along the same way as presented in (Savchuk, 1983; Kovalenko et al., 1986). At first one shows the convergence of measures induced by random processes (8) in the space of functions without unremovable discontinuities equipped with the Skorokhod topology to a measure corresponding to a solution of a stochastic differential equation, after that the form of the solution and its characteristics are derived. Limit processes turn out to be continuous with probability 1 and the convergence in the space of functions without unremovable discontinuities equipped with the uniform topology takes place. A general scheme of the method and results from the theory of random processes refferred to are provided in section 7.1 of (Kovalenko et al., 1986).

Let us take points $t_{N_k} = k/N,\ k = 0, \dots, N$, as nonrandom times. A sequence of arrays of vectors

$$\mu_{N0}, \mu_{N1}, \dots, \mu_{NN}, \qquad N = 1, 2, \dots, \tag{15}$$

where $\mu_{Nk} = \mu_N(t_{Nk})$ is such that each array is a Markov chain and $\mathbf{P}\{\mu_{N0} = 0\} = 1$. The values of transition probabilities are uniquely defined by the first $l + n$ coordinates of random vectors μ_{Nk}, i.e., by vectors $\bar{\varkappa}_{Nk} = (\varkappa^1_{Nk}, \dots, \varkappa^{l+n}_{Nk})$. The first and second conditional moments of increments of processes $\mu_N(t)$ can be represented by formulas

$$\mathbf{E}\left\{(\Delta\mu_{Nk} - \alpha_{Nk}\Delta t_{Nk})(\Delta\mu_{Nk} - \alpha_{Nk}\Delta t_{Nk})^T \,|\mathcal{F}_{Nk}\right\} = \beta^2_{Nk}\Delta t_{Nk},$$

$$\mathbf{E}\left\{\Delta\mu_{Nk}|\mathcal{F}_{Nk}\right\} = \alpha_{Nk}\Delta t_{Nk}, \tag{16}$$

where $\Delta\mu_{Nk} = \mu_{N,k+1} - \mu_{N,k}$, $\Delta t_{Nk} = t_{N,k+1} - t_{N,k}$, $\mathcal{F}_{Nk}$ is a σ-algebra generated by random vectors $\mu_{N0}, \dots, \mu_{Nk}$, is the transposition symbol, α_{Nk} are random vectors (columns), β^2_{Nk} is a symmetric positive semidefinite matrix. There exist nonrandom Borel functions $a_{Nk}(x)$ and $B^2_{Nk}(x)$ such that $\alpha_{Nk} = a_{Nk}(\bar{\varkappa}_{Nk})$ and $\beta^2_{Nk} = B^2_{Nk}(\bar{\varkappa}_{Nk})$. Here $x = (x_1, \dots, x_{n+l}) \in \mathbf{R}^{l+n}$, function $a_{Nk}(x)$ takes on values in $\mathbf{R}^{l+n+1}$ and $B^2_{Nk}(x)$ are matrix-valued functions:

$$a_{Nk}(x) = \left(a^1_{Nk}(x), \dots, a^{l+n+1}_{Nk}(x)\right), \quad B^2_{Nk}(x) = \|b^{ij}_{Nk}\|_1^{l+n+1}. \tag{17}$$

In view of Lemma 7.1.3 from (Kovalenko et al., 1986) it suffices to prove Theorem 3 when $m/N = \lambda + o(1/\sqrt{N})$. Let us define a discrete set of points $X = X(N, m, l, k) = \{x\colon \mathbf{P}\{\bar{\varkappa}_{Nk} = x\} > 0\}$. The asymptotic formula (7.3.7) from (Kovalenko et al., 1986) for the conditional probability that an arbitrary cell $j > Nt_{Nk}$ contains exactly r balls from the given batches if $\bar{\varkappa}_{Nk} = x \in X$ makes it possible to find that under the same condition the conditional probability R' that the $(k+1)$th cell contains no less balls from the first l batches than from the last n batches has the form

$$R' = R + Q/\sqrt{N}(1 - t_{Nk})$$

$$+O\left(\frac{|x|^2}{N(1 - t_{Nk})^2}\right) + o\left(\frac{x}{N(1 - t_{Nk})}\right) + o\left(\frac{1}{\sqrt{N}}\right), \tag{18}$$

where

$$Q = \sum_{h=0}^{l}\sum_{r=0}^{h} b_h(l, \lambda) b_r(n, \lambda)\lambda^{-1}(1 - \lambda)^{-1}\left[\left(\lambda - \frac{h}{l}\right)\sum_{i=1}^{l} x_i + \left(\lambda - \frac{r}{n}\right)\sum_{i=l+1}^{l+n} x_i\right].$$

Let $\mathbf{E}_x$, Var_x, Cov_x respectively be symbols of conditional mathematical expectations, variance and covariance under the condition that $\bar{\varkappa}_{Nk} = x$. For any $0 \le k \le N(1 - \delta)$, $0 < \delta = \mathrm{const}$, $x \in X$ relations (16)–(18) imply provided

$\Delta Z_{N,k}, i = 1, \ldots, l+n$, are Boolean random variables (cf. proof of Theorem 7.3.1 in (Kovalenko et al., 1986)):

$$\begin{aligned}
a^i_{Nk}(x) &= N\mathbf{E}_x \Delta \varkappa^i_{Nk} = \sqrt{N}\mathbf{E}_x\{\Delta Z^i_{Nk} - \Delta \mathbf{E} Z^i_{Nk}\} \\
&= -\frac{x_i}{1-t_{Nk}} + o(1), \qquad i = 1, \ldots, l+n, \\
a^{l+n+1}_{Nk}(x) &= N\mathbf{E}_x \Delta \xi_{Nk} = \sqrt{N}\mathbf{E}_x(\Delta Y_{Nk} - \Delta \mathbf{E} Y_{Nk}) \\
&= \sqrt{N}\left(R' - R + o(1/\sqrt{N})\right) \\
&= Q + O\left(\frac{|\, x\,|^2}{\sqrt{N}(1-t_{Nk})^2}\right) + o\left(\frac{x}{\sqrt{N}(1-t_{Nk})}\right) + o(1), \\
b^{ii}_{Nk}(x) &= \mathrm{Var}_x \Delta Z^i_{Nk} = \lambda(1-\lambda) \\
&\quad + O\left(\frac{x_i}{\sqrt{N}(1-t_{Nk})}\right) + o\left(\frac{1}{\sqrt{N}}\right), \qquad i = 1, \ldots, l+n, \\
b^{l+n+1,l+n+1}_{Nk}(x) &= \mathrm{Var}_x \Delta Y_{Nk} = R'(1-R') \\
&= R(1-R) + O\left(\frac{x}{\sqrt{N}(1-t_{Nk})}\right) + o\left(\frac{1}{\sqrt{N}}\right), \\
b^{i,l+n+1}_{Nk}(x) &= \mathrm{Cov}_x\, (\Delta Z^i_{Nk}, \Delta Y_{Nk}) \\
&= \lambda(R_1 - R) + O\left(\frac{x}{\sqrt{N}(1-t_{Nk})}\right) + o\left(\frac{1}{\sqrt{N}}\right), \qquad i = 1, \ldots, l, \\
b^{i,l+n+1}_{Nk}(x) &= \lambda(R_2 - R) \\
&\quad + O\left(\frac{x}{\sqrt{N}(1-t_{Nk})}\right) + o\left(\frac{1}{\sqrt{N}}\right), \qquad i = l+1, \ldots, l+n, \\
b^{ij}_{Nk}(x) &= \mathrm{Cov}_x\left(\Delta Z^i_{Nk}, \Delta Z^j_{Nk}\right) = 0 \qquad 1 \le i \ne j \le l+n.
\end{aligned}$$

Let us introduce the vector

$$a(t,x) = (-x_1, \ldots, -x_{l+n}, Q)(1-t)^{-1} \tag{19}$$

and the symmetric matrix

$$B^2 = \|b_{ij}\|_1^{l+n+1}, \tag{20}$$

where $b_{ij} = \lambda(1-\lambda)\delta_{ij}$, $i,j = 1, \ldots, l+n$,

$$b_{l+n+1,l+n+1} = R(1-R), \quad b_{i,l+n+1} = \lambda(R_1 - R), \qquad i = 1, \ldots, l,$$

$$b_{j,l+n+1} = \lambda(R_2 - R), \qquad j = l+1, \ldots, l+n.$$

Then for $x \in X(N,m,l,n,k)$ uniformly in $t \in [0, 1-\delta]$

$$a_{Nk}(x) = a(t_{Nk}, x) + \overbrace{(1, \ldots, 1)}^{l+n+1} \times \left(O(\,|\, x\,|^2 / \sqrt{N}) + o(x/\sqrt{N} + 1)\right),$$

$$B^2_{Nk}(x) = B^2\left(1 + O\left(\frac{x}{\sqrt{N}}\right) + o\left(\frac{1}{\sqrt{N}}\right)\right).$$

For $x \notin X(N,m,l,n,k)$ we set by definition

$$a_{Nk}(x) = a(t_{Nk},x), \quad B^2_{Nk}(x) = B. \tag{21}$$

Introduce also two functions

$$a_N(t,x) = a_{Nk}(x), \quad B^2_N(t,x) = B^2_{Nk}(x) \qquad \text{if} \quad t \in [t_{Nk}, t_{Nk+1}). \tag{22}$$

The verification of conditions 1)–7) of Lemma 7.1.1 from (Kovalenko et al., 1986) (or conditions of Theorem 13 from Section 3, Chapter 2 in (Gikhman and Skorokhod, 1975)) for the sequence of arrays of random vectors (15) and functions (19)–(22) shows that the sequence of random processes $\mu(t)$ for any $\delta > 0$ converges weakly on the segment $[0, 1-\delta]$ in the space of $(l+n+1)$-dimentional functions without unremovable discontinuities equipped with the Skorokhod topology to a solution of the stochastic differential equation

$$d\mu(t) = a(t,\mu(t)) + Bdw(t), \qquad \mu(0) = 0,$$

having trend vector (19) and diffusion operator (20).
(Kovalenko et al., 1986) (or as in Theorem 6 from (Savchuk, 1983)). By Lemma 7.4.1 from (Kovalenko et al., 1986) weak convergence $\mu_N(t) \Rightarrow \mu(t)$ holds on the segment $[0,1]$ in $\mathbf{D}^{l+n+1}$.
The distribution of the $(l+n+1)$th coordinate of limit process $\mu(t)$ with $t = s = 1$ yields Corollary 4 (cf. (11)).

REFERENCES

Gikhman, I. I. and Skorokhod, A. V. (1979). *Theory of Random Processes*. Vol. 3, Springer-Verlag, Berlin–New York.

Ivanov, V. A. (1981). Randomized divisible statistics. In: *Theory and Practice of Application of Mathematical Methods and Computer Science* (Yu. Yermolyev and I. Kovalenko, Eds.). Cybernetic Center of the Ukrainian Academy of Sciences, Kiev, pp. 67–75 (in Russian).

Ivanov, V. A. (1982). Limiting theorems in allocation scheme with random levels. *Math. Notes* **31**, 619–631 (in Russian).

Kovalenko, I. N., Levitskaya, A. A., and Savchuk, M. N. (1986). *Selected Chapters of Probabilistic Combinatorics*. Naukova Dumka, Kiev (in Russian).

Savchuk, M. N. (1983). Some limiting theorems in equiprobable scheme of ball batch allocation. *Theory Probab. Math. Statist.* **28**, 122–130 (in Russian).

Savchuk, M. N. (1986). Limiting theorems in batch allocation scheme with random levels. In: *Mathematical Methods of Analysis and Optimization of Complex Systems Operating Under Uncertainty Conditions* (Yu. Yermolyev and I. Kovalenko, Eds.). Cybernetic Center of the Ukrainian Academy of Sciences, Kiev, pp. 40–46 (in Russian).

Probabilistic Methods in Discrete Mathematics, pp. 437 – 450
V. F. Kolchin et al. (Eds.)
1993 TVP/VSP

RUNS IN FINITE MARKOV CHAINS

L. Ya. SAVELYEV
Institute of Mathematics, Siberian Branch of Russia Academy of Sciences, Universitetskii Prospect 4, 630090 Novosibirsk-90 Russia

ABSTRACT

Joint distributions of some characteristics of runs in finite Markov chains are considered. A system of difference equations for the probabilities is written out in scalar and matrix form. The equation for the corresponding generating function is given. Some special cases and few examples are considered.

1. INTRODUCTION

There is a lot of papers and monographs about Markov chains and processes. The book (Kemeny and Snell, 1960) is devoted to finite Markov chains. For Markov processes the necessary definitions, formulations and proofs of statements used in the present paper can be found in (Ventsel, 1975).

1.1. Let $(\Omega, \mathcal{F}, \mathbf{P})$ be a probability space and C be a finite set. Consider a Markov process $\xi_n: \Omega \to C$, $n \in \mathbf{Z}$, with distributions $P^n = (p_\delta^n)$, $p_\delta^n = \mathbf{P}\{\xi_n = \delta\}$ and transition matrices $Q^{on} = (q_{\gamma\delta}^{0n})$, $q_{\gamma\delta}^{0n} = \mathbf{P}\{\xi_n = \delta \mid \xi_{n-1} = \gamma\}$, provided $\mathbf{P}\{\xi_{n-1} = \gamma\} > 0$ $(\gamma, \delta \in C)$.
We assume that $\xi_n^{-1}(A) \in \mathcal{F}$ for any $n \in \mathbf{Z}$ and $A \subseteq C$, and $C = C^+ \cup \{\theta\}$, $\xi_n(\omega) = \theta$, $n < 0$, $\omega \in \Omega$, $\xi_n(\omega) \in C^+$, $n \geq 0$, $\omega \in \Omega$. Let $C = \sum A_\nu$, $\nu \in \Gamma^+$, $A_\theta = \{\theta\}$, $\Gamma = \Gamma^+ \cup \{\theta\}$, $A_\nu \neq 0$, $A_\mu A_\nu = 0$, $\mu \neq \nu$, $\mu \in \Gamma$, $\nu \in \Gamma$, $B_\nu = C - A_\nu$, $\nu \in \Gamma$. Part of a set we will identify with the indicator function of the corresponding subset.
The random variable $u_{n,\nu} = A_\nu(\xi_n)$ with values

$$u_{n,\nu}(\omega) = A_\nu(\xi_n(\omega)) = 1, \qquad \xi_n(\omega) \in A_\nu,$$

$$u_{n,\nu}(\omega) = 0, \qquad \xi_n(\omega) \in B_\nu,$$

describes the event $\xi_n \in A_\nu$. For example, $u_{n,\theta} = 1$, $n < 0$, $u_{n,\theta} = 0$, $n \geq 0$, and $u_{n,\nu} = 0$, $n < 0$, $\nu \neq \theta$.

1.2. The random variables ξ_n, $n \geq 0$, form a finite Markov chain with the initial distribution $P = P^0$ and the transition matrices $Q^{(n)} = Q^{0n}$, $n > 0$. By definition $p_\delta = p_\delta^0 = \mathbf{P}\{\xi_0 = \delta\}$ and $p_\theta = 0$. It is clear that each Markov chain ξ_n, $n \geq 0$, can be embedded into a Markov process ξ_n, $n \in \mathbf{Z}$.

1.3. We investigate runs of the events $\xi_n \in A_\nu$ for $\nu \neq \theta$.
The number of ν-values, i.e., the events $\xi_l \in A_\nu$ in the sequence ξ_l, $l \leq n$, is described by the random variable

$$x_{n\nu} = \sum_{l \leq n} u_{l,\nu}.$$

The number of ν-runs in this sequence is described by the random variable

$$y_{n\nu} = \sum_{l \leq n} \max(0, u_{l,\nu} - u_{l-1,\nu}).$$

The maximum of lengths of these ν-runs is described by the random variable

$$z_{n\nu} = \max_{l \leq n} \Big(\sum_{l \leq m \leq n} \prod_{l \leq k \leq m} u_{k,\nu} \Big).$$

By definition $x_{n\theta} = y_{n\theta} = z_{n\theta} = 0$, $n < 0$, $\nu \neq \theta$, and $x_{n\theta} = \infty$, $y_{n\theta} = 1$, $z_{n\theta} = \infty$, $n \in \mathbf{Z}$.

1.4. Let $x_n = (x_{n\nu})$, $y_n = (y_{n\nu})$, $z_n = (z_{n\nu})$ and $i = (i(\nu))$, $j = (j(\nu))$, $k = (k(\nu))$, where $\nu \in \Gamma$ and $i(\nu), j(\nu), k(\nu) \in \mathbf{Z}$. Suppose further that $k(\nu) \geq 0$, $\nu \in \Gamma$. The sets $R, S, T \subseteq \Gamma$ define the families $Rx_n = (R(\nu)x_{n\nu})$, $Sy_n = (S(\nu)y_{n\nu})$, $Tz_n = (T(\nu)z_{n\nu})$ and $Ri = (R(\nu)i(\nu))$, $Sj = (S(\nu)j(\nu))$, $Tk = (T(\nu)k(\nu))$. We assume that $\theta \notin R$, $\theta \notin S$, $\theta \notin T$, and $0 \cdot \infty = \infty \cdot 0 = 0$. The operations and the order for the families are defined by the operations and the order for their values. The correctness of the equalities

$$\mathbf{R}x_n = Ri, \quad Sy_n = Sj,$$

and the inequality $Tz_n \leq Tk$ depends on the correctness of $x_{n\rho} = i(\rho)$, $y_{n\sigma} = j(\sigma)$ and $z_{n\tau} \leq k(\tau)$ for $\rho \in R$, $\sigma \in S$, $\tau \in T$ only. For example, these equalities and inequality are true always for $R = S = T = 0$. There exists a unique index $\mu = \mu(\delta) \in \Gamma$ such that $\delta \in A_\mu$. Let $m = k(\mu)$. Then $\mu(\theta) = \theta$ and $m = k(\theta)$. We will consider also the families $e = (e(\nu))$ and $e_\mu = (e_\mu(\nu))$ with values $e_\mu(\nu) = 1$, $\nu = \mu$, $e_\mu(\nu) = 0$, $\nu \neq \mu$, $e(\nu) = 1$, $\nu \in \Gamma$. The sets R, S, T define the families Re_μ, Se_μ, Te_μ, Re, Se, Te.

1.5. Consider the probabilities

$$f_\delta(i,j,k,n) = \mathbf{P}\{Rx_n = Ri,\ Sy_n = Sj,\ Tz_n \leq Tk,\ \xi_n = \delta\},$$
$$f(i,j,k,n) = \mathbf{P}\{Rx_n = Ri,\ Sy_n = Sj,\ Tz_n \leq Tk\}.$$

It is clear that

$$f(i,j,k,n) = \sum_\delta f_\delta(i,j,k,n).$$

The probabilities $f(i,j,k,n)$ define the joint distribution of the random variables $x_{n\rho}$, $y_{n\sigma}$, $z_{n\tau}$, $\rho \in R$, $\sigma \in S$, $\tau \in T$.

The problem is to find the probabilities $f(i,j,k,n)$. In the present paper this problem is solved with the help of a system of difference equations and a generating function.

2. SCALAR EQUATIONS

We compose a system of difference equations for the probabilities $f_\delta(i,j,k,n)$ with the help of the Markov property. We assume that the sets R, S, T are given.

2.1. Notice that $z_{n-1} \le z_n$ and

$$\{z_n \le k\} = \{z_{n-1} \le k\} - \{z_{n-1} \le k, z_n \le k\}.$$

Therefore,

$$U_\delta(i,j,k,n) = V_\delta(i,j,k,n) - W_\delta(i,j,k,n), \tag{1}$$

where

$$U_\delta(i,j,k,n) = \{Rx_n = Ri,\ Sy_n = Sj,\ Tz_n \le Tk,\ \xi_n = \delta\},$$
$$V_\delta(i,j,k,n) = \{Rx_n = Ri,\ Sy_n = Sj,\ Tz_{n-1} \le Tk,\ \xi_n = \delta\},$$
$$W_\delta(i,j,k,n) = \{Rx_n = Ri,\ Sy_n = Sj,\ Tz_{n-1} \le Tk,\ Tz_n \le Tk,\ \xi_n = \delta\}.$$

It is clear that

$$V_\delta(i,j,k,n) = \sum_\gamma U_\gamma(i - e_\mu, j - B_\mu(\gamma)e_\mu, k, n-1)\{\xi_n = \delta\}, \tag{2}$$

$$W_\delta(i,j,k,n) = T(\mu)\sum_\gamma U_\gamma(i - (m+1)e_\mu, j - e_{\mu'}, k, n-m-1)$$
$$\times B_\mu(\gamma)\{\xi_{n-m} \in A, \dots, \xi_n \in A\}\{\xi_n = \delta\}. \tag{3}$$

Let $n < 0$. Then

$$U_\delta(i,j,k,n) = V_\delta(i,j,k,n) = 0, \qquad \delta \ne \theta;$$
$$U_\theta(i,j,k,n) = V_\theta(i,j,k,n) = \Omega \qquad [(Ri = 0) \wedge (Sj = 0)],$$
$$U_\theta(i,j,k,n) = V_\theta(i,j,k,n) = 0 \qquad [(Ri \ne 0) \vee (Sj \ne 0)];$$
$$W_\delta(i,j,k,n) = 0.$$

Let $n = 0$. Then

$$U_\delta(i,j,k,n) = (\xi_0 = \delta) \qquad [(Ri = Re_\mu) \wedge (S_j = Se_\mu) \wedge (Tk \ge Te_\mu)],$$
$$U_\delta(i,j,k,n) = 0 \qquad [(Ri \ne Re_\mu) \vee (Sj \ne Re_\mu) \vee (Tk \not\ge Te_\mu)];$$
$$V_\delta(i,j,k,n) = (\xi_0 = \delta) \qquad [(Ri = Re_\mu) \wedge (Sj = Se_\mu)],$$
$$V_\delta(i,j,k,n) = 0 \qquad [(Ri \ne Re_\mu) \vee (Sj \ne Se_\mu)];$$
$$W_\delta(i,j,k,n) = (\xi_0 = \delta) \qquad [(Ri = Re_\mu) \wedge (Sj = Se_\mu) \wedge (Tk \not\ge Te_\mu)],$$
$$W_\delta(i,j,k,n) = 0 \qquad [(Ri \ne Re_\mu) \vee (Sj \ne Se_\mu) \vee (Tk \ge Te_\mu)].$$

In the two last equalities we take into account that

$$Tk \ge Te_\mu \iff T(\mu)m \ge T(\mu); \quad Tk \ge Te_\mu \iff T(\mu) = 1, \quad m = 0.$$

Let $n \ge 0$. Then $\xi_n \ne \theta$ and

$$U_\theta(i,j,k,n) = V_\theta(i,j,k,n) = W_\theta(i,j,k,n) = 0.$$

2.2. Consider the probabilities

$$f_\delta(i,j,k,n) = \mathbf{P}\{U_\delta(i,j,k,n)\},$$
$$g_\delta(i,j,k,n) = \mathbf{P}\{V_\delta(i,j,k,n)\},$$
$$h_\delta(i,j,k,n) = \mathbf{P}\{W_\delta(i,j,k,n)\},$$

and the transition probabilities

$$q_{\gamma\delta}^{m,n}(A) = \mathbf{P}\{\xi_{n-m} \in A, \dots, \xi_n \in A, \xi_n = \delta \mid \xi_{n-m-1} = \gamma\}, \qquad A \subseteq C.$$

It is easy to show that the equations

$$q_{\gamma\delta}^{0,n}(A) = q_{\gamma\delta}^{0,n} A(\delta),$$
$$q_{\gamma\delta}^{m,n}(A) = \sum_\beta q_{\gamma\beta}^{0,n-m}(A) q_{\beta\delta}^{m-1,n}(A), \qquad m \geq 1. \tag{4}$$

are valid. Notice that $q_{\gamma\delta}^{m,n}(A) = 0$, where $\delta \notin A$.

THEOREM 1. *The probabilities $f_\delta(i,j,n,k)$ are the unique solution of the system of equations*

$$f_\delta(i,j,k,n) = \sum_\gamma f_\gamma(i - e_\mu, j - B_\mu(\gamma)e_\mu, k, n-1) q_{\gamma\delta}^{0,n} \tag{5}$$
$$- T(\mu) \sum_\gamma f_\gamma(i - (m+1)e_\mu, j - e_\mu, k, n-m-1) B_\mu(\gamma) q_{\gamma\delta}^{m,n}(A_\mu),$$

which satisfies the following conditions:

$$\begin{aligned} f_\delta(i,j,k,n) &= 0 \qquad ([n<0] \wedge [\delta \neq \theta]),\\ f_\theta(i,j,k,n) &= 1 \qquad ([n<0] \wedge [(Ri = 0) \wedge (S_j = 0)]),\\ f_\theta(i,j,k,n) &= 0 \qquad ([n<0] \wedge [(Ri \neq 0) \vee (Sj \neq 0)]),\\ f_\delta(i,j,k,n) &= 0 \qquad ([n \geq 0] \wedge [(Ri \ngeq 0) \vee (Sj \ngeq 0)]). \end{aligned} \tag{6}$$

Proof. It follows from (1) that

$$f_\delta(i,j,k,n) = g_\delta(i,j,k,n) - h_\delta(i,j,k,n).$$

From (2)–(3) and the Markov property we obtain that

$$g_\delta(i,j,k,n) = \sum_\gamma f_\gamma(i - e_\mu, j - B_\mu(\gamma)e_\mu, k, n-1) q_{\gamma\delta}^{0,n},$$
$$h_\delta(i,j,k,n) = T(\mu) \sum_\gamma f_\gamma(i - (m+1)e_\mu, j - e_\mu, k, n-m-1) B_\mu(\gamma) q_{\gamma\delta}^{m,n}(A_\mu).$$

Thus (5) is true. The equalities from the end of 2.1 imply (6). The family of probabilities for which (5) and (6) are true is unique. This assertion can be proved by induction.
From (6) it follows that

$$f_\delta(i,j,k,n) = 0 \qquad [(Ri \ngeq 0) \vee (Sj \ngeq 0)].$$

Indeed, $f_\delta(i,j,k,n) \neq 0$ for $n < 0$ only if $Ri = 0$ and $Sj = 0$.

2.3. The probabilities $f(i,j,k,n)$ can be found as the sums of $f_\delta(i,j,k,n)$. The probabilities

$$p(i,j,k,n) = \mathbf{P}\{Rx_n = Ri,\ Sy_n = Sj,\ Tz_n = Tk\}$$

can be found with the help of $f(i,j,k,n)$. Let $L \subseteq T$, $M = T - L$ and $|M|$ be the number of elements of a set M. Notice that

$$\{Tz_n = Tk\} = \prod\{z_{n\tau} = k(\tau)\} = \prod\left[\{z_{n\tau} \le k(\tau)\} - \{z_{n\tau} \le k(\tau) - 1\}\right]$$

$$= \sum(-1)^{|M|}\{z_{n\lambda} \le k(\lambda)\}\{z_{n\nu} \le k(\nu) - 1\}, \qquad \lambda \in L,\ \nu \in M,\ \tau \in T,$$

$$\{Tz_n = Tk\} = \sum(-1)^{|M|}\{z_n \le k - Me\}, \qquad M \subseteq T.$$

From this equality for the indicator functions it follows that

$$p(i,j,k,n) = \sum(-1)^{|M|}f(i,j,k - Me, n), \qquad M \subseteq T.$$

In particular, if $T = \{1\}$; $k = k(1)$; $M = \varnothing, T$; $|M| = 0, 1$, then

$$p(i,j,k,n) = f(i,j,k,n) - f(i,j,k-1,n),$$

and if $T = \{1,2\}$, $k = (k(1), k(2))$; $M = \varnothing, \{1\}, \{2\}, T$; $|M| = 0,1,1,2$, then

$$\begin{aligned} p(i,j,k,n) = f(i,j,(k(1),k(2)),n) &- f(i,j,(k(1)-1,k(2)),n) \\ - f(i,j,(k(1),k(2)-1,n)) &+ f(i,j,(k(1)-1,k(2)-1,n). \end{aligned}$$

2.4. Let the random variables ξ_n, $n \ge 0$, be independent and identically distributed. Then

$$q_{\gamma d}^{0,n} = \mathbf{P}\{\xi_0 = \delta\} = p_\delta, \qquad n \ge 0.$$

Such a sequence $\{\xi_n\}$ is called a *Laplace sequence.* For $|C^+| = 2$, the Laplace sequence (ξ_n) is a *Bernoulli sequence.*
Put

$$a_\nu = \mathbf{P}\{\xi_0 \in A_\nu\} = \sum_\delta p_\delta A_\nu(\delta).$$

LEMMA 1. (1) *Let ξ_n, $n \ge 0$, be a Laplace sequence. Then*

$$q_{\gamma\delta}^{m,n}(A_\mu) = a_\mu^m p_\delta, \qquad 0 \le m \le n.$$

(2) *Let ξ_n, $n \in \mathbf{Z}$, be a Markov process. Then*

$$q_{\gamma\delta}^{m,n}(A_\mu) = 0, \qquad 0 \le n \le m.$$

Proof. (1) From the definitions it follows that

$$q_{\gamma\delta}^{0,n}(A_\mu) = q_{\gamma\delta}^{0,n} A_\mu(\delta) = p_\delta = a_\mu^o p_\delta.$$

Let $q^{m-1,n}_{\beta\delta}(A_\mu) = a^{m-1}_\mu p_\delta$, $1 \le m \le n$. Then it follows from (4) that

$$q^{mn}_{\gamma\delta}(A_\mu) = \sum_\beta p_\beta A_\mu(\beta) a^{m-1}_\mu p_\delta = a_\mu a^{m-1}_\mu p_\delta = a^m_\mu p_\delta, \qquad 1 \le m \le n.$$

The first assertion of the lemma is proved.
(2) Let $\beta \ne \theta$. Then $q^{0,n-m}_{\gamma\beta} = 0$ for $n < m$. Therefore,

$$q^{0,n-m}_{\gamma\beta}(A_\mu) = q^{0,n-m}_{\gamma\beta} A_\mu(\beta) = 0, \qquad \beta \ne \theta.$$

In addition,

$$q^{1,0}_{\gamma\delta}(A_\mu) = \sum_\beta q^{0,-1}_{\gamma\beta}(A_\mu) q^{00}_{\beta\delta}(A_\mu) = 0,$$

since

$$q^{0,-1}_{\gamma\beta} A_\mu(\beta) = 0, \quad \mu \ne \theta; \qquad q^{0,0}_{\beta\delta} = 0, \quad \mu = \theta = \delta.$$

Let $q^{m-1,n}_{\beta\delta}(A_\mu) = 0$, $0 \le n \le m-1$. Then

$$q^{m,n}_{\gamma\delta}(A_\mu) = \sum_\beta q^{0,n-m}_{\gamma\beta}(A_\mu) q^{m-1,n}_{\beta\delta}(A_\mu) = 0, \qquad 0 \le n < m.$$

The second assertion of the lemma is proved.

2.5. The systems of difference equations in (Savelyev, 1985, 1991) for the case where C is a finite set are special cases of system (5).

Example 1. The system from Section 2.2 of (Savelyev, 1985) is obtained from (5) with $\Gamma^+ = \{1,2\}$, $A_1 = A$, $A_2 = B$, $R = 0$, $S = 0$, $T = \{1\}$, $k(1) = k$, $f(\cdot,\cdot,k,n) = f(k,n)$.

Example 2. The system from Section 2 of (Savelyev, 1991) is obtained from (5) with $\Gamma^+ = \{1,2\}$, $A_1 = A$, $A_2 = B$, $R = \{1\}$, $S = \{1\}$, $T = \{1,2\}$, $i(1) = i$, $j(1) = j$, $k(1) = k$, $k(2) = l$, $f(i(1), j(1), (k(1), k(2)), n) = f(i,j,k,l,n)$.

Example 3. Let $C^+ = \{0,1\} = \Gamma^+$, $\theta = -1$, $C = \{-1,0,1\} = \Gamma$, $A_0 = \{0\}$, $A_1 = \{1\}$, $R = 0$, $S = 0$, $T = \{1\}$, $k(1) = k$, $z_{n1} = z_n$, and let ξ_n, $n \ge 0$, be a Bernoulli sequence, $p_1 = p$, $p_0 = q = 1-p$, $f(\cdot,\cdot,k(1),n) = f(k,n)$. It follows from (5) and (6) that

$$f(k,n) = 1, \qquad n < k,$$
$$f(k,n) = f(k,n-1) - q^{1\wedge(n-k)} p^{k+1} f(k, n-k-2), \qquad n \ge k.$$

Here $1 \wedge (n-k) = \min\{1, n-k\}$.
In exactly the same way we can obtain a similar system for a Laplace sequence with finite C^+ and $\Gamma^+ = \{0,1\}$.

3. A MATRIX EQUATION

Equalities (4)–(6) can be written in a matrix form.

3.1. Let

$$F(i,j,k,n) = (f_\delta(i,j,k,n)),$$
$$G(i,j,k,n) = (g_\delta(i,j,k,n)),$$
$$H(i,j,k,n) = (h_\delta(i,j,k,n));$$

let E be the unit C-row, E^* be the unit C-column, I be the unit $C \times C$ matrix, A be the diagonal $C \times C$ matrix with the diagonal elements $A(\delta,\delta) = A(\delta)$ for $A \subseteq C$, and let $Q^{m,n}(A) = (q_{\gamma\delta}^{m,n}(A))$.
The matrix A is called an indicator matrix, and we identify it with the set A. Notice that the equality $A+B = C$ for sets is equivalent to the equality $A+B = I$ for indicator matrices.
It follows from (4) that

$$Q^{0,n}(A) = Q^{0,n}A, \quad Q^{m,n}(A) = Q^{0,n-m}(A)Q^{m-1,n}(A). \tag{7}$$

3.2. The following assertion is a corollary to Theorem 1.

THEOREM 2. *The row F is the unique solution of the matrix equation*

$$F(i,j,k,n) = \sum_\mu (F(i-e_\mu,j,k,n-1)A_\mu + F(i-e_\mu,j-e_\mu,k,n-1)B_\mu)Q^{0,n}(A_\mu)$$
$$- \sum_\mu T(\mu)F(i-(m+1)e_\mu, j-e_\mu, k, n-m-1)B_\mu Q^{m,n}(A_\mu), \tag{8}$$

which satisfies the following conditions:

$$\begin{aligned} F(i,j,k,n) &= EA_\theta && ([n<0] \wedge [(Ri=0) \wedge (S_j=0)]),\\ F(i,j,k,n) &= 0 && ([n<0] \wedge [(Ri \neq 0) \vee (Sj \neq 0)]),\\ F(i,j,k,n) &= 0 && ([n \geq 0] \wedge [(Ri \not\geq 0) \vee (Sj \not\geq 0)]). \end{aligned} \tag{9}$$

Notice that equality (8) is equivalent to the equalities

$$F(i,j,k,n) = G(i,j,k,n) - H(i,j,k,n),$$
$$G(i,j,k,n) = \sum_\mu F(i-e_\mu,j,k,n-1)A_\mu Q^{0,n}A_\mu$$
$$+ \sum_\mu F(i-e_\mu,j-e_\mu,k,n-1)B_\mu Q^{0,n}A_\mu,$$
$$H(i,j,k,n) = \sum_\mu T(\mu)F(i-(m+1)e_\mu, j-e_\mu, k, n-m-1)B_\mu Q^{m,n}(A_\mu).$$

It follows from (9) that

$$F(i,j,k,n) = 0 \qquad [(Ri \not\geq 0) \vee (Sj \not\geq 0)].$$

3.3. Let $|i| = \sum |i(\nu)|$, $|j| = \sum |j(\nu)|$, $|k| = \sum k(\nu)$, $\chi(i,j,k,n) = 1 - [1 \wedge (|Ri| + |Sj| + |m| + |n|)]$ for families i, j, k and integers m, n.
From Theorem 2 we obtain the following assertion.

COROLLARY 1. *The equalities*

$$F(i,j,k,0) = \sum_{\mu} (1 - (1 - (1 \wedge m))T(\mu))\chi(i - e_\mu, j - e_\mu, 0, 0)PA_\mu,$$
$$G(i,j,k,0) = \sum_{\mu} \chi(i - e_\mu, j - e_\mu, 0, 0)PA_\mu, \tag{10}$$
$$H(i,j,k,0) = \sum_{\mu} T(\mu)\chi(i - e_\mu, j - e_\mu, m, 0)PA_\mu$$

are true.

Proof. Since $f_\delta(i,j,k,-1) = 0$, $\delta \neq \theta$, $f_\theta(i,j,k,-1) = 1$, $Ri = Sj = 0$, and $S(j - e_\theta) = Sj$, $\theta \notin S$, we have

$$F(i,j,k,-1) = \chi(i,j,0,0)EA_\theta, \quad \chi(i - e_\theta, j, 0, 0) = \chi(i - e_\theta, j - e_\theta, 0, 0).$$

In addition,

$$EA_\theta Q^{0,0} A_\mu = EA_\theta(E^* P)A_\mu = (EA_\theta E^*)PA_\mu = PA_\mu.$$

Therefore,

$$\begin{aligned} G(i,j,k,0) &= \sum_{\mu} \chi(i - e_\mu, j, 0, 0)EA_\theta A_\mu Q^{0,0} A_\mu \\ &\quad + \sum_{\mu} \chi(i - e_\mu, j - e_\mu, 0, 0)EA_\theta B_\mu Q^{0,0} A_\mu \\ &= \chi(i - e_\theta, j, 0, 0)PA_\theta + \sum_{\mu \neq \theta} \chi(i - e_\mu, j - e_\mu, 0, 0)PA_\mu \\ &= \sum_{\mu} \chi(i - e_\mu, j - e_\mu, 0, 0)PA_\mu. \end{aligned}$$

The second equality from (10) is proved.
Since $p_\theta = 0$ and

$$Q^{0,0}(A_\theta) = Q^{0,0} A_\theta = (E^* P)A_\theta = E^*(PA_\theta) = 0,$$
$$Q^{0,-1}(A_\mu) = Q^{0,-1} A_\mu = (E^* EA_\theta)A_\mu = (E^* E)(A_\theta A_\mu) = 0, \qquad \mu \neq \theta,$$

we have

$$Q^{1,0}(A_\mu) = Q^{0,-1}(A_\mu)Q^{0,0}(A_\mu) = 0$$

for every $\mu \in \Gamma$. By induction it is easy to show that $Q^{m,0}(A_\mu) = 0$, $m > 0$. Therefore the third equality from (10) is true.
Since $F = G - H$ and

$$\chi(i,j,m,0) = (1 - (1 \wedge m))\chi(i,j,0,0),$$

the second and the third equality imply the first equality from (10).

3.4. Notice that

$$Q^{0,l} = Q^{0,-1} = E^*EA_\theta, \qquad l < 0, \qquad Q^{0,0} = E^*P.$$

In what follows we assume that

$$Q^{0,n} = Q^{0,1} = Q, \qquad n > 0.$$

Denote

$$A_{\mu\mu} = A_\mu Q A_\mu, \quad B_{\mu\mu} = B_\mu Q A_\mu.$$

Since $A_\mu A_\mu = A_\mu$, we have

$$Q^{1,n}(A_\mu) = Q^{0,n-1}A_\mu Q^{0,n}A_\mu = QA_\mu A_\mu QA_\mu = QA_\mu A_{\mu\mu}$$

for $n > 1$. By induction we obtain that

$$Q^{m,n}(A_\mu) = Q^{0,n-m}A_\mu Q^{m-1,n}(A_\mu) = QA_\mu(QA_\mu A_{\mu\mu}^{m-1}) = QA_\mu A_{\mu\mu}^m$$

for $n > m \geq 1$. In exactly the same way we obtain that for $m \geq 1$

$$Q^{m,n}(A_\mu) = Q^{0,0}A_\mu A_{\mu\mu}^m.$$

Since $A_{\mu\mu}^0 = I$, this equality is true for $m = 0$.
Let $0 < n < m$. Then for $\mu \neq \theta$

$$Q^{m,n}(A_\mu) = E^*EA_\theta A_\mu Q^{m-1,n}(A_\mu) = 0.$$

Since $T(\theta) = 0$, we have $T(\mu)B_\mu Q^{m,n}(A_\mu) = 0$.
Let $0 < n = m$. Then

$$B_\mu Q^{m,n}(A_\mu) = B_\mu Q^{0,0}A_\mu A_{\mu\mu}^m.$$

Let $0 \leq m < n$. Then

$$B_\mu Q^{m,n}(A_\mu) = B_{\mu\mu}A_{\mu\mu}^m.$$

In the Laplace case

$$Q^{0,0} = Q = E^*P, \quad A_{\mu\mu} = A_\mu(E^*P)A_\mu, \quad B_{\mu\mu} = B_\mu(E^*P)A_\mu.$$

3.5. Consider an example from Section 5 of (Savelyev, 1985) which concerns the addition of random binary numbers. Let $C = \{0,1,2,3\}$, $\Gamma = \{0,1,2\}$, $A_0 = \{0\}$, $A_1 = \{1\}$, $A_2 = \{2,3\}$, $P = (0,0,3/4,1/4)$, $R = S = 0$, $T = \{1\}$;

$$Q^{0l} = \begin{pmatrix} 1 & 0 & 0 & 0 \\ 1 & 0 & 0 & 0 \\ 1 & 0 & 0 & 0 \\ 1 & 0 & 0 & 0 \end{pmatrix}, \quad Q^{0,0} = \begin{pmatrix} 0 & 0 & 3/4 & 1/4 \\ 0 & 0 & 3/4 & 1/4 \\ 0 & 0 & 3/4 & 1/4 \\ 0 & 0 & 3/4 & 1/4 \end{pmatrix},$$

$$Q^{0,n} = \begin{pmatrix} 1 & 0 & 0 & 0 \\ 0 & 1/2 & 1/4 & 1/4 \\ 0 & 0 & 3/4 & 1/4 \\ 0 & 1/2 & 1/4 & 1/4 \end{pmatrix}, \qquad l < 0, \quad n > 0.$$

Notice that

$$A_0 Q^{0,n} A_0 = A_0,$$

$$A_{11} = \begin{pmatrix} 0 & 0 & 0 & 0 \\ 0 & 1/2 & 0 & 0 \\ 0 & 0 & 0 & 0 \\ 0 & 0 & 0 & 0 \end{pmatrix}, \qquad B_{11} = \begin{pmatrix} 0 & 0 & 0 & 0 \\ 0 & 0 & 0 & 0 \\ 0 & 0 & 0 & 0 \\ 0 & 1/2 & 0 & 0 \end{pmatrix},$$

$$A_{22} = \begin{pmatrix} 0 & 0 & 0 & 0 \\ 0 & 0 & 0 & 0 \\ 0 & 0 & 3/4 & 1/4 \\ 0 & 0 & 1/4 & 1/4 \end{pmatrix}, \qquad B_{22} = \begin{pmatrix} 0 & 0 & 0 & 0 \\ 0 & 0 & 1/4 & 1/4 \\ 0 & 0 & 0 & 0 \\ 0 & 0 & 0 & 0 \end{pmatrix}.$$

By induction we obtain from (7) that

$$B_{11} A_{11}^m = \begin{pmatrix} & 0 & & 0 \\ \begin{pmatrix} 0 & 0 \\ 0 & (1/2)^{m+1} \end{pmatrix} & & & 0 \end{pmatrix},$$

$$B_{22} A_{22}^m = \begin{pmatrix} 0 & \begin{pmatrix} 0 & 0 \\ 1/4 & 1/4 \end{pmatrix} \begin{pmatrix} 3/4 & 1/4 \\ 1/4 & 1/4 \end{pmatrix}^m \\ 0 & 0 \end{pmatrix}.$$

Thus the matrix coefficients of equation (8) in this example are easily calculated.

4. FUNCTIONAL EQUATIONS

In what follows we assume that the random variables ξ_n, $n \geq 0$, form a homogeneous Markov chain: $Q^{0,n} = Q^{0,1} = Q$, $n > 0$. In this case the equation for the required generating function can be easily deduced from (8)–(10).

4.1. Let $s = (s_\nu)$, $t = (t_\nu)$ be families of complex numbers such that $0 < |s_\nu|$, $|t_\nu| \leq 1$, $\nu \in \Gamma$, and

$$s^{Ri} = \prod s_\nu^{Ri(\nu)} = \prod s_\rho^{i(\rho)}, \quad t^{Sj} = \prod t_\nu^{Sj(\nu)} = \prod t_\sigma^{j(\sigma)}, \qquad \rho \in R, \ \sigma \in S.$$

Consider the vector generating functions

$$\widehat{F}(s,t,k,u) = \sum_{i,j,n} F(i,j,k,n) s^{Ri} t^{Sj} u^n,$$

$$\widehat{G}(s,t,k,u) = \sum_{i,j,n} G(i,j,k,n) s^{Ri} t^{Sj} u^n,$$

$$\widehat{H}(s,t,k,u) = \sum_{i,j,n} H(i,j,k,n) s^{Ri} t^{Sj} u^n,$$

where $n \geq 0,\ u \in \mathbf{C},\ 0 < |u| < 1$. The inequalities

$$\sum_{ij} |f_\delta(i,j,k,n)s^{Ri}t^{Sj}| \leq \mathbf{P}\{z_n \leq k, \xi_n = \delta\} \leq 1,$$
$$\sum_{ij} |g_\delta(i,j,k,n)s^{Ri}t^{Sj}| \leq \mathbf{P}\{z_{n-1} \leq k, \xi_n = \delta\} \leq 1,$$
$$\sum_{ij} |h_\delta(i,j,k,n)s^{Ri}t^{Sj}| \leq \mathbf{P}\{z_{n-1} \leq k, z_n \leq k, \xi_n = \delta\} \leq 1$$

ensure the summability for the families under consideration. The relation $F = G - H$ implies $\widehat{F} = \widehat{G} - \widehat{H}$.
The scalar generating function

$$\widehat{f}(s,t,k,u) = \sum_{i,j,n} f(i,j,k,n)s^{Ri}t^{Sj}u^n$$

is obtained with the help of the vector generating function $\widehat{F}(s,t,k,u)$. It follows from the definitions that

$$\widehat{f}(s,t,k,u) = \widehat{F}(s,t,k,u)E^*.$$

4.2. From Theorem 2 we obtain the following assertion.

THEOREM 3. *Let ξ_n, $n \geq 0$, form a homogeneous Markov chain. Then*

$$\widehat{F}(I - Ku) = L, \tag{11}$$

where

$$K = \sum_\mu (A_{\mu\mu}s^{Re_\mu} + B_{\mu\mu}\Delta_\mu),$$
$$L = \sum_\mu PA_\mu\Delta_\mu, \tag{12}$$
$$\Delta_\mu = I \cdot T(\mu) \cdot A^m_{\mu\mu}s^{mRe_\mu}t^{Se_\mu}u^m.$$

Proof. It follows from the definitions that

$$\widehat{G} = \sum_\mu \left[\sum_{i,j,n} F(i - e_\mu, j, k, n-1)s^{Ri}t^{Sj}u^n\right] A_\mu Q^{0,n}A_\mu$$
$$+ \sum_\mu \left[\sum_{i,j,n} F(i - e_\mu, j - e_\mu, k, n-1)s^{Ri}t^{Sj}u^n\right] B_\mu Q^{0,n}A_\mu,$$
$$\widehat{H} = \sum_\mu \left[\sum_{i,j,n} F(i - (m+1)e_\mu, j - e_\mu, k, n - m - 1)s^{Ri}t^{Sj}u^n\right] T(\mu)B_\mu Q^{m,n}(A_\mu).$$

In the inner sums we consider the term for $n = 0$. Using (10) we obtain that

$$\sum_{i,j} G(i,j,k,0)s^{Ri}t^{Sj} = \sum_{\mu} PA_{\mu}s^{Re_{\mu}}t^{Se_{\mu}}.$$

In addition the equalities

$$\sum_{i,j,n} F(i-e_{\mu},j,k,n-1)s^{Ri}t^{Sj}u^{n} = \widehat{F}s^{Re_{\mu}}u,$$

$$\sum_{i,j,n} F(i-e_{\mu},j-e_{\mu},k,n-1)s^{Ri}t^{Sj}u^{n} = \widehat{F}s^{Re_{\mu}}t^{Se_{\mu}}u$$

are true for $n > 1$. Thus

$$\widehat{G} = \sum_{\mu} PA_{\mu}s^{Re_{\mu}}t^{Se_{\mu}} + \widehat{F}\sum_{\mu}[A_{\mu\mu} + B_{\mu\mu}t^{Se_{\mu}}]s^{Re_{\mu}}u.$$

Now consider the terms for $0 \le n \le m$. Making use of (9), we obtain that

$$F(i-(m+1)e_{\mu},j-e_{\mu},k,n-m-1) = F(0,0,k,-1) = EA_{\theta},$$

$$EA_{\theta}B_{\mu}Q^{m,m}(A_{\mu}) = EA_{\theta}B_{\mu}E^{*}PA_{\mu}A^{m}_{\mu\mu} = PA_{\mu}A^{m}_{\mu\mu}$$

for $n = m$. In addition, the equalities

$$\sum_{i,j,n} F(i-(m+1)e_{\mu},j-e_{\mu},k,n-m-1)s^{Ri}t^{Sj}u^{n} = \widehat{F}s^{(m+1)Re_{\mu}}t^{Se_{\mu}}u^{m+1}$$

are true for $n > m$. Hence

$$\widehat{H} = \sum_{\mu} T(\mu)PA_{\mu}A^{m}_{\mu\mu}s^{(m+1)Re_{\mu}}t^{Se_{\mu}}u^{m} + \widehat{F}\sum_{\mu} T(\mu)B_{\mu\mu}A^{m}_{\mu\mu}s^{(m+1)Re_{\mu}}t^{Se_{\mu}}u^{m+1}.$$

Substituting the obtained values

$$\widehat{G} = \widehat{G}(s,t,k,u), \quad \widehat{H} = \widehat{H}(s,t,k,u)$$

into

$$\widehat{F} = \widehat{F}(s,t,k,u) = \widehat{G}(s,t,k,u) - \widehat{H}(s,t,k,u) = \widehat{G} - \widehat{H},$$

we see that

$$\begin{aligned}\widehat{F} &= \sum_{\mu}[PA_{\mu}s^{Re_{\mu}}t^{Se_{\mu}}T(\mu)PA_{\mu}A^{m}_{\mu\mu}s^{(m+1)Re_{\mu}}t^{Se_{\mu}}u^{m}] \\ &\quad + \widehat{F}\sum_{\mu}[(A_{\mu\mu} + B_{\mu\mu}t^{Se_{\mu}})s^{Re_{\mu}}u - T(\mu)B_{\mu\mu}A^{m}_{\mu\mu}s^{(m+1)Re_{\mu}}t^{Se_{\mu}}u^{m+1}] \\ &= L + \widehat{F}Ku.\end{aligned}$$

These equalities are equivalent to (11)–(12). The theorem is proved.

4.3. Consider the normed space $\mathfrak{L}^1 = \mathfrak{L}^1(C)$ and a linear operator $M\colon \mathfrak{L}^1 \to \mathfrak{L}^1$, which is represented by a $C \times C$ matrix $M = (a_{\gamma\delta})$ with the norm $|M| = \max_\gamma \sum_\delta |a_{\gamma\delta}|$. Notice that $|A_\mu| = |B_\mu| = |Q| = 1$. With the help of the Neumann series it is easy to prove the following assertion on the solution of equation (11).

THEOREM 4. *Let* $|u| < 1/(3|\Gamma|)$. *Then equation* (11) *is well-defined and has the solution*

$$\widehat{F} = L(I - Ku)^{-1} = L \sum K^n u^n, \tag{13}$$

where the matrices K, L *are defined in* (12).

Proof. It follows from the definitions that

$$|K| \le \sum (|A_{\mu\mu}| + |B_{\mu\mu}|\,|\Delta_\mu|),$$

$$|A_{\mu\mu}| \le |A_\mu|\,|Q|\,|A_\mu| = 1, \quad |B_{\mu\mu}| = 1,$$

$$|\Delta_\mu| \le |I| + |A_{\mu\mu}|^m \le 2.$$

Consequently, $|K| \le 3|\Gamma|$ and for $|u| < 1/(3|\Gamma|)$

$$|K|\,|u| < 1$$

This implies (see Chapter 4, Section 5 of (Kolmogorov and Fomin, 1989)) that the theorem is true.

4.4. This theorem leads us to the following assertion.

COROLLARY 2. *Let* $|u| < 1/(3|\Gamma|)$. *Then*

$$\widehat{f} = L(I - Ku)^{-1}E^* = L \sum (K^n E^*) u^n. \tag{14}$$

Proof. Since E^* is a continuous linear functional,

$$(L \sum K^n u^n) E^* = L \sum (K^n E^*) u^n.$$

It follows from this equality and (13) that the assertion is true.
The scalar generating function $\widehat{f}$ can be obtained from the system of equations

$$\widehat{f} \cdot 1 - \widehat{F} \cdot E^* = 0, \quad \widehat{f} \cdot 0 - \widehat{F} \cdot (I - Ku) = L.$$

This system can be written in the block form:

$$(\widehat{f}, \widehat{F}) \begin{pmatrix} 1 & 0 \\ -E^* & I - Ku \end{pmatrix} = (0, L). \tag{15}$$

THEOREM 5. *Let* $|u| < 1/(3|\Gamma|)$. *Then*

$$\widehat{f} = \det \begin{pmatrix} 1 & 0 \\ -E^* & I - Ku \end{pmatrix} (\det(I - Ku))^{-1}. \tag{16}$$

Proof. It follows from Theorem 4 that $I - Ku$ is a nonsingular matrix and $\det(I - Ku) \ne 0$. Equality (16) is obtained from (15) by the Cramer rule.

4.5. The solution of the formulated problem is given by equalities (4)–(16). Consider a few examples.

Example 1. The equations and the generating function from Section 4.3 of (Savelyev, 1985) follows from (15)–(16) with $\Gamma^+ = \{1,2\}$, $R = 0$, $S = O, T = \{1\}$ and the notation changed respectively (see 2.5).

Example 2. The equations and the generating function from Section 4.5 of (Savelyev, 1991) follows from (15)–(16) with $\Gamma^+ = \{1,2\}$, $R = \{1\}$, $S = \{1\}$, $T = \{1,2\}$ and the notations changed respectively (see 2.5). Let $0 \le p$, $\alpha, \beta \le 1$ and

$$P = (p, 1-p), \qquad Q = \begin{pmatrix} \alpha & 1-\alpha \\ 1-\beta & \beta \end{pmatrix}.$$

Then

$$K = \begin{pmatrix} \alpha s & (1-\alpha)(1-\beta^l u^l) \\ (1-\beta)st(1-\alpha^k s^k u^k) & \beta \end{pmatrix},$$

$$L = (p(1-\alpha^k s^k u^k)st, (1-p)(1-\beta^l u^l)),$$

$$\widehat{f}(s,t,k,u) = d^{-1}\{p(1-\alpha^k s^k u^k)[(1-\beta u) + (1-\alpha)(1-\beta^l u^l)u]st$$
$$+(1-p)(1-\beta^l u^l)[(1-\alpha s u) + (1-\alpha^k s^k u^k)(1-\beta)stu]\},$$
$$d = (1-\alpha s u)(1-\beta u) - (1-\alpha)(1-\beta)(1-\alpha^k s^k u^k)(1-\beta^l u^l)stu^2.$$

If $|u| < 1/9$, then $d \neq 0$ for $|s|, |t| \le 1$.
The expression for the generating function is simplified if the Bernoulli case is considered and $\alpha = p$, $\beta = 1-p$. This expression is particularly simple if $p = \alpha = \beta = 1/2$.

Example 3. In Section 3.5 the generating function has the form

$$\widehat{f}(k,n) = \frac{1-(1/2)u-(1/2)^{k+3}u^{k+1}+(1/2)^{k+4}u^{k+2}}{1-(3/2)u+(1/2)u^2+(1/2)^{k+3}u^{k+2}-(1/2)^{k+4}u^{k+3}}.$$

This equality is obtained in (Dedukhin, 1979).

REFERENCES

Dedukhin, A. D. (1979). Carry of digits in the addition of random numbers. In: *Proceedings of the 17th Conference on Mathematics*, Novosibirsk Univ. Press, Novosibirsk, pp. 25–30 (in Russian).

Kemeny, J. G. and Snell, J. J. (1960). *Finite Markov Chains*. Van Nostrand, Princeton.

Kolmogorov, A. N. and Fomin, S. V. (1989). *Elements of the Theory of Functions and the Functional Analysis*. Nauka, Moscow (in Russian).

Savelyev, L. Y. (1985). Long runs in Markov sequences. In: *Proceedings Inst. Math. Siberian Branch Acad. Sci. USSR*, 5, pp. 137–144 (in Russian).

Savelyev, L. Y. (1991). Runs in Markov chains. *Siberian Math. J.* 32, 116–132 (in Russian).

Ventsel, A. D. (1975). *Theory of Random Processes*. Nauka, Moscow (in Russian).

Probabilistic Methods in Discrete Mathematics, pp. 451 – 453
V. F. Kolchin et al. (Eds.)
1993 TVP/VSP

HOW OFTEN IS ADDITION EQUAL TO BITWISE ADDITION?

V. I. SHERSTNEV
Steklov Mathematical Institute, Vavilov Street 42, 117966 Moscow GSP-1, Russia

ABSTRACT

We estimate the probability that the addition modulo 2^n coincides with the bitwise addition for independent random summands uniformly distributed on the set of integers from 0 to $2^n - 1$. In particular, if the number of summands is congruent with 3 modulo 4 then this probability equals 2^{1-n}.

1. Let V^n be the set of integers from 0 to $2^n - 1$. We may regard the following two computer-based kinds of addition on V^n. The first of them is the addition modulo 2^n denoted below as $+$. (We shall denote the addition of real numbers also as $+$, but it can be done without ambiguity.) The second is the bitwise addition denoted below as $\oplus$. Algebraic connection between them is known to be complicated. The aim of the paper is to answer the title question. More precisely, we wish to know how many collections $v_1, \dots, v_k \in V^n$ exist such that $v_1 + \dots + v_k = v_1 \oplus \dots \oplus v_k$.
2. Let $X_{i,j}(\Gamma)$ $(i = 1, \dots, k,\ j = 1, \dots, n)$ be independent random variables with $\mathbf{P}\{X_{i,j} = 1\} = \mathbf{P}\{X_{i,j} = 0\} = 1/2$, and $Y_i = \sum_{j=1}^{n} X_{i,j} 2^{j-1}$. Evidently, Y_i $(i = 1, \dots, k)$ are independent random variables with the uniform distribution on V^n. Obviously, $X_{i,n}, \dots, X_{i,1}$ is the binary representation of Y_i. In what follows we estimate

$$q(n,k) = \mathbf{P}\{Y_1 + \dots + Y_k = Y_1 \oplus \dots \oplus Y_k\}.$$

3. Note that the lowest-order bit of $Y_1 + \dots + Y_k$ and the same bit of $Y_1 \oplus \dots \oplus Y_k$ being equal to $X_{1,1} \oplus \dots \oplus X_{k,1}$ are identical. Hence, we may pay no attention to this bit. Let $]x[$ denote the right shift of $x \in V^n$, i. e., if x has the binary representation $x_n, \dots, x_1$, then the binary representation of $]x[$ is $x_n, \dots, x_2$. Now let us consider the sum $X_{1,1} + \dots + X_{k,1}$. A moments consideration shows that this sum converges in distribution to an uniform on V^n random variable, as k tends to infinity. It follows that the distribution of $](Y_1 + \dots + Y_k) \oplus Y_1 \oplus \dots \oplus Y_k[$ converges to the uniform distribution on V^{n-1}, as $k \to \infty$. In particular,

$$q(n,k) \to 1/2^{n-1}, \quad k \to \infty, \quad \text{for any fixed } n.$$

4. The sum $Y_1 + \dots + Y_k$ equals $Y_1 \oplus \dots \oplus Y_k$ if and only if the carry from any bit excluding the highest-order one is even.

We construct the random variables $Z_0, Z_1, \ldots$ in the following way. Let $Z_0 = 0$ and let

$$Z_i = \begin{cases} [(Z_{i-1} + \sum_{j=1}^{k} X_{i,j})/2], & \text{if it is even and } Z_{i-1} \geq 0, \\ -1, & \text{otherwise,} \end{cases}$$

for any $i \geq 1$ (square brackets denote integral part). Note that Z_i is equal to the carry from the ith bit if all preceding carries were even. So, it is not hard to see that

$$q(n,k) = \mathbf{P}\{Z_{n-1} \geq 0\}. \tag{1}$$

The sequence $Z = (Z_n)_{n\geq 0}$ is a Markov chain having the state space $\{-1, 0, 2, 4, \ldots\}$, which we number anew as $\{-1, 0, 1, \ldots\}$. The state -1 is absorbing, therefore one-step transition matrix P of this chain is as follows

$$P = \begin{pmatrix} 1 & 0 \cdots 0 \\ d & R \end{pmatrix},$$

where d is some non-negative vector. The sum $\sum_{j=1}^{k} X_{i,j}$ has a binomial distribution with parameters $(k, 1/2)$, hence an element r_{lm} of the matrix R can be represented as

$$r_{lm} = \mathbf{P}\{Z_i = m \mid Z_{i-1} = l\} = \left(\binom{k}{2(2m-l)} + \binom{k}{2(2m-l)+1} \right) 2^{-k}, \tag{2}$$

for any $l, m = 0, 1, \ldots, [k/2]$. From this equality one can deduce that the even rows of the matrix R differ by shift only. We denote by $a(k)$ the sum of elements for any even row of R. Similarly, let $b(k)$ denote the sum of elements for any odd row of R. From (2) we obtain that $0 < a(k)$, $b(k) < 1$ and $a(k) + b(k) = 1$.
The l-step transition matrix is

$$P^l = \begin{pmatrix} 1 & 0 \ldots 0 \\ d^{(l)} & R^l \end{pmatrix}$$

with some vector $d^{(l)}$ and $R^l = (r_{ij}^{(l)})$ being the l-th power of the matrix R. Hence, we obtain

$$\mathbf{P}\{Z_n \geq 0\} = \sum_j r_{0j}^{(n-1)}. \tag{3}$$

5. Now we estimate $q(n,k)$ for a fixed k. Let $M(n,k) = \max_i \sum_j r_{ij}^{(n)}$ and $m(n,k) = \min_i \sum_j r_{ij}^{(n)}$. For $n = 1$ we have $M(1,k) = \max(a(k), b(k))$ and $m(1,k) = \min(a(k), b(k))$. The equality (3) implies that

$$m(n,k) \leq q(n,k) \leq M(n,k),$$

but $m(n,k) \geq (m(1,k))^n$ and $M(n,k) \leq (M(1,k))^n$, therefore we obtain

$$\big(m(1,k)\big)^{n-1} \leq q(n,k) \leq \big(M(1,k)\big)^{n-1}. \tag{4}$$

The spectral radius $\rho(R)$ of a non-negative matrix R satisfies the inequality (see, e. g., (Horn and Johnson, 1986))

$$m(1,k) \le \rho(R) \le M(1,k)$$

for any k. Besides, $\rho(R)$ is known (Gantmaher, 1967, Chapter XIII.4) to be the maximal eigenvalue with a positive eigenvector $\pi = (\pi_i)$. This gives us the estimate

$$(\min \pi_i / \max \pi_i)\rho(R)^{n-1} \le q(n,k) \le (\max \pi_i / \min \pi_i)\rho(R)^{n-1}.$$

6. Next we estimate $q(n,k)$ in more explicit terms. From (2) we derive the equalities

$$a(k) = 3/8 + b(k-4)/4, \quad b(k) = 3/8 + a(k-4)/4.$$

Let us consider the sequence $d(t) = 3/8 + d(t-1)/4$, $t = 1, \dots$ For this sequence we have $d(t) < 1/2$ or $d(t) = 1/2$ or $d(t) > 1/2$ for any t. If $d(t) < 1/2$, then the sequence is monotonically increasing and if $d(t) > 1/2$, then the sequence is monotonically decreasing. Therefore,

$$M(1,k) = 3/8 + M(1,k-4)/4 \quad \text{and} \quad m(1,k) = 3/8 + m(1,k-4)/4.$$

The solutions of this recurrences are

$$M(1,k) = 1/2 + \big(M(1,k_0) - 1/2\big)/4^{[k/4]},$$

$$m(1,k) = 1/2 + \big(m(1,k_0) - 1/2\big)/4^{[k/4]},$$

where $k \equiv k_0 \pmod 4$, $0 \le k_0 \le 3$. It is easy to check that $M(1,0) = M(1,1) = 1$, $M(1,2) = 3/4$, $M(1,3) = 1/2$ and $m(1,0) = m(1,1) = 0$, $m(1,2) = 1/4$, $m(1,3) = 1/2$. From these formulas and inequality (4) we deduce that

$$\left(\frac{1}{2} - \frac{M(1,k_0) - 1/2}{2\cdot 4^{[k/4]}}\right)^{n-1} \le q(n,k) \le \left(\frac{1}{2} + \frac{M(1,k_0) - 1/2}{2\cdot 4^{[k/4]}}\right)^{n-1}.$$

Consequently,

$$q(n,k) = 2^{1-n}\big(1 + O(n2^{-k/2})\big) \quad \text{as} \quad k \to \infty.$$

7. For $k \equiv 3 \pmod 4$ we obtain from (2) that $a(k) = b(k) = 1/2$ and formula (4) implies that in this case

$$q(n,k) = 2^{1-n}.$$

REFERENCES

Gantmaher, F. R. (1967). *Matrix Theory*. Nauka, Moscow (in Russian).

Horn, R. and Johnson, C. (1986). *Matrix Analysis*. Cambridge Univ. Press, Cambridge.

Probabilistic Methods in Discrete Mathematics, pp. 454 – 458
V. F. Kolchin et al. (Eds.)
1993 TVP/VSP

BOOLEAN FUNCTIONS AND APPROXIMATION OF VECTORS

V. M. SIDELNIKOV and O. V. SIPACHEVA
Moscow State University, Vorobjevy Gory, 119899 Moscow, Russia

ABSTRACT

The approximation of vectors on the sphere S^{n-1} in Euclidean space $\mathbf{R}^n$ by discrete vectors generated by Boolean functions is considered.

Let $(\mathbf{F}_2)^m$ denote the m-dimensional linear space over the field $\mathbf{F}_2 = \{0,1\}$, $n = 2^m$. The addition in the field $\mathbf{F}_2$ is denoted by $\oplus$.
Let f be an arbitrary Boolean function defined on $(\mathbf{F}_2)^m$, and all points of $(\mathbf{F}_2)^m$ be numbered such that

$$(\mathbf{F}_2)^m = \{\alpha^1, \alpha^2, \ldots, \alpha^n\}, \qquad n = 2^m.$$

The point

$$\mathbf{v}_f = ((-1)^{f(\alpha^1)}, \ldots, (-1)^{f(\alpha^n)})$$

on the sphere $S^{n-1}_{\sqrt{n}}$ of radius $\sqrt{n}$ in the n-dimensional Euclidean space $\mathbf{R}^n$ corresponds to the function f. Choose a basis $e^1, \ldots, e^n$ in $\mathbf{R}^n$ such that coordinates of each point of $\mathbf{R}^n$ correspond to a vector in $(\mathbf{F}_2)^m$.

LEMMA 1. *Let f and g be binary functions on $(\mathbf{F}_2)^m$. The vectors $\mathbf{v}_f$ and $\mathbf{v}_g$ are mutually orthogonal if and only if the function $f \oplus g$ is equivalued, i. e., takes the values 0 and 1 in the same number of points.*

LEMMA 2. *Each linear nonconstant binary function on $(\mathbf{F}_2)^m$ is equivalued.*

Proof is obvious.

LEMMA 3. *Let L^0 be the space of all linear binary functions on $(\mathbf{F}_2)^m$ (including the function identically equal to zero on $(\mathbf{F}_2)^m$). Then the set $\{\mathbf{v}_l\colon\ l \in L^0\}$ represents an orthogonal basis of Euclidean space $\mathbf{R}^n$ consisting of vectors of length $\sqrt{n}$.*

Proof. Obviously, the length of each $\mathbf{v}_f$ vector is equal to $\sqrt{n}$. Let $l_1, l_2 \in L^0$ and $l_1 \neq l_2$. Since l_1 and l_2 take the same value at zero, the linear function $l_1 \oplus l_2$ is not constant. This function is equivalued by virtue of Lemma 2. It follows from Lemma 1 that the vectors $\mathbf{v}_{l_1}$ and $\mathbf{v}_{l_2}$ are mutually orthogonal. It remains to note that the cardinality of the set L^0 is exactly $n = 2^m$, since each linear binary function l with the value 0 at zero is defined in some fixed basis of $(\mathbf{F}_2)^m$ by an equation of the form

$$l(\alpha) = a_1\alpha_1 \oplus a_2\alpha_2 \oplus \cdots \oplus a_m\alpha_m = 0, \qquad \alpha \in (\mathbf{F}_2)^m,$$

where α_i are the coordinates of the vector α and a_i are elements of the field $\mathbf{F}_2$. We conclude that the cardinality of the set $\{\mathbf{v}_l\colon\ l \in L^0\}$ is equal to n and the set consists of the pairwise orthogonal vectors of the same length. Therefore the set is a basis.

THEOREM 1. *Let f be a Boolean function on $(\mathbf{F}_2)^m$, and let there exist a basis $\{\beta^1,\ldots,\beta^m\}$ of $(\mathbf{F}_2)^m$ such that the equality*

$$f = g^{(k)} \oplus l^{(f)}$$

holds. Here $g^{(k)}$ is an arbitrary Boolean function dependent on the first k coordinates only and $l^{(f)} \in L^0$. Then the vector $\mathbf{v}_f \in \mathbf{R}^n$ has at most 2^k nonzero coordinates in the basis $\{\mathbf{v}_l\colon\ l \in L^0\}$.

Proof. Let $l \in L^0$. The coordinate of the vector $\mathbf{v}_f$ corresponding to the basis vector $\mathbf{v}_l$ is zero if and only if the vectors $\mathbf{v}_f$ and $\mathbf{v}_l$ are mutually orthogonal or if the function $f \oplus l$ is equivalued, what is the same according to Lemma 1. The function $l \oplus l^{(f)}$ can be represented as the sum

$$l_1 \oplus l_2, \tag{1}$$

where $l_1 \oplus l_2 \in L^0, l_1$ depends on the first k coordinates only, and l_2 depends on the last $m-k$ coordinates.

LEMMA 4. *If $l_2 \neq 0$ in (1), the function $f \oplus l$ is equivalued.*

Proof. Fix an arbitrary binary vector β of length k and consider the subset $\{b\} \times (\mathbf{F}_2)^{m-k}$ in $(\mathbf{F}_2)^m$. Each vector from the subset can be written in the basis $\{\beta^1,\ldots,\beta^m\}$ as follows: the first k coordinates are fixed and coincide with the coordinates of the vector β in the basis $\{\beta^1,\ldots,\beta^k\}$, and the last $m-k$ ones are arbitrary. Clearly,

$$(\mathbf{F}_2)^m = \bigcup\{\{\beta\} \times (\mathbf{F}_2)^{m-k}\colon\ \beta \in (\mathbf{F}_2)^k\}.$$

The function $f \oplus l$ can be represented in the form

$$f \oplus l = g^{(k)} \oplus l_1 \oplus l_2.$$

The functions $g^{(k)}$ and l_1 are independent of the last $m-k$ coordinates. It follows that l_2 is nonconstant and, because of its linearity, is equivalued on

$$\{\beta\} \times (\mathbf{F}_2)^{m-k}$$

for each β from $(\mathbf{F}_2)^k$. The restriction of the function $f \oplus l$ onto the set $\{\beta\} \times (\mathbf{F}_2)^{m-k}$ is therefore equivalued for each $\beta \in (\mathbf{F}_2)^k$. Since

$$(\mathbf{F}_2)^m = \bigcup \{\{\beta\} \times (\mathbf{F}_2)^{m-k} \colon \beta \in (\mathbf{F}_2)^k\},$$

the function $f \oplus l$ is equivalued on the whole space $(\mathbf{F}_2)^m$.

We now have to determine what is the number of $l \in L^0$ with $l_2 = 0$, i.e., what is the number of l from L^0 such that the function $l^{(f)} \oplus l$ depends on the first k coordinates only. Clearly, this number coincides with the number of all functions from L^0 that depend on the first k coordinates only and is equal to 2^k.

We conclude that if the nonlinear part of a Boolean function f in some basis depends only on a small number of coordinates then the number of nonzero coordinates of the corresponding vector $\mathbf{v}_f$ on the sphere $S^{n-1}_{\sqrt{n}}$ in $\mathbf{R}^n$ in the basis $\{\mathbf{v}_l \colon l \in L^0\}$ is not large too. In particular, if f is a quadratic binary surface with the canonical form

$$f(\alpha) = \bigoplus_{j \leq k} y_{2j-1} y_{2j} + l + c,$$

where y_i is the ith coordinate of the vector α in the basis in which f assumes the canonical form, $l \in L^0$ and c is a constant, then the number of nonzero coordinates of the vector $\mathbf{v}_f$ in the basis $\{\mathbf{v}_l \colon l \in L^0\}$ is equal to 2^k.

To turn from quadratic binary surfaces with nonlinear parts depending only on a small number of coordinates in some basis to arbitrary ones additional transformations can be applied. Here the formalism of discrete differentiation is helpful: many usual properties of ordinary derivatives have Boolean analogies. In what follows we will use the discrete differentiation.

LEMMA 5. *Let f be a quadratic binary function on $(\mathbf{F}_2)^m$. Then for each $\alpha \in (\mathbf{F}_2)^m$ the discrete derivative f_α of the function f along the direction α defined as*

$$f_\alpha(x) = f(x) \oplus f(x \oplus \alpha)$$

is linear.

Proof. Choose an arbitrary basis in $(\mathbf{F}_2)^m$. The function f can be represented in this basis as the Zhegalkin polynomial

$$f(x) = \Big(\bigoplus_{i<j} a_{ij} x_i x_j\Big) \oplus \Big(\bigoplus_{i \leq m} a_i x_i\Big) \oplus a_0,$$

where $a_i, a_{ij} \in \mathbf{F}_2$ and x_i is the ith coordinate of the vector x. Calculating the sum of the functions in this form results in a linear function.

We have agreed that the coordinates of all points of the Euclidean space $\mathbf{R}^n$ are numbered by vectors from $(\mathbf{F}_2)^m$. The additive group $(\mathbf{F}_2)^m$ acts on the space $\mathbf{R}^n$ as follows: for $\alpha \in (\mathbf{F}_2)^m$ and $x \in \mathbf{R}^n$, $x = (x_1, \dots, x_n) = (x_{\alpha^1}, \dots, x_{\alpha^n})$

$$a * x = (x_{\alpha+\alpha^1}, \dots, x_{\alpha+\alpha^n}) = (y_1, \dots, y_n)$$

(it should be kept in mind that $(\mathbf{F}_2)^m = \{\alpha^1, \alpha^2, \dots, \alpha^n\}$).

If f, g are Boolean functions on $(\mathbf{F}_2)^m$ and, for some fixed α, $g(x) = f(x \oplus \alpha)$ for all x from $(\mathbf{F}_2)^m$, then it is easy to see that $\mathbf{v}_g = \alpha * \mathbf{v}_f$. It is also clear that for any Boolean functions f_1 and f_2 the coordinates of the vector $\mathbf{v}_{f_1 \oplus f_2}$ are the products of the coordinates of the vectors $\mathbf{v}_{f_1}$ and $\mathbf{v}_{f_2}$. Therefore the discrete differentiation of the function f in $(\mathbf{F}_2)^m$ along the direction α in the Euclidean space corresponds to the operation $T_\alpha(\mathbf{v}_f)$ which implies the action of the element α on the vector $\mathbf{v}_f$ followed by the multiplication of the coordinates of the vector $\mathbf{v}_f$ by those of the vector $\alpha * v_f$. Since multiplying the coordinates of two vectors is a bilinear mapping, T_α is in a canonical correspondence with the tensor $\widehat{T}_\alpha$ of the type $(2,1)$: for any $x, y \in \mathbf{R}^n$ and linear functional A on $\mathbf{R}^n$

$$\widehat{T}_\alpha(x, y, A) = A(x \cdot (\alpha * y))$$

(the coordinatewise multiplication).
Therefore we obtain the following theorem.

THEOREM 2. *For an arbitrary Boolean function f and any $\alpha \in (\mathbf{F}_2)^m$*

$$\mathbf{v}_{f_\alpha} = T_\alpha(\mathbf{v}_f)$$

(*here f_α is the discrete derivative of the function f along the direction α*).

Clearly, the mapping $T_\alpha(\mathbf{v}_f)$ is independent of the basis in $\mathbf{R}^n$.
If a binary function f is quadratic, the vector $T_\alpha(\mathbf{v}_f)$ should either coincide with one of the basis vectors $\mathbf{v}_l$ or have the form $(-1) \cdot \mathbf{v}_l$ for any α. This follows from linearity of discrete derivative of a quadratic function, and for this reason this derivative either belongs to L^0 or has the form $\mathbf{1} \oplus l$ where $\mathbf{1}$ is the function identically equal to unity and $l \in L^0$.
Consequently, for any quadratic Boolean function f and all α from $(\mathbf{F}_2)^m$, the vector $\mathbf{v}_{f_\alpha}$ has exactly one nonzero coordinate in the basis $\{\mathbf{v}_l\colon\ l \in L^0\}$. This makes it possible to determine the proximity of any point of the sphere $S^{n-1}_{\sqrt{n}}$ to the family of all points corresponding to the quadratic Boolean functions: one should calculate the vectors $T_\alpha(x)$ and estimate to what extent these vectors are close to the set of all vectors on the sphere such that have exactly one nonzero coordinate in the basis $\{\mathbf{v}_l\colon\ l \in L^0\}$. Of course, $T_\alpha(x)$ should be considered not for all α from $(\mathbf{F}_2)^m$ but only for the basis set $\{\alpha\}$ of the space $(\mathbf{F}_2)^m$.
The same technique is applicable to estimating the proximity of point x to the set BQ, where B is a nondegenerate linear transformation and $Q \subset S^{n-1}_{\sqrt{n}}$ is the set of all points of the form $\mathbf{v}_f$ corresponding to the quadratic bilinear surfaces f on $(\mathbf{F}_2)^m$. Let $\mathbf{v}_f$ be a point on the sphere corresponding to a binary quadratic surface f, and

$$\mathbf{v}_f = \sum_{i=1}^{n} v_i e^i,$$

where $e^1, \ldots, e^n$ is the initial fixed basis of $\mathbf{R}^n$. Then

$$B\mathbf{v}_f = \sum_{i=1}^{n} v_i B e^i$$

and if $Be^i = (\varepsilon_1^i, \dots, \varepsilon_n^i)$ for $i \le n$, then

$$\begin{aligned} B\mathbf{v}_f &= \sum_{i=1}^{n} v_i B e^i = \sum_{i=1}^{n} v_i \Big(\sum_{j=1}^{n} \varepsilon_j^i e^i\Big) = \sum_{i=1}^{n} v_i e^j \sum_{j=1}^{n} \varepsilon_j^i \\ &= \Big(\mathbf{v}_1 \sum_{j=1}^{n} \varepsilon_j^1, \mathbf{v}_2 \sum_{j=1}^{n} \varepsilon_j^2, \dots, \mathbf{v}_n \sum_{j=1}^{n} \varepsilon_j^n\Big) \end{aligned}$$

or

$$B\mathbf{v}_f = \Big(\mathbf{v}_{\alpha_1} \sum_{j=1}^{n} \varepsilon_j^{\alpha_1}, \mathbf{v}_{\alpha_2} \sum_{j=1}^{n} \varepsilon_j^{\alpha_2}, \dots, \mathbf{v}_{\alpha_n} \sum_{j=1}^{n} \varepsilon_j^{\alpha_n}\Big)$$

what is the same.

Thus for any α,

$$\begin{aligned} T_\alpha(B\mathbf{v}_f) = \Big(&\mathbf{v}_{\alpha_1}\mathbf{v}_{\alpha_1\oplus\alpha}\Big(\sum_{j=1}^{n} \varepsilon_j^{\alpha_1}\Big)\Big(\sum_{j=1}^{n} \varepsilon_j^{\alpha_1\oplus\alpha}\Big), \\ &\dots, \mathbf{v}_{\alpha_n}\mathbf{v}_{\alpha_n\oplus\alpha}\Big(\sum_{j=1}^{n} \varepsilon_j^{\alpha_n}\Big)\Big(\sum_{j=1}^{n} \varepsilon_j^{\alpha_n\oplus\alpha}\Big)\Big). \end{aligned}$$

Since

$$T_\alpha(B\mathbf{v}_f) = (\mathbf{v}_{\alpha_1}\mathbf{v}_{\alpha_1\oplus\alpha}, \dots, \mathbf{v}_{\alpha_n}\mathbf{v}_{\alpha_n\oplus\alpha}),$$

$T_\alpha(B\mathbf{v}_f)$ should have a single nonzero coordinate in the basis $\{\mathbf{v}_l\colon\ l \in L^0\}$.

Probabilistic Methods in Discrete Mathematics, pp. 459 – 469
V. F. Kolchin et al. (Eds.)
1993 TVP/VSP

LIMIT THEOREMS FOR RANDOM A-PERMUTATIONS

A. L. YAKYMIV
Steklov Mathematical Institute, Vavilov Street 42, 117966 Moscow GSP-1, Russia

ABSTRACT

We consider the set T_n of permutations of degree n whose cycle lengths belong to a fixed subset A of the set $\mathbf{N} = \{1,2,\ldots\}$. We assume that

$$A = \{m: m \in \mathbf{N},\ g(m) \in \Delta\},$$

where $g(t)$ is a real-valued function of t, Δ is a finite union of intervals from $[0,1]$, and $\{a\}$ is the fractional part of a. Earlier we have considered the case $g(t) = t^{\alpha} l(t)$, where $l(t)$ is a function slowly varying at infinity, and $\alpha > 0$, $\alpha \notin \mathbf{N}$. In the present paper we investigate the case where α is a natural number.

1. INTRODUCTION

Let A be a fixed subset of $\mathbf{N} = \{1,2,3,\ldots\}$. An A-permutation is a permutation whose cycle lengths belong to A (see (Sachkov, 1978)). Let T_n denote the set of all A-permutations of degree n, and $|X|$ denote the number of elements of the finite set X or the Lebesque measure of the infinite set X. Let ζ_n and ζ_{nm} be respectively the total number of cycles and the number of cycles of the length $m \in A$ of the random permutation, uniformly distributed on T_n. We consider the following general problems for A-permutations:

1) the asymptotics of $|T_n|$ as $n \to \infty$,

2) the asymptotical behavior of the distributions of ζ_n and ζ_{nm} as $n \to \infty$ and m is fixed.

Some results in this direction were obtained in (Sachkov, 1978; Bolotnikov, Sachkov, and Tarakanov, 1979; Kolchin, 1991; Pavlov, 1984; 1986 a, b; 1991; Yakymiv, 1989, 1991). The following three theorems were proved in (Yakymiv, 1991).

THEOREM 1. *Let* $n \to \infty$,

$$|k: k \le n,\ k \in A|/n \to \sigma > 0 \tag{1}$$

and for $m \ge n,\ m = O(n)$

$$|k: k \le n,\ k \in A,\ m-k \in A|/n \to \sigma^2. \tag{2}$$

Then

$$|T_n| \sim n! n^{\sigma-1} L(n) e^{-\sigma\gamma}/\Gamma(\sigma), \qquad n \to \infty,$$

where

$$L(n) = \exp\Big(\sum_{m \le n, m \in A} 1/m - \sigma \log n \Big)$$

is a slowly varying function, γ is the Euler constant and $\Gamma(\cdot)$ is the gamma-function.

THEOREM 2. *Let $n \to \infty$ and the set A satisfies conditions (1) and (2). Then the distribution of random variable*

$$\zeta_n' = (\zeta_n - a_n)/\sqrt{\sigma \log n}$$

converges weakly as $n \to \infty$ to the standard normal law, where

$$a_n = \sum_{m \in A, m \le n} 1/m.$$

THEOREM 3. *Let $n \to \infty$, the set A satisfies conditions (1), (2), and $m \in A$ is fixed. Then the distribution of random variable ζ_{nm} converges weakly as $n \to \infty$ to the Poisson distribution with parameter $1/m$.*

Now the question is: in what cases does the set A satisfy conditions (1) and (2)? A particular answer was given in (Yakymiv, 1991, 1992) and this article contains further investigations in this direction.

We assume that

$$A = \{m: m \in \mathbf{N}, g(m) \in \Delta\}, \tag{3}$$

where $g(t)$ is a real-valued function of argument $t \ge 0$ and Δ is a finite union of intervals from $[0,1]$. It is easy to see that if for arbitrary intervals $\Delta_1, \Delta_2 \subseteq [0,1]$

$$|k: k \le n, \{g(k)\} \in \Delta_1, \{g(m-k)\} \in \Delta_2|/n \to |\Delta_1||\Delta_2| \tag{4}$$

as $n \to \infty, m \ge n, m = O(n)$, then (1) and (2) are valid with $\sigma = |\Delta|$.

A sequence $(g(n))$ is said to be uniformly distributed (u.d.) if for an arbitrary interval $\Delta \subseteq [0,1]$

$$|k: k \le n, \{g(k)\} \in \Delta|/n \to |\Delta| \tag{5}$$

as $n \to \infty$ (see (Kuipers and Niderreiter, 1974)). Note that (4) implies (5) but the converse is not true. A counterexample is given by the function $g(n) = n\theta$, where θ is irrational. The sequence $(g(n))$ satisfying (4) is called strongly uniformly distributed (s.u.d.).

Thus, to verify (1) and (2) it suffices to check that the sequence $(g(n))$ is s.u.d. The following statement gives an analytical base for such a verification.

LEMMA 1. *The sequence $(g(n))$ is s.u.d. if and only if for all integers $a, b \ne 0$*

$$\sum_{t=1}^{n} \exp\Big(2\pi i\big(ag(t) + bg(m-t)\big)\Big) = o(n) \tag{6}$$

as $n \to \infty$, $m \geq n$, $m = O(n)$.

It is shown in (Yakymiv, 1991) that if

$$g(t) = t^{\alpha} l(t), \tag{7}$$

where $\alpha \in (0,1)$, $l(t)$ is a function slowly varying at infinity and

$$\frac{d^n}{dt^n} l(t) = o\big(t^{-n} l(t)\big), \qquad t \to \infty, \tag{8}$$

for $n = 1, 2, 3$, then the sequence $(g(n))$ is s.u.d. In (Yakymiv, 1992) it is proved that $(g(n))$ is s.u.d. if (7) is valid for some $\alpha > 1, \alpha \notin \mathbf{N}$, and (8) is valid for all $n = 1, \ldots, [\alpha] + 2$.

For every function slowly varying at infinity there exists an equivalent function $l(t)$ satisfying (8) for all $n \in \mathbf{N}$ (Yakymiv, 1982). For example, the functions $\log t$, $\log\log t$, $\exp(\log^{\beta} t)$ $(\beta < 1)$, their various products and powers satisfy (8) for all $n \in \mathbf{N}$.

In the present paper we investigate the case where α in (7) belongs to $\mathbf{N}$ and $l(t)$ is a slowly varying function.

2. THE MAIN RESULTS

Denote by S the set of all functions slowly varying at infinity. Let I_m $(m \in \mathbf{N})$ be the set of all functions from S satisfying (8) for $n = 1, \ldots, m$, and let $I = \bigcap_{m \in \mathbf{N}} I_m$. For $b \in \mathbf{R}$ and $n \in \mathbf{N}$, let $b^{[n]} = b(b-1)\cdots(b-n+1)$.

THEOREM 4. *Let $l(t) \in S$ and $\alpha \in \mathbf{N}$ in (7). If for every $t \geq 1$*

$$l'(t) = \delta t^{-1} l_1(t), \tag{9}$$

where $\delta = 1$ or $\delta = -1$, $l_1(t) \in I_{\alpha+1}$ and

$$l_1(t) = o\big(l(t)\big), \qquad t \to \infty, \tag{10}$$

then $(g(n))$ is s.u.d.

COROLLARY 1. *The sequence $(n^{\alpha} \log^{\beta} n)$ is s.u.d. for arbitrary $\alpha \in \mathbf{N}$ and $\beta \neq 0$.*

COROLLARY 2. *The sequence $(n^{\alpha} \log^{\gamma} \log n)$ is s.u.d. for arbitrary $\alpha \in \mathbf{N}$ and $\gamma \neq 0$.*

COROLLARY 3. *The sequence $(n^{\alpha} \log^{\beta} n \log^{\gamma} \log n)$ is s.u.d. for arbitrary $\alpha \in \mathbf{N}$ and $\beta, \gamma \neq 0$.*

COROLLARY 4. *The sequence $(n^{\alpha} \exp(c \log^{\beta} n))$ is s.u.d. for arbitrary $c \neq 0, \beta < 1$ and $\alpha \in \mathbf{N}$.*

COROLLARY 5. *Let all the assumptions of Theorem 4 are valid. Then Theorems 1–3 are true for the set A defined by (3).*

3. AUXILIARY RESULTS

Further we will assume that $\delta = 1$. (The case $\delta = -1$ is considered in a similar way).

LEMMA 2. *Let $g(t) = t^\alpha l(t)$, where $\alpha \notin N \cup \{0\}$. If $l(t) \in I_m$ for some $m \in \mathbf{N}$, then*

$$g^{(m)}(t) \sim \alpha^{[m]} t^{\alpha-m} l(t), \qquad t \to \infty. \tag{11}$$

LEMMA 3. *Let all the assumptions of Theorem 4 are valid. Then $l(t) \in I_{\alpha+2}$.*

LEMMA 4. *Under the assumptions of Theorem 4, as $t \to \infty$,*

$$g^{(\alpha)}(t) \sim \alpha! l(t), \tag{12}$$

$$g^{(\alpha+1)}(t) \sim \alpha! t^{-1} l_1(t), \tag{13}$$

$$g^{(\alpha+2)}(t) \sim -\alpha! t^{-2} l_1(t). \tag{14}$$

For $m \in \mathbf{N}$ and $h_1, \ldots, h_m \in \mathbf{N}$ we put $g(h_1, t) = (g(t + h_1) - g(t), \ldots, g(h_1, \ldots, h_m, t) = g(h_1, \ldots, h_{m-1}, t + h_m) - g(h_1, \ldots, h_{m-1}, t)$.

LEMMA 5. *Let $h_1, \ldots, h_m \in \mathbf{N}$, $m = \alpha - 1$, and $g(h_1, \ldots, h_m, t) = g_m(t)$. Under the assumptions of Theorem 4, $g'_m(t) \sim h\alpha! l(t)$, $g''_m(t) \sim h\alpha! t^{-1} l_1(t)$, $g'''_m(t) \sim -h\alpha! t^{-2} l_1(t)$ as $t \to \infty$, where $h = h_1, \ldots, h_m$.*

LEMMA 6. *Let $l(t) \in S$. Then there exist functions $r(t)$ and $s(t)$ such that $r(t) \to \infty$, $s(t) \to 0$ and*

$$l(\lambda t)/l(t) \to 1, \qquad t \to \infty, \tag{15}$$

uniformly with respect to $\lambda \in [s(t), r(t)]$.

LEMMA 7. *Let (12), (13), (14) are valid for some function $g(t)$, $\alpha = 1$, and $l(t), l_1(t) \in S$. Then the sequence $(g(n))$ is s.u.d.*

LEMMA 8. *If the sequence $(g(h_1, \ldots, h_k, n),\ n \in N)$ is s.u.d. for some $k \in \mathbf{N}$ and arbitrary $h_1, \ldots, h_k \in \mathbf{N}$, then the sequence $(g(n))$ is s.u.d.*

LEMMA 9. *A function $l(t) \in I_m$, $m \in \mathbf{N}$, if and only if for all $n = 1, \ldots, m$*

$$\varphi^{(n)}(t) = o(1), \quad \text{as} \quad t \to \infty, \tag{16}$$

where $\varphi(t) = \log l(e^t)$.

LEMMA 10. *Let $l(t) \in I_m$, where $m \in \mathbf{N}$. Then $l^\beta(t) \in I_m$ for any $\beta \in \mathbf{R}$.*

LEMMA 11. *Let functions $l_1(t)$ and $l_2(t)$ belong to I_m, where $m \in \mathbf{N}$. Then the function $l(t) = l_1(t)l_2(t)$ also belongs to I_m.*

LEMMA 12. *Functions $\log^\beta t$, $\log^\gamma \log t$ belong to I for any $\beta, \gamma \in \mathbf{R}$.*

4. PROOFS

Proof of Lemma 2. Differentiate $g(t)$ m times. We have $g^{(m)}(t) = \sum_{k=0}^{m} C_m^k \alpha^{[k]} t^{\alpha-k} l^{(m-k)}(t)$ here $\alpha^{[0]} = 1, l^{(0)}(t) = l(t), C_m^k = \binom{m}{k}$. For every $k < m$ we have

$$t^{\alpha-k} l^{(m-k)}(t) = o(t^{\alpha-m} l(t)), \qquad t \to \infty;$$

therefore (11) is true. Lemma 2 is proved.

Proof of Lemma 3. From (9) and (10) we obtain that $l'(t) = t^{-1} l_1(t) = o(t^{-1} l(t)), t \to \infty$. By Lemma 2, for $m = 2, \ldots, \alpha + 2$, $n = m - 1$,

$$\frac{d^m}{dt^m} l(t) = \frac{d^n}{dt^n}(t^{-1} l_1(t)) \sim (-1)^n n! t^{-n-1} l_1(t) = o(t^{-m} l(t)) \tag{17}$$

as $t \to \infty$. Therefore, $l(t) \in I_{\alpha+2}$. Lemma 3 is proved.

Proof of Lemma 4. Differentiate (7) m times, $m \leq \alpha$. Then $g^{(m)}(t) = \sum_{k=0}^{m} C_\alpha^k \alpha^{[m-k]} t^{\alpha-(m-k)} l^{(k)}(t)$. For $m = \alpha$

$$g^{(\alpha)}(t) = \sum_{k=0}^{\alpha} C_\alpha^k \alpha^{[\alpha-k]} t^k l^{(k)}(t). \tag{18}$$

Lemma 3 implies that for $k = 1, \ldots, \alpha$ we have

$$l^{(k)}(t) = o(t^{-k} l(t)), \qquad t \to \infty.$$

Thus from (18) we obtain that

$$g^{(\alpha)}(t) \sim \alpha^{[\alpha]} C_\alpha^0 l^{(0)}(t) = \alpha! l(t), \qquad t \to \infty.$$

Therefore, (12) is proved. Differentiating (18), we obtain

$$g^{(\alpha+1)}(t) = \sum_{k=0}^{\alpha} C_{\alpha+1}^{k+1} \alpha^{[\alpha-k]} t^k l^{(k+1)}(t). \tag{19}$$

From (17) and (19) it follows that

$$g^{(\alpha+1)}(t) \sim \frac{l_1(t)}{t} \sum_{k=0}^{\alpha} (-1)^k C_{\alpha+1}^{k+1} \alpha^{[\alpha-k]} k!$$

$$= \frac{l_1(t)}{t}\alpha! \sum_{k=0}^{\alpha}(-1)^k C_{\alpha+1}^{k+1} = l_1(t)t^{-1}\alpha!\lambda, \tag{20}$$

where $\lambda = -\sum_{k=0}^{\alpha}(-1)^k C_{\alpha+1}^{k+1}$. Let us calculate the value λ. It is clear that $\lambda = -\sum_{i=1}^{\alpha+1}(-1)^i C_{\alpha+1}^i = -\sum_{i=1}^{\beta}(-1)^i C_\beta^i$, where $\beta = \alpha + 1$. For an arbitrary real s we have $\sum_{i=0}^{\beta} C_\beta^i s^i = (1+s)^\beta$. Thus $\sum_{i=0}^{\beta} C_\beta^i(-1)^i = (-1+1)^\beta = 0$.
From this we have that $1 - \lambda = 0$ and $\lambda = 1$. Now (13) follows from (20). Differentiating (19), we obtain $g^{(\alpha+2)}(t) = \sum_{k=0}^{\alpha} C_{\alpha+2}^{k+2}\alpha^{[\alpha-k]}t^k l^{(k+2)}(t)$. Taking into account (17), we have

$$g^{(\alpha+2)}(t) \sim \sum_{k=0}^{\alpha} C_{\alpha+2}^{k+2}\alpha^{[\alpha-k]}t^{-k}t^{k-2}(-1)^{k+1}(k+1)!l_1(t)$$

$$= \alpha!t^{-2}l_1(t)\sum_{k=0}^{\alpha} C_{\alpha+2}^{k+2}(-1)^{k+1}(k+1) = \alpha!t^{-2}l_1(t)\mu, \tag{21}$$

where $\mu = \sum_{k=0}^{\alpha} C_{\alpha+2}^{k+2}(-1)^{k+1}(k+1) = -\sum_{i=2}^{\gamma} C_\gamma^i(-1)^i(i-1)$, $\gamma = \alpha+2$. For the calculation of the value μ we note that $\sum_{i=0}^{\gamma} C_\gamma^i s^{i+1} = s(s+1)^\gamma$. Differentiating this equation, we obtain

$$\sum_{i=0}^{\gamma} C_\gamma^i(i+1)s^i = (s+1)^{\gamma-1}(1+(1+\gamma)s),$$

and

$$\sum_{i=0}^{\gamma} C_\gamma^i(i-1)s^i = (s+1)^\gamma(1+(1+\gamma)s) - 2\sum_{i=0}^{\gamma} C_\gamma^i s^i$$

$$= (s+1)^{\gamma-1}(1+(1+\gamma)s) - 2(1+s)^\gamma.$$

Thus

$$-\sum_{i=0}^{\gamma} C_\gamma^i(i-1)(-1)^i = 0$$

and

$$\mu = -\sum_{i=2}^{\gamma} C_\gamma^i(i-1)(-1)^i = -1 + 0 = -1. \tag{22}$$

Relations (21) and (22) imply (14). Lemma 4 is proved.
The proof of Lemma 5 is analogous to the proof of Lemma 3 in (Yakymiv, 1992).

Proof of Lemma 6. According to a theorem on the integral representation of the slowly varying functions there exist a real number $b > 0$ and functions $\eta(t)$ and $\varepsilon(t)$ defined for $t \geq b$ such that $\eta(t) \to c$, $\varepsilon(t) \to 0$ as $t \to \infty$, and

$$l(t) = \exp\left\{\eta(t) + \int_b^t \frac{\varepsilon(u)}{u}du\right\}, \qquad t \geq b.$$

Without loss of generality we can suppose that $\lambda \geq 1$. Then for $t \geq b$

$$\frac{l(\lambda t)}{l(t)} = \exp\Big(\eta(\lambda t) - \eta(t) + \int_t^{\lambda t} \frac{\varepsilon(u)}{u} du\Big). \tag{23}$$

Put $r(t) = \exp(1/\sqrt{\delta(t)})$, where $\delta(t) = \sup_{u \geq t} |\varepsilon(u)|$. Then

$$\Big|\int_t^{\lambda t} \frac{\varepsilon(u)}{u} du\Big| \leq \delta(t) \log_e \lambda \leq \sqrt{\delta(t)} \to 0, \qquad t \to \infty,$$

uniformly with respect to $\lambda \in [1, r(t)]$. Relation (15) follows from (23) and the last inequality. Lemma 6 is proved.

Proof of Lemma 7. According to Lemma 1 we have to verify (6). We put $\theta = |b/a|, f(t) = ag(t) + bg(m-t)$. Let us consider the case $b/a > 0$. According to Lemma 6, we choose a sequence $s(n) \to 0$ as $n \to \infty$ such that (15) is fulfilled and $ns(n) \in \mathbf{N}$. Making use of (12) for $\alpha = 1$ and Lemma 6, we obtain that for $t \in [ns(n), n - ns(n)]$

$$f'(t) = a\big(g'(t) - \theta g'(m-t)\big) = o\big(l(t)\big) + O\big(l(m-t)\big) = O\big(l(n)\big) \tag{24}$$

as $n \to \infty, m \geq n, m = O(n)$. As it follows from (13) and (14), there exists a constant M such that for $m \geq n \geq M$ and $t \in [ns(n), n - ns(n)]$

$$|f''(t)| = a\big(g''(t) + \theta g''(m-t)\big) \geq |a| g''(t) \geq |a| g''(n) \asymp n^{-1} l(n) \tag{25}$$

as $n \to \infty$. From (24), 25 and Theorem 2.7 of we obtain that

$$\Big|\sum_{t=ns(n)}^{n-ns(n)} \exp\big(2\pi i f(t)\big)\Big| \leq \Big(\big|f'(ns(n))\big| + \big|f'(n - ns(n))\big| + 2\Big)(4/\sqrt{\rho} + 3)$$

$$= O\big(l(n)\big) \cdot O\Big(1/\sqrt{n^{-1} l_1(n)}\Big) = O\Big(\sqrt{n} l(n)/\sqrt{l_1(n)}\Big) = o(n)$$

as $n \to \infty, m \geq n, m = O(n)$, where $\rho = \inf_{ns(n) \geq t \geq n - s(n)} \big|f''(t)\big|$. Hence

$$\sum_{t=1}^{n} \exp\big(2\pi i f(t)\big) = o(n) + O\big(ns(n)\big) = o(n)$$

as $n \to \infty, m \geq n, m = O(n)$. Thus (6) is true in this case. For $b/a < 0$ the proof is similar to the proof of Lemma 4 from (Yakymiv, 1992). Lemma 7 is proved. Lemma 8 is proved in (Yakymiv, 1992).

Proof of Lemma 9. Let $l(t) \in I_m$ and $\psi(t) = l(e^t)$. According to formula (0.10) from (Gradshtein and Ryzhik, 1965) for $n = 1, \ldots, m$

$$\frac{d^n}{dt^n} \psi(t) = \sum_{i_1, \ldots, i_n} l^{(k)}(e^t) C_{i_1, \ldots, i_n} \prod_{j=1}^{n} (e^t)^{i_j},$$

where the summation is taken over all ordered sets of integers $i_1, ..., i_n \geq 0$ such that $\sum_{j=1}^{n} j i_j = n$, $k = \sum_{j=1}^{n} i_j$, $C_{i_1,...,i_n}$ are some constants. From this we have

$$\psi^{(n)}(t) = \sum_{i_1,...,i_n} o(e^{-tk} l(e^t)) e^{tk} = o(l(e^t)) = o(\psi(t))$$

as $t \to \infty$. Hence for any $j = 1, \ldots, m$ the asymptotic relation

$$\psi^{(i)}(t) = o(\psi(t)), \qquad t \to \infty, \tag{26}$$

holds. Applying formula (0.10) from (Gradshtein and Ryzhik, 1965) for function $\varphi(t) = \log \psi(t)$, we obtain that for $n = 1, \ldots, m$

$$\varphi^{(n)}(t) = \sum_{i_1,...,i_n} \log^{(k)} (\psi(t)) C_{i_1,...,i_n} \prod_{j=1}^{n} \left(\psi^{(j)}(t)\right)^{i_j}.$$

Using (26), we obtain that

$$\varphi^{(n)}(t) = \sum_{i_1,...,i_n} (-1)^{k+1}(k-1)! \left(\psi(t)\right)^{-k} C_{i_1,...,i_n} o\left(\psi(t)^k\right) + o(1)$$

as $t \to \infty$. Relation (16) is proved. A proof of the converse statement is contained in (Yakymiv, 1981).

Proof of Lemma 10. We put $\varphi(t) = \log l(e^t), \Phi(t) = \log l^{\beta}(e^t)$. According to Lemma 9, relation (16) is valid for any $n = 1, \ldots, m$. Thus, for any $n = 1, \ldots, m$

$$\Phi^{(n)}(t) = o(1), \qquad t \to \infty.$$

Consequently, from Lemma 9 we obtain that $l^{\beta}(t) \in I_m$. Lemma 10 is proved.

Proof of Lemma 11. Let $n = 1, \ldots, m$. Differentiating $l(t)$ n times we obtain that

$$\frac{d^n}{dt^n} l(t) = \sum_{i=0}^{n} C_n^i l_1^{(i)}(t) l_2^{(n-i)}(t),$$

and

$$l^{(n)}(t) = \sum_{i=1}^{n-1} o\left(t^{-i} l_1(t)\right) o\left(t^{-(n-i)} l_2(t)\right) + o\left(l_1(t) t^{-n} l_2(t)\right)$$

$$= o\left(t^{-n} l_1(t) l_2(t)\right), \qquad t \to \infty.$$

Therefore, $l(t) \in I_m$. Lemma 11 is proved.

Proof of Lemma 12. Note that for an arbitrary $n \in \mathbf{N}$

$$\frac{d^n}{dt^n} \log_e t = (-1)^{n-1} ! t^{-n} = o(t^{-n} \log t), \qquad t \to \infty.$$

Thus, $\log t \in I$. For $\log \log t$ we have

$$\frac{d}{dt} \log \log t = \frac{1}{t \log t} = o(t^{-1} \log \log t) \tag{27}$$

and for $m \geq 2, n = m-1$ according to (17)

$$\frac{d^m}{dt^m}\log\log t = \frac{d^n}{dt^n}(t^{-1}\log^{-1} t) \sim (-1)^n n! t^{-n-1}\log^{-1} t = o(t^{-m}\log\log t) \quad (28)$$

as $t \to \infty$. From (27) and (28) it follows that $\log\log t \in I$. Lemma 10 now implies that $\log^\gamma \log t \in I$. Lemma 12 is proved.

Proof of Theorem 4. If $\alpha = 1$, then due to Lemma 7, the Theorem 4 is true. Let $\alpha > 1$. Put

$$l(t) = h\alpha! l(t), \quad l_1(t) = h\alpha! l_1(t), \quad m = \alpha - 1, \quad g(t) = g_m(t).$$

From Lemma 5 and Lemma 7 we obtain that for any $h_1, \ldots, h_m \in \mathbf{N}$ the sequence $(g(h_1, \ldots, h_m, n), n \in \mathbf{N})$ is s.u.d. Now from Lemma 8 it follows that $(g(n))$ is s.u.d. Theorem 4 is proved.

Proof of Corollary 1. For $l(t) = \log^\beta t$

$$l'(t) = \beta \log^{\beta-1} t/t.$$

Due to Lemma 12 the function

$$l_1(t) = |\beta| \log^{\beta-1} t \in I.$$

Besides, it is clear that $l_1(t) = o(l(t))$ as $t \to \infty$. Thus, Corollary 1 is proved.

Proof of Corollary 2. Put $L(t) = \log\log t$. Then $l(t) = \log^\gamma \log t = L^\gamma(t)$. Note that

$$l'(t) = \gamma L^{\gamma-1} L'(t) = \frac{\gamma L^{\gamma-1}(t)}{t \log t}.$$

Thus

$$l_1(t) = |\gamma| L^{\gamma-1}(t) \log^{-1}(t).$$

From Lemma 10 it follows that $\log^{-1}(t) \in I$. Lemma 12 implies that $L^{\gamma-1}(t) \in I$. According to Lemma 11 we have $l_1(t) \in I$. Corollary 2 is proved.

Proof of Corollary 3. We consider only the case $\alpha > 0, \beta > 0$ (other cases can be treated similarly). We have

$$l'(t) = \frac{\beta \log^{\beta-1} t \log^\gamma t}{t} + \gamma \frac{\log^\beta t \log^{\gamma-1} \log t}{t \log t}.$$

Thus

$$l_1(t) = \beta \log^{\beta-1} t \log^\gamma \log t + \gamma \log^{\beta-1} t \log^{\gamma-1} \log t$$
$$\sim \beta \log^{\beta-1} t \log^\gamma \log t, \qquad t \to \infty.$$

It is clear that $l_1(t) \in S$ and $l_1(t) = o(l(t))$ as $t \to \infty$. Note that $l_1(t) = l_2(t) + l_3(t)$, where

$$l_2(t) = \beta \log^{\beta-1} t \log^\gamma \log t, \quad l_3(t) = \gamma \log^{\beta-1} t \log^{\gamma-1} \log t.$$

According to Lemmas 11 and 12 the functions $l_2(t)$ and $l_3(t)$ belong to I. Thus for an arbitrary $n \in \mathbf{N}$

$$l_1^{(n)}(t) = l_2^{(n)}(t) + l_3^{(n)}(t) = o\big(t^{-n} l_2(t)\big) + o\big(t^{-n} l_3(t)\big)$$

$$= o\big(t^{-n} l_1(t)\big), \qquad t \to \infty.$$

Therefore, $l_1(t) \in I$. Corollary 3 is proved.

Proof of Corollary 4. Note that

$$l'(t) = c\beta l(t) \log^{\beta-1} t/t$$

and $l_1(t) = |c\beta| l(t) \log^{\beta-1} t$. For $\beta < 1$

$$\frac{tl'(t)}{l(t)} = c\beta \log^{\beta-1} t \to 0, \qquad t \to \infty.$$

Thus, $l(t) \in S$. According to Lemma 12 we have $\log^{\beta-1} t \in I$. Let us check that $l(t) \in I$. Indeed,

$$\varphi(t) = \log l(e^t) = ct^\beta$$

and, for any $n \in \mathbf{N}$,

$$\varphi^{(n)}(t) = c\beta^{[n]} t^{\beta-n} = o(1), \qquad t \to \infty.$$

Therefore, by Lemma 9 we have $l(t) \in I$. The function $l_1(t)$ belongs to a product of functions from I (Lemma 11). Corollary 4 is proved.

REFERENCES

Bolotnikov, Yu. V., Sachkov, V. N., and Tarakanov, V. E. (1979). On some classes of random variables on the cycles of random permutations. *Math. USSR Sb.* **124**, 91–104 (in Russian).

Gradshtein, I. S. and Ryzhik, I. M. (1962). *Tables of Integrals, Sums, Series and Products.* Academic Press, New York (1965).

Kolchin, V. F. (1991). On the number of permutations with constraints on the cycle lengths *Discrete Math. Appl.* **1**, 179–194.

Kuipers, L. and Niderreiter, H. (1974). *Uniform Distribution of Sequences.* John Wiley, New York.

Pavlov, A. I. (1984). On the number and cycle structure of the permutations in some classes, *Math. USSR Sb.* **124**, 536–556 (in Russian).

Pavlov, A. I. (1986 a). On some classes of permutations with number-theoretic restrictions on the lengths of cycles. *Math. USSR Sb.* **129**, 252–283 (in Russian).

Pavlov, A. I. (1986 b). On substitutions with cycle lengths from a given set. *Theory Probab. Appl.* **31**, 618–619.

Pavlov, A. I. (1991). The number of permutations with cycle lengths from a given set. *Discrete Math. Appl.* **3**, 109–123 (in Russian).

Sachkov, V. N. (1978). *Probabilistic Methods on Combinatorial Analysis.* Nauka, Moscow (in Russian).

Yakymiv, A. L. (1981). *Limit Theorems for Branching Processes.* Ph. D. Thesis, Moscow (in Russian).

Yakymiv, A. L. (1982). Multidimensional Tauberian theorems and their application to Bellman–Harris branching processes. *Math. USSR Sb.* **43**, 413–425 (in Russian).

Yakymiv, A. L. (1989). On the number of A-permutations. *Math. USSR Sb.* **180**, 294–303 (in Russian).

Yakymiv, A. L. (1991). Substitutions with cycle lengths from a fixed set. *Discrete Math. Appl.* **1**, 105–116 (in Russian).

Yakymiv, A. L. (1992). On some classes of permutations with cycle lengths from a given set *Discrete Math. Appl.* **4**, 128–134 (in Russian).

LIST OF CONTRIBUTORS

ORGANIZING COMMITTEE

Bei Fragen zur Produktsicherheit wenden Sie sich bitte an:
If you have any questions regarding product safety,
please contact:

Walter de Gruyter GmbH
Genthiner Straße 13
10785 Berlin
productsafety@degruyterbrill.com